MONOGRAPHS ON STATISTICS AND APPLIED PROBABILITY

General Editors

D.R. Cox, V. Isham, N. Keiding, N. Reid and H. Tong

1 Stochastic Population Models in Ecology and Epidemiology *M.S. Bartlett* (1960)
2 Queues *D.R. Cox and W.L. Smith* (1961)
3 Monte Carlo Methods *J.M. Hammersley and D.C. Handscomb* (1964)
4 The Statistical Analysis of Series of Events *D.R. Cox and P.A.W. Lewis* (1966)
5 Population Genetics *W.J. Ewens* (1969)
6 Probability, Statistics and Time *M.S. Bartlett* (1975)
7 Statistical Inference *S.D. Silvey* (1975)
8 The Analysis of Contingency Tables *B.S. Everitt* (1977)
9 Multivariate Analysis in Behavioural Research *A.E. Maxwell* (1977)
10 Stochastic Abundance Models *S. Engen* (1978)
11 Some Basic Theory for Statistical Inference *E.J.G. Pitman* (1979)
12 Point Processes *D.R. Cox and V. Isham* (1980)
13 Identification of Outliers *D.M. Hawkins* (1980)
14 Optimal Design *S.D. Silvey* (1980)
15 Finite Mixture Distributions *B.S. Everitt and D.J. Hand* (1981)
16 Classification *A.D. Gordon* (1981)
17 Distribution-free Statistical Methods, 2nd edition *J.S. Maritz* (1995)
18 Residuals and Influence in Regression *R.D. Cook and S. Weisberg* (1982)
19 Applications of Queueing Theory, 2nd edition *G.F. Newell* (1982)
20 Risk Theory, 3rd edition *R.E. Beard, T. Pentikainen and E. Pesonen* (1984)
21 Analysis of Survival Data *D.R. Cox and D. Oakes* (1984)
22 An Introduction to Latent Variable Models *B.S. Everitt* (1984)
23 Bandit Problems *D.A. Berry and B. Fristedt* (1985)
24 Stochastic Modelling and Control *M.H.A. Davis and R. Vinter* (1985)
25 The Statistical Analysis of Compositional Data *J. Aitchison* (1986)
26 Density Estimation for Statistics and Data Analysis *B.W. Silverman*
27 Regression Analysis with Applications *G.B. Wetherill* (1986)

28 Sequential Methods in Statistics, 3rd edition
G.B. Wetherill and K.D. Glazebrook (1986)
29 Tensor Methods in Statistics *P. McCullagh* (1987)
30 Transformation and Weighting in Regression *R.J. Carroll and D. Ruppert* (1988)
31 Asymptotic Techniques for Use in Statistics *O.E. Barndorff-Nielsen and D.R. Cox* (1989)
32 Analysis of Binary Data, 2nd edition *D.R. Cox and E.J. Snell* (1989)
33 Analysis of Infectious Disease Data *N.G. Becker* (1989)
34 Design and Analysis of Cross-Over Trials *B. Jones and M.G. Kenward* (1989)
35 Empirical Bayes Methods, 2nd edition *J.S. Maritz and T. Lwin* (1989)
36 Symmetric Multivariate and Related Distributions *K.-T. Fang S. Kotz and K.W. Ng* (1990)
37 Generalized Linear Models, 2nd edition *P. McCullagh and J.A. Nelder* (1989)
38 Cyclic and Computer Generated Designs, 2nd edition
J.A. John and E.R. Williams (1995)
39 Analog Estimation Methods in Econometrics *C.F. Manski* (1988)
40 Subset Selection in Regression *A.J. Miller* (1990)
41 Analysis of Repeated Measures *M.J. Crowder and D.J. Hand* (1990)
42 Statistical Reasoning with Imprecise Probabilities *P. Walley* (1991)
43 Generalized Additive Models *T.J. Hastie and R.J. Tibshirani* (1990)
44 Inspection Errors for Attributes in Quality Control
N.L. Johnson, S. Kotz and X. Wu (1991)
45 The Analysis of Contingency Tables, 2nd edition *B.S. Everitt* (1992)
46 The Analysis of Quantal Response Data *B.J.T. Morgan* (1993)
47 Longitudinal Data with Serial Correlation: A State-space Approach
R.H. Jones (1993)
48 Differential Geometry and Statistics *M.K. Murray and J.W. Rice* (1993)
49 Markov Models and Optimization *M.H.A. Davis* (1993)
50 Networks and Chaos - Statistical and Probabilistic Aspects
O.E. Barndorff-Nielsen, J.L. Jensen and W.S. Kendall (1993)
51 Number-theoretic Methods in Statistics *K.-T. Fang and Y. Wang* (1994)
52 Inference and Asymptotics *O.E. Barndorff-Nielsen and D.R. Cox* (1994)
53 Practical Risk Theory for Actuaries *C.D. Daykin, T. Pentikäinen and M. Pesonen* (1994)

54 Biplots *J.C. Gower and D.J. Hand* (1996)
55 Predictive Inference: An Introduction *S. Geisser* (1993)
56 Model-Free Curve Estimation *M.E. Tarter and M.D. Lock* (1993)
57 An Introduction to the Bootstrap *B. Efron and R.J. Tibshirani* (1993)
58 Nonparametric Regression and Generalized Linear Models *P.J. Green and B.W. Silverman* (1994)
59 Multidimensional Scaling *T.F. Cox and M.A.A. Cox* (1994)
60 Kernel Smoothing *M.P. Wand and M.C. Jones* (1995)
61 Statistics for Long Memory Processes *J. Beran* (1995)
62 Nonlinear Models for Repeated Measurement Data *M. Davidian and D.M. Giltinan* (1995)
63 Measurement Error in Nonlinear Models *R.J. Carroll, D. Ruppert and L.A. Stefanski* (1995)
64 Analyzing and Modeling Rank Data *J.I. Marden* (1995)
65 Time Series Models - In econometrics, finance and other fields *D.R Cox, D.V. Hinkley and O.E. Barndorff Nielsen* (1996)
66 Local Polynomial Modeling and its Applications *J. Fan and I. Gijbels* (1996)
67 Multivariate Dependencies - Models, analysis and interpretation *D.R. Cox and N. Wermuth* (1996)
68 Statistical Inference - Based on the likelihood *A. Azzalini* (1996)
69 Bayes and Empirical Bayes Methods for Data Analysis *B.P. Carlin and T.A. Louis* (1996)
70 Hidden Markov and Other Models for Discrete-valued Time Series *I.L. Macdonald and W. Zucchini* (1997)
71 Statistical Evidence: A Likelihood paradigm *R. Royall* (1997)
72 Analysis of Incomplete Multivariate Data *J.L. Schafer* (1997)
73 Multivariate Models and Dependence Concepts *H. Joe* (1997)
74 Theory of Sample Surveys *M.E. Thompson* (1997)
75 Retrial Queues *G.I. Falin and J.G.C. Templeton* (1997)
76 Theory of Dispersion Models *B. Jørgensen* (1997)
77 Mixed Poisson Processes *J. Grandell* (1997)

(Full details concerning this series are available from the Publishers).

Mixed Poisson Processes

Jan Grandell

Department of Mathematics
Royal Institute of Technology
Stockholm
Sweden

CHAPMAN & HALL

London · Weinheim · New York · Tokyo · Melbourne · Madras

Published by Chapman & Hall, 2–6 Boundary Row, London SE1 8HN, UK

Chapman & Hall, 2–6 Boundary Row, London SE1 8HN, UK

Chapman & Hall GmbH, Pappelallee 3, 69469 Weinheim, Germany

Chapman & Hall USA., 115 Fifth Avenue, New York, NY 10003, USA

Chapman & Hall Japan, ITP-Japan, Kyowa Building, 3F, 2-2-1 Hirakawacho, Chiyoda-ku, Tokyo 102, Japan

Chapman & Hall Australia, 102 Dodds Street, South Melbourne, Victoria 3205, Australia

Chapman & Hall India, R. Seshadri, 32 Second Main Road, CIT East, Madras 600 035, India

First edition 1997

Printed in Great Britain by St Edmundsbury Press, Bury St Edmunds, Suffolk.

ISBN 0 412 78700 8

A catalogue record for this book is available from the British Library

∞ Printed on permanent acid-free text paper, manufactured in accordance with ANSI/NISO Z39.48 - 1992 and ANSI/NISO Z39.48 - 1984 (Permanence of Paper).

Contents

Preface

More than fifty years ago, Ove Lundberg presented his thesis *On random processes and their application to sickness and accident statistics* about Markov point processes and – more particularly – mixed Poisson processes; a class of processes which was introduced by Dubourdieu (1938). The thesis appeared 1940, with a second edition 1964. Lundberg talked about a **compound** Poisson process, but since that nowadays means a different process we will use the word **mixed**. As a synonym of 'mixed', one finds 'weighted' in the literature. The mixed Poisson processes have ever since played a prominent rôle in actuarial mathematics, but they have also sometimes caused misunderstandings.

Ove Lundberg's starting point was a practical problem. He, like Dubourdieu, wanted to describe the number of claims, in fixed periods, occurring in sickness and accident insurance. It was known from investigations by Pólya (1930), Eggenberger (1924), and Newbold (1927) that the negative binomial distribution is – in this kind of applications – more realistic than the Poisson distribution. His wish to understand these questions more theoretically led to his thesis. Ove Lundberg has, both personally and in his thesis, acknowledged the importance of the influence of William Feller, whom at that time belonged to Harald Cramér's group of probabilists in Stockholm.

The present book can be looked upon as a detailed survey, and contains no essential new results. It differs from other surveys – as those given by Albrecht (1985), Haight (1967), and Jung and Lundberg (1969) – in the sense that we have tried to combine 'insurance mathematics' and 'point process theory'. Mixed Poisson processes have mainly been studied by scientists with their primary interests in either insurance or point processes, and often those two groups seem not to have been aware of each other. Not too seldom, results originating in Ove Lundberg's thesis have been rediscovered much later.

In the introduction – Chapter 1 – we try to give a rather intuitive background to the subject. Especially we discuss the relation between accident proneness and contagion.

Chapter 2 deals with the mixed Poisson distribution, which means that the mean, or intensity, of a Poisson distribution is regarded as a random variable Λ. A usual assumption is that Λ is Γ-distributed, which is the only case when the best estimate of the outcome of Λ is linear. We further consider thinning and infinite divisibility. Some examples are given when Λ has other distributions than the Γ-distribution.

In Chapter 3 we have collected some basic definitions and results about point processes, Markov processes, and martingales.

Chapter 4 is, to a great extent, a slightly modernized summary of Lundberg's work. A careful reading of Lundberg's thesis shows that it is amazingly modern, which is reflected by the words 'slightly modernized'. Most – or almost all – of the modernization lies in the definition of a mixed Poisson process as a Poisson process with random intensity. Although very natural, this definition is based on distributions of point processes – a concept belonging to the more recent development of stochastic processes.

Certain results are more easily proved, or can be more satisfactory stated, for mixed Poisson processes than for mixed Poisson distributions. One explanation is that Cox processes, where the Poisson intensity is varying and random, is the natural process correspondence to the mixed Poisson distribution. Cox processes are discussed in Chapter 5 together with Gauss–Poisson processes and mixed renewal processes. The latter class of processes seems not to have been much considered in the literature. Although we are not at all convinced about their practical relevance, mixed renewal processes are a natural extension of mixed Poisson processes.

Chapter 6 deals with characterization of mixed Poisson processes. Here it is of the utmost importance to specify the class within which the characterization holds. Lundberg gave such several characterizations within the class of birth processes. Although the concept of martingales was unknown in 1940, Lundberg gave a characterization, which now would be called a martingale characterization. Some of Lundberg's characterizations essentially hold also within the framework of stationary and general point processes. Newer characterizations in terms of thinning, random translations, Palm probabilities, and symmetric distributions are given.

The compound mixed Poisson distribution is obtained as a random sum of independent identically distributed random variables. In Chapter 8 we consider bounds, asymptotic behaviour, and recursive evaluation of probabilities related to such sums. It turns out that the bounds are related to reliability properties of the underlying random variable Λ. These properties are discussed in Chapter 7.

Models for the risk fluctuations in an insurance business are considered in Chapter 9. The fundamental difference between mixed Poisson processes and Ammeter processes – introduced by Ammeter (1948) – is emphasized.

One of the most important and difficult questions in an application of the mixed Poisson process is to estimate the distribution of the random variable Λ from data. From a purely mathematical point of view this is not at all a trivial problem, but we will not discuss it at all. The reason is that we have nothing to add to Albrecht (1982), a paper which we strongly advice a reader to consult.

The end of proofs is marked by ■ and the end of examples and remarks is marked by □.

It is a pleasure to thank Boualem Djehiche, Paul Embrechts, Jan Enger, Bronius Grigelionis, Jan Jung, Vladimir Kalashnikov, Claudia Klüppelberg, Ove Lundberg, Thomas Mikosch, Hanspeter Schmidli, Torbjörn Thedéen, Olof Thorin, and Gordon Willmot for their substantial help in the preparation of this monograph. I am also highly indebted to Nicos Yannaros who allowed me to freely use his unpublished results, leading to Proposition 2.3, and to Gunnar Karlsson who many times helped me with fine details in the typesetting system TEX.

Jan Grandell
Stockholm, March 1997

CHAPTER 1

Introduction

In every person's life, now and then, rare unpredictable events occur. The word 'unpredictable' is here meant to be understood in an 'every day' sense and not strictly mathematically.

Examples of those kinds of rare events we have in mind are:

to win a lot of money in a game;
to have a housebreaking or burglary at one's home;
to become unemployed;
to be involved in a traffic accident;
to become seriously sick;
to die.

Only the first example may be regarded as 'pleasant'. The other examples are 'unpleasant', and one naturally wants to avoid them or at least to reduce their consequences.

One way to control the consequences is to insure oneself and in some of the examples above insurance may be compulsory. Typical for events which it is possible to insure against is a certain unpredictability or randomness. Although this book is not going to treat insurance mathematics, we will often have actuarial applications in mind. Consequently the events will mostly be 'unpleasant'.

Other kinds of rare events of great importance in a person's life are:

to inherit a lot of money;
to marry;
to separate;
to have children.

These events, although they sometimes may be regarded as unpredictable, are of a different kind, which we do not have in mind. Especially they are not possible to insure 'against', at least if we disregard certain speculative insurance contracts which might be accepted by Lloyd's.

We will mainly be interested in events where randomness enters in several ways:

the number of events;
the times of the events;
the consequences of the events.

Naturally the situation may differ if the events are looked at from an individual point of view or from the point of view of a society or an insurance company. A drastic example of this difference is the event 'to die'.

In order to be more concrete, let N be a random variable describing the number of non-fatal traffic accidents in which an individual is involved during a year. During that year the individual is exposed to many risky situations, of which only a few – in most years hopefully none – develop into accidents. A natural model for the completely random situation is that N is **Poisson distributed** with mean α, i.e.

$$P\{N = k\} = \frac{\alpha^k}{k!} e^{-\alpha}, \quad k = 0, 1, \ldots \tag{1.1}$$

The value of α depends on factors like:

the distance travelled;
the traffic environment;
the skillfulness of the individual as a driver.

In any case, we will regard α as a parameter characterizing the individual's **accident proneness**. Suppose now that we consider the same individual during several years. Naturally the value of α may be different from year to year, due to changes in the individual's behaviour. We will, however, disregard this and assume that the number of accidents in different years are independent and Poisson distributed with the same α. A slight generalization of this assumption is to assume that the accident pattern is described by a **Poisson process** with intensity α, see Definition 3.4 for details.

Let us now consider the number of traffic accidents for several individuals during a year, all driving about the same distance and in a similar traffic environment. In spite of these similar features, the numbers of accidents will most probably vary more than they 'ought' to do, if they all were regarded as independent observations from a

Poisson distribution. A natural explanation is that different individuals have different accident proneness, i.e. each individual corresponds a 'personal' α-value. Naturally it is difficult – or more or less impossible – to determine such an individual α-value, unless the individual accident pattern is known. Since traffic accidents – luckily enough – are rare events, this means that an individual has to be observed for a long time before reasonably certain conclusions about α can be drawn. Despite this difficulty, we suppose that, to each individual, there corresponds a personal Poisson process, describing future accidents.

Although the personal α is unknown, we may very well have a general idea about its variation from individual to individual, at least if all individuals form a reasonably homogeneous group. Therefore it seems natural to regard a personal α as the outcome of a random variable Λ with known distribution U. The distribution U is called the **structure distribution** (of the group) and may be regarded as a **prior distribution**. The notion of a prior distribution usually refers to a description of a subjective opinion or belief, but we will not discuss the philosophical questions connected with subjective probabilities and Bayesian statistics. We may then interpret (1.1) as the conditional distribution of N, given the outcome $\Lambda = \alpha$. The unconditional distribution of N is in that case given by

$$P\{N = k\} = \int_{0-}^{\infty} \frac{\ell^k}{k!} e^{-\ell} \, dU(\ell), \quad k = 0, 1, \ldots \tag{1.2}$$

This distribution is called a **mixed Poisson distribution** and we have, see Proposition 2.1,

$$E[N] = E[\Lambda] \quad \text{and} \quad \mathrm{Var}[N] = E[\Lambda] + \mathrm{Var}[\Lambda] \geq E[\Lambda].$$

These relations may be compared with the Poisson case, which corresponds to $\mathrm{Var}[\Lambda] = 0$, where $E[N] = \mathrm{Var}[N]$. The mixed Poisson distribution is thus **over-dispersed** relative to the Poisson distribution. Intuitively this is quite natural, since – see Remark 2.1 – $\mathrm{Var}[\Lambda]$ may be interpreted as a measure of the intensity variation and $E[\Lambda]$ as related to the Poisson variation. The latter interpretation comes from $\mathrm{Var}[N - \Lambda] = E[\Lambda]$.

Our discussion has, up to now, closely followed the pioneering works of Greenwood and Yule (1920) and Newbold (1926). They regarded the individual accident proneness as the outcome of a Γ-distributed random variable with density u given by

$$u(\ell) = \frac{\beta^\gamma}{\Gamma(\gamma)} \ell^{\gamma-1} e^{-\beta\ell}, \quad \ell \geq 0, \tag{1.3}$$

where γ and β are positive parameters. In this case N is **negative binomially distributed** with

$$P\{N = k\} = \binom{\gamma + k - 1}{k} \left(\frac{\beta}{\beta+1}\right)^\gamma \left(\frac{1}{\beta+1}\right)^k, \quad k = 0, 1, \ldots.$$

Naturally the mixed Poisson distribution is, in general, more 'complicated' than the Poisson distribution. It may, however, be noticed that $\gamma = 1$, which means that Λ is **exponentially distributed**, implies

$$P\{N = k\} = \left(\frac{\beta}{\beta+1}\right) \left(\frac{1}{\beta+1}\right)^k, \quad k = 0, 1, \ldots.$$

In this case N has a **geometric distribution**; a distribution which might be regarded as even 'simpler' than the Poisson distribution.

Similarly we can define the **mixed Poisson process**, by considering personal – or conditional – Poisson processes. The mixed Poisson process, with its structure distribution given by (1.3) is called a **Pólya process**. Consider now an individual, who is insured against traffic accidents. The individual was insured at time $t = 0$. Up to time t there have been $N(t)$ claims, and the insurance company wants to predict the number of claims during the 'next' premium period $(t, t+h]$, i.e. the company wants to predict $N(t+h) - N(t)$, by using its information particular for this individual. A natural predictor is

$$\begin{aligned} &E[N(t+h) - N(t) \mid N(t) = n] \\ &= E[E[N(t+h) - N(t) \mid \Lambda,\ N(t) = n] \mid N(t) = n] \\ &= hE[\Lambda \mid N(t) = n]. \end{aligned}$$

In the Pólya case we have

$$E[\Lambda \mid N(t) = n] = \frac{\gamma + n}{\beta + t}. \tag{1.4}$$

The given motivation for the Pólya or – more generally – the mixed Poisson process may be summarized in the following three postulates, where 'N' stands for Newbold.

N1: Individuals differ from each other in accident proneness.

N2: No contagion, i.e. accidents incurred in the past do not influence accidents in the future.

N3: The individual accident proneness remains constant in time.

Pólya (1930) considered a very different scheme, which becomes more natural in the process setting. Let us think of the number of sicknesses instead of accidents, and we now assume that all individuals have the same sickness proneness at time $t = 0$. One may think of a rather homogeneous group of individuals, taking a sickness insurance. Let $N(t)$ describe the number of sicknesses in $(0, t]$ for an individual member of the group.

Suppose first that at time $t = 0$ we want to predict $N(h)$, where h is small. For some known value $\kappa_0(0)$, equal for all individuals in the group, we have

$$\begin{aligned} P\{N(h) = 0\} &\approx 1 - \kappa_0(0) \cdot h \\ P\{N(h) = 1\} &\approx \kappa_0(0) \cdot h. \end{aligned} \tag{1.5}$$

For small values of h we have

$$P\{N(h) = 1\} \approx E[N(h)] = hE[\Lambda],$$

and therefore to predict $N(h)$ is (almost) the same as to compute the probability $P\{N(h) = 1\}$.

Now we go forward to time t. To predict $N(t + h) - N(t)$ is then, as discussed above, almost the same as to compute the conditional probability

$$P\{N(t + h) - N(t) = 1 \mid \text{the sickness pattern in } (0, t]\}.$$

The dependence on the sickness pattern may be looked upon as caused by contagion. Let us postulate that the precise times of the sicknesses do not influence the conditional probability above. This is the same as to postulate that $N(\cdot)$ is a **Markov process**. In fact, we assume, for $t > 0$, that

$$P\{N(t + h) - N(t) = k \mid N(t) = n\}$$

$$= \begin{cases} 1 - \kappa_n(t)h + o(h), & k = 0, \\ \kappa_n(t)h + o(h), & k = 1, \\ o(h), & k > 1, \end{cases}$$

which means that $N(\cdot)$ is a **birth process**. As usual $o(h)$ denotes a quantity which, for small h, is of lower order than h, i.e.

$$\lim_{h \downarrow 0} \frac{o(h)}{h} = 0.$$

It is natural to suppose that the **transition intensity** $\kappa_n(t)$ is increasing in in n, meaning that many diseases make the individual more liable to new diseases, and decreasing in t. A simple model, fulfilling these requirements is

$$\kappa_n(t) = \frac{\gamma + an}{\beta + bt}, \quad \gamma,\, a,\, \beta,\, b \geq 0.$$

If $a = b = 0$, i.e. if there is no contagion present, then $N(\cdot)$ is a Poisson process with intensity γ/β. Assume that $a > 0$. We may then reduce with a and reparameterize the model, so there is no restriction to let $a = 1$. Therefore we consider

$$\kappa_n(t) = \frac{\gamma + n}{\beta + bt}, \quad \gamma,\, \beta,\, b \geq 0. \tag{1.6}$$

Assume now – this is a stationarity assumption – that

$$P\{N(h) = 1\} = P\{N(t+h) - N(t) = 1\}, \tag{1.7}$$

which is highly natural, at least for t so small that the fact that the individual has become older has no essential influence. Let, in fact, $t = h$. Then (1.7) can be written as

$$P\{N(h) = 1\} = P\{N(2h) - N(h) = 1\}$$

$$= P\{N(h) = 0\}P\{N(t+h) - N(t) = 1 \mid N(h) = 0\}$$
$$+ P\{N(h) = 1\}P\{N(t+h) - N(t) = 1 \mid N(h) = 1\},$$

or, equivalently,

$$\kappa_0(0)h = (1 - \kappa_0(0)h)\kappa_0(h)h + \kappa_0(0)h\kappa_1(h)h + o(h^2).$$

Using (1.6) we get, after multiplication with $(\beta + bh)\beta/(\gamma h)$,

$$\beta + bh = \left(1 - \frac{\gamma}{\beta}h\right)\beta + (\gamma + 1)h + o(h) = \beta + h + o(h).$$

Thus we must have $b = 1$. Comparing (1.4) and (1.6), it follows that $N(\cdot)$ is a Pólya process. This is the first non-elementary result used here, and we have to rely on the remarks after Definition 3.6 and Proposition 4.1.

The given motivation for the Pólya process is a modernized version of the scheme considered by Pólya (1930). Pólya's original scheme was formulated in terms of drawings from an urn, and that led to the assumption (1.6) with $b = 1$. To summarize, the Pólya scheme is based on the following three postulates:

P1: Individuals have identical sickness proneness.

P2: Contagion is present as described above.

P3: The sickness proneness depends on time.

Although P2 and P3 are related, we see that N1–N3 and P1–P3 are contradictory. Going back to traffic accidents, P3 may be interpreted as if the accident proneness decreases as the driver's experience increases – practice makes perfect. In spite of that, the two sets of postulates lead to the same model.

Certainly it may seem surprising that two contradictory sets of postulates give rise to the same model. This was first observed by Ove Lundberg in his thesis *On random processes and their application to sickness and accident statistics* from 1940 and further discussed by Feller (1943). As mentioned in the preface, a second edition of Lundberg's thesis appeared 1964, and we have chosen to refer to that edition since we believe that it is more easily available. In this thesis, mixed Poisson **processes** were thoroughly studied for the first time. Lundberg based his treatment of mixed Poisson processes on the theory of birth processes. In fact, he also extended the theory of Markov processes, in order to be able to consider sufficiently general birth processes. He defined a mixed Poisson process $N(\cdot)$ as a birth process fulfilling

$$P\{N(t) = k\} = \int_{0-}^{\infty} \frac{(\ell t)^k}{k!} e^{-\ell t} \, dU(\ell), \quad t \geq 0, \; k = 0, 1, \ldots,$$

for some distribution U. This was a very natural definition at that time. Although the mixed Poisson distribution, as a mixture of Poisson distributions, was well understood, the theory of stochastic processes was not enough developed to allow a 'similar' definition of mixed Poisson processes. When we indicated that definition above, we relied on more recent theory. These questions will be discussed in detail in Chapter 4.

Let, as above, Λ be a random variable with distribution U. It is not too difficult to see that the corresponding mixed Poisson process is a birth process with its transition intensities given by

$$\kappa_n(t) = E[\Lambda \mid N(t) = n]. \tag{1.8}$$

The converse problem, to characterize those birth processes which are mixed Poisson processes, is much more difficult. Lundberg has given several such characterizations, see Theorem 6.1, of which we here only mention the following:

A birth process with transition intensities $\kappa_n(t)$ is a mixed Poisson process if and only if

$$\kappa_{n+1}(t) = \kappa_n(t) - \frac{\kappa_n'(t)}{\kappa_n(t)}. \tag{1.9}$$

Consider $\kappa_n(t)$ given by (1.6) again. Simple calculations yield

$$\kappa_n(t) - \frac{\kappa_n'(t)}{\kappa_n(t)} = \frac{\gamma + n + b}{\beta + bt}.$$

Thus a birth process with $\kappa_n(t)$ given by (1.6) is not a mixed Poisson process unless $b = 1$. This indicates that (1.9) is a restrictive condition.

In order to understand how restrictive (1.9) really is, let us go back to (1.4), or (1.6) with $b = 1$, and suppose that we are dealing with contagion. A first generalization of (1.4) may be to consider $\kappa_n(t) = \kappa_n/(\beta + t)$, i.e. to allow for a non-linear dependence. A further generalization is to allow for a more general dependence of time. A natural extension of (1.4) might then be

$$\kappa_n(t) = \kappa_n \cdot v(t), \tag{1.10}$$

for some sequence $\{\kappa_n\}$ and some function $v(t)$. However, see Theorem 4.1, the so-defined birth process will be a mixed Poisson process, if and only if $\kappa_n(t)$ is given be (1.4).

A very interesting discussion about accident proneness versus contagion is found in Bates and Neyman (1952a, b). Much of our discussion has been inspired by these two papers. Surprisingly, they seem to have been little noticed in actuarial literature. It seems natural to base a model for accident proneness on an assumption of the distribution U of Λ, and a model for contagion on the intensities $\kappa_n(t)$. Bates and Neyman (1952b) came to the conclusion that it is 'highly unlikely' that a model for contagion shall 'happen' to coincide with a model for accident proneness. Within the process version, this is the same as saying that it is highly unlikely that intensities $\kappa_n(t)$, motivated by theoretical or empirical considerations about contagion, shall satisfy (1.9).

Bates and Neyman (1952a, b) discuss the relation between light and severe accidents. A light accident is an accident which the victim surely survives, while a severe accident is an accident which may be fatal. In these terms our discussion has been about light accidents.

Following Bates and Neyman, let us think about an airline company which is going to employ a pilot. The airline company knows that the applicant, during a time period of length t, has had N_L light accidents of a kind which ought to be related to his proneness for aircraft crashes. Regarding an aircraft crash as certainly fatal, a natural quantity to estimate is

$$P\{N_S = 0 \mid N_L = n\}, \tag{1.11}$$

where N_S is the number – 0 or 1 – of future aircraft crashes during the planned length s of the employment.

Let, as before, α denote the individual proneness for the light accidents. The individual proneness for aircraft crashes is denoted by α_S. The relation between the proneness for light accidents and for future aircraft crashes is assumed to be

$$\alpha_S = q \cdot \alpha,$$

for some, probably small, q. This assumption is certainly very strong. Bates and Neyman did, however, from empirical investigations, reach the conclusion that the assumption may lead to reasonable consequences.

Now we regard α as the outcome of a random variable Λ with distribution U. Let $N(\cdot)$ be the corresponding mixed Poisson process. The conditional probability in (1.11) is then equivalent to

$$P\{N(t+qs) - N(t) = 0 \mid N(t) = n\}. \tag{1.12}$$

Put $\kappa_n(t) = E[\Lambda \mid N(t) = n]$, as before. Let us first consider the case $t = 0$ and $n = 0$, and put

$$p_n(s) = P\{N(s) = n\}.$$

By the 'usual' arguments for Markov processes, see (4.2), we get

$$p_0'(s) = -\kappa_0(s)p_0(s),$$

which has the solution

$$p_0(s) = e^{-\int_0^s \kappa_0(\tau)\, d\tau}.$$

Assume now that $N(t) = n$ and consider the point process $N^*(\cdot)$ defined by $N^*(s) = N(t+s) - N(t)$. Then $N^*(\cdot)$ is a birth process, for details we refer to Chapter 4, with

$$\kappa_0^*(s) = \kappa_n(t+s).$$

Thus we get

$$P\{N(t+qs) - N(t) = 0 \mid N(t) = n\} = e^{-\int_0^{qs} \kappa_n(t+\tau)\, d\tau}$$

$$= e^{-\int_t^{t+qs} \kappa_n(\tau)\, d\tau} = e^{-\int_t^{t+qs} E[\Lambda | N(\tau)=n]\, d\tau}.$$

A similar relation has been shown by Rubin (1972, p.549) to hold also for point processes which are not birth processes. In that case $E[\Lambda \mid N(\tau) = n]$ has to be replaced by the so called $\boldsymbol{F}^N$**-intensity** of N, see Definition 3.7.

Bates and Neyman continue this discussion further and discuss more systematically 'selection models', which mathematically means that they consider the conditional probabilities

$$P\{N_S = 0 \mid N_L \leq n\}.$$

For further results in this direction, we refer to Haight (1965).

As indicated above, light and severe accidents have different consequences. When we look upon consequences, we will have actuarial applications in mind. An 'accident' will then correspond to a claim and its consequence to the cost for the insurance company to cover its liability. The costs of the claims will always be described by a sequence $\{Z_k\}_1^\infty$ of independent and identically distributed random variables, which further is independent of $N(\cdot)$. The total amount of claims paid by the company in the interval $(0, t]$ is then described by the **claim process**

$$Y(t) = \sum_{k=1}^{N(t)} Z_k, \quad \left(\sum_{k=1}^{0} Z_k \stackrel{\text{def}}{=} 0 \right).$$

Let us first consider a fixed value of t. The random variable $Y = Y(t)$ is said to be **compound mixed Poisson distributed**. Let $F_Y(y) = P\{Y \leq y\}$. For the insurance company, upper bounds for and asymptotic behaviour of $1 - F_Y(y)$ are of great importance, especially for large values of y. For moderate values of y the distribution of Y can often be found by recursive evaluation. These questions will be discussed in Chapter 8.

Chapter 9 is devoted to the 'process version', i.e. to the claim process. An important part of insurance mathematics is **ruin theory**. Roughly speaking, we then consider a model of the risk business of an insurance company and study the probability of ruin, i.e. the probability that the risk business ever will be below some specified (negative) value. The risk business is described by the **risk process**, $X(\cdot)$, defined by

$$X(t) = ct - Y(t),$$

where ct is the premium income in the interval $(0, t]$. More precisely, the **ruin probability** $\Psi(u)$ of a company facing the risk process $X(\cdot)$ and having initial capital u is defined by

$$\Psi(u) = P\{u + X(t) < 0 \text{ for some } t > 0\}.$$

Although a reader is assumed to be acquainted with classical ruin theory, i.e. where $N(\cdot)$ is a Poisson process, some basic facts will be given in that case.

Apart from the case where $N(\cdot)$ is a mixed Poisson process, we will consider a very different, although related, model due to Ammeter (1948). We will define the **Ammeter process** within the framework of **Cox processes**, see Chapter 5. Loosely speaking, an Ammeter process is built up by independent mixed Poisson process in the following way:

Let $\Delta > 0$ be fixed. Originally Ammeter considered the random variable $N(\Delta)$. With similar arguments as in Greenwood and Yule (1920) and Newbold (1926), he let $N(\Delta)$ be mixed Poisson distributed with U given by (1.3). He then considered the discrete parameter process $\{N(k \cdot \Delta)\}$, where the random variables

$$N(\Delta),\ N(2\Delta) - N(\Delta),\ N(3\Delta) - N(2\Delta), \ldots$$

are independent and identically distributed.

An Ammeter process is a continuous time version of the original Ammeter model where U is not necessarily a Γ-distribution. We refer to Chapters 5 and 9 for the precise definition.

CHAPTER 2

The mixed Poisson distribution

Most of the historical misunderstandings about mixed Poisson processes, mentioned in the Preface, have been due to confusion of distributions of random variables and processes. In this chapter we will only consider distributions of random variables. As a convention, we will understand that a discrete random variable is non-negative integer-valued.

Definition 2.1. *A discrete random variable N is said to be* **Poisson distributed**, Po(α), *if, for some $\alpha \geq 0$,*

$$\pi_{\alpha,k} \stackrel{\text{def}}{=} P\{N = k\} = \frac{\alpha^k}{k!}\, e^{-\alpha}, \quad k = 0, 1, \ldots$$

Let U be the distribution function of a non-negative random variable $\varLambda$.

Definition 2.2. *A discrete random variable N is said to be* **mixed Poisson distributed**, MP(t, U), *with* **structure distribution** U *if*

$$p_k(t) \stackrel{\text{def}}{=} P\{N = k\} = E\left[\frac{(\varLambda t)^k}{k!}\, e^{-\varLambda t}\right]$$

$$= \int_{0-}^{\infty} \frac{(\ell t)^k}{k!}\, e^{-\ell t}\, dU(\ell), \quad k = 0, 1, \ldots$$

The t appearing in Definition 2.2 may, at the moment, seem confusing but it will turn out to be practical. We do, however, sometimes denote MP($1, U$) by MP(U). Let U_t be the distribution function of $\varLambda t$, i.e. $U_t(\ell) = U(\ell/t)$. It is easy to see that MP(t, U) = MP(U_t).

It is often convenient to discuss the random variables $\varLambda$ and N simultaneously. Formally this means that $\varLambda$ and N must be defined on the same probability space, or – which is the same – that we consider the random vector ($\varLambda$, N). The distribution of that vector is, with a slight extension of Definition 2.2, given by

$$P\{\Lambda \le x,\, N = n\} \stackrel{\text{def}}{=} \int_{0-}^{x} \frac{(\ell t)^n}{n!}\, e^{-\ell t}\, dU(\ell), \tag{2.1}$$

for $x \ge 0$ and $n = 0, 1, \ldots$

We collect some simple and well-known properties of the mixed Poisson distribution in Proposition 2.1.

Proposition 2.1. *Let N be $\mathrm{MP}(t, U)$, where U is the distribution of a non-negative random variable Λ with mean μ_Λ and variance σ^2_Λ. Then*

(i) $E[N] = t\mu_\Lambda$;

(ii) $\mathrm{Var}[N] = t\mu_\Lambda + t^2\sigma^2_\Lambda$;

(iii) $\frac{1}{t}P\{N > n\} = \int_0^\infty \frac{(\ell t)^n}{n!}\, e^{-\ell t}(1 - U(\ell))\, d\ell$.

(iv) $$P\{\Lambda \le x \mid N = n\} = \frac{\int_{0-}^{x} \ell^n e^{-\ell t}\, dU(\ell)}{\int_{0-}^{\infty} \ell^n e^{-\ell t}\, dU(\ell)}; \tag{2.2}$$

(v) $$E[\Lambda \mid N = n] = \frac{\int_{0-}^{\infty} \ell^{n+1} e^{-\ell t}\, dU(\ell)}{\int_{0-}^{\infty} \ell^n e^{-\ell t}\, dU(\ell)}; \tag{2.3}$$

(vi) *The generating function $G_N(s)$ of N is given by*

$$G_N(s) \stackrel{\text{def}}{=} E[s^N] = \hat{u}(t(1-s)), \quad s \le 1;$$

where $\hat{u}(v) = \int_{0-}^{\infty} e^{-\ell v}\, dU(\ell)$ is the Laplace transform of Λ.

Proof. We have

$$E[N] = E[E[N \mid \Lambda]] = E[\Lambda t] = t\mu_\Lambda$$

and

$$\begin{aligned}\mathrm{Var}[N] &= E[\mathrm{Var}[N \mid \Lambda]] + \mathrm{Var}[E[N \mid \Lambda]] \\ &= E[\Lambda t] + \mathrm{Var}[\Lambda t] = t\mu_\Lambda + t^2\sigma^2_\Lambda,\end{aligned}$$

which prove (i) and (ii).

Integration by parts yields

$$P\{N > n\} = \int_0^\infty \sum_{k=n+1}^{\infty} \frac{(\ell t)^k}{k!}\, e^{-\ell t}\, dU(d\ell)$$

$$= \int_0^\infty \sum_{k=n+1}^\infty \left(\frac{t(\ell t)^{k-1}}{(k-1)!} e^{-\ell t} - \frac{t(\ell t)^k}{k!} e^{-\ell t} \right) (1 - U(\ell))\, d\ell$$

$$= t \int_0^\infty \frac{(\ell t)^n}{n!} e^{-\ell t}(1 - U(\ell))\, d\ell,$$

which proves (iii).

From (2.1) we get

$$P\{\Lambda \le x \mid N = n\} = \frac{P\{\Lambda \le x,\, N = n\}}{P\{N = n\}}$$

$$= \frac{\int_{0-}^x \frac{(\ell t)^n}{n!} e^{-\ell t}\, dU(\ell)}{\int_{0-}^\infty \frac{(\ell t)^n}{n!} e^{-\ell t}\, dU(\ell)} = \frac{\int_{0-}^x \ell^n e^{-\ell t}\, dU(\ell)}{\int_{0-}^\infty \ell^n e^{-\ell t}\, dU(\ell)},$$

which proves (iv). Item (v) follows from

$$E[\Lambda \mid N = n] \stackrel{\text{def}}{=} \int_0^\infty x\, dP\{\Lambda \le x \mid N = n\}.$$

Further

$$G_N(s) = \sum_{k=0}^\infty s^k p_k(t) = \int_{0-}^\infty \sum_{k=0}^\infty s^k \frac{(\ell t)^k}{k!} e^{-\ell t}\, dU(\ell)$$

$$= \int_{0-}^\infty e^{s\ell t} e^{-\ell t}\, dU(\ell) = \hat{u}(t(1-s)),$$

which proves (vi). ■

Remark 2.1. Items (i) and (ii) in Proposition 2.1 also follow by differentiation of $G_N(s)$. More generally we then get, cf. Ottestad (1944, p.34), the factorial moments

$$E[N(N-1)\dots(N-k+1)] = t^k E[\Lambda^k], \quad k = 1,\ 2, \dots$$

Another way to understand (ii) is to consider

$$N = (N - t\Lambda) + t\Lambda,$$

and to observe that

$$\mathrm{Cov}[N - t\Lambda, t\Lambda] = E[(N - t\Lambda)\, t\Lambda] = E[E[(N - t\Lambda)\, t\Lambda \mid \Lambda]] = 0.$$

We may interpret $t\Lambda$ as a 'signal' and $N - t\Lambda$ as 'noise'. Notice, however, that

$$\mathrm{Var}[N - t\Lambda \mid \Lambda] = t\Lambda,$$

which implies that $t\Lambda$ and $N - t\Lambda$ are not independent.

In this context it is natural to interpret $\mathrm{Var}[t\Lambda] = t^2\sigma_\Lambda$ as the **intensity variation** and $\mathrm{Var}[N - t\Lambda] = t\mu_\Lambda$ as the **Poisson variation**.

□

The following theorem, in which (i) was noticed by Feller (1943, p.392), is essentially a corollary to Proposition 2.1 (vi). Let $\stackrel{d}{=}$ mean 'equality in distribution' and let $\stackrel{d}{\to}$ mean 'convergence in distribution'. With the notation in the theorem, $\Lambda_1 \stackrel{d}{=} \Lambda_2$ means that $U_1(\ell) = U_2(\ell)$ for all ℓ and $\Lambda_n \stackrel{d}{\to} \Lambda$ means that

$$\lim_{n\to\infty} U_n(\ell) = U(\ell) \text{ at each point of continuity of } U.$$

Theorem 2.1. *Let Λ, Λ_1, $\Lambda_2, \ldots$ be non-negative random variables with distributions U, U_1, $U_2, \ldots$ and let N, N_1, $N_2, \ldots$ be* MP(U), MP(U_1), MP(U_2)... *respectively. Then*

(*i*) $N_1 \stackrel{d}{=} N_2$ *if and only if* $\Lambda_1 \stackrel{d}{=} \Lambda_2$;

(*ii*) $N_n \stackrel{d}{\to} N$ *if and only if* $\Lambda_n \stackrel{d}{\to} \Lambda$;

(*iii*) *If* $N_n \stackrel{d}{\to} \widetilde{N}$, *where* $\widetilde{N}$ *is some discrete random variable, then* $\widetilde{N}$ *is* MP.

Proof. (i) If $\Lambda_1 \stackrel{d}{=} \Lambda_2$ then $N_1 \stackrel{d}{=} N_2$ follows by definition.

If $N_1 \stackrel{d}{=} N_2$ it follows from Proposition 2.1 (vi) that $\hat{u}_1(v) = \hat{u}_2(v)$ for $0 \le v \le 1$. Since a Laplace transform is uniquely determined by its values on $[0, 1]$, cf. Feller (1971, p.430), $\Lambda_1 \stackrel{d}{=} \Lambda_2$ follows from the uniqueness theorem for Laplace transforms.

(ii) If $\Lambda_n \stackrel{d}{\to} \Lambda$ then, by Proposition 2.1 (vi) and the continuity theorems for Laplace transforms and generating functions, $\hat{u}_n(v) \to \hat{u}(v)$ and thus $G_{N_n}(s) \to G_N(s)$ which implies $N_n \stackrel{d}{\to} N$.

If $N_n \stackrel{d}{\to} N$ we get, as above, $\hat{u}_n(v) \to \hat{u}(v)$ for $0 \le v \le 1$. By a slight generalization of the continuity theorem for Laplace transforms, cf. Kallenberg (1983, p.167), $\Lambda_n \stackrel{d}{\to} \Lambda$ follows.

(iii) We have $G_{N_n}(s) \to G_{\tilde{N}}(s)$ and thus $\hat{u}_n(v) \to G_{\tilde{N}}(1-v)$ for $0 \le v \le 1$. Since $G_{\tilde{N}}(1-t) \to G_{\tilde{N}}(1) = 1$ as $t \to 0$, the desired result follows from (ii). ■

Example 2.1. The most common choice of the structure distribution U is certainly the **Γ-distribution**. Then

$$u(\ell) \stackrel{\text{def}}{=} U'(\ell) = \frac{\beta^\gamma}{\Gamma(\gamma)} \ell^{\gamma-1} e^{-\beta\ell}, \quad \ell \ge 0,$$

where γ and β are positive parameters and $\Gamma(\gamma)$ is the Γ-function, defined by

$$\Gamma(\gamma) \stackrel{\text{def}}{=} \int_0^\infty x^{\gamma-1} e^{-x}\, dx, \quad \gamma > 0.$$

γ is called the shape parameter and β the scale parameter. In this case we say that Λ is $\Gamma(\gamma, \beta)$.

Well-known properties of the Γ-function are

$$\Gamma(n+1) = n!, \quad \Gamma(\gamma+1) = \gamma\Gamma(\gamma) \quad \text{and} \quad \Gamma(\tfrac{1}{2}) = \sqrt{\pi}. \tag{2.4}$$

For $\gamma = 1$ the random variable Λ is **exponentially distributed.** Further, when γ is an integer, Λ has the same distribution as the sum of γ independent and exponentially distributed random variables. Those Γ-distributions are often referred to as **Erlang distributions.**

We have, see for instance Bühlmann (1970, pp.18–19),

$$\mu_\Lambda = \frac{\gamma}{\beta}, \quad \sigma_\Lambda^2 = \frac{\gamma}{\beta^2}, \quad \hat{u}(v) = \left(1 + \frac{v}{\beta}\right)^{-\gamma}$$

and

$$p_k(t) = \binom{\gamma+k-1}{k} \left(\frac{\beta}{\beta+t}\right)^\gamma \left(\frac{t}{\beta+t}\right)^k$$

for $k = 0, 1, \ldots$, i.e. N has a **negative binomial distribution.** For $\gamma = 1$ this reduces to the **geometric distribution.**

Table 2.1. *Distributions illustrated in Figures 2.1–2.5.*

	β	γ	t	μ_Λ	σ_Λ^2	$E[N]$	Var[N]
Fig. 2.1:	0.5	0.5	10	1	2	10	210
Fig. 2.2:	1	1	10	1	1	10	110
Fig. 2.3:	2	2	10	1	0.5	10	60
Fig. 2.4:	10	10	10	1	0.1	10	20
Fig. 2.5:	Poisson		10	1	0	10	10

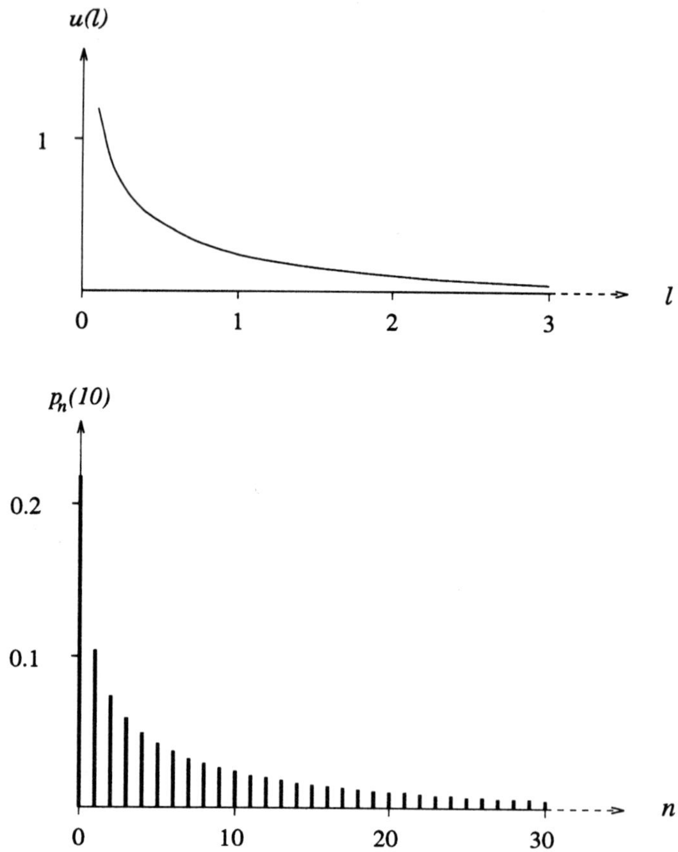

Figure 2.1. *Distributions of Λ and N for $\beta = 0.5$, $\gamma = 0.5$, and $t = 10$.*

In Figures 2.1–2.5 we illustrate the relation between the distributions of Λ and N in the cases given in Table 2.1. We have chosen the relatively large value $t = 10$, in order to be able to see how $p_k(t)$ behaves for small values of k. In Figures 2.1–2.3 the intensity variation dominates the Poisson variation, due to the choice of t. In Figure 2.4 the two kinds of variations are the same, cf. Remark 2.1. The pure Poisson variation is illustrated in Figure 2.5, which formally corresponds to $\beta = \gamma = \infty$.

It follows from (2.2) that Λ conditioned upon the outcome $N = n$

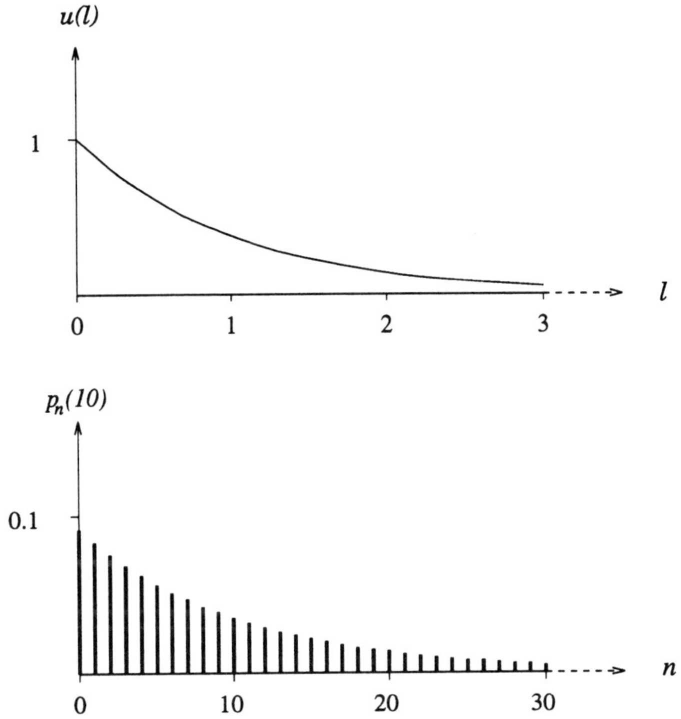

Figure 2.2. *Distributions of Λ and N for $\beta = 1$, $\gamma = 1$, and $t = 10$.*

is $\Gamma(\gamma + n,\ \beta + t)$, and thus

$$E[\Lambda \mid N = n] = \frac{\gamma + n}{\beta + t}. \tag{2.5}$$

Using Proposition 2.1 (v) and (2.5) we get

$$\frac{\gamma + n}{\beta + t} = \frac{p_{n+1}(t)(n+1)!/t^{n+1}}{p_n(t)n!/t^n} = \frac{p_{n+1}(t)(n+1)}{p_n(t)\,t}$$

or

$$p_{n+1}(t) = \frac{(\gamma + n)\,t}{(\beta + t)(n+1)}\,p_n(t), \quad n = 0, 1, 2, \ldots \tag{2.6}$$

Since $p_0(t) = \left(\frac{\beta}{\beta+t}\right)^\gamma$ we get by (2.6) a simple recursive algorithm for the distribution of N. Usually (2.6) is written in the form

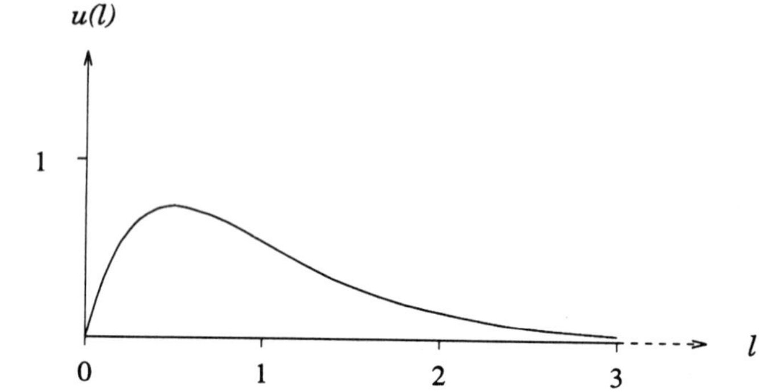

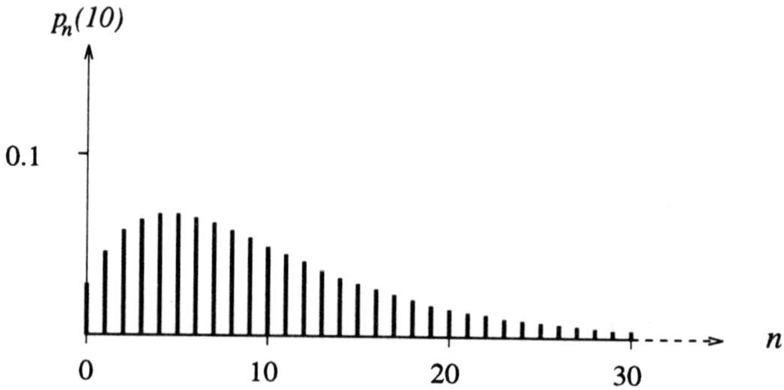

Figure 2.3. *Distributions of Λ and N for $\beta = 2$, $\gamma = 2$, and $t = 10$.*

$$p_n(t) = \left(a + \frac{b}{n}\right) p_{n-1}(t), \quad n = 1, 2, \dots, \tag{2.7}$$

where $a = \frac{t}{\beta+t}$ and $b = \frac{(\gamma-1)t}{\beta+t}$. This algorithm is the **Panjer assumption**, used in the well-known Panjer recursion, cf. Panjer (1981). The example will be continued. □

Let N_t be MP(t, U). The index t is here only meant to emphasize that the distribution of N_t depends on t. From (2.3) it seems natural to interpret $\Lambda_B^* \stackrel{\text{def}}{=} E[\Lambda \mid N_t]$ as the **best estimate** – or the Bayes estimate – of Λ. This interpretation will be discussed further in Chapter 4.

The simplest, and probably most obvious, estimate of Λ is $\frac{1}{t}N_t$. For $\frac{1}{t}N_t$ we have

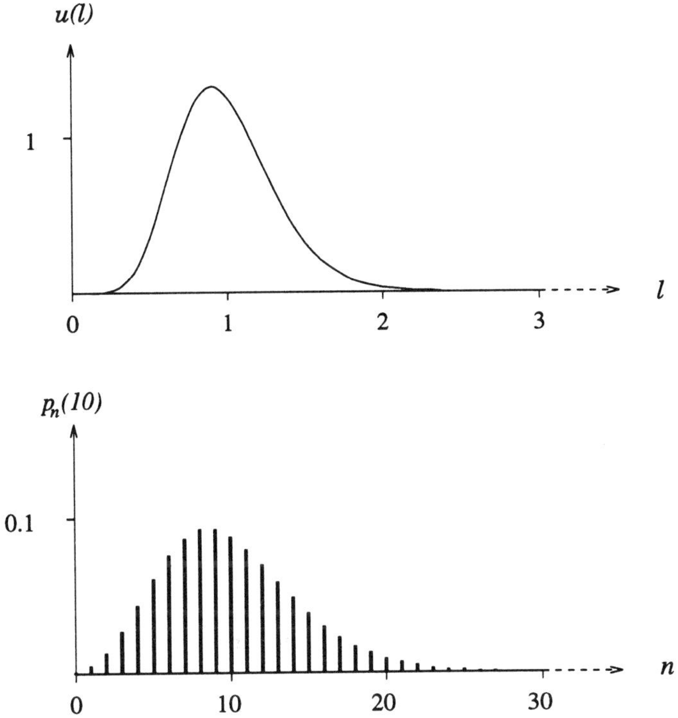

Figure 2.4. *Distributions of* Λ *and* N *for* $\beta = 10$, $\gamma = 10$, *and* $t = 10$.

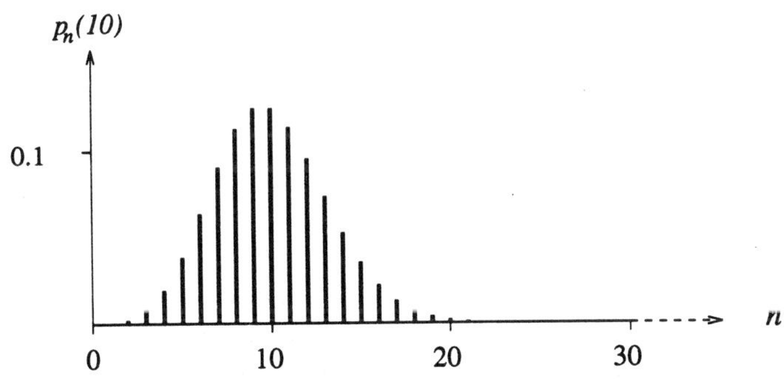

Figure 2.5. *The* Po(10)*-distribution.*

$$E[\tfrac{1}{t}N_t \mid \Lambda] = \Lambda$$

and, cf. Remark 2.1,

$$E[(\tfrac{1}{t}N_t - \Lambda)^2] = E\big[E[(\tfrac{1}{t}N_t - \Lambda)^2 \mid \Lambda]\big] = \frac{\mu_\Lambda}{t}.$$

Thus, for large values of t, we have $\frac{1}{t}N_t \approx \Lambda$. Although it is slightly close to the process setting, one might say that $\frac{1}{t}N_t \to \Lambda$, provided the convergence is interpreted in mean square, in probability, or in distribution, since these notions of convergence are only based on the distribution of (Λ, N_t). Under general assumptions Adell and de la Cal (1993) have obtained the order of convergence as $t \to \infty$. The question whether the convergence holds almost surely, on the other hand, would require a simultaneous distribution of $(\Lambda, N_t, t \geq 0)$, i.e. a process setting, cf. Proposition 4.2 below. The question is therefore meaningless in this setting.

The second simplest estimate ought to be the **best linear estimate**, which in actuarial literature is often called the **credibility mean**, Λ_L^* of Λ. This means that we consider the estimate $\Lambda_L^* = a + b\frac{1}{t}N_t$, where a and b are chosen so that $E[(\Lambda_L^* - \Lambda)^2]$ is minimized. Routine calculations, see for instance Bühlmann (1970, p.101) or Grandell (1976, p.129), yield

$$\Lambda_L^* = \frac{\mu_\Lambda^2 + \sigma_\Lambda^2 N_t}{\mu_\Lambda + \sigma_\Lambda^2 t} = \frac{\mu_\Lambda}{\mu_\Lambda + \sigma_\Lambda^2 t} \cdot \mu_\Lambda + \frac{\sigma_\Lambda^2 t}{\mu_\Lambda + \sigma_\Lambda^2 t} \cdot \tfrac{1}{t}N_t, \tag{2.8}$$

where $\frac{\sigma_\Lambda^2 t}{\mu_\Lambda + \sigma_\Lambda^2 t}$ is called the **credibility**, and

$$E[(\Lambda_L^* - \Lambda)^2] = \frac{\mu_\Lambda \sigma_\Lambda^2}{\mu_\Lambda + \sigma_\Lambda^2 t}.$$

One may notice that (2.8) also follows from (2.5) since Λ_L^* only depends on U via μ_Λ and σ_Λ^2.

Conversely, Λ_B^* is linear in N_t only when U is a Γ-distribution, as will be shown in Theorem 2.2, which is due to Johnson (1957) and (1967). The proof to be given is based on Johnson (1957). In the process setting, see Theorem 4.1, the corresponding result goes back to Lundberg (1964). Recall, here and in the sequel, that Lundberg's original thesis appeared 1940, while we have chosen to refer to the second edition.

Theorem 2.2. *Let N be* MP(U) *with* $\mu_\Lambda < \infty$. *The distribution U is a Γ-distribution or a one-point distribution if and only if Λ_B^* is linear in N.*

Proof. If U is a Γ-distribution it is already established that Λ_B^* is linear. If U is a one-point distribution, i.e. if $\Lambda = \mu_\Lambda$ almost surely, we have $\Lambda_B^* = \mu_\Lambda$.

Now we assume that Λ_B^* linear, i.e.

$$E[\Lambda \mid N = n] = a + bn \quad \text{for some } a \text{ and } b \text{ and for all } n.$$

Since $E[\Lambda \mid N = n] \geq 0$ for all n it follows that $a,\ b \geq 0$.

If $a = 0$ we have $E[\Lambda \mid N = 0] = 0$, which implies $\mu_\Lambda = 0$ since $P\{N = 0\} > 0$. Thus the theorem holds in this case, and we can assume that $a > 0$ and that $\mu_\Lambda > 0$.

Further $a + b\mu_\Lambda = \mu_\Lambda$, since $E[E[\Lambda \mid N]] = E[\Lambda] = E[N]$, and thus $a = (1-b)\mu_\Lambda$. Putting this together, it follows that we can assume

$$E[\Lambda \mid N = n] = a + bn \quad \text{for some } a > 0 \text{ and } b \in [0,1).$$

It follows from (2.3) that

$$a + bn = \frac{E[X^{n+1}]}{E[X^n]},$$

where X has distribution U_X given by

$$dU_X(x) = e^{-x}\, dU(x)/\hat{u}(1).$$

Thus

$$E[X] = a,\ E[X^2] = a(a+b), \dots,\ E[X^{n+1}] = a(a+b)\cdots(a+nb).$$

Hence $E[X^n] < (a+nb)^n$ and so

$$\limsup_{n\to\infty} \frac{1}{n} E[X^n]^{1/n} < a + b.$$

Thus, see Feller (1971, p.514), these moments determine U_X uniquely, i.e. X is $\Gamma(\frac{a}{b}, \frac{1}{b})$ if $b \in (0,1)$. From the relation between U and U_X, it then follows that Λ is $\Gamma(\frac{a}{b}, \frac{1-b}{b})$. For $b = 0$ it follows that $X = \Lambda = a$ almost surely. ■

We refer to Goel (1982) for a broader discussion on best and best linear estimation of random variables.

We will now consider (vi) in Proposition 2.1.

Definition 2.3. *A function $h(s)$ whose all nth derivatives $h^{(n)}(s)$ exist for $s \in (a,b)$ is* **absolutely monotone** *in (a,b) if*

$$h^{(n)}(s) \geq 0 \quad n = 0, 1, \ldots \text{ and } s \in (a,b)$$

and is **completely monotone** *in (a,b) if*

$$(-1)^n h^{(n)}(s) \geq 0 \quad n = 0, 1, \ldots \text{ and } s \in (a,b).$$

A function $h(s)$ is called a generating function if $h(s) = E[s^N]$ for some discrete random variable N and a Laplace transform if $h(s) = E[e^{-sX}]$ for some non-negative random variable X.

It is well known, see for instance Feller (1971, pp.223, 439), that a function $h(s)$ is a generating function if and only if $h(s)$ is absolutely monotone in $(0,1)$ and $h(1) = 1$ and that a function $h(s)$ is a Laplace transform if and only if $h(s)$ is completely monotone in $(0,\infty)$ and $h(0) = 1$.

The following proposition, due to Puri and Goldie (1979, pp.140–141), gives an analytical definition of the mixed Poisson distribution, which may at first sight look somewhat artificial. It will, however, turn out to have a natural probabilistic interpretation.

Proposition 2.2. *A discrete random variable N with generating function G_N is mixed Poisson distributed if and only G_N is absolutely monotone in $(-\infty, 1)$.*

Proof. Assume that N is MP(U) for some distribution U. From Proposition 2.1 (vi) it follows that $G_N(s) = \hat{u}(1-s)$. Thus we have

$$G^{(n)}(s) = (-1)^n \hat{u}^{(n)}(1-s) \geq 0$$

for $0 < 1 - s < \infty$ or for $s \in (-\infty, 1)$.

Assume now that $G_N(s)$ is absolutely monotone in $(-\infty, 1)$. Then it follows that $G_N(s+1)$ is absolutely monotone in $(-\infty, 0)$ and that $G_N(-s+1)$ is completely monotone in $(0,\infty)$. Since $G_N(-0+1) = 1$ we have

$$G_N(-s+1) = \int_{0-}^{\infty} e^{-\ell t}\, dU(\ell) = \hat{u}(s)$$

for some distribution U. Thus $G_N(s) = \hat{u}(1-s)$ which implies that N is MP(U), since the generating function determines the distribution. ■

Assume now that some discrete random variable N is used as a model for the number of claims in, let us say, the time-interval $(0,t]$.

Definition 2.4. *Let N be a discrete random variable, let ξ_1, $\xi_2, \ldots$ be independent and identically distributed Bernoulli variables, i.e.*

$$P\{\xi_i = 0\} = 1 - p \quad and \quad P\{\xi_i = 1\} = p$$

for some $p \in (0,1]$ and all $i = 1, 2, \ldots$, and let N be independent of the ξ_is. The discrete random variable

$$N_p = \xi_1 + \xi_2 + \ldots + \xi_N, \ (N_p = 0 \text{ if } N = 0)$$

is called an independent p-**thinning** *of N.*

It follows immediately from Definition 2.4 that N_p, conditioned upon N, is binomially distributed, $\text{Bin}(N, p)$, i.e.

$$P\{N_p = k \mid N = n\} = \binom{n}{k} p^k (1-p)^{n-k},$$

for $n = 0, 1, \ldots$ and $k = 0, \ldots, n$.

The notion of p-thinning is more natural in the process setting. Its implication on risk modeling is discussed in Grandell (1991, pp.45–55). Here we will only give one example. Suppose we are interested in the number of claims N_p in $(0, t]$, which will cost the company more than a certain amount. If the claim costs are independent and if the probability of a claim cost to exceed that amount is p, then N_p is a p-thinning.

Consider now N and N_p as above. Then

$$G_{N_p}(s) = E\big[E[s^{N_p} \mid N]\big] = E\left[\sum_{k=0}^{N} \binom{N}{k} s^k p^k (1-p)^{N-k}\right]$$

$$= E\big[(1 - p + ps)^N\big] = G_N(1 - p(1-s)).$$

Conversely, we say that N_p is obtained by p-thinning, and we have

$$G_N(s) = G_{N_p}(1 - \tfrac{1-s}{p}). \tag{2.9}$$

If N is $\text{MP}(t, U)$ we have

$$G_{N_p}(s) = \hat{u}(t(p(1-s))) = \hat{u}(pt(1-s)),$$

i.e. N_p is $\text{MP}(pt, U)$, and N can be obtained by p-thinning from $N_{1/p}$ which is $\text{MP}(t/p, U)$. In general a random variable N cannot necessarily be obtained by p-thinning.

Proposition 2.3. *A discrete random variable N with generating function G_N can be obtained by independent p-thinning for every $p \in (0,1)$ if and only G_N is absolutely monotone in $(-\infty, 1)$.*

Proof. Let N be any random variable with generating function G_N, and let p be fixed. Consider the function G defined by $G(s) = G_N(1 - \frac{1-s}{p})$. Simple calculations yield

$$G^{(n)}(s) = p^{-n} G_N^{(n)}(1 - \tfrac{1-s}{p}).$$

Since G is a generating function if and only if it is absolutely monotone on $(0,1)$ it follows that G is a generating function if and only if G_N is absolutely monotone on $(1 - \frac{1}{p}, 1) = (-\frac{1-p}{p}, 1)$. Thus the proposition follows. ∎

Combining Propositions 2.2 and 2.3, we get the following theorem, which is a very special case of a characterization of Cox processes due to Mecke (1968), see Theorem 5.1 below.

Theorem 2.3. *A discrete random variable N is mixed Poisson distributed if and only if it can be obtained by independent p-thinning for every $p \in (0,1)$.*

We will now consider infinite divisibility.

Definition 2.5. *A discrete random variable N is called* **discretely infinitely divisible**, *DID, if for each n there exists a discrete random variable N_n such that N has the same distribution as the sum of n independent copies of N_n.*
A random variable Λ is called **infinitely divisible**, *ID, if for each n there exists a random variable Λ_n such that Λ has the same distribution as the sum of n independent copies of Λ_n.*

An equivalent analytical definition is that N is DID if for each n there exists a generating function G_n such that $G_N(s) = G_n^n(s)$. In the same way Λ, with Laplace transform $\hat{u}$, is ID if for each n there exists a Laplace transform $\hat{u}_n$ such $\hat{u}(v) = \hat{u}_n^n(v)$.

Remark 2.2. Generally a discrete random variable is called ID when it is, according to Definition 2.5, DID. The trivial variable $N = 1$ is ID but not DID. On the other hand, an ID discrete variable N is DID if and only if $P\{N = 0\} > 0$, see Kallenberg (1983, p.55). We will give probabilistic arguments.

If N is DID it is obviously ID, and $P\{N = 0\} > 0$ follows from the representation to be given. Assume therefore that N is ID and that $P\{N = 0\} > 0$. For each n there exist independent and identically distributed random variables $\Lambda_{n,1}, \ldots, \Lambda_{n,n}$ so that

$$N \stackrel{\mathrm{d}}{=} \Lambda_{n,1} + \Lambda_{n,2} + \ldots + \Lambda_{n,n}.$$

Since, as is easily shown, $\Lambda_{n,1}, \ldots, \Lambda_{n,n}$ are non-negative, it follows that $P\{N = 0\} > 0$ implies $P\{\Lambda_{n,k} = 0\} > 0$. If $\Lambda_{n,1}, \ldots, \Lambda_{n,n-1} = 0$, which holds with positive probability, we have $N = \Lambda_{n,n}$. Thus $\Lambda_{n,n}$ is discrete, i.e. N is DID. □

Definition 2.6. *A discrete random variable N is said to be* **compound Poisson distributed** *if*

$$N = \xi_1 + \xi_2 + \ldots + \xi_{\tilde{N}}, \quad (N = 0 \text{ if } \tilde{N} = 0),$$

where $\xi_1, \xi_2, \ldots$ are independent and identically distributed discrete variables, $\tilde{N}$ is Po$(\tilde{\alpha})$, *and $\tilde{N}$ is independent of the ξ_is.*

For N, as in Definition 2.6, we have

$$G_N(s) = \sum_{k=0}^{\infty} \frac{\tilde{\alpha}^k}{k!} e^{-\tilde{\alpha}} G_\xi^k(s) = e^{\tilde{\alpha}(G_\xi(s)-1)}. \tag{2.10}$$

The following proposition, see for instance Feller (1968, p.290), is well known.

Proposition 2.4. *A discrete random variable N is* DID *if and only if it is compound Poisson distributed.*

Proposition 2.5 (Maceda, 1948). *Let N be* MP(U), *where U is the distribution of a non-negative random variable Λ. N is* DID *if Λ is* ID.

Proof. Assume that Λ, with Laplace transform $\hat{u}$, is ID and choose n. Then there exists a Laplace transform $\hat{u}_n$ so that $\hat{u}(v) = \hat{u}_n^n(v)$. Put, see Proposition 2.1 (vi), $G_n(s) = \hat{u}_n(1 - s)$. Then

$$G_N(s) = \hat{u}(1 - s) = \hat{u}_n^n(1 - s) = G_n^n(s),$$

and the proposition follows. ■

Proposition 2.5 is very natural from the construction of mixed Poisson distributed random variables. More surprising is certainly that the proposition does not hold 'if and only if'. Kallenberg (1983, p.71) has given an explicit counter-example, which we will consider in some detail. For further discussion on mixtures and infinite divisibility, we refer to Godambe (1977), Huff (1978), Puri and Goldie (1979), Shanbhag and Westcott (1977), and Steutel (1970).

Kallenberg's counter-example. Consider the polynomial

$$g(s) = s - \varepsilon s^2 + s^3 + s^4$$

and the function

$$G(s) = e^{g(s)-g(1)}. \tag{2.11}$$

If $\varepsilon \leq 0$ it follows from (2.10) that $G(s)$ is the generating function of a compound Poisson distribution and therefore DID. As noticed by Raikov and Lévy, $G(s)$ can be a generating function also for $\varepsilon > 0$. In order to show that, we – following Lukacs (1960, pp.178–179) – write $G(s)$ on the form

$$G(s) = e^{-(1+g(1))}e^{1+g(s)} = e^{-(1+g(1))}\sum_{k=0}^{\infty}\frac{(1+g(s))^k}{k!}. \tag{2.12}$$

Consider $(1+g(s))^2$ and $(1+g(s))^3$. By routine computations we get

$$(1+g(s))^2 = 1 + 2s + (1-2\varepsilon)s^2 + 2(1-\varepsilon)s^3 + (4+\varepsilon^2)s^4$$
$$+ 2(1-\varepsilon)s^5 + (1-2\varepsilon)s^6 + 2s^7 + s^8$$

and

$$(1+g(s))^3 = 1 + 3s + 3(1-\varepsilon)s^2 + 2(2-3\varepsilon)s^3 + 3(3-\varepsilon+\varepsilon^2)s^4$$
$$+ 3(3-2\varepsilon+\varepsilon^2)s^5 + (6-12\varepsilon-\varepsilon^3)s^6 + 3(3-2\varepsilon+\varepsilon^2)s^7$$
$$+ 3(3-\varepsilon+\varepsilon^2)s^8 + 2(2-3\varepsilon)s^9 + 3(1-\varepsilon)s^{10} + 3s^{11} + s^{12}.$$

If all coefficients in $(1+g(s))^2$ and $(1+g(s))^3$ are non-negative, it follows that $(1+g(s))^{2k} = [(1+g(s))^2]^k$ and $(1+g(s))^{2k+1} = (1+g(s))^3[(1+g(s))^2]^{k-1}$ also have non-negative coefficients. Considering the expressions above, this is seen to be the case, provided that $\varepsilon \leq 1/2$, except for the coefficient $6-12\varepsilon-\varepsilon^3$ of s^6 of $(1+g(s))^3$.

The equation

$$6 - 12\varepsilon - \varepsilon^3 = 0$$

has the real solution

$$\varepsilon_0 \stackrel{\text{def}}{=} \frac{\left(3+\sqrt{73}\right)^{2/3} - 4}{\left(3+\sqrt{73}\right)^{1/3}} \approx 0.4902.$$

Consider now $G(s)$, as given by (2.12), for $0 < \varepsilon \le \varepsilon_0$. It follows that only s^2 can have a negative coefficient. The coefficient of s^2 is $e^{-g(1)}(\frac{1}{2} - \varepsilon)$, and thus $G(s)$ is a generating function.

Kallenberg's counter-example is to let Λ be a discrete random variable with generating function $G(s)$ for $0 < \varepsilon \le \varepsilon_0$. It follows from (2.10) that Λ cannot be compound Poisson distributed, and therefore not DID. Since $G(0) = P\{\Lambda = 0\} \neq 0$ it can neither be ID.

Let N be MP(t, U). Since $\hat{u}(v) = G(e^{-v})$ it follows from Proposition 2.1 (vi) that

$$G_N(s) = \exp\left\{e^{t(s-1)} - \varepsilon e^{2t(s-1)} + e^{3t(s-1)} + e^{4t(s-1)} - 3 + \varepsilon\right\}.$$

It follows from Proposition 2.4 that N is DID if and only if all coefficients in the MacLaurin expansion of

$$e^{t(s-1)} - \varepsilon e^{2t(s-1)} + e^{3t(s-1)} + e^{4t(s-1)}$$

are non-negative. The coefficient of s^k is readily seen to be

$$\left(e^{-t} - \varepsilon e^{-2t}2^k + e^{-3t}3^k + e^{-4t}4^k\right)/k!,$$

which certainly is positive if $\varepsilon < e^{-t}$. Thus, for each t, we can construct a counter-example. On the other hand, it is impossible to construct a counter-example which holds for all t, cf. Theorem 4.2.

The given counter-example may look rather complicated. It is, however, not possible to construct a simpler one of the form (2.11), since Lévy (1937) has shown that necessary for a function $G(s)$, of the form (2.11), to be a generating function is that a term in the polynomial $g(s)$ with a negative coefficient is preceded by one term and followed by at least two terms with positive coefficients. Puri and Goldie (1979, pp.150–151) have, on the other hand, given the counter-example

$$u(\ell) = 2e^{-\ell} - 6e^{-3\ell} + 5e^{-5\ell}, \quad \ell \geq 0, \tag{2.13}$$

for $t = 1$. □

We will take the Puri–Goldie counter-example (2.13) as a basis of a short discussion on mixture and infinite divisibility. The most well-known result in this direction is probably that mixtures of exponential distributions are ID; a result due to Goldie (1967) and mentioned in Feller (1971, p.452). More precisely, a random variable X is said to be **mixed exponentially distributed**, ME(U), if its density function f_X is given by

$$f_X(x) = \int_0^\infty \ell e^{-\ell x}\, dU(\ell), \quad x \geq 0,$$

for some distribution U with $U(0) = 0$. The distribution U may, as in the mixed Poisson case, be regarded as a prior distribution. It will be seen from Proposition 4.5 that the class of mixed exponential distributions is highly relevant in connection with mixed Poisson processes. In particular it follows that a finite mixture, i.e.

$$f_X(x) = \sum_{j=1}^{m} u_j \ell_j e^{-\ell_j x}, \quad x \geq 0, \tag{2.14}$$

where $u_j > 0$, $\sum_{j=1}^m u_j = 1$ and $\ell_j > 0$ is ID. Without restriction we may assume that

$$0 < \ell_1 < \ell_2 < \cdots < \ell_m.$$

Let X and Y be independent and exponentially distributed with densities

$$f_X(x) = \ell_1 e^{-\ell_1 x} \quad \text{and} \quad f_Y(x) = \ell_2 e^{-\ell_2 x}, \quad x \geq 0,$$

respectively and let $0 < \ell_1 < \ell_2$. The density of $X + Y$ is then given by

$$f_{X+Y}(x) = \frac{\ell_2}{\ell_2 - \ell_1} \ell_1 e^{-\ell_1 x} - \frac{\ell_1}{\ell_2 - \ell_1} \ell_2 e^{-\ell_2 x}, \quad x \geq 0, \tag{2.15}$$

i.e. f_{X+Y} is a linear combination of exponential distributions, LCE. From the construction it is obvious that f_{X+Y} is a density, i.e.

$$f_{X+Y}(x) \geq 0 \text{ for all } x \quad \text{and} \quad \int_0^\infty f_{X+Y}(x)\, dx = 1, \tag{2.16}$$

and that $X + Y$ is ID. The distribution can, however, not be obtained by a 'Bayesian argument' since $-\ell_1/(\ell_2 - \ell_1)$ is negative. Therefore we prefer to call f_{X+Y} a linear combination, instead of a mixture, of exponential distributions.

Let us now consider a function f_X which is an LCE of the form (2.14) where we only require that $u_j \neq 0$, $\sum_{j=1}^m u_j = 1$ and $\ell_j > 0$, i.e. the sequence $\{u_j\}$ need not be a probability distribution. A necessary condition for f_X to be a density, cf. Steutel (1967), is

$$u_1 > 0 \quad \text{and} \quad \sum_{j=1}^{m} u_j \ell_j \geq 0.$$

A simple sufficient condition for f_X to be a density, due to Bartholomew (1969), is

$$\sum_{j=1}^{r} u_j \ell_j \geq 0, \qquad r = 1, 2, \ldots, m.$$

Although the necessary and sufficient condition are far from each other, they coincide for $m = 2$ and it follows that an LCE of order 2 is a density if and only if $0 < u_1 \leq \ell_2/(\ell_2 - \ell_1)$. Thus (2.15) is on the 'boundary' to be a density. The sufficient condition does not apply to (2.13), and neither does a more refined sufficient condition also due to Bartholomew (1969). Nevertheless (2.13) is a density, see Steutel (1967).

A sufficient condition, due to Steutel (1967), for an LCE-distributed random variable to be ID is that in the sequence $u_1, \ldots, u_m$ there is at most one change of sign. Thus, for $m = 2$ any LCE variable is ID and it follows that (2.13) is of lowest possible order. Both to prove that (2.13) **is not** ID and to prove that the corresponding mixed Poisson distribution **is** ID, require more advanced theory of infinite divisibility than Kallenberg's counter-example.

As for exponential distributions, one may of course consider linear combinations of Poisson distributions, LCP. We do, however, not know any situation where those distributions turn up or why they would be interesting other than – possibly – as a mathematical construction. We therefore restrict ourselves to $m = 2$. Let N_t be a random variable with

$$P\{N_t = k\} = u_1 \frac{(t\ell_1)^k}{k!} e^{-\ell_1 t} + u_2 \frac{(t\ell_2)^k}{k!} e^{-\ell_2 t}, \quad k = 0, 1, \ldots,$$

where $u_j \neq 0$, $u_1 + u_2 = 1$ and $0 < \ell_1 < \ell_2$. Simple manipulations yield

$$P\{N_t = k\} \geq 0 \quad \text{if and only if} \quad u_2 \geq (u_2 - 1) \left(\frac{\ell_1}{\ell_2}\right)^k e^{(\ell_2 - \ell_1)t}.$$

By letting $k \to \infty$ this implies that $u_2 \geq 0$, and therefore we may restrict consideration to $u_2 > 1$, since $0 < u_2 < 1$ is no problem and $u_2 = 1$ is not allowed. Since $\ell_1/\ell_2 < 1$ it readily follows that the distribution of N_t is well-defined if and only if

$$0 < u_2 \leq \frac{e^{-\ell_1 t}}{e^{-\ell_1 t} - e^{-\ell_2 t}}.$$

Notice that the right-hand side tends to 1 as $t \to \infty$, which implies that the distribution of N_t is well-defined for all t if and only if u_1 and u_2 are probabilities. We will return to this case in Example 4.3.

Let Λ with Laplace transform $\hat{u}$ be ID and let N be MP(U). It is well known, see for instance Feller (1971, p.450), that $\hat{u}$ then has a representation which may be looked upon as generalization of (2.10) or – which is the same – that

$$-\frac{d \log \hat{u}(v)}{dv} \quad \text{is completely monotone in } (0, \infty).$$

By Propositions 2.1 (vi), 2.4, and 2.5 it follows that

$$\log \hat{u}(v) = \tilde{\alpha}(G_\xi(1-v) - 1),$$

and therefore $G'_\xi(s)$ must be absolutely monotone in $(-\infty, 1)$. Conversely, let N be DID and assume that $G'_\xi(s)$ is absolutely monotone in $(-\infty, 1)$. Then, cf. Godambe (1977, p.299) or Puri and Goldie (1979, p.142), N must be MP(U) where U is the distribution of some ID random variable Λ. If further $G_\xi(s) \geq 0$ for all $s \in (-\infty, 1)$ it follows from Proposition 2.2 that ξ itself must be mixed Poisson distributed. In the representation of a DID variable as a random sum, cf. Definition 2.6, it is natural to require that $P\{\xi = 0\} = 0$, since otherwise there may be many zeros involved in the sum. In that case $G_\xi(0) = 0$ and therefore $G_\xi(s) < 0$ for $s < 0$ unless $G_\xi(s) \equiv 0$; this is a trivial case which we disregard since it corresponds to $P\{\Lambda = 0\} = 1$. In the representation (2.10) $\tilde{\alpha}$ and G_ξ are not uniquely determined. In fact we may always assign an arbitrary probability p to the event $\{\xi = 0\}$ or, more precisely, consider a random variable ζ with

$$G_\zeta(s) = p + (1-p)G_\xi(s) \quad \text{or} \quad G_\xi(s) = \frac{G_\zeta(s) - p}{1-p}.$$

Then

$$\tilde{\alpha}(G_\xi(1-v) - 1) = \frac{\tilde{\alpha}}{1-p}(G_\zeta(1-v) - 1)$$

which has a natural probabilistic interpretation; the Poisson intensity is increased in order to compensate for the zeros. If $G_\xi(-\infty) > -\infty$ we may always choose a p such that

$$0 = G_\zeta(-\infty) = p + (1-p)G_\xi(-\infty) \quad \text{or} \quad p = \frac{|G_\xi(-\infty)|}{1 + |G_\xi(-\infty)|}.$$

In this case it readily follows that ζ is mixed Poisson distributed. From the representation of $\log \hat{u}(v)$ it follows that $G_\xi(-\infty) > -\infty$ if and only if $P\{\Lambda = 0\} > 0$. If $P\{\Lambda = 0\} = 0$, which may be regarded as the 'usual' case, it is not possible to express N as a random sum of mixed Poisson distributed random variables. For some further discussion we refer to Puri and Goldie (1979, pp.142–143).

Example 2.1 (continued). Recall that Λ is $\Gamma(\gamma, \beta)$-distributed. From its Laplace transform

$$\hat{u}(v) = \left(1 + \frac{v}{\beta}\right)^{-\gamma}$$

it follows that Λ is ID and that Λ_n is $\Gamma(\gamma/n, \beta)$. Let N be MP(t, U). From Proposition 2.5 it follows that N is DID, and thus compound Poisson distributed. We have, cf. Bühlmann (1970, p.71),

$$\begin{aligned} G_N(s) = \hat{u}(t(1-s)) &= \left(1 + \frac{t(1-s)}{\beta}\right)^{-\gamma} \\ &= \exp\left\{-\gamma\left[\log\left(\frac{\beta+t}{\beta}\right) + \log\left(1 - \frac{ts}{\beta+t}\right)\right]\right\} \\ &= \exp\left\{\gamma\log\left(\frac{\beta+t}{\beta}\right)\left(-\frac{\log\left(1-\frac{ts}{\beta+t}\right)}{\log\left(\frac{\beta+t}{\beta}\right)} - 1\right)\right\}. \end{aligned}$$

Thus, see (2.10), N is compound Poisson distributed with

$$\tilde{\alpha} = \gamma\log\left(\frac{\beta+t}{\beta}\right) \quad \text{and} \quad G_\xi(s) = -\frac{\log\left(1-\frac{ts}{\beta+t}\right)}{\log\left(\frac{\beta+t}{\beta}\right)}. \tag{2.17}$$

It is not at all obvious that $G_\xi(s)$ is a generating function. A series expansion yields

$$G_\xi(s) = \frac{1}{\log\left(\frac{\beta+t}{\beta}\right)} \sum_{n=1}^{\infty} \frac{1}{n}\left(\frac{ts}{\beta+t}\right)^n,$$

which, since $G_\xi(1) = 1$, implies

$$P\{\xi = n\} = \frac{\left(\frac{t}{\beta+t}\right)^n}{n \log\left(\frac{\beta+t}{\beta}\right)}, \quad n = 1, 2, \ldots$$

Thus $G_\xi(s)$ is a generating function. The distribution of ξ is sometimes called the **logarithmic distribution**. It is worth noting that $\tilde{\alpha}$ is **not** proportional to t and that the distribution of ξ depends on t. □

Naturally one may in certain applications have good reasons to use other structure distributions than the Γ-distribution. Often it is then not possible to get a closed – or a reasonably simple – expression for $p_k(t)$, but in many cases recursive evaluation is possible. Such recursions are, for instance, discussed by Willmot (1986). Even for the Γ-distribution, as was seen in Example 2.1, recursive algorithms may be numerically useful.

Willmot (1993) considers a more general approach, which applies to many structure distributions; some of them we will discuss below. We will, however, illustrate the idea behind those recursions for a shifted Γ-distribution, before discussing Willmot's general method.

Example 2.2. Let Λ have a shifted Γ-distribution with $u(\ell)$ given by

$$u(\ell) = \frac{\beta^\gamma}{\Gamma(\gamma)}(\ell - \alpha)^{\gamma-1} e^{-\beta(\ell-\alpha)}, \quad \ell \geq \alpha.$$

This seems to be a natural distribution, where α can be interpreted as a 'basic risk'.

Let N be MP(t, U). This model was proposed by Delaporte (1960). Ruohonen (1988) fitted it to several data encountered in the literature, while Willmot and Sundt (1989a) and Schröter (1990) studied it from a more theoretical point of view. N is said to be **Delaporte distributed**.

Since Λ is ID it follows from Proposition 2.5 that N is DID.

One may notice that N is the sum of a Poisson variable and an independent negative binomial variable. Although this implies, cf. Example 2.1,

$$p_m(t) = \sum_{n=0}^{m} \frac{(\alpha t)^{m-n}}{(m-n)!} e^{-\alpha t} \binom{\gamma + n - 1}{n} \left(\frac{\beta}{\beta + t}\right)^\gamma \left(\frac{t}{\beta + t}\right)^n,$$

it may be tempting to try to determine the distribution in a recursive way.

We have

$$\hat{u}(v) = e^{-\alpha v}\left(1+\frac{v}{\beta}\right)^{-\gamma}$$

and thus

$$G_N(s) = e^{-\alpha t(1-s)}\left(1+\frac{t(1-s)}{\beta}\right)^{-\gamma}.$$

Differentiation leads, after simple calculations, to

$$(\beta + t - ts)G_N'(s) = (\gamma t + \alpha t(\beta + t) - \alpha t^2 s)G_N(s).$$

Using

$$G_N(s) = \sum_{m=0}^{\infty} p_m(t)s^m \quad \text{and} \quad G_N'(s) = \sum_{m=0}^{\infty}(m+1)\,p_{m+1}(t)s^m$$

and equating the coefficients of s^m yields, with $p_{-1}(t) = 0$, the recursion

$$(\beta + t)(m+1)\,p_{m+1}(t)$$
$$= \big(\gamma + \alpha(\beta + t) + m\big)tp_m(t) - \alpha t^2 p_{m-1}(t), \quad m = 0,\,1,\,2\ldots \quad (2.18)$$

Since $p_0(t) = G_N(0) = e^{-\alpha t}\left(\frac{\beta}{\beta+t}\right)^{\gamma}$ we can easily compute $p_m(t)$ for any m. □

The idea behind the derivation of the recursion in Example 2.2 was simply to derive a differential equation for G_N, and than to make a MacLaurin expansion. In Example 2.2 the generating function G_N had a simple form, so the differential equation was easily found by derivation. This is, however, not always the case. Further the recursion followed easily since G_N and G_N' were multiplied by polynomials of low order.

Willmot (1993) considered the following class of absolutely continuous structure distributions:

(i) $U(\ell_0) = 0$ and $U(\ell_1) = 1$ for $0 \le \ell_0 < \ell_1 \le \infty$;

(ii) $\displaystyle \frac{d}{d\ell}\log u(\ell) = \frac{\eta(\ell)}{\theta(\ell)} = \frac{\sum_{n=0}^{k}\eta_n\ell^n}{\sum_{n=0}^{k}\theta_n\ell^n}$ for $\ell_0 \le \ell \le \ell_1$.

In Example 2.2 we had $\ell_0 = \alpha > 0$ and we will in Chapter 9 see that also $\ell_1 < \infty$ may be of interest, so we gain not only formal generality by introducing ℓ_0 and ℓ_1. The assumption in (ii) that $\eta(\ell)$ and $\theta(\ell)$ are of the same order is no restriction, since some of the η_n or θ_n may be zero. Further, for notational reasons, we define $\eta_n = \theta_n = 0$ for $n \notin \{0, 1, \dots, k\}$.

Let N be MP(t, U) and recall from Proposition 2.1 (vi) that

$$G_N(s) = \int_{\ell_0}^{\ell_1} e^{-t(1-s)\ell} u(\ell)\, d\ell.$$

Notice that

$$G_N^{(n)}(s) = \int_{\ell_0}^{\ell_1} t^n \ell^n e^{-t(1-s)\ell} u(\ell)\, d\ell = \sum_{m=0}^{\infty} (m+1)^{\overline{n}} p_{m+n}(t) s^m,$$

where $G_N^{(n)}(s)$ denotes the nth derivative of $G_N(s)$ and

$$(m+1)^{\overline{n}} \stackrel{\text{def}}{=} \frac{(m+n)!}{m!} = \begin{cases} 1 & \text{if } n = 0, \\ (m+1)(m+2)\cdots(m+n) & \text{if } n > 0. \end{cases}$$

We have

$$\begin{aligned} &\frac{d}{d\ell}\Big(e^{-t(1-s)\ell} u(\ell)\theta(\ell)\Big) \\ &= e^{-t(1-s)\ell}\big(t(s-1)u(\ell)\theta(\ell) + u'(\ell)\theta(\ell) + u(\ell)\theta'(\ell)\big) \\ &= e^{-t(1-s)\ell} u(\ell)\big(t(s-1)\theta(\ell) + \eta(\ell) + \theta'(\ell)\big), \end{aligned} \tag{2.19}$$

where obviously

$$t(s-1)\theta(\ell) + \eta(\ell) + \theta'(\ell) = \sum_{n=0}^{k} \big(ts\theta_n - t\theta_n + \eta_n + (n+1)\theta_{n+1}\big)\ell^n.$$

Integration of (2.19) over (ℓ_0, ℓ_1) and multiplication with t^k leads to the desired differential equation

$$\begin{aligned} &\sum_{n=0}^{k} \big(ts\theta_n - t\theta_n + \eta_n + (n+1)\theta_{n+1}\big) t^{k-n} G_N^{(n)}(s) \\ &= t^k e^{-t(1-s)\ell_1} u(\ell_1)\theta(\ell_1) - t^k e^{-t(1-s)\ell_0} u(\ell_0)\theta(\ell_0). \end{aligned}$$

MacLaurin expansion of $G_N^{(n)}(s)$ and $e^{ts\ell}$ and equating the coefficients of s^m yield, cf. Willmot (1993, p.116),

$$\sum_{n=-1}^{k} \left(\theta_{n+1} m^{\overline{n+1}} + \left(\eta_n - \theta_n t + (n+1)\theta_{n+1}\right)(m+1)^{\overline{n}}\right) \cdot t^{k-n} p_{m+n}(t)$$

$$= t^k \pi_{\ell_1 t,m} u(\ell_1)\theta(\ell_1) - t^k \pi_{\ell_0 t,m} u(\ell_0)\theta(\ell_0), \tag{2.20}$$

where $\pi_{\ell t,m}$ is the Poisson probability and $p_{-1}(t) = 0$. Recall that $\theta_{k+1} = \theta_{-1} = \eta_{-1} = 0$.

In order to apply the **Willmot recursion** we need to know

$$p_0(t), \dots, p_{k-1}(t).$$

Sometimes those probabilities can be determined analytically or otherwise by numerical integration.

In any numerical algorithm, round-off errors are inevitable. We say that a recursion is **stable** if those errors do not add up in a 'dangerous' way. For a detailed discussion of these questions we refer to Panjer and Wang (1993) and to Wang and Panjer (1993). We will, however, point out that the analysis of numerical errors may be misleading if it is done in terms of absolute errors and not in terms of relative errors.

Example 2.2 (continued). We will now illustrate how Willmot's general approach works for the Delaporte distribution.

We have

$$\frac{d}{d\ell} \log u(\ell) = \frac{d}{d\ell}\left(\text{const.} + (\gamma - 1)\log(\ell - \alpha) - \beta(\ell - \alpha)\right)$$

$$= \frac{\gamma - 1}{\ell - \alpha} - \beta = \frac{\gamma - 1 + \alpha\beta - \beta\ell}{-\alpha + \ell},$$

and (2.20) applies with $k = 1$,

$$\begin{matrix} \ell_0 = \alpha \\ \ell_1 = \infty \end{matrix} \quad \text{and} \quad \begin{matrix} \eta_0 = \gamma - 1 + \alpha\beta & \eta_1 = -\beta \\ \theta_0 = -\alpha & \theta_1 = 1. \end{matrix}$$

Then we get, since $\theta(\ell_0) = 0$,

$$-\theta_0 t^2 p_{m-1}(t) + \left(\theta_1 m + \eta_0 - \theta_0 t + \theta_1\right) t p_m(t)$$

$$+ (\eta_1 - \theta_1 t)(m+1)p_{m+1}(t) = 0$$

or

$$- \alpha t^2 p_{m-1}(t) + (m + \gamma + \alpha\beta + \alpha t)\, t p_m(t)$$
$$- (\beta + t)(m+1)\, p_{m+1}(t) = 0$$

which is in agreement with (2.18).

Considering (2.18) we see that $p_m(t)$ and $p_{m-1}(t)$ appear with different signs, which indicates that the recursion is unstable. This does not mean that it is useless, but that it should be used with care. □

Consider (2.18) with $\alpha = 0$, which means that we are back in the Panjer assumption (2.7). The problem of different signs in (2.18) does then disappear, and (2.7) is known to be stable. Using the arguments which led to (2.6) 'backwards', it follows that

$$E[\Lambda \mid N = n] = \frac{p_{n+1}(t)(n+1)}{p_n(t)\, t} = \frac{a+b}{t} + \frac{an}{t}, \quad n = 0,\ 1, \ldots,$$

i.e. $E[\Lambda \mid N = n]$ is linear in n. Thus it follows from Theorem 2.2 that the Panjer recursion – among mixed Poisson distributions – only applies to the Poisson and negative binomial distributions.

Sundt and Jewell (1981) have shown that, besides these distributions, (2.7) also applies to the binomial distribution, and that this is all. The Panjer recursion has therefore been extended in different ways. In connection with mixed Poisson distributions, the most interesting extension seems to be the approach by Willmot, discussed above.

Example 2.3. Let Λ have a **generalized inverse Gaussian distribution**, GIGD, with

$$u(\ell) = \frac{(\psi/\chi)^{\gamma/2}}{2K_\gamma(\sqrt{\chi\psi})} \ell^{\gamma-1} e^{-(\chi\ell^{-1}+\psi\ell)/2}, \quad \ell > 0, \tag{2.21}$$

where K_γ is the modified Bessel function of the third kind. The GIGD was introduced by Good (1953, p.249).

The domain of variation for the parameters is

$$\gamma \in \mathbb{R}, \quad (\chi, \psi) \in \Theta_\gamma,$$

where

$$\Theta_\gamma = \begin{cases} \{(\chi,\psi):\ \chi \geq 0,\ \psi > 0\} & \text{if } \gamma > 0, \\ \{(\chi,\psi):\ \chi > 0,\ \psi > 0\} & \text{if } \gamma = 0, \\ \{(\chi,\psi):\ \chi > 0,\ \psi \geq 0\} & \text{if } \gamma < 0. \end{cases}$$

The corresponding mixed Poisson distribution is called the **Sichel distribution**. Before discussing this distribution, we will give some basic facts about the GIGD. The class of GIGDs is flexible and includes several important special cases. We will in Chapters 8 and 9 see that it is useful not only as a structure distribution.

It is often useful to introduce the parameters

$$\omega = \sqrt{\chi\psi} \quad \text{and} \quad \eta = \sqrt{\frac{\chi}{\psi}},$$

and then (2.21) takes the form

$$u(\ell) = \frac{\eta^{-\gamma}}{2K_\gamma(\omega)} \ell^{\gamma-1} e^{-\omega(\eta\ell^{-1}+\eta^{-1}\ell)/2}, \quad \ell > 0,$$

where (for γ fixed) ω is a concentration parameter and η a scale parameter.

At the first sight the GIGD may look very complicated, at least compared with the Γ-distribution. Certainly it is more complicated since it – as we will see – contains the Γ-distribution as a special case. However, probably it looks complicated mostly due to the norming factor $K_\gamma(\omega)$, which is much less usual than the Γ-function. Therefore we will start by giving some properties of $K_\gamma(\omega)$. These properties are taken from Jørgensen (1982, pp.170–171), which is a standard reference for the GIGD.

K_γ is defined by

$$K_\gamma(\omega) \stackrel{\text{def}}{=} \frac{1}{2}\int_0^\infty x^{\gamma-1} e^{-\omega(x^{-1}+x)/2}\, dx, \quad \omega > 0,$$

which is one of its many integral representations, and it satisfies the relations

$$K_\gamma(\omega) = K_{-\gamma}(\omega) \tag{2.22}$$

and

$$K_{\gamma+1}(\omega) = \frac{2\gamma}{\omega} K_\gamma(\omega) + K_{\gamma-1}(\omega). \tag{2.23}$$

When γ is a ‘half-integer’ $K_{n+1/2}$ has the representation

$$K_{n+1/2}(\omega) = \sqrt{\frac{\pi}{2\omega}}\, e^{-\omega} \sum_{i=0}^{n} \frac{(n+i)!}{(n-i)!i!}(2\omega)^{-i}, \tag{2.24}$$

for $n = 0, 1, 2\ldots$

Properties (2.23) and (2.24) may be looked upon as simplifying relations for K_γ, similar to (2.4) for the Γ-function.

Consider (2.21) in the case where $\chi = 0$. It is seen from Θ_γ that this is possible for $\gamma > 0$ and $\chi > 0$. Since

$$\lim_{\omega\to 0} \omega^\gamma K_\gamma(\omega) = \Gamma(\gamma)2^{\gamma-1}, \quad \gamma > 0, \tag{2.25}$$

it readily follows that Λ is $\Gamma(\gamma, \chi/2)$-distributed. Since this case – which from the point of view of mixed Poisson distributions is the most important case – was treated in detail in Example 2.1, we suppose for the rest of this example that $\chi > 0$.

For $\psi = 0$ it follows from Θ_γ that $\gamma < 0$ and $\chi > 0$, and we say that Λ is **reciprocal Γ-distributed.** Using (2.22) and (2.25) we get

$$u(\ell) = \frac{(\chi/2)^{-\gamma}}{\Gamma(-\gamma)} \ell^{\gamma-1} e^{-\chi/(2\ell)}, \quad \ell > 0.$$

For $\psi > 0$ all moments exist and are given by, see Jørgensen (1982, pp.12–13),

$$E[\Lambda^k] = \frac{K_{\gamma+k}(\omega)}{K_\gamma(\omega)} \eta^k.$$

For $\psi = 0$ we have

$$E[\Lambda^k] = \begin{cases} \frac{\Gamma(-\gamma+k)}{\Gamma(-\gamma)} \left(\frac{\chi}{2}\right)^k, & k < -\gamma, \\ \infty, & k \geq -\gamma. \end{cases}$$

The Laplace transform is given by

$$\hat{u}(v) = \frac{K_\gamma\left(\omega\sqrt{1+2v/\psi}\right)}{K_\gamma(\omega)(1+2v/\psi)^{\gamma/2}}, \quad \psi > 0,$$

and

$$\hat{u}(v) = \frac{2K_\gamma(\sqrt{2\chi v})}{\Gamma(-\gamma)(\chi v/2)^{\gamma/2}}, \quad \psi = 0.$$

A particular important case is $\gamma = -1/2$, in which case we say that Λ has an **inverse Gaussian distribution**, IGD. It then follows from (2.22) that $E[\Lambda] = \eta$. With

$$\mu_\Lambda = \eta = \sqrt{\frac{\chi}{\psi}} \quad \text{and} \quad \beta = \frac{\omega}{2\eta} = \frac{\psi}{2}$$

we get, by (2.24),

$$u(\ell) = \frac{\mu_\Lambda}{\sqrt{\pi \ell^3/\beta}} e^{-\beta(\ell-\mu_\Lambda)^2/\ell},$$

which is a convenient parameterization. The Laplace transform is reduced to

$$\hat{u}(v) = e^{-2\beta\mu_\Lambda\left(\sqrt{1+(v/\beta)}-1\right)},$$

and it follows that Λ is ID. For $\psi = 0$, i.e. when Λ is both IGD and reciprocal Γ-distributed, we have $\mu_\Lambda = \infty$ and $\beta = 0$. Then

$$u(\ell) = \frac{\sqrt{\chi}}{\sqrt{2\pi\ell^3}} e^{-\chi/(2\ell)} \quad \text{and} \quad \hat{u}(v) = e^{-\sqrt{2\chi v}},$$

which is a **stable distribution of order** 1/2, see for instance Feller (1971, p.52).

Consider now N being MP(t, U). In the IGD case, first treated by Holla (1967), N is said to be **inverse Gaussian–Poisson distributed**. This case was further studied by Sichel (1971) and Willmot (1987), and extended to Λ being GIGD. In this generality the corresponding mixed Poisson distribution is called the Sichel distribution.

It follows from (2.21) that

$$p_m(t) = \left(\frac{\psi}{\psi+2t}\right)^{\gamma/2} \left(\frac{\chi t^2}{\psi+2t}\right)^{m/2} \frac{K_{\gamma+m}\left(\sqrt{\chi(\psi+2t)}\right)}{m! K_\gamma\left(\sqrt{\chi\psi}\right)}.$$

Although this can be simplified in the IGD case, it is much easier to derive the distribution recursively. Using (2.23) we get, cf. Sichel (1974, p.27),

$$(\psi+2t)m(m+1)\,p_{m+1}(t)$$
$$= 2tm(\gamma+m)\,p_m(t) + \chi t^2 p_{m-1}(t), \quad m = 1, 2, \ldots \tag{2.26}$$

It is also possible to deduce (2.26) by Willmot's general approach.

We have

$$\frac{d}{d\ell}\log u(\ell) = \frac{d}{d\ell}\left(\text{const.} + (\gamma-1)\log\ell - \frac{\chi}{2\,\ell} - \frac{\psi\ell}{2}\right)$$

$$= \frac{\gamma-1}{\ell} + \frac{\chi}{2\,\ell^2} - \frac{\psi}{2} = \frac{\chi + 2(\gamma-1)\ell - \psi\ell^2}{2\,\ell^2},$$

and (2.20) applies with $k = 2$,

$$\begin{matrix} \ell_0 = 0 & \text{and} & \eta_0 = \chi & \eta_1 = 2(\gamma-1) & \eta_2 = -\psi \\ \ell_1 = \infty & & \theta_0 = 0 & \theta_1 = 0 & \theta_2 = 2. \end{matrix}$$

Using $\theta(\ell_0) = 0$, (2.26) follows by simple routine calculations.

In order to apply (2.26), we need to know $p_0(t)$ and $p_1(t)$. If γ is a 'half-integer' they are easily found by using (2.24). For general γ they can be approximately computed by using a connection between K_γ and the modified Bessel function of the first kind, cf. Sichel (1975) and Jørgensen (1982, pp.170–171).

For $\gamma > -1$ the probabilities $p_0(t)$ and $p_1(t)$ appear with the same sign in (2.26), so the recursion is stable, cf. Panjer and Wang (1993, p.245). If $\gamma < -1$, the safest way to use (2.26) may be to compute $p_{m_\gamma - 1}(t)$ and $p_{m_\gamma}(t)$ for $m_\gamma - 1 \le -\gamma < m_\gamma$, and to use (2.26) for $m \ge m_\gamma$. For $m \le m_\gamma - 1$, we then use the recursion backwards:

$$\chi t^2 p_{m-1}(t) = -2tm(\gamma + m)\,p_m(t) + (\psi + 2t)m(m+1)\,p_{m+1}(t).$$

□

Example 2.4. Let Λ be **beta distributed** with

$$u(\ell) = \frac{\ell^{a-1}(\ell_1 - \ell)^{b-1}}{B(a,b)\ell_1^{a+b-1}}, \quad 0 < \ell < \ell_1, \tag{2.27}$$

where $a > 0$ and $b > 0$. The **beta function** $B(a,b)$ is defined by

$$B(a,b) \stackrel{\text{def}}{=} \int_0^1 x^{a-1}(1-x)^{b-1}\,dx.$$

A well-known relation is

$$B(a,b) = \frac{\Gamma(a)\Gamma(b)}{\Gamma(a+b)},$$

from which it follows that

$$\mu_\Lambda = \frac{a\ell_1}{a+b} \quad \text{and} \quad \sigma_\Lambda^2 = \frac{ab\ell_1^2}{(a+b)^2(a+b+1)}.$$

This structure distribution may at first seem a bit artificial, and we mainly consider it since, as mentioned, the case $\ell_1 < \infty$ will be of interest. A simple way to motivate this choice, is to say that the beta distributions are a flexible class with bounded support. A somewhat more convincing argument is to consider 'mixed independent p-thinning' of a $\text{Po}(\ell_1)$-distributed random variable. More precisely, let $\widetilde{N}$ be $\text{Po}(\ell_1)$-distributed and let N be obtained by independent p thinning, see Definition 2.4. Then N is $\text{Po}(p\ell_1)$-distributed, which follows since N, conditioned upon $\widetilde{N}$, is $\text{Bin}(\widetilde{N}, p)$. Assume now that p may be regarded as the out-come of a random variable. As structure, or prior, distribution in connection with the binomial distribution, the beta distribution (with $\ell_1 = 1$) is the most common choice. With this construction, N is $\text{MP}(U)$, where u is given by (2.27).

Let now N be $\text{MP}(t, U)$. It is obvious that the distribution of N depends on t and ℓ_1 only through $t \cdot \ell_1$, so without any restrictions we may put $t = 1$. We say that N is **beta–Poisson distributed.** Although we regarded this distribution as a bit artificial, it is natural in certain biological applications, where the random p corresponds to a survival probability, cf. Gurland (1958), and its value may depend on the environment and other external conditions.

Quinkert (1957) studied the beta–Poisson distribution as a model for the number of claims in a time-interval. His reasons for studying the model was merely that the beta–Poisson distribution is a flexible distribution and that $\ell_1 < \infty$ might be natural. Under the normalization $\mu_\Lambda = 1$, obtained by the requirements $\ell_1 > 1$ and $b = a(\ell_1 - 1)$, Quinkert (1957, pp.12–15) showed that the beta–Poisson distribution as limiting cases contains the Poisson distribution, for $a \to \infty$ and ℓ_1 fixed, and the negative binomial distribution, for a fixed and $\ell_1 \to \infty$.

We will consider the two limiting cases mentioned above. We do, however, regard μ_Λ as a free parameter, i.e. we consider the parameters a, μ_Λ and $\ell_1 > \mu_\Lambda$ and let $b = a(\ell_1 - \mu_\Lambda)/\mu_\Lambda$. Since μ_Λ always will be fixed, we let u_{a,ℓ_1} denote the density given by (2.27). Similarly Λ_{a,ℓ_1} denotes a random variable with that density and N_{a,ℓ_1} the corresponding beta–Poisson variable.

Let $a \to \infty$, μ_Λ and ℓ_1 be fixed. Then

$$\sigma_\Lambda^2 = \frac{\mu_\Lambda(\ell_1 - \mu_\Lambda)}{a\ell_1 + \mu_\Lambda} \to 0, \quad a \to \infty.$$

Thus $\Lambda_{a,\ell_1} \xrightarrow{\mathrm{P}} 0$ as $a \to \infty$ which, cf. Theorem 2.1, implies that $N_{a,\ell_1} \xrightarrow{\mathrm{d}} N$, where N is $\mathrm{Po}(\mu_\Lambda)$.

Let now a and μ_Λ be fixed and $\ell_1 \to \infty$. Notice that $b/\ell_1 \to a/\mu_\Lambda$. We have

$$\lim_{\ell_1\to\infty} u_{a,\ell_1}(\ell) = \lim_{\ell_1\to\infty} \frac{\ell^{a-1}(\ell_1-\ell)^{b-1}}{B(a,b)\ell_1^{a+b-1}}$$

$$= \frac{\ell^{a-1}}{\Gamma(a)} \lim_{\ell_1\to\infty} \frac{\Gamma(a+b)}{\ell_1^a\Gamma(b)}\left(1-\frac{\ell}{\ell_1}\right)^{b-1}$$

$$= \frac{\ell^{a-1}}{\Gamma(a)} e^{-(a/\mu_\Lambda)\ell} \lim_{\ell_1\to\infty} \frac{\Gamma(a+b)}{\ell_1^a\Gamma(b)}.$$

By Stirling's formula,

$$\Gamma(x) \sim \sqrt{2\pi}\, x^{x-\frac{1}{2}} e^{-x}, \quad x\to\infty,$$

i.e. the ratio of the two sides tends to one as $x \to \infty$, we get

$$\lim_{\ell_1\to\infty} \frac{\Gamma(a+b)}{\ell_1^a\Gamma(b)} = \lim_{\ell_1\to\infty} \frac{(a+b)^{a+b-\frac{1}{2}}e^{-a}}{\ell_1^a b^{b-\frac{1}{2}}}$$

$$= e^{-a} \lim_{\ell_1\to\infty} \left(\frac{a+b}{\ell_1}\right)^a \left(\frac{a+b}{b}\right)^{b-\frac{1}{2}} = e^{-a}\left(\frac{a}{\mu_\Lambda}\right)^a e^a = \left(\frac{a}{\mu_\Lambda}\right)^a.$$

Thus

$$\lim_{\ell_1\to\infty} u_{a,\ell_1}(\ell) = \frac{(a/\mu_\Lambda)^a}{\Gamma(a)} \ell^{a-1} e^{-(a/\mu_\Lambda)\ell},$$

which is the density of the $\Gamma(a, a/\mu_\Lambda)$-distribution. By Scheffé's theorem, see for instance Billingsley (1968, p.224), it follows that $\Lambda_{a,\ell_1} \xrightarrow{\mathrm{d}} \Lambda$ as $\ell_1 \to \infty$, where Λ is $\Gamma(a, a/\mu_\Lambda)$. From Theorem 2.1 $N_{a,\ell_1} \xrightarrow{\mathrm{d}} N$ follows, where N is negatively binomially distributed.

By MacLaurin expansion of $e^{-\ell}$ we get, cf. McNolty (1964, p.288),

$$p_n(1) = \frac{\ell_1^n}{B(a,b)n!} \sum_{k=0}^{\infty} \frac{(-\ell_1)^k B(a+n+k,b)}{k!}. \tag{2.28}$$

This probability can by expressed in terms of the **confluent hypergeometric function** ${}_1F_1$:

$${}_1F_1(\alpha,\gamma,x) \stackrel{\text{def}}{=} \sum_{k=0}^{\infty} \frac{\Gamma(\alpha+k)}{\Gamma(\alpha)} \frac{\Gamma(\gamma)}{\Gamma(\gamma+k)} \frac{x^k}{k!}, \quad \gamma \neq 0, -1, -2, \ldots$$

More strictly than above, (2.28) follows from the integral representation, see for instance Gurland (1958, p.247),

$$ {}_1F_1(\alpha,\gamma,x) = \frac{1}{B(\alpha,\gamma-\alpha)} \int_0^1 t^{\alpha-1}(1-t)^{\gamma-\alpha-1} e^{tx}\, dt, $$

$\gamma > \alpha > 0$, and the observation, cf. Albrecht (1984), that

$$ p_n(1) = \frac{\ell_1^n B(n+a,b)}{n!B(a,b)} {}_1F_1(n+a, n+a+b, -\ell_1). $$

Although the representation of $p_n(1)$ in terms of ${}_1F_1$ is helpful, we will restrict ourselves to recursive determination by Willmot's approach.

We have $\ell_0 = 0$ and

$$ \frac{d}{d\ell}\log u(\ell) = \frac{d}{d\ell}(\text{const.} + (a-1)\log\ell + (b-1)\log(\ell_1-\ell)) $$

$$ = \frac{a-1}{\ell} - \frac{b-1}{\ell_1-\ell} = \frac{(a-1)\ell_1 - (a+b-2)\ell}{\ell_1\ell - \ell^2} $$

and (2.20) applies with $k = 2$,

$$ \begin{array}{lll} \eta_0 = (a-1)\ell_1 & \eta_1 = -(a+b-2) & \eta_2 = 0 \\ \theta_0 = 0 & \theta_1 = \ell_1 & \theta_2 = -1. \end{array} $$

Since $\theta(\ell_0) = \theta(\ell_1) = 0$, we get by (2.20),

$$ m(m+1)\, p_{m+1}(1) $$

$$ = m(m-1+a+b+\ell_1)\, p_m(1) - \ell_1(m+a-1)\, p_{m-1}(1), $$

for $m = 1, 2, \ldots$. This is in agreement with a recursion due to Gurland (1958, p.233). The recursion is unstable, and shall therefore be used with care. Gurland (1958, p.233) recommends that (2.28) is used and that the recursion is merely used for checking the computed values.

We now simplify the model, and consider the case where $a = 1$, i.e. we consider

$$ u(\ell) = \frac{b(\ell_1-\ell)^{b-1}}{\ell_1^b}, \quad 0 < \ell < \ell_1, $$

where $b > 0$. This model was introduced by Beall and Rescia (1953), in connection with biological data. The recursion is simplified to

$$(m+1)\,p_{m+1}(1) = (m+b+\ell_1)\,p_m(1) - \ell_1 p_{m-1}(1).$$

Notice that the recursion is still unstable and of the same order as before. Let us, however, start again with Willmot's approach. Then we have

$$\frac{d}{d\ell}\log u(\ell) = -\frac{b-1}{\ell_1 - \ell}$$

and (2.20) applies with $k = 1$,

$$\begin{aligned} \eta_0 &= 1 - b & \eta_1 &= 0 \\ \theta_0 &= \ell_1 & \theta_1 &= -1, \end{aligned}$$

which may seem to be a much more impressive simplification. However, the recursion is still the one given by (2.27), so no further simplification is gained.

Let us finally consider also $b = 1$, which implies that Λ is uniformly distributed on $[0, \ell_1]$. We then say that N is **uniform–Poisson distributed**. This distribution has been used by Bhattacharya and Holla (1965) in connection with accident proneness.

The recursion is not much simplified, but since

$$\begin{aligned} p_0(1) &= \frac{1}{\ell_1}\left(1 - e^{-\ell_1}\right), \\ p_1(1) &= \frac{1}{\ell_1}\left(1 - (1+\ell_1)e^{-\ell_1}\right), \end{aligned}$$

it is easy to apply, provided its numerical instability is taken into account. □

Example 2.5. Let X be a normally distributed random variable with mean μ_G and variance σ_G^2; the index G stands for Gaussian.

Let Λ be **truncated normally distributed**, i.e.

$$U(\ell) = P\{X \le \ell \mid X \ge 0\} = \frac{\Phi\left(\frac{\ell - \mu_G}{\sigma_G}\right) - \Phi\left(-\frac{\mu_G}{\sigma_G}\right)}{1 - \Phi\left(-\frac{\mu_G}{\sigma_G}\right)}$$

or

$$u(\ell) = \frac{\frac{1}{\sigma_G}\varphi\left(\frac{\ell - \mu_G}{\sigma_G}\right)}{\Phi\left(\frac{\mu_G}{\sigma_G}\right)},$$

where, as usual,

$$\varphi(w) \stackrel{\text{def}}{=} \frac{1}{\sqrt{2\pi}} e^{-w^2/2} \quad \text{and} \quad \Phi(w) = \int_{-\infty}^{w} \varphi(z)\, dz.$$

Recall that $\Phi(-w) = 1 - \Phi(w)$.

We have, cf. Kupper (1962b, p.126),

$$\mu_\Lambda = \mu_G + u(0)\sigma_G^2,$$

$$\sigma_\Lambda^2 = \sigma_G^2 \left(1 - u(0)\mu_G - u(0)^2\sigma_G^2\right) = \sigma_G^2 \left(1 - u(0)\mu_\Lambda\right)$$

and

$$\hat{u}(v) = \frac{\Phi\left(\frac{\mu_G - \sigma_G^2 v}{\sigma_G}\right)}{\Phi\left(\frac{\mu_G}{\sigma_G}\right)} \exp\{-\mu_G v + \tfrac{1}{2}\sigma_G^2 v^2\}.$$

Let N be MP(t, U). We say that N is **truncated normal–Poisson distributed**. This distribution was introduced independently by Berljand *et al.* (1962), Kupper (1962a, b) and Patil (1964). Although both Kupper and Patil have expressed $p_n(t)$ explicitly by a series, recursive determination is more tractable.

We have

$$\frac{d}{d\ell} \log u(\ell) = \frac{d}{d\ell}\left(\text{const.} - \frac{1}{2}\left(\frac{\ell - \mu_G}{\sigma_G}\right)^2\right) = -\frac{\ell - \mu_G}{\sigma_G^2}$$

and (2.20) applies with $k = 1$,

$$\begin{aligned} \ell_0 &= 0 & & \eta_0 = \mu_G & \eta_1 &= -1 \\ \ell_1 &= \infty & \text{and} \quad & \theta_0 - \sigma_G^2 & \theta_1 &= 0. \end{aligned}$$

Thus we get, cf. Willmot (1993, p.126),

$$\begin{aligned} p_0(t) &= \hat{u}(t), \\ p_1(t) &= (\mu_G - \sigma_G^2 t)\, t p_0(t) + \sigma_G^2 t u(0), \\ (m+1)\, p_{m+1}(t) &= (\mu_G - \sigma_G^2 t)\, t p_m(t) + \sigma_G^2 t^2 p_{m-1}(t), \quad m \geq 1. \end{aligned}$$

This is in agreement with a recursion due to Kupper (1962a, p.462). The recursion is stable for $\mu_G \geq \sigma_G^2 t$.

In the case $\mu_G = \sigma_G^2 t$ we have, as pointed out by both Kupper and Patil,

$$p_n(t) = \frac{e^{-\mu_G t/2}}{\mu_G t \Phi(\sqrt{\mu_G t}\,)} \frac{\left(\frac{\mu_G t}{2}\right)^{\frac{n}{2}+1}}{\Gamma\left(\frac{n}{2}+1\right)}. \qquad \square$$

Example 2.6. Let Λ be **lognormally distributed**, i.e. let $\log \Lambda$ be normally distributed with mean μ_G and variance σ_G^2. Like in Example 2.5 the index G stands for Gaussian, while the relation between Λ and the normal distribution is quite different in the two examples. It is well known that

$$u(\ell) = \frac{1}{\sqrt{2\pi}\sigma_G \ell} e^{-(\log \ell - \mu_G)^2/(2\sigma_G^2)}$$

and that

$$\mu_\Lambda = e^{\mu_G + \sigma_G^2/2}, \quad \sigma_\Lambda^2 = e^{2\mu_G + \sigma_G^2}(e^{\sigma_G^2} - 1).$$

Let N be $\mathrm{MP}(t, U)$. We say that N is **lognormal–Poisson distributed**. Explicit forms neither of the probabilities $p_k(t)$ nor of the generating function $G_N(s)$ seem to be known. Further $u(\ell)$ is not of the form required in the Willmot recursion. A table of the probabilities $p_k(1)$ is, however, given by Brown and Holgate (1971) for

$$\begin{matrix} \mu_G = 0.1, 0.2, \ldots, 0.9 \\ \sigma_G^2 = 0.1, 0.2, \ldots, 1.0 \end{matrix} \quad \text{and} \quad \begin{matrix} \mu_G = 1, 2, \ldots, 9 \\ \sigma_G^2 = 1, 2, \ldots, 10 \end{matrix} \quad \text{respectively.}$$

Different approximations of of $p_k(t)$ have, however, been proposed. We will here only consider an approximation due to Bulmer (1974). For a survey of other approximations, together with a historical background and applications of the lognormal–Poisson distribution, we refer to Shaban (1988).

Let X be a $\Gamma(k, t)$-distributed random variable, cf. Example 2.1. We have, for $k > 0$,

$$\begin{aligned} p_k(t) &= \int_0^\infty \frac{(\ell t)^k}{k!} e^{-\ell t} \frac{1}{\sqrt{2\pi}\sigma_G \ell} e^{-(\log \ell - \mu_G)^2/(2\sigma_G^2)}\, d\ell \\ &= \frac{1}{k\sqrt{2\pi}\sigma_G} \int_0^\infty e^{-(\log \ell - \mu_G)^2/(2\sigma_G^2)} \frac{t^k \ell^{k-1}}{(k-1)!} e^{-t\ell}\, d\ell \\ &= \frac{1}{k\sqrt{2\pi}\sigma_G} E\left[e^{-(\log X - \mu_G)^2/(2\sigma_G^2)}\right]. \end{aligned} \tag{2.29}$$

Bulmer's idea is to expand the function $e^{-(\log x - \mu_G)^2/(2\sigma_G^2)}$ in a Taylor series at $x = E[X] = k/t$ and to ignore powers higher than the second. Routine calculations yield

$$e^{-\frac{(\log x-\mu_G)^2}{2\sigma_G^2}} \approx e^{-\frac{(\log(k/t)-\mu_G)^2}{2\sigma_G^2}} (1 + C_1(x - k/t) + C_2(x - k/t)^2),$$

where C_1 is irrelevant and

$$C_2 = \frac{t^2}{2k^2\sigma_G^2} \left(\frac{(\log(k/t) - \mu_G)^2}{\sigma_G^2} + \log(k/t) - \mu_G - 1 \right).$$

Combining this with (2.29) leads to

$$p_k(t) \approx \frac{1}{k\sqrt{2\pi}\sigma_G} e^{-(\log(k/t)-\mu_G)^2/(2\sigma_G^2)}$$

$$\times \left\{ 1 + \frac{1}{2k\sigma_G^2} \left(\frac{(\log(k/t) - \mu_G)^2}{\sigma_G^2} + \log(k/t) - \mu_G - 1 \right) \right\} \quad (2.30)$$

which is the approximation obtained by Bulmer (1974, p.103). Notice that (2.30) is undefined for $k = 0$.

This approximation – as stated by Bulmer – has a relative error less than 10^{-3} when $k \geq 10$ and $t = 1$, for values of μ_G and σ_G^2 likely to be encountered in practice. Unfortunately Bulmer does not specify those values any further.

For large values of k it is tempting to simplify (2.30) to

$$p_k(t) \approx \frac{1}{k\sqrt{2\pi}\sigma_G} e^{-(\log(k/t)-\mu_G)^2/(2\sigma_G^2)} = (1/t)\, u(k/t)$$

and to consider

$$P\{N > n\} \approx \sum_{k=n+1}^{\infty} u(k/t)\,(1/t) \approx \int_{n/t}^{\infty} u(\ell)\, d\ell = 1 - U(n/t). \quad (2.31)$$

Whether (2.31) is mathematically stringent as a limit theorem, i.e. if $\lim_{n\to\infty} P\{N > n\}/(1 - U(n/t)) = 1$, seems to be an open question. Some remarks related to this question will be given after Proposition 8.6 below. □

Further examples of special structure distributions are, for instance, found in Albrecht (1984), Philipson (1960), and Willmot (1986 and 1993). In Philipson (1960) – reviewed by Haight (1967, pp.41–42) – special emphasis is given to the importance of confluent hypergeometric functions. Kemp (1968) studies the **tail-truncated Γ-distribution**, i.e.

$$u(\ell) = \frac{\beta^\gamma}{\Gamma(\gamma)} \ell^{\gamma-1} e^{-\beta\ell} \Bigg/ \int_0^{\ell_1} \frac{\beta^\gamma}{\Gamma(\gamma)} x^{\gamma-1} e^{-\beta x}\, dx, \quad 0 \le \ell \le \ell_1,$$

and points at its relation to confluent hypergeometric functions.

In Chapter 8 we will discuss some relations between the tail behaviour of the distributions of Λ and N.

CHAPTER 3

Some basic concepts

In this chapter we will collect some basic definitions and results about point processes, Markov processes, and martingales. The reader is recommended to scan this chapter hastily and reread if necessary.

Let $(\Omega, \mathcal{F}, P)$ be a complete probability space on which all stochastic quantities are defined. A **filtration** $\boldsymbol{F} = (\mathcal{F}_t;\ t \geq 0)$ is a non-decreasing family of sub-σ-algebras of $\mathcal{F}$. For any process $Y = \{Y(t);\ t \geq 0\}$, the **natural** filtration $\boldsymbol{F}^Y = (\mathcal{F}_t^Y;\ t \geq 0)$ is defined by $\mathcal{F}_t^Y = \sigma\{Y(s);\ s \leq t\}$. Thus $\mathcal{F}_t^Y$ is the σ-algebra generated by Y up to time t, and represents the **internal history** of Y up to time t. All filtrations $\boldsymbol{F}$ will be **completed**, i.e. all P-null sets of $\mathcal{F}$ are added to $\mathcal{F}_0$.

Definition 3.1. *Y is called a* **Markov process** *if*

$$P\{Y(t) \leq y \mid \mathcal{F}_s^Y\} = P\{Y(t) \leq y \mid Y(s)\} \qquad P\text{-a.s.}$$

for all $t \geq s \geq 0$ and all $y \in \mathbb{R}$.

Let $\mathcal{N}$ be set of all functions $\nu = \{\nu(t);\ t \geq 0\}$ such that:

(i) $\nu(0) = 0$;

(ii) $\nu(t)$ is an integer for all $t < \infty$;

(iii) $\nu(\cdot)$ is non-decreasing and right-continuous.

A function $\nu \in \mathcal{N}$ is called a **counting function.**

Let $\mathcal{B}(\mathcal{N})$ be the σ-algebra generated by projections, i.e.

$$\mathcal{B}(\mathcal{N}) = \sigma\{\nu(t) \leq y;\ \nu \in \mathcal{N},\ t \geq 0,\ y < \infty\}.$$

Definition 3.2. *A* **point process** *is a measurable mapping from the probability space $(\Omega, \mathcal{F}, P)$ into $(\mathcal{N}, \mathcal{B}(\mathcal{N}))$.*

Usually a point process is denoted by N. The **distribution** of N is a probability measure Π on $(\mathcal{N}, \mathcal{B}(\mathcal{N}))$. In Figure 3.1 some notation used in a point process is illustrated. The random process $N(\cdot)$ is sometimes called the **counting process** and $N(t)$ denotes the number of points in $(0,t]$. The position of the points are denoted by T_1, $T_2, \ldots$ It is sometimes convenient to illustrate a point process by just marking the points on the time-axis, as shown at the bottom of Figure 3.1.

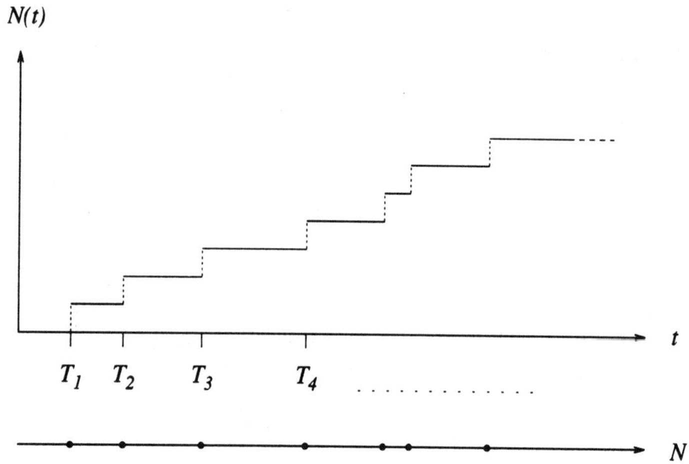

Figure 3.1. *Illustration of notation in a point process.*

Put $\mathcal{N}_S = \{\nu \in \mathcal{N};\ \nu(t) - \nu(t-) = 0 \text{ or } 1\}$, i.e. a $\nu \in \mathcal{N}_S$ increases exactly one unit at its epochs of increase. A point process is called **simple** if $\mathcal{N}$ in Definition 3.2 is replaced by $\mathcal{N}_S$ or if $\Pi\{\mathcal{N}_S\} = 1$, which is the same since $\mathcal{N}_S \in \mathcal{B}(\mathcal{N})$. The point process illustrated in Figure 3.1 is, in fact, simple.

Definition 3.3. *A point process N is said to be (strictly)* **stationary** *if N_s, defined by $N_s(t) = N(s+t) - N(s)$, has the same distribution for all $s \geq 0$.*

To say that N is stationary is the same as to say that the (counting) process $N(\cdot)$ has stationary increments.

A realization of a stationary point process N has P-a.s. no point or infinitely many points, i.e. $P\{N(\infty) = 0 \text{ or } \infty\} = 1$.

The **shift operator** $T_s : \mathcal{N} \to \mathcal{N}$ is defined by

$$(T_s\nu)(t) = \nu(s+t) - \nu(s), \quad s \geq 0.$$

Let $T_s^{-1}B = \{\nu \in \mathcal{N};\ T_s\nu \in B\}$ for any $B \in \mathcal{B}(\mathcal{N})$. A set $B \in \mathcal{B}(\mathcal{N})$ is said to be **invariant**, if $B = T_s^{-1}B$ for all $s \geq 0$

A stationary point process N with distribution Π is **ergodic** if $\Pi\{B\} = 0$ or 1 for all invariant sets $B \in \mathcal{B}(\mathcal{N})$.

Let N be stationary. Then $E[N(t)] = \alpha \cdot t$, where α is called the **intensity** of N. There always exists a, possibly infinite, random variable $\overline{N}$ with $E[\overline{N}] = \alpha$, called the **individual intensity**, such that $N(t)/t \to \overline{N}$ P-a.s. as $t \to \infty$. If $\alpha < \infty$ it follows that $\overline{N}$ is finite P-a.s. Since $\{\nu \in \mathcal{N};\ \overline{\nu} \leq x\}$ is invariant for each x it follows that $\overline{N}$ is deterministic if N is ergodic.

Definition 3.4. *A point process N is called a* **Poisson process with intensity** α, *and its distribution is denoted by Π_α, if*

(*i*) *$N(t)$ has independent increments;*

(*ii*) *$N(t) - N(s)$ is* Po$(\alpha \cdot (t-s))$*-distributed.*

For $n = 1, 2, \ldots, 0 = t_0 < t_1 < \ldots < t_n$, $0 = k_0 \leq k_1 \leq \ldots \leq k_n$, and

$$B = \{\nu \in \mathcal{N};\ \nu(t_1) = k_1, \nu(t_2) = k_2, \ldots, \nu(t_n) = k_n\} \in \mathcal{B}(\mathcal{N}) \quad (3.1)$$

we have

$$\Pi_\alpha\{B\} = \prod_{j=1}^{n} \frac{(\alpha \cdot (t_j - t_{j-1}))^{k_j - k_{j-1}}}{(k_j - k_{j-1})!} e^{-(t_j - t_{j-1})\alpha}.$$

Any probability measure Π on $(\mathcal{N}, \mathcal{B}(\mathcal{N}))$ is determined by $\Pi\{B\}$ for all B of the above kind. Thus $\Pi_\alpha\{B\}$ is determined for all $B \in \mathcal{B}(\mathcal{N})$, i.e. Π_α is well-defined by Definition 3.4. A Poisson process is stationary, ergodic and simple.

The distribution of a simple point process is, in fact, essentially determined by the one-dimensional distributions. Let N be a simple point process. For an interval $I = (s, t]$ we put $N\{I\} = N(t) - N(s)$, so that N then is interpreted as a **random measure**. Let $\mathcal{I}_U$ be the set of finite unions of intervals in $[0, \infty)$. For any $A \in \mathcal{I}_U$ we denote by $N\{A\}$ the number of points in in A.

Theorem 3.1. *The distribution Π of a simple point process is determined by $\Pi\{B\}$ for all B of the form*

$$B = \{\nu \in \mathcal{N};\ \nu\{A\} = 0\}, \quad A \in \mathcal{I}_U.$$

Remark 3.1. Theorem 3.1 is not true if $\mathcal{I}_U$ is the set of unions of any fixed number of intervals. In fact, see Szász (1970), there exists for any fixed value of n a non-Poisson process with distribution Π so that $\Pi\{B\} = \Pi_\alpha\{B\}$ for B of the form given in (3.1). □

Let $A(t)$ be a right-continuous non-decreasing function with $A(0) = 0$ and $A(t) < \infty$ for each $t < \infty$. A point process N is called a (non-homogeneous) **Poisson process with intensity measure** A if (ii) in Definition 3.4 is replaced by

(*ii*) *$N(t) - N(s)$ is Poisson distributed with mean $A(t) - A(s)$.*

A Poisson process with intensity measure A is simple if and only if A is continuous. Without much loss of generality we may in that case assume that A has the representation $A(t) = \int_0^t \alpha(s)\, ds$, where $\alpha(\cdot)$ is called the **intensity function.**

Definition 3.5. *An* $\boldsymbol{F}$**-martingale** $M = \{M(t);\ t \geq 0\}$ *is a real valued process such that:*

(*i*) *$M(t)$ is $\mathcal{F}_t$-measurable for $t \geq 0$;*

(*ii*) *$E[|M(t)|] < \infty$ for $t \geq 0$;*

(*iii*) *$E[M(t) \mid \mathcal{F}_s] = M(s)$ $\quad$ P-a.s. for $t \geq s$.*

An $\boldsymbol{F}$-martingale M is called **right continuous** *if:*

(*i*) *The trajectories $M(t)$ are right continuous;*

(*ii*) *The filtration $\boldsymbol{F}$ is right continuous, i.e.*

$$\mathcal{F}_t = \bigcap_{s>t} \mathcal{F}_s \quad \text{for } t \geq 0.$$

All processes which we consider have right continuous trajectories and the filtrations are always such that the condition (ii) of the right continuity definition is satisfied.

Let X be a random variable with finite mean, and put $M(t) \stackrel{\text{def}}{=} E[X \mid \mathcal{F}_t]$. Then

$$E[M(t) \mid \mathcal{F}_s] = E\big[E[X \mid \mathcal{F}_t] \mid \mathcal{F}_s\big] = M(s),$$

i.e. M is an $\boldsymbol{F}$-martingale.

The following well-known proposition, which goes back to Lévy, may thus be regarded as a martingale convergence result.

Proposition 3.1. *Let X with $E[|X|] < \infty$ be a random variable. Then $E[X \mid \mathcal{F}_t] \to E[X \mid \mathcal{F}_\infty]$ P-a.s., where $\mathcal{F}_\infty$ is the smallest σ-algebra containing $\mathcal{F}_t$, $t \geq 0$.*

If X is $\mathcal{F}_\infty$-measurable it follows that $E[X \mid \mathcal{F}_t] \to X \quad P$-a.s.

Definition 3.6. *Let N be a simple point process and let*

$$\hat{\Lambda} = \{\hat{\Lambda}(t);\ t \geq 0\}$$

be a random process with P-a.s. non-decreasing and continuous realizations. $\hat{\Lambda}$ is called the **$\boldsymbol{F}$-compensator** *of N if $\hat{\Lambda}(t)$ is $\mathcal{F}_t$-measurable and if $N(t) - \hat{\Lambda}(t)$ is an $\boldsymbol{F}$-martingale.*

The $\boldsymbol{F}^N$-compensator is uniquely (up to an equivalence) determined by the point process, and, conversely, the point process is uniquely determined by its $\boldsymbol{F}^N$-compensator. With a certain extension of Definition 3.6 every simple point process has a compensator. The important restriction is that $\hat{\Lambda}$ is assumed to have continuous realizations.

Let N be a non-homogeneous Poisson process with continuous intensity measure A. Then it is easy to see that $N(t) - A(t)$ is an $\boldsymbol{F}^N$-martingale. The following characterization result is generally regarded as the first important result linking point processes and martingales.

Theorem 3.2 (Watanabe, 1964). *A simple point process N is a Poisson process with continuous intensity measure A if and only if $N(t) - A(t)$ is an $\boldsymbol{F}^N$-martingale.*

In other words, Watanabe's theorem states that a simple point process with continuous $\boldsymbol{F}^N$-compensator is a Poisson process if and only if its $\boldsymbol{F}^N$-compensator is deterministic.

It is, at least intuitively, obvious that to require that a point process has a deterministic $\boldsymbol{F}^N$-compensator is (almost) the same as to require that it has independent increments. Therefore Watanabe's theorem may be looked upon as a martingale variant of Prekopa's characterization of the Poisson process. In order to formulate that theorem we need the following definition: A point process N is said to be **non-atomic** if

$$P\{N(t) - N(t-) = 0\} = 1 \text{ for all } t.$$

A Poisson process with intensity measure A is non-atomic if and only if A is continuous, i.e. if and only if it is simple.

Theorem 3.3 (Prekopa, 1958). *A simple non-atomic point process N is a (non-homogeneous) Poisson process if and only if it has independent increments.*

In many cases the $\boldsymbol{F}$-compensator $\hat{A}$ is not only continuous but also absolutely continuous.

Definition 3.7. *Let N be a simple point process with $\boldsymbol{F}$-compensator $\hat{A}$. A non-negative right continuous random process*

$$\hat{\lambda} = \{\hat{\lambda}(t);\ t \geq 0\}$$

satisfying

$$\hat{A}(t) = \int_0^t \hat{\lambda}(s)\, ds$$

is called the **$\boldsymbol{F}$-intensity** *of N.*

The intuitive interpretation of the $\boldsymbol{F}$-intensity is

$$\hat{\lambda}(t) = E[dN(t) \mid \mathcal{F}_t].$$

It is not easy to find simple conditions on a point process to ascertain that it has an $\boldsymbol{F}$-intensity.

Let N be a point process and let T_k denote the position of the kth point, as illustrated in Figure 3.1.

Definition 3.8. *A point process N is called an ordinary* **renewal process** *(with inter-occurrence time distribution K) if the variables $T_1, T_2 - T_1, T_3 - T_2, \ldots$ are independent and have a common distribution K.*

Further it will be tacitly understood that $K(0) = 0$, which is equivalent to N being simple.

A fine account of renewal theory is given by Feller (1971, pp.358–388), from which most of the material below is taken.

A Poisson process is an ordinary renewal process with exponentially distributed inter-occurrence times. The Poisson process is the only ordinary renewal process which is also stationary.

Assume now that $E[T_1] < \infty$ and let $1/\alpha$ denote that mean, i.e.

$$\frac{1}{\alpha} \stackrel{\text{def}}{=} \int_0^\infty t\, dK(t) = \int_0^\infty (1 - K(t))\, dt.$$

Let the variables $T_2 - T_1$, $T_3 - T_2, \ldots$ have distribution K, while the distribution of T_1 is modified to

$$P\{T_1 \le t\} = \alpha \int_0^t (1 - K(s))\, ds.$$

The resulting point process is called a **stationary** renewal process and is a stationary point process in the sense of Definition 3.3. In the sequel we do, however, by a 'renewal process' mean an 'ordinary renewal process'.

Suppose that we are interested in the **renewal measure** $A(t) \stackrel{\text{def}}{=} E[N(t)]$. It follows from the 'renewal property' that

$$A(t) = \int_0^t (1 + A(t-s))\, dK(s) = K(t) + \int_0^t A(t-s)\, dK(s). \quad (3.2)$$

Equation (3.2), which is an example of a **renewal equation**, has the unique solution

$$A(t) = K(t) + A * K(t) = K(t) + (K + A * K) * K(t)$$

$$= K(t) + K^{2*}(t) + A * K^{2*}(t) = \ldots = \sum_{n=1}^{\infty} K^{n*}(t), \quad (3.3)$$

where K^{n*} denotes the nth convolution of K with itself.

Since, cf. Figure 3.1, $\{N(t) > n\} = \{T_{n+1} \le t\}$ we have

$$P\{N(t) = n\} = K^{n*}(t) - K^{(n+1)*}(t), \quad n = 1, 2, \ldots,$$

and thus (3.3) also follows directly.

Assume now that K is **non-arithmetic**, i.e. there exists no number d such that K is concentrated on d, $2d, \ldots$. Then

$$\lim_{t \to \infty} A(t+h) - A(t) = h\alpha \quad \text{for every } h > 0; \quad (3.4)$$

a result known as Blackwell's theorem.

Let K be continuous with density k and let

$$\alpha(t) \stackrel{\text{def}}{=} \sum_{n=1}^{\infty} k^{n*}(t),$$

where k^{n*} denotes the density of K^{n*}. Obviously $\alpha(\cdot)$ satisfies the equation

$$\alpha(t) = k(t) + \int_0^t \alpha(t-s)k(s)\,ds. \tag{3.5}$$

Conversely, since (3.5) is a renewal equation, it has a unique solution $\alpha(\cdot)$ whenever k is bounded. Since $A(t) = \int_0^t \alpha(s)\,ds$ satisfies (3.2) it follows that $\alpha(\cdot)$ is a density for $A(\cdot)$.

Let $T_0 = 0$. The $\boldsymbol{F}^N$-intensity of N is given by

$$\hat{\lambda}(t) = \frac{k(t - T_{N(t)})}{1 - K(t - T_{N(t)})} = r(t - T_{N(t)}), \tag{3.6}$$

where $r(\cdot)$ is the failure rate for K, see (7.1) below.

CHAPTER 4

The mixed Poisson process

Let N be a Markov point process. Its marginal distributions and transition probabilities are defined by

$$p_n(t) \stackrel{\text{def}}{=} P\{N(t) = n\}$$

and

$$p_{m,n}(s,t) \stackrel{\text{def}}{=} P\{N(t) = n \mid N(s) = m\}, \quad 0 \le s \le t,$$

respectively.

Since N is a point process we have $p_0(0) = 1$, $p_n(t) = p_{0,n}(0,t)$ and $p_{m,n}(s,t) = 0$ for $n < m$.

The transition probabilities satisfy the Chapman–Kolmogorov equations

$$p_{m,n}(s,t) = \sum_{k=m}^{n} p_{m,k}(s,\tau)\, p_{k,n}(\tau,t), \quad \text{for } s \le \tau \le t,\ m \le n.$$

Definition 4.1. *A simple Markov point process N is called a* **birth process** *with* **transition intensities** $\kappa_n(t)$ *if, for* $t > 0$,

$$p_{m,n}(t, t+h) = \begin{cases} 1 - \kappa_m(t)h + o(h), & n = m, \\ \kappa_m(t)h + o(h), & n = m+1, \\ o(h), & n > m+1, \end{cases}$$

as $h \downarrow 0$.

A birth process is thus a simple Markov point process where the jumps are caused by the intensities $\kappa_n(t)$. (The trivial simple Markov point process $N(t) = \lfloor t \rfloor$, where $\lfloor\cdot\rfloor$ means 'integer part', is not a birth process.) The intensities must be non-negative. Further we assume – although it is not necessary – that they are continuous in t

on $(0,\infty)$. We do, however, allow for the possibility that $\kappa_n(0)=\infty$, but we require that $\int_0^t \kappa_0(s)\,ds<\infty$ for all $t<\infty$ in order to ensure that $N(0)=0$. It is more or less obvious that a birth process is determined by its intensities.

On the other hand, if $\kappa_n(t)$ are non-negative continuous functions in t such that

$$\sum_{n=k}^{\infty} \frac{1}{\max\limits_{0\le s<t} \kappa_n(s)} = \infty, \quad \text{for any } t \text{ and } k, \tag{4.1}$$

then the intensities define a birth process, cf. Lundberg (1964, pp.46–47). Condition (4.1) is to ascertain that the process does not explode, which – with our definition of a point process – is not permitted. We do not at all claim that the given conditions are the best possible, but we will not need any sophisticated set of conditions in this direction.

Assume now that a birth process N is given. From the Chapman–Kolmogorov equations with $m=0$, $s=0$, $\tau=t$, and $t=t+h$ we get

$$p_n(t+h)$$

$$= p_n(t)(1-\kappa_n(t)h+o(h)) + p_{n-1}(t)(\kappa_{n-1}(t)h+o(h)) + \sum_{k=0}^{n-2} p_k(t)o(h).$$

Since we only sum over finitely many values of k, we get

$$p_n(t+h)-p_n(t) = -p_n(t)\kappa_n(t)h + p_{n-1}(t)\kappa_{n-1}(t)h + o(h),$$

and thus, cf. Lundberg (1964, p.18),

$$\begin{aligned} p_0'(t) &= -\kappa_0(t)\,p_0(t), \\ p_n'(t) &= -\kappa_n(t)\,p_n(t) + \kappa_{n-1}(t)\,p_{n-1}(t), \quad n>0. \end{aligned} \tag{4.2}$$

It follows from (4.2) that the (one-dimensional) marginal distributions determine the intensities, and thus characterize the distribution of the birth process.

Remark 4.1. Equations (4.2) are really **forward equations**. By exactly the same arguments, only that arbitrary m and s are considered, the forward equations for the transition probabilities,

$$\begin{aligned} \frac{\partial p_{n,n}(s,t)}{\partial t} &= -\kappa_n(t)\,p_{n,n}(s,t), \\ \frac{\partial p_{m,n}(s,t)}{\partial t} &= -\kappa_n(t)\,p_{m,n}(s,t) + \kappa_{n-1}(t)\,p_{m,n-1}(s,t), \end{aligned} \tag{4.3}$$

for $m < n$, follow.

The **backward equations** follow from the Chapman–Kolmogorov equations with $\tau = s + h$. Then we get, for $m < n$,

$$\begin{aligned} & p_{m,n}(s,t) \\ &= (1 - \kappa_m(s)h)\, p_{m,n}(s+h,t) + \kappa_m(s) h p_{m+1,n}(s+h,t) + o(h) \\ &= p_{m,n}(s+h,t) - \kappa_m(s) h p_{m,n}(s,t) + \kappa_m(s) h p_{m+1,n}(s,t) + o(h). \end{aligned}$$

By the 'usual' arguments we get,

$$\begin{aligned} \frac{\partial p_{m,m}(s,t)}{\partial s} &= \kappa_m(s)\, p_{m,m}(s,t), \\ \frac{\partial p_{m,n}(s,t)}{\partial s} &= \kappa_m(s)[\, p_{m,n}(s,t) - p_{m+1,n}(s,t)], \end{aligned} \tag{4.4}$$

for $m < n$. □

Ove Lundberg, essentially, defined a mixed Poisson process, cf. Lundberg (1964, p.72), in the following way:

Lundberg's definition of a mixed Poisson process. *A birth process N is called a* **mixed Poisson process**, MPP(U), *if $N(t)$ is MP(t, U)-distributed for all $t \geq 0$ and some distribution U.*

The natural question, of course settled by Lundberg (1964), is:

Does there exist a mixed Poisson process to each distribution U, and is the process uniquely defined?

We have already discussed the uniqueness, so only the existence remains to be established. Consider the Laplace transform

$$\hat{u}(t) = \int_{0-}^{\infty} e^{-\ell t}\, dU(\ell),$$

and notice that

$$p_n(t) = (-1)^n \frac{t^n}{n!} \hat{u}^{(n)}(t). \tag{4.5}$$

Differentiation of (4.5) yields

$$p_0'(t) = \hat{u}'(t) = \frac{\hat{u}'(t)}{\hat{u}(t)}\, p_0(t)$$

and, for $n > 0$,

$$p'_n(t) = (-1)^n \frac{t^n}{n!} \hat{u}^{(n+1)}(t) + (-1)^n \frac{t^{(n-1)}}{(n-1)!} \hat{u}^{(n)}(t)$$

$$= \frac{\hat{u}^{(n+1)}(t)}{\hat{u}^{(n)}(t)} p_n(t) - \frac{\hat{u}^{(n)}(t)}{\hat{u}^{(n-1)}(t)} p_{n-1}(t).$$

It follows from (4.2), cf. Lundberg (1964, p.72), provided N is a birth process and $U(0) < 1$, that,

$$\kappa_n(t) = -\frac{\hat{u}^{(n+1)}(t)}{\hat{u}^{(n)}(t)} = \frac{\int_{0-}^{\infty} \ell^{n+1} e^{-\ell t} \, dU(\ell)}{\int_{0-}^{\infty} \ell^{n} e^{-\ell t} \, dU(\ell)} . \tag{4.6}$$

It follows from (4.6) that the intensities are positive, continuous and differentiable for $t > 0$. Further

$$\int_0^t \kappa_0(s) \, ds = -\int_0^t \frac{d}{ds} \log \hat{u}(s) \, ds = -\log \hat{u}(t) < \infty \quad \text{for all } t < \infty.$$

Naturally we may change $\kappa_n(t)$ in isolated points, but it will be tacitly understood that we have chosen the natural version, i.e. the one given by (4.6). Since $\sum_{n=0}^{\infty} p_n(t) = 1$ for all t there can be no explosion, so it is no problem that (4.1) does not necessarily hold, cf. Lundberg (1964, p.75). Thus there exists a well-defined birth process with its intensities given by (4.6).

If $U(0) = 1$, i.e. if $P\{\Lambda = 0\} = 1$, we have P-a.s. no points. Then $\kappa_0(t) \equiv 0$, while $\kappa_n(t)$, for $n > 0$, is irrelevant and may be undefined. We will disregard this trivial case, except in some cases where it is required for completeness.

Assume that $N(s) = m$ and consider the conditional point process N^* defined by $N^*(t) = N(s+t) - N(s)$. Obviously N^* is a birth process with, cf. Lundberg (1964, p.91) and Willmot and Sundt (1989b),

$$\kappa_n^*(t) = \kappa_{m+n}(s+t) \quad \text{and} \quad p_n^*(t) = p_{m,m+n}(s, s+t). \tag{4.7}$$

From (4.6) we get

$$\kappa_n^*(t) = \frac{\int_{0-}^{\infty} \ell^{m+n+1} e^{-\ell(s+t)} \, dU(\ell)}{\int_{0-}^{\infty} \ell^{m+n} e^{-\ell(s+t)} \, dU(\ell)} = \frac{\int_{0-}^{\infty} \ell^{n+1} e^{-\ell t} \, dU^*(\ell)}{\int_{0-}^{\infty} \ell^{n} e^{-\ell t} \, dU^*(\ell)},$$

where

$$U^*(x) = \frac{\int_{0-}^{x} \ell^m e^{-s\ell}\, dU(\ell)}{\int_{0-}^{\infty} \ell^m e^{-s\ell}\, dU(\ell)}. \tag{4.8}$$

Thus N^* is an MPP(U^*) and we have, cf. Lundberg (1964, p.85),

$$\begin{aligned}
p_{m,n}(s,t) &= p^*_{n-m}(t-s) \\
&= \int_{0-}^{\infty} \frac{(\ell\cdot(t-s))^{n-m}}{(n-m)!} e^{-\ell\cdot(t-s)}\, dU^*(\ell) \\
&= \frac{\int_{0-}^{\infty} \frac{\ell^n\cdot(t-s)^{n-m}}{(n-m)!}\, e^{-\ell t}\, dU(\ell)}{\int_{0-}^{\infty} \ell^m e^{-\ell s}\, dU(\ell)} \\
&= \binom{n}{m} \left(\frac{s}{t}\right)^m \left(1-\frac{s}{t}\right)^{n-m} \frac{p_n(t)}{p_m(s)}.
\end{aligned} \tag{4.9}$$

Let Π_U denote the distribution of N, and consider the set

$$B = \{\nu \in \mathcal{N};\ \nu(t_1) = k_1,\ \nu(t_2) = k_2, \ldots,\ \nu(t_n) = k_n\}.$$

From the Markov property it follows that

$$\begin{aligned}
\Pi_U\{B\} &= p_{k_1}(t_1)\, p_{k_1,k_2}(t_1,t_2)\, p_{k_2,k_3}(t_2,t_3) \cdots p_{k_{n-1},k_n}(t_{n-1},t_n) \\
&= p_{k_1}(t_1) \prod_{j=2}^{n} \binom{k_j}{k_{j-1}} \left(\frac{t_{j-1}}{t_j}\right)^{k_{j-1}} \left(1-\frac{t_{j-1}}{t_j}\right)^{k_j-k_{j-1}} \frac{p_{k_j}(t_j)}{p_{k_{j-1}}(t_{j-1})} \\
&= p_{k_n}(t_n) \frac{k_n!\, t_1^{k_1}}{k_1!\, t_n^{k_n}} \prod_{j=2}^{n} \frac{(t_j-t_{j-1})^{k_j-k_{j-1}}}{(k_j-k_{j-1})!} \\
&= \int_{0-}^{\infty} \Pi_\ell\{B\}\, dU(\ell),
\end{aligned} \tag{4.10}$$

where Π_ℓ denotes the distribution of a Poisson process with intensity ℓ, cf. Definition 3.4. As pointed out in Chapter 3, Π_U is determined by (4.10). For each $B \in \mathcal{B}(\mathcal{N})$ the mapping $\ell \mapsto \Pi_\ell\{B\}$ is measurable (Grandell 1976, pp.5–7). Thus we can define an MPP(U) in the following way:

Definition 4.2. *A point process N is called a* **mixed Poisson process**, MPP(U), *if its distribution Π_U is given by*

$$\Pi_U\{B\} = \int_{0-}^{\infty} \Pi_\ell\{B\}\, dU(\ell), \quad B \in \mathcal{B}(\mathcal{N}).$$

Since any Poisson process is stationary, the same also holds for a mixed Poisson process.

Remark 4.2. In the literature, both Lundberg's definition and Definition 4.2 have been used. In some publications based on point process theory – Grandell (1976) is a typical such example – Definition 4.2 is used and, without any comments on the difference in definitions, results by Lundberg (1964) are quoted. Albrecht (1981) filled this gap, and showed that the two definitions are equivalent. □

Definition 4.2 leads to the following intuitive interpretation of a mixed Poisson process N:

First a realization ℓ of a non-negative random variable Λ is generated and conditioned upon that realization N is a Poisson process with intensity ℓ.

More precisely, we consider (Λ, N) as a measurable mapping from the probability space $(\Omega, \mathcal{F}, P)$ into $(\mathbb{R} \times \mathcal{N}, \mathcal{B}(\mathbb{R} \times \mathcal{N}))$, where

$$\mathcal{B}(\mathbb{R} \times \mathcal{N}) = \sigma\{(\ell \le x, \nu(t)) \le y;\ \nu \in \mathcal{N},\ t \ge 0,\ x, y < \infty\},$$

with its distribution determined by

$$P\{\Lambda \le x,\ N \in B\} = \int_{0-}^{x} \Pi_\ell\{B\}\, dU(\ell), \quad B \in \mathcal{B}(\mathcal{N}).$$

With this definition it is natural to interpret the mixed Poisson process as the Bayesian version of the Poisson process and U as the prior distribution of the intensity. This interpretation is essentially not new at all, cf. Lundberg (1964, p.92). For any $B \in \mathcal{B}(\mathcal{N})$ with $\Pi_U\{B\} > 0$ we can then regard

$$U^B(x) \stackrel{\text{def}}{=} P\{\Lambda \le x \mid N \in B\} = \frac{\int_{0-}^{x} \Pi_\ell\{B\}\, dU(\ell)}{\int_{0-}^{\infty} \Pi_\ell\{B\}\, dU(\ell)}$$

as the **posterior distribution** of the intensity given B or, more precisely, $N \in B$. Notice that, while $B \in \mathcal{B}(\mathcal{N})$, the event $\{N \in B\} \in \mathcal{F}$.

Assume that we have observed $N(t) = n$. The posterior distribution is then given by, see (2.2),

$$U^*(x) \stackrel{\text{def}}{=} P\{\Lambda \leq x \mid N(t) = n\} = \frac{\int_{0-}^{x} \ell^n e^{-\ell t}\, dU(\ell)}{\int_{0-}^{\infty} \ell^n e^{-\ell t}\, dU(\ell)},$$

which is in agreement with (4.8). The **best estimate** – or the Bayes estimate – of Λ is given by, see (2.3),

$$\Lambda_B^* \stackrel{\text{def}}{=} E[\Lambda \mid N(t) = n] = \int_{0-}^{\infty} x\, dU^*(x)$$

$$= \frac{\int_{0-}^{\infty} \ell^{n+1} e^{-\ell t}\, dU(\ell)}{\int_{0-}^{\infty} \ell^n e^{-\ell t}\, dU(\ell)} = \kappa_n(t), \tag{4.11}$$

where the last equality follows from (4.6).

The best linear estimate Λ_L^* of Λ is given by (2.8) with N replaced by $N(t)$.

Remark 4.3. Instead of basing Λ_B^* and Λ_L^* on $N(t)$ it would be natural to base them on the whole realization of $N(s)$ for $s \leq t$. For the best estimate that would mean $\Lambda_B^* = E[\Lambda \mid \mathcal{F}_t^N]$. Due to the structure of the mixed Poisson process, the estimates are the same, see Grandell (1976, pp.95, 129–130). □

Proposition 4.1. *Let N be* MPP(U) *with $E[\Lambda] < \infty$. N has $\boldsymbol{F}^N$-intensity $\hat{\lambda}(t) = E[\Lambda \mid \mathcal{F}_t^N] = \kappa_{N(t)}(t)$.*

Proof. Consider the filtration $\boldsymbol{F}$ with

$$\mathcal{F}_t = \sigma\{\Lambda,\ N(s);\ s \leq t\}.$$

This means that Λ is known all the time, and thus N is, essentially, a Poisson process with intensity Λ. The $\boldsymbol{F}$-compensator is then, cf. Theorem 3.2, $\Lambda \cdot t$. Thus, cf. Definition 3.7, the $\boldsymbol{F}$-intensity is Λ. This implies, see Brémaud (1981, pp.42–43) for details, that $E[\Lambda \mid \mathcal{F}_t^N]$ is the $\boldsymbol{F}^N$-intensity. The proposition now follows from (4.11) and Remark 4.3. ■

For 'kind' birth processes $\kappa_{N(t)}(t)$ is the $\boldsymbol{F}^N$-intensity. We will not use this, but only point out that what 'kind' shall really mean is not a trivial question. For a, more general, discussion of related matters we refer to Aven (1985).

Sometimes it is desirable to have Λ explicitly in the definition. We can then choose the following definition. A **standard Poisson process** is a Poisson process with intensity 1.

Definition 4.3. *Let a random variable Λ and a standard Poisson process $\widetilde{N}$ be independent of each other. The point process $N = \widetilde{N} \circ \Lambda$, where $\widetilde{N} \circ \Lambda(t) \stackrel{\text{def}}{=} \widetilde{N}(\Lambda t)$, is called a mixed Poisson process.*

Strictly speaking, in Definition 4.3 it is enough to require that N and $\widetilde{N} \circ \Lambda$ are equal in distribution. Moreover, we ought to show that the mapping $(\widetilde{N}, \Lambda) \mapsto \widetilde{N}(\Lambda t)$ is measurable. For these questions we refer to Grandell (1976, pp.9–16).

With Definition 4.3 the following result is a triviality, although the whole question about almost sure convergence was meaningless in the distributional setting.

Proposition 4.2. *Let N be* MPP(U). *Then*

$$\lim_{t\to\infty} \tfrac{1}{t}N(t) = \Lambda \quad P\text{-a.s.}$$

Proof. On $\{\Lambda = 0\}$ we have

$$\lim_{t\to\infty} \tfrac{1}{t}N(t) = \lim_{t\to\infty} \tfrac{1}{t}\widetilde{N}(0) = 0.$$

On $\{\Lambda > 0\}$ we have

$$\lim_{t\to\infty} \tfrac{1}{t}N(t) = \lim_{t\to\infty} \tfrac{1}{t}\widetilde{N}(\Lambda t) = \Lambda \lim_{t\to\infty} \tfrac{1}{\Lambda t}\widetilde{N}(\Lambda t)$$
$$= \Lambda \lim_{s\to\infty} \tfrac{1}{s}\widetilde{N}(s) = \Lambda.$$

■

Proposition 4.2 does in fact hold for all stationary point processes, cf. for instance Grandell (1991, p.128), provided Λ is interpreted as the individual intensity. That is also true for the following corollary, cf. Albrecht (1983) and Gerber (1983).

Corollary 4.1. *Let N be* MPP(U) *with $U(0) = 0$ and let T_n denote the time of the nth jump. Then*

$$\lim_{n\to\infty} \frac{T_n}{n} = \Lambda^{-1} \quad P\text{-a.s.}$$

Proof. It follows from Proposition 4.2 that

$$\Lambda = \lim_{t\to\infty} \tfrac{1}{t}N(t) = \lim_{n\to\infty} \frac{N(T_n)}{T_n} = \lim_{n\to\infty} \frac{n}{T_n},$$

and the corollary follows. ■

Consider now $\kappa_{N(t)}(t)$ as a random process. The following proposition, cf. Lundberg (1964, p.93), holds.

Proposition 4.3. *Let N be* MPP(U) *with $E[\Lambda] < \infty$ or, equivalently, with $\kappa_0(0) < \infty$. Then*

$$\lim_{t\to\infty} \kappa_{N(t)}(t) = \Lambda \quad P\text{-a.s.}$$

Proof. Recall from Proposition 4.1 that

$$\kappa_{N(t)}(t) = E[\Lambda \mid \mathcal{F}_t^N].$$

It follows from Proposition 4.2 that Λ is $\mathcal{F}_\infty^N$-measurable, and then Proposition 3.1 yields the desired result. ■

Proposition 4.3 is intuitively very natural, when compared with Proposition 4.2, since $\kappa_{N(t)}(t)$, in comparison with $N(t)/t$, ought to be 'closer' to Λ. In a way Lundberg (1964, p.93) used this idea, and considered $\kappa_n(t)$ when n and t tend to infinity in such a way that n/t tends to a fixed value.

The following lemma, cf. Lundberg (1964, p.72), will be useful. As will be seen in Theorem 6.1, it does, in fact, characterize those birth processes which are also mixed Poisson processes.

Lemma 4.1. *Let N be* MPP(U) *with $U(0) < 1$. Its intensities $\kappa_n(t)$ satisfy the recurrence relation*

$$\kappa_{n+1}(t) = \kappa_n(t) - \frac{\kappa_n'(t)}{\kappa_n(t)}, \qquad n = 0, 1, \ldots$$

Proof. The lemma follows from (4.6) since

$$\kappa_n'(t) = -\frac{\hat{u}^{(n+2)}(t)\hat{u}^{(n)}(t) - \hat{u}^{(n+1)}(t)^2}{\hat{u}^{(n)}(t)^2}$$

$$= -\kappa_{n+1}(t)\kappa_n(t) + \kappa_n(t)^2.$$

■

Example 4.1. An MPP(U) N is called a **Pólya process** if U is a Γ-distribution, cf. Example 2.1. It follows from (2.5) that if Λ is $\Gamma(\gamma, \beta)$, i.e. if

$$u(\ell) \stackrel{\text{def}}{=} U'(\ell) = \frac{\beta^\gamma}{\Gamma(\gamma)} \ell^{\gamma-1} e^{-\beta\ell}, \quad \ell \geq 0,$$

then

$$\kappa_n(t) = \frac{\gamma + n}{\beta + t}.$$

Suppose that we have observed $N(s) = m$. The density u^* of the posterior distribution U^*, cf. (4.8), is then given by

$$u^*(\ell) = \text{const.}\ \ell^m e^{-s\ell} \cdot \ell^{\gamma-1} e^{-\beta\ell}.$$

Thus U^* is a $\Gamma(\gamma + m, \beta + s)$-distribution. This is, of course, well known from Bayesian statistics. The fact that U and U^* belong to the same class of distributions is generally expressed by saying that Γ-distributions form a family of **conjugate prior distributions** relative to the Poisson distribution.

An alternative way to determine the posterior distribution is to rely on (4.7). Then we find

$$\kappa_n^*(t) = \kappa_{m+n}(s+t) = \frac{\gamma + (m+n)}{\beta + (s+t)} = \frac{(\gamma + m) + n}{(\beta + s) + t},$$

which yields the same posterior distribution. □

In Proposition 4.4, cf. Lundberg (1964, pp.76–84), we list various properties of the intensities $\kappa_n(t)$, although some of them have already been mentioned.

Proposition 4.4. *Let N be* MPP(U) *with $U(0) < 1$. The transition intensities $\kappa_n(t)$ have the following properties:*

(*i*) *$\kappa_n(t)$ is positive, continuous, differentiable, non-increasing in $t > 0$, and non-decreasing in n;*

(*ii*) *$\kappa_n(0) < \infty$ if and only if $E[\Lambda^{n+1}] < \infty$;*

(*iii*) *$\int_0^\infty \kappa_0(s)\,ds = \infty$ if $U(0) = 0$,*

$\int_0^\infty \kappa_0(s)\,ds < \infty$ if $U(0) > 0$,

$\int_t^\infty \kappa_n(s)\,ds = \infty$ for all t if $n > 0$;

(*iv*) *for fixed n, there exist $t_1, t_2, \ldots \to \infty$ such that*

$$\kappa_n(t_k) > \frac{n}{t_k};$$

(*v*) *for fixed t, there exist $n_1, n_2, \ldots$ such that*

$$\kappa_{n_k}(t) < \frac{n_k + 1}{t};$$

(*vi*) *for fixed* t *and* $\varepsilon > 0$, *there exists an* n_0, *depending on* t *and* ε, *such that*

$$\kappa_n(t) < \frac{1+\varepsilon}{t} n \log n, \quad n \geq n_0.$$

Proof. (i) As already mentioned, it follows from (4.6) that the intensities are positive, continuous and differentiable for $t > 0$. Further, it follows from (4.6), by Schwartz' inequality applied to the variables $\Lambda^{n/2}e^{-\Lambda t/2}$ and $\Lambda^{(n+2)/2}e^{-\Lambda t/2}$, that

$$\begin{aligned}\kappa_{n+1}(t) - \kappa_n(t) &= \frac{\hat{u}^{(n+1)}(t)}{\hat{u}^{(n)}(t)} - \frac{\hat{u}^{(n+2)}(t)}{\hat{u}^{(n+1)}(t)} \\ &= \frac{\hat{u}^{(n+1)}(t)^2 - \hat{u}^{(n+2)}(t)\hat{u}^{(n)}(t)}{\hat{u}^{(n+1)}(t)\hat{u}^{(n)}(t)} \geq 0,\end{aligned}$$

i.e. $\kappa_n(t)$ is non-decreasing in n. By Lemma 4.1 it follows that $\kappa_n(t)$ is non-increasing in t.
(ii) We will show this by induction. By (4.6) we get $\kappa_0(0) = E[\Lambda]$, so (ii) holds for $n = 0$.

Assume that (ii) holds for $n-1$. If $E[\Lambda^n] < \infty$ if follows from (4.6) that $\kappa_n(0) = E[\Lambda^{n+1}]/E[\Lambda^n]$, and so (ii) holds. If $E[\Lambda^n] = \infty$ then $E[\Lambda^{n+1}] = \infty$. By the induction assumption, $\kappa_{n-1}(0) = \infty$ follows. Since, by (i), $\kappa_n(t) \geq \kappa_{n-1}(t)$, it follows that $\kappa_n(0) - \infty$. Thus (i) holds also for n.
(iii) Using (4.2) we get $p_0(t) = \exp\{-\int_0^t \kappa_0(s)\,ds\}$. Thus

$$\int_0^\infty \kappa_0(s)ds = -\lim_{t\to\infty} \log p_0(t) = -\lim_{t\to\infty} \log \hat{u}(t) = -\log U(0),$$

and the statements about $\kappa_0(\cdot)$ follow. Assume now that $N(t) = n > 0$ and consider $N^*(s) = N(t+s) - N(t)$. Since, see (4.7) and (4.8), $\kappa_0^*(s) = \kappa_n(s+t)$ and $U^*(0) = 0$, also the statement about $\kappa_n(\cdot)$ is true.
(iv) Since (iv) is trivially true for $n = 0$, we consider $n > 0$. Differentiation of (4.5) yields, cf. (4.6),

$$p_n'(t) = (-1)^n \frac{t^{(n-1)}}{(n-1)!}\hat{u}^{(n)}(t) + (-1)^n \frac{t^n}{n!}\hat{u}^{(n+1)}(t)$$

$$= p_n(t)\left(\frac{n}{t} - \kappa_n(t)\right).$$

Since $p_n(t)$ is positive and tends to zero as $t \to \infty$, we cannot have $p_n'(t) \geq 0$ for **all** $t > t_0$, however large t_0 may be. Therefore $\kappa_n(t) \leq n/t$ for all $t > t_0$ is impossible, and (iv) follows.
(v) From (4.5) and (4.6) we get

$$\frac{p_{n+1}(t)}{p_n(t)} = \frac{t}{n+1}\left(-\frac{u^{(n+1)}(t)}{u^{(n)}(t)}\right) = \frac{t}{n+1} \cdot \kappa_n(t).$$

Since, for fixed t, $\{p_n(t)\}$ is a probability distribution, and therefore summable, we must have $p_{n+1}(t)/p_n(t) < 1$ for some sequence $n_1, n_2, \ldots$ of n-values, and (v) follows.
(vi) The proof is very technical, and we refer to Lundberg (1964, pp.78–81). ■

The following characterization of the Pólya process is due to Lundberg (1964, pp.97–99)

Theorem 4.1. *Let N be* MPP(U) *with intensities $\kappa_n(t)$. The following three statements are equivalent:*

(i) N is a Pólya or a Poisson process.

(ii) $\kappa_n(t)$ is, for any fixed t, linear in n.

(iii) $\kappa_n(t)$ is a product of two factors, one depending on n only and the other on t only.

Proof. If N is a Poisson process with intensity α, then $\kappa_n(t) \equiv \alpha$, and (ii) and (iii) hold. If N is a Pólya process (ii) and (iii) follow from Example 4.1.

Assume that (ii) holds, i.e. that $\kappa_n(t) = a(t) + b(t)n$ for some functions $a(t)$ and $b(t)$. From Lemma 4.1 it follows that

$$b(t) = -\frac{a'(t) + b'(t)n}{a(t) + b(t)n}$$

and hence

$$a'(t) = -a(t)b(t) \quad \text{and} \quad b'(t) = -b^2(t).$$

The latter equation has the solution

$$b(t) = \frac{1}{b + t},$$

where $b = \infty$ is interpreted as $b(t) \equiv 0$. Since $\kappa_n(t) \geq 0$ for all n it follows that $b \geq 0$. Inserting this expression in the former equation yields

$$a(t) \equiv \alpha \text{ for } b = \infty \quad \text{and} \quad a(t) = \frac{a}{b+t} \text{ for } b < \infty.$$

Thus $b = \infty$ implies a Poisson process, and we assume $b < \infty$. Then we have

$$\kappa_n(t) = \frac{a+n}{b+t}. \tag{4.12}$$

Since $a = 0$ means that $\kappa_0(t) \equiv 0$, which implies $N(t) \equiv 0$ P-a.s. – formally this may be regarded as a Poisson process with intensity 0 – we assume that $a > 0$. Further $b = 0$ would imply $\int_0^t \kappa_0(s)\, ds = \infty$ for all t. This would imply, see (4.2), $p_0(t) \equiv 0$, which is impossible. Thus the intensities given by (4.12) defines a Pólya process.

Assume that (iii) holds, i.e. that $\kappa_n(t) = \kappa_n \cdot b(t)$ for some sequence $\{\kappa_n\}$ and some function $b(t)$. From Lemma 4.1 it follows that

$$\kappa_{n+1} - \kappa_n = -\frac{b'(t)}{b^2(t)}.$$

Since this relation must hold for all $n \geq 0$ and $t > 0$ we get

$$\kappa_{n+1} - \kappa_n = c \quad \text{or} \quad \kappa_n = \kappa_0 + cn,$$

for some constant c. Hence $\kappa_n(t)$ is linear in n, and (ii) applies. ■

Theorem 4.1 (iii) has no correspondence for mixed Poisson distributions, while (ii) follows from Theorem 2.2, at least if we disregard the condition $\mu_\Lambda < \infty$. We have given the proof of Theorem 4.1 (ii) partly for historical reasons – it is seventeen years older – and partly since it shows that processes sometimes are easier to handle than distributions. Note that Theorem 2.2 does not follow from Theorem 4.1 (ii).

We shall now consider infinite divisibility, which also is simplified in the process setting.

Definition 4.4. *A point process N is called* **infinitely divisible** *if for each n there exists a point process N_n such that N has the same distribution as the sum of n independent copies of N_n.*

Definition 4.4 is the natural point process version of DID, as given by Definition 2.5. The standard reference about infinitely divisible point processes is Matthes *et al.* (1978).

The following nicer 'process version' of Proposition 2.5 is due to Bühlmann and Buzzi (1971).

Theorem 4.2. *Let N be* MPP(U), *where U is the distribution of a non-negative random variable Λ. N is infinitely divisible if and only if Λ is* ID.

Proof. Assume that Λ is ID. For each n the random variable Λ has the same distribution as the sum of n independent and identically distributed variables. To each of those variables we construct an MPP, and it easy to see that N has the same distribution as the sum of those MPPs. Thus N is infinitely divisible.

Assume now that N is infinitely divisible. Then $N(t)$ is DID for each t, and thus $N(t)/t$ is ID. Since, see Proposition 4.2, $N(t)/t \to \Lambda$ and since the limit of ID variables is ID, it follows that Λ is ID. ∎

A detailed analysis of the Pólya process, from the point of view of infinite divisibility, is given by Waymire and Gupta (1983).

The random variable T_1 is the waiting time to the first event in a point process N, and is of interest in, for instance, reliability models. If N describes the break-downs of some machinery, T_1 is the time to the first failure.

Proposition 4.5. *For any* MPP(U) *with* $U(0) = 0$, *the waiting time* T_1 *is* ID.

Proof. Let N be MPP(U) with the waiting time T_1 to the first event. Then

$$P\{T_1 > t\} = P\{N(t) = 0\} = \int_0^\infty e^{-\ell t}\, dU(\ell),$$

i.e. the distribution of T_1 is a mixture of exponential distributions. Such a mixture is infinitely divisible as pointed out in the discussion on page 30. ∎

We will, from a different point of view, discuss reliability properties in Chapter 7. Here we just notice that $\kappa_0(t)$ is the failure rate, see (7.1), of T_1.

Example 4.2. We will now mention a class of infinitely divisible mixed Poisson processes which was proposed by Hofmann (1955). Recall from Feller (1971, p.450) that

Λ is infinitely divisible if and only if $\hat{u}(v) = e^{-\tau(v)}$ where $\tau(v)$ has completely monotone derivative and $\tau(0) = 0$.

The above representation may be compared with (2.10) and Proposition 2.4 for the discrete case.

For a Poisson process with intensity α, we have

$$\hat{u}(v) = e^{-\alpha v} \Rightarrow \tau_0(v) = \alpha v \quad \text{and} \quad \tau_0'(v) = \alpha.$$

The index put on $\tau(\cdot)$ will be explained a few lines below.

If N is a Pólya process with $\hat{u}(v) = \left(1 + \frac{v}{\beta}\right)^{-\gamma}$, cf. Example 2.1, we have

$$\tau_1(v) = \gamma \log\left(1 + \frac{v}{\beta}\right) \quad \text{and} \quad \tau_1'(v) = \gamma \frac{\frac{1}{\beta}}{1 + \frac{v}{\beta}}.$$

In order to get notation closer to Hofmann's, we put $c = 1/\beta$ and $\alpha = E[\Lambda] = \gamma/\beta$. Then we have

$$\hat{u}(v) = \left(\frac{1}{1 + cv}\right)^{\alpha/c},$$

$$\tau_1(v) = \frac{\alpha}{c} \log(1 + cv) \quad \text{and} \quad \tau_1'(v) = \alpha \left(\frac{1}{1 + cv}\right).$$

Hofmann (1955, p.524) proposed the class of infinitely divisible mixed Poisson processes – sometimes called **Hofmann processes** – defined by

$$\tau_\delta'(v) = \alpha \left(\frac{1}{1 + cv}\right)^\delta, \quad c > 0, \ \alpha > 0, \ \delta \geq 0. \tag{4.13}$$

Since $\tau_1'(v)/\alpha$ is the Laplace transform of the exponential distribution and since that distribution is ID it follows that also $\tau_\delta'(v)/\alpha$ is a Laplace transform. Thus $\tau_\delta'(v)$, defined by (4.13), is completely monotone.

Notice, see (4.6), that $\kappa_0(t) = \tau_\delta'(t)$. By applying Lemma 4.1 we can – at least in principle – easily derive $\kappa_1(t), \kappa_2(t), \ldots$ Doing this, we get by routine calculations:

$$\kappa_0(t) = \frac{\alpha}{(1 + ct)^\delta},$$

$$\kappa_1(t) = \kappa_0(t) + \frac{\delta c}{1 + ct},$$

$$\kappa_2(t) = \kappa_1(t) + \frac{\delta c}{1 + ct} \cdot \frac{\alpha + c(1 + ct)^{\delta - 1}}{\alpha + \delta c(1 + ct)^{\delta - 1}},$$

and so on.

Hofmann (1955) derived some properties of this class. Since its practical relevance, unless $\delta = 0$ or 1, is unclear, we will just give the reference. □

Let us now consider computation of transition intensities for some of the examples considered in Chapter 2.

Example 2.2 continued. Let N be a Delaporte process, i.e. let Λ be shifted Γ-distributed, cf. Example 2.2, with

$$\hat{u}(v) = e^{-\alpha v}\left(1 + \frac{v}{\beta}\right)^{-\gamma}.$$

By (4.6), or by (4.6) and Lemma 4.1 as for the Hofmann process considered above, we get

$$\kappa_0(t) = \alpha + \frac{\gamma}{\beta + t},$$

$$\kappa_1(t) = \kappa_0(t) + \frac{\gamma}{\kappa_0(t)(\beta + t)^2},$$

$$\kappa_2(t) = \kappa_1(t) + \frac{2\kappa_1(t) - \kappa_0(t) - \alpha - (\kappa_1(t) - \kappa_0(t))^2(\beta + t)}{\kappa_1(t)(\beta + t)}.$$

A Delaporte process may be looked upon as the sum of a Poisson process, $N^{(\alpha)}$, and an independent Pólya process, $N^{(\beta,\gamma)}$. If we had been able to observe those two processes separately, then the transition intensities would have been much simplified. In order to realize this, we recall from Proposition 4.1 that $\kappa_{N(t)}(t)$ is the $\boldsymbol{F}^N$-intensity of N, i.e.

$$\kappa_{N(t)}(t) = E[\Lambda \mid \mathcal{F}_t^N].$$

To observe both $N^{(\alpha)}$ and $N^{(\beta,\gamma)}$ is the same as to use the filtration $\boldsymbol{F}$ with

$$\mathcal{F}_t = \sigma\{N^{(\alpha)}(s),\, N^{(\beta,\gamma)}(s);\ s \le t\}.$$

It is easy to see that

$$E[\Lambda \mid \mathcal{F}_t] = \alpha + \frac{\gamma + N^{(\beta,\gamma)}(t)}{\beta + t}.$$

Since $E[\Lambda \mid \mathcal{F}_t^N] = E[E[\Lambda \mid \mathcal{F}_t] \mid \mathcal{F}_t^N]$, it follows by Markovity that

$$\kappa_n(t) = \alpha + \frac{\gamma + E[N^{(\beta,\gamma)}(t) \mid N(t) = n]}{\beta + t}.$$

We do not claim that this representation leads to a more efficient way to derive $\kappa_n(t)$ than the more analytical method used above. We will, however, illustrate how $\kappa_n(t)$ may be explicitly expressed by the representation for $n = 0$ and 1.

For $n = 0$ we have $E[N^{(\beta,\gamma)}(t) \mid N(t) = 0] = 0$, and $\kappa_0(t)$ follows.

For $n = 1$ we have

$$\begin{aligned} E[N^{(\beta,\gamma)}(t) \mid N(t) = 1] &= P\{N^{(\beta,\gamma)}(t) = 1 \mid N(t) = 1\} \\ &= \frac{P\{N^{(\alpha)}(t) = 0\}P\{N^{(\beta,\gamma)}(t) = 1\}}{P\{N(t) = 1\}}. \end{aligned}$$

From Example 2.1 we get

$$P\{N^{(\beta,\gamma)}(t) = 0\} = \left(\frac{\beta}{\beta + t}\right)^{\gamma}$$

and

$$P\{N^{(\beta,\gamma)}(t) = 1\} = \frac{\gamma t}{\beta + t}\left(\frac{\beta}{\beta + t}\right)^{\gamma}.$$

Using these probabilities we get

$$E[N^{(\beta,\gamma)}(t) \mid N(t) = 1] = \frac{\gamma}{\gamma + \alpha(\beta + t)} = \frac{\gamma}{\kappa_0(t)(\beta + t)^2},$$

and $\kappa_1(t)$ follows.

With, hopefully, obvious notation we get

$$E[N^{(\beta,\gamma)}(t) \mid N(t) = n] = \frac{\sum_{k=1}^{n} k p_{n-k}^{(\alpha)}(t)\, p_k^{(\beta,\gamma)}(t)}{\sum_{k=0}^{n} p_{n-k}^{(\alpha)}(t)\, p_k^{(\beta,\gamma)}(t)}.$$

The actual computations seem, however, tedious. □

Example 2.3 (continued). Let Λ be inverse Gaussian–Poisson distributed, see Example 2.3, with

$$\hat{u}(v) = e^{-2\beta\mu_\Lambda\left(\sqrt{1+(v/\beta)}-1\right)}.$$

As for the Delaporte process above, we get

$$\kappa_0(t) = \frac{\mu_\Lambda}{\sqrt{1+\frac{t}{\beta}}},$$

$$\kappa_1(t) = \kappa_0(t) + \frac{1}{2(\beta+t)},$$

$$\kappa_2(t) = \kappa_1(t) + \frac{(\kappa_1(t)-\kappa_0(t))(2\kappa_1(t)-\kappa_0(t))}{\kappa_1(t)}.$$

□

Example 2.4 (continued). Consider the uniform–Poisson process, see Example 2.4 and Lundberg (1964, p.82), where Λ is uniformly distributed on $[0, \ell_1)$.

It follows by (4.6) that

$$\kappa_n(t) = \frac{\int_0^{\ell_1} \ell^{n+1} e^{-\ell t} \cdot \frac{1}{\ell_1}\, d\ell}{\int_0^{\ell_1} \ell^n e^{-\ell t} \cdot \frac{1}{\ell_1}\, d\ell} = \frac{\int_0^{\ell_1} \ell^{n+1} e^{-\ell t}\, d\ell}{\int_0^{\ell_1} \ell^n e^{-\ell t}\, d\ell},$$

i.e. $\kappa_n(t)$ is the ratio of two incomplete Γ-functions, cf. Lundberg (1964, p.82), and we get

$$\kappa_0(t) = \frac{1}{t} \cdot \frac{1-(t\ell_1+1)e^{-t\ell_1}}{1-e^{-t\ell_1}},$$

$$\kappa_1(t) = \frac{1}{t} \cdot \frac{2-(t^2\ell_1^2+2t\ell_1+2)e^{-t\ell_1}}{1-(t\ell_1+1)e^{-t\ell_1}},$$

$$\kappa_2(t) = \frac{1}{t} \cdot \frac{6-(t^3{\ell_1}^3+3t^2{\ell_1}^2+6t\ell_1+6)e^{-t\ell_1}}{2-(t^2\ell_1^2+2t\ell_1+2)e^{-t\ell_1}}.$$

□

A common feature of the transition intensity in the studied cases is that the complexity increases with increasing n. This may indicate that it is fairly difficult to find examples, besides the Pólya process, where $\kappa_n(t)$ is reasonably simple. Taking Theorem 4.1 into regard, this is perhaps not too surprising.

One case where the transition intensities really are simple will be considered in Example 4.3 below.

Example 4.3. Let Λ have a two-point distribution, U, with masses u_1 and u_2 at ℓ_1 and ℓ_2 respectively, i.e.

$$P\{\Lambda = \ell_1\} = u_1, \quad P\{\Lambda = \ell_2\} = u_2 \quad \text{and} \quad u_1 + u_2 = 1.$$

For convenience we assume $\ell_1 \leq \ell_2$. We have

$$\mu_\Lambda = u_1\ell_1 + u_2\ell_2 \quad \text{and} \quad \sigma_\Lambda^2 = u_1u_2(\ell_2 - \ell_1)^2.$$

This example was considered by Lundberg (1964, pp.94–95), and he called N a **double Poisson process.**

It follows from (4.11) that

$$\kappa_n(t) = \frac{u_1\ell_1^{n+1}e^{-\ell_1 t} + u_2\ell_2^{n+1}e^{-\ell_2 t}}{u_1\ell_1^n e^{-\ell_1 t} + u_2\ell_2^n e^{-\ell_2 t}}. \tag{4.14}$$

Assume now that we have observed $N(t)$, and recall that we may interpret $\Lambda_B^* = \kappa_{N(t)}(t)$ as the best estimate of Λ. In Chapter 2 we further considered the best linear estimate Λ_L^* of Λ. In this case, see (2.8), we have

$$\Lambda_L^* = \frac{(u_1\ell_1 + u_2\ell_2)^2 + u_1u_2(\ell_2 - \ell_1)^2 N(t)}{u_1\ell_1 + u_2\ell_2 + u_1u_2(\ell_2 - \ell_1)^2 t}$$

and

$$E[(\Lambda_L^* - \Lambda)^2] = \frac{(u_1\ell_1 + u_2\ell_2)u_1u_2(\ell_2 - \ell_1)^2}{u_1\ell_1 + u_2\ell_2 + u_1u_2(\ell_2 - \ell_1)^2 t}$$

Illustrations of Λ_B^* and Λ_L^* in the case $u_1 = u_2 = 1/2$, $\ell_1 = 0.5$ and $\ell_2 = 1.5$ by means of random generations are given by Grandell (1976, pp.147–148).

Let us simplify the example further by putting $\ell_1 = 0$. Then (4.14) reduces to

$$\Lambda_B^* = \begin{cases} \frac{u_2\ell_2 e^{-\ell_2 t}}{u_1 + u_2 e^{-\ell_2 t}}, & N(t) = 0, \\ \ell_2, & N(t) > 0. \end{cases}$$

In terms of the transition intensities, this means that $\kappa_n(t) = \ell_2$ for $n > 0$. This is very natural, since on $\{\Lambda = 0\}$ we have $N(t) = 0$ P-a.s. and thus we know that $\Lambda = \ell_2$ if $N(t) > 0$. Thus we have a case where $\kappa_n(t)$ becomes simpler with increasing n. Further we have,

$$E[(\Lambda_B^* - \Lambda)^2] = u_1 E[(\Lambda_B^* - \Lambda)^2 \mid \Lambda = 0] + u_2 E[(\Lambda_B^* - \Lambda)^2 \mid \Lambda = \ell_2]$$

$$= u_1 \left(\frac{u_2\ell_2 e^{-\ell_2 t}}{u_1 + u_2 e^{-\ell_2 t}} \right)^2 + u_2 \cdot e^{-\ell_2 t} \left(\frac{u_2\ell_2 e^{-\ell_2 t}}{u_1 + u_2 e^{-\ell_2 t}} - \ell_2 \right)^2$$

$$= \frac{u_1 u_2 \ell_2^2 e^{-\ell_2 t}}{u_1 + u_2 e^{-\ell_2 t}}$$

and

$$E[(\Lambda_L^* - \Lambda)^2] = \frac{u_1 u_2 \ell_2^2}{1 + u_1 \ell_2 t}.$$

In this simple case, which is the only 'non-Pólya' case where we have managed to express $E[(\Lambda_B^* - \Lambda)^2]$ explicitly, it is seen that Λ_B^* is much 'better' than Λ_L^* when $\ell_2 t$ is large. □

Let us return to birth processes. In the literature a birth process is often understood to be **homogeneous**, i.e. to have time-independent transition intensities κ_n. This is a quite reasonable assumption in a model of a population of members which can give birth to new members but cannot die. If all individuals give birth to new individuals independently of each other, a natural model is a **linear birth process** with

$$\kappa_n = v + n\kappa,$$

where κ is the individual birth intensity and v an external birth intensity which can be interpreted as immigration. For $v = \kappa$ an alternative interpretation is that we start with one individual and consider the population with that individual as its ancestor. The corresponding birth process is called a **Yule process**.

A natural question is when – or if – a mixed Poisson process is a homogeneous birth process. The following simple proposition gives the formal answer.

Proposition 4.6. *A mixed Poisson process N is a homogeneous birth process if and only if it is a Poisson process.*

Proof. Obviously a Poisson process with intensity α is a homogeneous birth process with $\kappa_n \equiv \alpha$.

Assume now that N a homogeneous birth process with, let us say, $\kappa_0 = \alpha > 0$. By Lemma 4.1 it follows that $\kappa_n = \kappa_{n-1}$, and thus the proposition follows. ■

A more interesting question is when a mixed Poisson process after a time-transformation is a homogeneous birth process and – in that case – how the time-transformation must look like. A continuously differentiable and (strictly) increasing function $A(t)$, $t \geq 0$, with $A(0) = 0$, $A(t) < \infty$ for each $t < \infty$ and $A(\infty) = \infty$ is called a **time-transformation**. Define the inverse B of A by

$$B(t) = \sup(s \mid A(s) \leq t).$$

The inverse B is also a time-transformation, and we have

$$A(B(t)) = B(A(t)) = t \text{ for all } t \geq 0.$$

Put $a(t) = A'(t)$ and $b(t) = B'(t)$ and notice that

$$a(B(t))b(t) = b(A(t))a(t) = 1 \text{ for all } t \geq 0.$$

Let N be a, not necessarily homogeneous, birth process. The **time-transformed** process N^A is defined by

$$N^A(t) \stackrel{\text{def}}{=} N(A(t)), \quad t \geq 0.$$

It is easy to see that N^A is a Markov process, and furthermore it is, according to our definition of a time-transformation, a birth process. By Definition 4.1 we get

$$\begin{aligned}
\kappa_n^A(t) &= \frac{P\{N^A(t+h) - N^A(t) = 1 \mid N^A(t) = n\}}{h} + o(1) \\
&= \frac{P\{N(A(t+h)) - N(A(t)) = 1 \mid N(A(t)) = n\}}{h} + o(1) \\
&= \kappa_n(A(t))\frac{A(t+h) - A(t)}{h} + o(1) = \kappa_n(A(t))a(t) + o(1).
\end{aligned}$$

Thus

$$\kappa_n^A(t) = \kappa_n(A(t))a(t),$$

and conversely

$$\kappa_n(t) = \kappa_n^A(B(t))b(t).$$

The following theorem is essentially due to Lundberg (1964, pp.57–58).

Theorem 4.3. *Let N be a mixed Poisson process. The following statements are equivalent:*

(*i*) *N is a Pólya process with $\kappa_n(t) = \frac{\gamma+n}{\beta+t}$ or a Poisson process.*

(*ii*) *There exists a time-transformation A such that N^A is a homogeneous birth process.*

In that case N^A is a linear birth process and the time-transformation A is of the form

$$A(t) = \begin{cases} at, & \textit{if } N \textit{ is a Poisson process,} \\ \beta\left(e^{at} - 1\right) & \textit{if } N \textit{ is a Pólya process,} \end{cases}$$

where $a \in (0, \infty)$ is some constant.

Proof. Assume that (ii) holds. Then there exists a time-transformation A, with inverse B, such that

$$\kappa_n(t) = \kappa_n^A(B(t))b(t) = \kappa_n^A b(t).$$

Thus (i) follows by Theorem 4.1 (iii). Furthermore, since we must have $\kappa_0(t) = \kappa_0^A b(t)$, it follows that

$$b(t) = c\kappa_0(t)$$

for some constant $c \in (0, \infty)$.

Assume that (i) holds. If N is a Poisson process with intensity α we must have $b(t) = c\alpha$. Thus $a(t) = 1/(c\alpha)$ and $A(t) = t/(c\alpha)$. Thus N^A is also a Poisson process, or – which is the same – a linear birth process with $v = 1/c$ and $\kappa = 0$.

If N is a Pólya process, we must have

$$b(t) = \frac{c\gamma}{\beta + t},$$

and it follows that

$$B(t) = c\gamma \log((\beta + t)/\beta)) \quad \text{and} \quad A(t) = \beta\left(e^{t/(c\gamma)} - 1\right),$$

i.e. of the form stated in the theorem. Thus we get

$$\kappa_n^A(t) = \kappa_n(A(t))a(t) = \frac{\gamma + n}{\beta e^{t/(c\gamma)}} \cdot \frac{1}{c\gamma} e^{t/(c\gamma)} = \frac{\gamma + n}{c\beta\gamma},$$

i.e. N^A is a linear birth process with $v = 1/(c\beta)$ and $\kappa = 1/(c\beta\gamma)$. Notice N^A is a Yule process when $\gamma = 1$, i.e. when Λ is exponentially distributed. ■

As mentioned in the Preface, mixed Poisson processes were introduced by Dubourdieu (1938) as models in sickness and accident insurance. Naturally, due to the state of art at that time, Dubourdieu's definition of a mixed Poisson process was rather similar to Lundberg's definition. Their approaches do, however, show – at least from a historical point of view – certain interesting differences.

Dubourdieu's starting point is to relate $N(t)$ with the times of the jumps $T_1, T_2, \ldots$. He assumes that T_n has a continuous distribution K_n with density k_n and derives the now well-known relations

$$p_0(t) = 1 - K_1(t) \text{ and } p_n(t) = K_n(t) - K_{n+1}(t), \quad n = 1, 2, \ldots \tag{4.15}$$

Those relations follow simply from $\{N(t) > n\} = \{T_{n+1} \le t\}$. Furthermore Dubourdieu shows that

$$E[N(t)] = \sum_{n=1}^{\infty} K_n(t).$$

Already the above relations make Dubourdieu (1938) a pioneering work in point process theory!

Next Dubourdieu (1938) assumes that $K_n = K^{n*}$ for some continuous distribution K with density k, where K^{n*} denotes the nth convolution of K with itself. In modern terminology we would say that N is an ordinary renewal process, see Definition 3.8. Let

$$\alpha(t) = \sum_{n=1}^{\infty} k^{n*}(t),$$

where – of course – k^{n*} denotes the density of K^{n*}. Dubourdieu shows that $\alpha(t)$ is the solution of the equation, cf. (3.5),

$$\alpha(t) = k(t) + \int_0^t k(t-s)\alpha(s)\, ds.$$

Today the above equation is known as a renewal equation and $\alpha(t)\,dt$ as the renewal measure. Under the assumption $p_0(t+s) = p_0(t) \cdot p_0(s)$, which essentially is an assumption of stationarity, Dubourdieu shows that N is a Poisson process, and points at the fundamental relation for the Poisson process:

$$p_n(t) = \sum_{m=0}^{n} p_m(s)\, p_{n-m}(t-s), \quad 0 \le s \le t. \tag{4.16}$$

In order to generalize the Poisson process, Dubourdieu goes back to to the general point process, and considers the following generalization of (4.16):

$$p_n(t) = \sum_{m=0}^{n} p_m(s)\, p_{m,n}(s,t), \quad 0 \le s \le t. \tag{4.17}$$

Notice that (4.17), although being a special case of the Chapman–Kolmogorov equations, holds quite generally.

Under the assumption that $p_0(t)$ is completely monotone in $(0, \infty)$, see Definition 2.3, Dubourdieu shows that

$$p_n(t) = (-1)^n \frac{t^n}{n!} p_0^{(n)}(t) \tag{4.18}$$

and

$$p_{m,n}(s,t) = (-1)^{n-m} \frac{(t-s)^{n-m}}{(n-m)!} \frac{p_0^{(n)}(t)}{p_0^{(m)}(s)} \tag{4.19}$$

satisfy (4.17). Furthermore Dubourdieu points out that $p_0(t)$ is completely monotone if and only if $p_0(t) = \int_{0-}^{\infty} e^{-\ell t}\, dU(\ell)$ for some distribution U, i.e. $p_0(t) = \hat{u}(t)$. In this generality, it is by no means obvious whether $p_n(t)$ and $p_{m,n}(s,t)$ uniquely determines any point process.

Assume now that N is a stationary Markov process. Since the distributions K_n are assumed to have densities, N is then a birth process. Dubourdieu claims that, in this case, it is easy to show that $p_n(t)$ and $p_{m,n}(s,t)$ must be of the form given above. This is a most remarkable claim, since it implies that the characterization by McFadden (1965), see Theorem 6.2 below, was already known by Dubourdieu. From Dubourdieu (1952, pp.269–282) it seems, however, as if an underlying assumption is that the location of the points in an interval, conditioned upon the number of points in that interval, is uniformly distributed. That assumption does in fact, characterize mixed Poisson processes, see Theorems 6.1 (vi), 6.3, and 6.6. In any case, when N is a birth process, it is uniquely determined by (4.18), and that definition agrees with Lundberg's definition as follows by (4.5). Simple computations show that (4.19) agrees with (4.9).

Assume now that $N(s) = m$ and consider the conditional point process N^* defined by $N^*(t) = N(s+t) - N(s)$. Recall from (4.7) that $p_n^*(t) = p_{m,m+n}(s, s+t)$. Let K_n^*, with density k_n^* and $\alpha^*(t)$, be defined by

$$p_0^*(t) = 1 - K_1^*(t) \quad \text{and} \quad p_n^*(t) = K_n^*(t) - K_{n+1}^*(t), \quad n = 1, 2, \ldots$$

and

$$\alpha^*(t) = \sum_{n=1}^{\infty} k_n^*(t).$$

Dubourdieu (1938, p.111) shows that

$$\alpha^*(t) = -\frac{p_0^{(m+1)}(s)}{p_0^{(m)}(s)} = -\frac{\hat{u}^{(m+1)}(s)}{\hat{u}^{(m)}(s)},$$

and thus, cf. (4.6), we have

$$\alpha^*(t) = \kappa_m(s), \quad t \geq 0. \tag{4.20}$$

The fact that $\alpha^*(0) = \kappa_m(s)$ is quite obvious, while (4.20) characterizes mixed Poisson processes, see Theorem 6.1 (iv).

CHAPTER 5

Some related processes

The mixed Poisson process is a generalization of the Poisson process. Naturally there are many other classes of point processes which have been used as generalizations of the Poisson process. The most well-known of such generalizations is probably the class of renewal processes. The only renewal process which is also a mixed Poisson process is, however, the Poisson process itself. We do therefore not regard that generalization of the Poisson process as 'related'. We will, however, in Chapter 9 rely on embedded renewal processes.

Cox processes, to be discussed in section 5.1 below, are very natural – and far reaching – generalizations, while Gauss–Poisson processes, treated in section 5.2, are 'related' but never mixed Poisson processes. Section 5.3 is devoted to mixed renewal processes.

5.1 Cox processes

Let us interpret the structure distribution U as a prior distribution, i.e. we look upon the mixed Poisson process in a Bayesian sense. A tempting generalization is then to let U, in some way, depend on t. We are here not thinking of the information gained by an observation of $N(t)$, as discussed in Chapter 4, but on cases where the intensity – or our opinion about it – changes with time. Such a process, perhaps after a time-transformation, ought to have marginal probabilities of the form

$$p_k(t) = \int_{0-}^{\infty} \frac{(\ell t)^k}{k!} e^{-\ell t} \, dU_t(\ell), \quad \text{for } k = 0, 1, \ldots \tag{5.1}$$

The generalization indicated above is, see for instance Philipson (1968, p.47), sometimes called 'a compound Poisson process in the wide sense', while the mixed Poisson process is called 'a compound Poisson process in the narrow sense'. The main drawback with this generalization is that it is, in the best case, a generalization of the one-dimensional marginal distributions and not of the process. We will now consider **Cox processes**, which seems to us to be the natural process correspondence of the above kind.

In order to do that, let $\Lambda = \{\Lambda(t);\ t \geq 0\}$ be a **random measure.** This means that the realizations of Λ belong to the set $\mathcal{M}$ of all functions $A = \{A(t);\ t \geq 0\}$ such that:

(i) $A(0) = 0$;

(ii) $A(t) < \infty$ for all $t < \infty$;

(iii) $A(\cdot)$ is non-decreasing and right-continuous.

The distribution U of a random measure Λ is a probability measure on $(\mathcal{M}, \mathcal{B}(\mathcal{M}))$, where $\mathcal{B}(\mathcal{M})$ is the σ-algebra generated by projections. Note that a point process is an integer-valued random measure, and that our definition of a random measure word for word equals the general definition of a point process, only that 'integer-valued' is extended to 'finite'.

We shall think of a Cox process N as generated in the following way. First a realization A of a random measure Λ is generated. Conditioned upon that realization, N is a non-homogeneous Poisson process with intensity measure A. A Poisson process with intensity measure A is defined as in Chapter 3, i.e.

(i) $N(t)$ has independent increments;

(ii) $N(t) - N(s)$ is Poisson distributed with mean $A(t) - A(s)$.

Its distribution is denoted by Π_A.

The formal definition looks very similar to Definition 4.2 of an MPP(U), although it is much more general.

Definition 5.1. *A point process N is called a* **Cox process** *if its distribution Π_U is given by*

$$\Pi_U\{B\} = \int_{\mathcal{M}} \Pi_A\{B\} U(dA), \quad B \in \mathcal{B}(\mathcal{N}). \tag{5.2}$$

For $B = \{\nu \in \mathcal{N};\ \nu(t) = n\}$ formula (5.2) reduces to (5.1) with $U_t(\ell) = P\{\Lambda(t)/t \leq \ell\}$.

The idea of Cox processes – sometimes also called 'doubly stochastic Poisson processes' – was introduced by Cox (1955). Pesonen (1968) proposed a process definition of 'compound Poisson processes in the wide sense', which – essentially – led to Cox processes. Cramér (1969) proposed Cox processes as models in risk theory. A detailed discussion of Cox processes and their impact on risk theory is to be found in Grandell (1991).

As indicated in Chapter 2, Cox processes are related to thinning. For later use, we will consider independent p-thinning a little more formally than needed for the moment. Let N be a point process with distribution Π, and denote by N_p the point process obtained by retaining (in the same location) every point of the process with probability p and deleting it with probability $q = 1-p$, independently of everything else. N_p is called the **p-thinning** of N. This notion is illustrated in Figure 5.1.

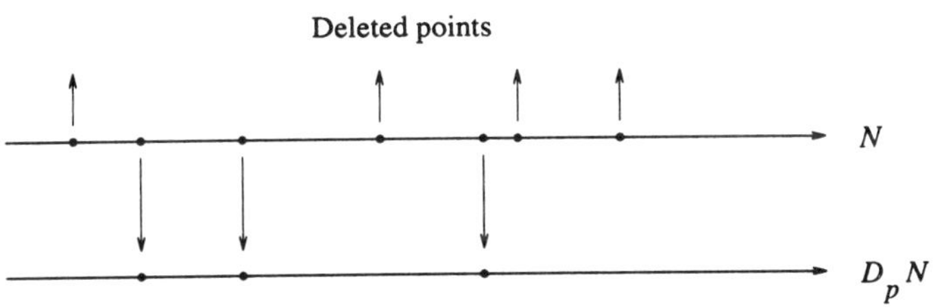

Figure 5.1. *Illustration of independent p-thinning.*

Let $\mathcal{P}$ be the set of distributions of point processes. Let $D_p : \mathcal{P} \to \mathcal{P}$ denote the p-thinning operator, i.e. for a point process N with distribution Π the distribution of N_p is $D_p\Pi$. In order to simplify the notation we also allow the operators to act on the point processes themselves. Then $N_p = D_pN$ just means that N_p has distribution $D_p\Pi$. Put

$$\mathcal{D}_p = \{D_p\Pi;\ \Pi \in \mathcal{P}\},$$

i.e. $\mathcal{D}_p$ is the set of distributions of point processes, which can be obtained by p-thinning.

Now we have the following characterization of Cox processes, of which Theorem 2.5 was a very special case.

Theorem 5.1 (Mecke, 1968). *A point process N is a Cox process if and only if it can be obtained by independent p-thinning for every $p \in (0,1)$.*

Theorems 2.5 and 5.1 indicate that Cox processes, and not mixed Poisson processes, really are the process correspondence to mixed

Poisson distributions. This may be an explanation why sometimes mixed Poisson processes were easier to handle than the distributions.

The following theorem, due to Kallenberg (1975), is an extension of Theorem 5.1. The notation $\xrightarrow{\mathrm{d}}$ means **convergence in distribution**, a notion here almost equivalent to 'convergence of finite-dimensional distributions'. This intuitive description is enough for our purposes – a discussion on the concept is found in Grandell (1991, p.42).

Theorem 5.2 (Kallenberg, 1975). *Let $N_1, N_2, \ldots$ be a sequence of point processes and let $p_1, p_2, \ldots \in (0,1)$ be real numbers such that $\lim_{n\to\infty} p_n = 0$. Then*

$$D_{p_n} N_n \xrightarrow{\mathrm{d}} \textit{some point process } N$$

if and only if

$$p_n N_n \xrightarrow{\mathrm{d}} \textit{some random measure } \Lambda.$$

If we have convergence it further holds that N is a Cox process with underlying random measure Λ.

Theorem 5.1 can be obtained a corollary of Theorem 5.2, see Grandell (1991, p.46).

We have given the most general definition of Cox processes in order to be able to formulate Theorems 5.1 and 5.2. In most models Λ has the representation

$$\Lambda(t) = \int_0^t \lambda(s)\, ds,$$

where $\lambda = \{\lambda(t);\ t \geq 0\}$ is called the **intensity process**. Obviously $\lambda(t) \geq 0$ P-a.s.

We note that a mixed Poisson process is a Cox process with underlying random measure $\Lambda\{dt\} = \Lambda \cdot dt$ or $\lambda(t) = \Lambda$, where Λ is a random variable.

Remark 5.1. Actually a mixed Poisson process ought to be defined as a Cox process with $\Lambda = \lambda \cdot A$, for some $A \in \mathcal{M}$ and the usual mixed Poisson ought to be called a 'stationary mixed Poisson process'. Using Definition 4.3 it readily follows that a time-transformed stationary mixed Poisson process, see Theorem 4.3, is a (non-stationary) mixed Poisson process. We will, however, except in section 6.3, only consider stationary mixed Poisson processes. □

Example 5.1. Let $\Delta > 0$ be a fixed value and $\{L_k;\ k = 0, 1, \dots\}$ a sequence of non-negative, independent, and identically distributed random variables. A rather simple Cox process is obtained by letting

$$\lambda(t) = L_k \quad \text{for } k\Delta \leq t < (k+1)\Delta.$$

The Cox process with this intensity is called an **Ammeter process**, since it is essentially the model considered by Ammeter (1948). We will return to this process in Chapter 9. The Ammeter process is technically related to the mixed Poisson process since it can be looked upon as built up by a sequence of independent mixed Poisson processes. However, besides from that relation, it is very different from a mixed Poisson process. This fundamental difference seems sometimes to have been overlooked, see e.g. Philipson (1968, p.52). □

5.2 Gauss–Poisson processes

The class of **Gauss–Poisson processes** was introduced by Newman (1970) and further studied by Milne and Westcott (1972) and Brillinger (1978). We will indicate the definition and some properties of this class of processes, mainly by considering 'Gauss–Poisson' distributions. These distributions have been studied by Kemp and Kemp (1965, 1966), who called them 'Hermite' distributions.

Let us suppose that a random variable N is Poisson distributed with unknown mean α. In some way we have an estimate of α at our disposal. Very often it is reasonable to assume that the estimate is Gaussian distributed. It would then be tempting to consider a random variable Λ being Gaussian distributed with mean μ_Λ and variance σ_Λ^2, and try to use that variable in a mixed Poisson process. The problem is naturally that Λ can be negative with positive probability. One might possibly say that if μ_Λ is large relative to σ_Λ^2, then the probability is so small for a negative Λ that we are willing to neglect that possibility.

One way to proceed is to consider the function

$$G_N(s) = \hat{u}(1-s),$$

where

$$\hat{u}(t) = E[e^{-t\Lambda}] = \exp\{-\mu_\Lambda t + \tfrac{1}{2}\sigma_\Lambda^2 t^2\}. \tag{5.3}$$

The question is now:

Is G_N a generating function?

The following proposition is due to Kemp and Kemp (1965, p.368), and was generalized to the process setting by Milne and Westcott (1972, p.153).

Proposition 5.1. *The function $G_N(s)$, determined by (5.3), is a generating function if and only if $\mu_\Lambda \geq \sigma_\Lambda^2$.*

Proof. We have

$$G_N(s) = \exp\{-\mu_\Lambda(1-s) + \tfrac{1}{2}\sigma_\Lambda^2(1-s)^2\}. \tag{5.4}$$

Assume that G_N is a generating function, i.e. that G_N is absolutely monotone in $(0,1)$. Then

$$0 \leq G_N'(s) = \{\mu_\Lambda - \sigma_\Lambda^2(1-s)\}G_N(s) \quad \text{for } 0 < s < 1, \tag{5.5}$$

which implies, as $s \downarrow 0$, $\mu_\Lambda \geq \sigma_\Lambda^2$.

Now assume that $\mu_\Lambda \geq \sigma_\Lambda^2$. We will show that N in fact is compound Poisson distributed. We have

$$\begin{aligned}
\log G_N(s) &= -\mu_\Lambda(1-s) + \tfrac{1}{2}\sigma_\Lambda^2(1-s)^2 \\
&= -\mu_\Lambda + \tfrac{1}{2}\sigma_\Lambda^2 + (\mu_\Lambda - \sigma_\Lambda^2)s + \tfrac{1}{2}\sigma_\Lambda^2 s^2 \\
&= (\mu_\Lambda - \tfrac{1}{2}\sigma_\Lambda^2)\left(\frac{(\mu_\Lambda - \sigma_\Lambda^2)s}{\mu_\Lambda - \frac{1}{2}\sigma_\Lambda^2} + \frac{\frac{1}{2}\sigma_\Lambda^2 s^2}{\mu_\Lambda - \frac{1}{2}\sigma_\Lambda^2} - 1\right).
\end{aligned}$$

From Definition 2.5 and (2.10) it follows that N is compound Poisson distributed with

$$\tilde{\alpha} = \mu_\Lambda - \tfrac{1}{2}\sigma_\Lambda^2, \quad P\{\xi_i = 1\} = \frac{\mu_\Lambda - \sigma_\Lambda^2}{\mu_\Lambda - \frac{1}{2}\sigma_\Lambda^2},$$

and

$$P\{\xi_i = 2\} = \frac{\frac{1}{2}\sigma_\Lambda^2}{\mu_\Lambda - \frac{1}{2}\sigma_\Lambda^2}. \tag{5.6}$$

■

Proposition 5.1 leads to the following definition.

Definition 5.2. *A discrete random variable N is said to be* **Gauss–Poisson distributed**, *$GP(\mu_\Lambda, \sigma_\Lambda^2)$, if its generating function is given by (5.4) with $\mu_\Lambda \geq \sigma_\Lambda^2$.*

We will now consider some properties of the Gauss–Poisson distribution, and compare with the mixed Poisson distribution.

Let N be $\mathrm{GP}(\mu_\Lambda, \sigma_\Lambda^2)$. Recall from (5.6) that N is compound Poisson distributed. It also follows form (5.6) that N has the representation

$$N = 2\widetilde{N}_1 + \widetilde{N}_2, \tag{5.7}$$

where $\widetilde{N}_1$ and $\widetilde{N}_2$ are independent random variables and $\mathrm{Po}(\frac{1}{2}\sigma_\Lambda^2)$ and $\mathrm{Po}(\mu_\Lambda - \sigma_\Lambda^2)$ respectively.

It follows from (5.7) that

$$E[N] = 2(\tfrac{1}{2}\sigma_\Lambda^2) + (\mu_\Lambda - \sigma_\Lambda^2) = \mu_\Lambda$$

and

$$\mathrm{Var}[N] = 4(\tfrac{1}{2}\sigma_\Lambda^2) + \mu_\Lambda - \sigma_\Lambda^2) = \mu_\Lambda + \sigma_\Lambda^2,$$

which is in agreement with Proposition 2.1 (i) and (ii). Further, it is well known that the Gaussian distribution is ID. Since N is compound Poisson distributed, it is DID, which is in agreement with Proposition 2.5.

Now we consider p-thinning, which is a fundamental concept for mixed Poisson processes. We have the following 'correspondence' to Theorem 2.5.

Theorem 5.3. *Let N be $GP(\mu_\Lambda, \sigma_\Lambda^2)$. N can be obtained by independent p-thinning if and only if $p \in [\sigma_\Lambda^2/\mu_\Lambda, 1]$.*

Proof. Assume that N is obtained by independent p-thinning from $N_{1/p}$. It follows from (2.9) and (5.4) that $N_{1/p}$ is $\mathrm{GP}(\mu_\Lambda/p, \sigma_\Lambda^2/p^2)$. Since $\mu_\Lambda/p \geq \sigma_\Lambda^2/p^2$ is equivalent to $p \geq \sigma_\Lambda^2/\mu_\Lambda$, the theorem follows. ■

Kemp and Kemp (1965) expressed $p_m = P\{N = m\}$ in terms of 'modified' Hermite polynomials. From (5.5) and the method used in Example 2.2 the recursion

$$\begin{aligned} p_0 &= \exp\{-\mu_\Lambda + \tfrac{1}{2}\sigma_\Lambda^2\}, \\ p_1 &= (\mu_\Lambda - \sigma_\Lambda^2)\, p_0, \\ (m+1)\, p_{m+1} &= (\mu_\Lambda - \sigma_\Lambda^2)\, p_m + \sigma_\Lambda^2 p_{m-1}, \quad m = 1, 2, \ldots \end{aligned}$$

follows very easily.

This is in agreement with a recursion due to Kemp and Kemp (1965, p. 383). Since $\mu_\Lambda \geq \sigma_\Lambda^2$ is required, the recursion is always stable. Notice that Willmot's general approach – which was so successfully used in Chapter 2 – is not applicable since, the Gauss–Poisson distribution is not a mixed Poisson distribution. Nevertheless, as pointed out by Kemp and Kemp (1966), it is natural to compare with the truncated normal–Poisson distribution with $\mu_G = \mu_\Lambda$ and $\sigma_G^2 = \sigma_\Lambda^2$, see Example 2.5. The recursion steps agree for $m \geq 1$, but of course the differences in p_0 and p_1 influence all probabilities.

A disadvantage with the Gauss–Poisson distribution is, of course, that it is not a mixed Poisson process. The truncated normal–Poisson distribution with $\mu_G = \mu_\Lambda$ and $\sigma_G^2 = \sigma_\Lambda^2$, on the other hand, may seem a bit *ad hoc* and further its mean and variance do not agree completely with μ_Λ and σ_Λ^2. One may therefore try to approximate a Gauss–Poisson distribution with some other mixed Poisson distribution where Λ is approximately normal. The simplest and most natural such case is certainly to let Λ be $\Gamma(\mu_\Lambda^2/\sigma_\Lambda^2, \mu_\Lambda/\sigma_\Lambda^2)$, in which case N is negatively binomially distributed, see Example 2.1. Denote the probabilities corresponding to these three distributions by p_m^{GP}, p_m^{TN}, and p_m^{NB} respectively. In order to compare the distributions we consider the **distance of variation**, DV, defined by

$$\mathrm{DV}(\mathrm{GP}, \mathrm{TN}) \stackrel{\mathrm{def}}{=} \frac{1}{2} \sum_{m=0}^{\infty} |p_m^{\mathrm{GP}} - p_m^{\mathrm{TN}}|.$$

For any two discrete distribution we have $0 \leq \mathrm{DV} \leq 1$ where $\mathrm{DV} = 0$ implies that the distributions agree while $\mathrm{DV} = 1$ implies that the distributions have disjoint support. In Table 5.1 the distance of variation is illustrated in the three cases above.

Table 5.1. *The distance of variation between the Gauss–Poisson distribution, the truncated normal distribution, and the negative binomial distribution in the case* $\mu_\Lambda = 10$.

σ_Λ^2	DV(GP, TN)	DV(GP, NB)	DV(TN, NB)
10	5.00×10^{-1}	5.00×10^{-1}	3.04×10^{-2}
5	1.26×10^{-5}	1.16×10^{-2}	1.16×10^{-2}
2	1.00×10^{-12}	2.53×10^{-3}	2.53×10^{-3}

In the case $\sigma_\Lambda^2 = 10$ the Gauss–Poisson distribution cannot be expected to be close to any mixed Poisson distribution, since in that case $p_m^{\text{GP}} = 0$ for m odd, cf. (5.7). The corresponding negative binomial distribution is illustrated in Figure 2.4. For the other values of σ_Λ^2 given, the Gauss–Poisson distribution and the truncated normal distribution are much closer to each other than to the negative binomial distribution.

Kemp and Papageorgiou (1982) discuss bivariate versions of the Gauss–Poisson distribution.

Let us now try to define a Gauss–Poisson process N, which would correspond to a mixed Poisson process. If N was MPP(U), where U is the distribution of a non-negative random variable Λ with $E[\Lambda] = \mu_\Lambda$ and $\text{Var}[\Lambda] = \sigma_\Lambda^2$, then the random variable $N(t)$ would be MP(U_t), where U_t is the distribution function of Λt. Since $E[\Lambda t] = \mu_\Lambda t$ and $\text{Var}[\Lambda t] = \sigma_\Lambda^2 t^2$ it seems natural to require that $N(t)$ is $\text{GP}(\mu_\Lambda t, \sigma_\Lambda^2 t^2)$. It follows, however, from Proposition 5.1 that this is possible only if $\mu_\Lambda t \geq \sigma_\Lambda^2 t^2$, i.e. only for $t \leq \mu_\Lambda / \sigma_\Lambda^2$.

Let μ_Λ and σ_Λ^2 be given and put $T = \mu_\Lambda / \sigma_\Lambda^2$ and let $\widetilde{N}$ be a Poisson process with intensity $\sigma_\Lambda^2/2$, defined on the square $[0, T]^2$. Following Brillinger (1978), we define a point process N on $[0, t]$ by

$$N(t) = \widetilde{N}\{[0,t] \times [0,T]\} + \widetilde{N}\{[0,T] \times [0,t]\}. \tag{5.8}$$

By splitting $[0,T]^2$ in the disjoint sets $[0,t]^2$, $[t,T] \times [0,t]$, $[0,t] \times [t,T]$, and $[t,T]^2$, it is seen that

$$N(t) = 2\underbrace{\widetilde{N}\{[0,t]^2\}}_{\stackrel{\text{def}}{=}\widetilde{N}_1} + \underbrace{\widetilde{N}\{[t,T] \times [0,t]\} + \widetilde{N}\{[0,t] \times [t,T]\}}_{\stackrel{\text{def}}{=}\widetilde{N}_2}.$$

Since the sets above are disjoint, it follows that $\widetilde{N}_1$ and $\widetilde{N}_2$ are independent and Poisson distributed. Further

$$E[\widetilde{N}_1] = \frac{\sigma_\Lambda^2}{2} t^2$$

and

$$E[\widetilde{N}_2] = \frac{\sigma_\Lambda^2}{2} 2t\,(T - t) = \sigma_\Lambda^2 t \left(\frac{\mu_\Lambda}{\sigma_\Lambda^2} - t \right) = \mu_\Lambda t - \sigma_\Lambda^2 t^2.$$

Thus, cf. (5.7), $N(t)$ is $\text{GP}(\mu_\Lambda t, \sigma_\Lambda^2 t^2)$ as required. It follows from Brillinger (1978) that N is a Gauss–Poisson process. We note that

$N(T) = 2\widetilde{N}\{[0,T]^2\}$ and thus $N(T)$ can not be obtained by any p-thinning. Consequently neither can the process N be obtained by p-thinning, and this is essentially due to the fact that we have chosen T as large as possible, cf. Theorem 5.3.

Remark 5.2. In order to indicate that this is as 'close' one can come to a mixed Poisson process, we consider the general definition given by Brillinger (1978). Put

$$I_{s,t} = \begin{cases} [s,t] & \text{if } t < \infty, \\ [s,\infty) & \text{if } t = \infty, \end{cases} \qquad I_t = I_{0,t} \text{ and } \bar{I}_{s,t} = I_T \setminus I_{s,t},$$

and let $\tilde{A}$ be a symmetric measure on I_T^2 such that

$$\tilde{A}\{I_t \times I_T\} < \infty \text{ for all } t \in I_T.$$

Let $\widetilde{N}$ be a (non-homogeneous) Poisson process on I_T^2 with intensity measure $\tilde{A}$ and let $\widetilde{N}_3$ be a (non-homogeneous) Poisson process on I_T with intensity measure $\tilde{A}_3$. A Gauss–Poisson process N is now defined, cf. (5.8), by

$$\begin{aligned} &N(t) - N(s) \\ &\quad = N\{I_{s,t}\} = \widetilde{N}\{I_{s,t} \times I_T\} + \widetilde{N}\{I_T \times I_{s,t}\} + \widetilde{N}_3\{I_{s,t}\}. \end{aligned} \tag{5.9}$$

It follows from Brillinger (1978) that (5.9) is the most general definition of a Gauss–Poisson process. By splitting I_T in disjoint sets, as above, we get

$$\begin{aligned} &N(t) - N(s) \\ &\quad = \underbrace{2\,\widetilde{N}\{I_{s,t}^2\}}_{\stackrel{\text{def}}{=}\widetilde{N}_1} + \underbrace{\widetilde{N}\{I_{s,t} \times \bar{I}_{s,t}\} + \widetilde{N}\{\bar{I}_{s,t} \times I_{s,t}\} + \widetilde{N}_3\{I_{s,t}\}}_{\stackrel{\text{def}}{=}\widetilde{N}_2}. \end{aligned}$$

where $\widetilde{N}_1$ and $\widetilde{N}_2$ are independent and Poisson distributed with

$$E[\widetilde{N}_1] = \tilde{A}\{I_{s,t}^2\} \quad \text{and} \quad E[\widetilde{N}_2] = 2\tilde{A}\{I_{s,t} \times \bar{I}_{s,t}\} + \tilde{A}_3\{I_{s,t}\}.$$

We now require that $N\{I_{s,t}\}$ is $\mathrm{GP}(\mu_A(t-s), \sigma_A^2(t-s)^2)$, which means that we have included a stationarity requirement. This implies that

$$\tilde{A}\{I_{s,t}^2\} = \frac{\sigma_\Lambda^2}{2}(t-s)^2$$

and

$$2\tilde{A}\{I_{s,t} \times \bar{I}_{s,t}\} + \tilde{A}_3\{I_{s,t}\} = \mu_\Lambda(t-s) - \sigma_\Lambda^2(t-s)^2.$$

Since $I_t^2 = I_t \times I_s \cup I_s \times I_t \cup I_{s,t}^2 \setminus I_s^2$ we get

$$2\tilde{A}\{I_t \times I_s\} = \tilde{A}\{I_t^2\} + \tilde{A}\{I_s^2\} - \tilde{A}\{I_{s,t}^2\}$$

$$= \frac{\sigma_\Lambda^2}{2}\left(t^2 + s^2 - (t-s)^2\right) = \sigma_\Lambda^2 ts.$$

Thus $\widetilde{N}$ is a (homogeneous) Poisson process with intensity $\sigma_\Lambda^2/2$. Further we must have

$$2\tilde{A}\{I_{s,t} \times \bar{I}_{s,t}\} + \tilde{A}_3\{I_{s,t}\} = \sigma_\Lambda^2(t-s)(T-t+s) + \tilde{A}_3\{I_{s,t}\}$$

$$= \mu_\Lambda(t-s) - \sigma_\Lambda^2(t-s)^2,$$

which yields

$$\tilde{A}_3\{I_t\} = \mu_\Lambda t - \sigma_\Lambda^2\left(t^2 + t(T-t)\right) = \left(\mu_\Lambda - \sigma_\Lambda^2 T\right) t.$$

Thus $\widetilde{N}_3$ is a (homogeneous) Poisson process with intensity $\mu_\Lambda - \sigma_\Lambda^2 T$ and we see that $T \le \mu_\Lambda/\sigma_\Lambda^2$. Only for $\sigma_\Lambda^2 = 0$ we can choose $T = \infty$, and then N is a Poisson process. □

Further discussion of Gauss–Poisson distributions and processes and generalizations of them is to be found in Milne and Westcott (1993).

5.3 Mixed renewal processes

Recall from Chapter 3 that a Poisson process is a renewal process with exponentially distributed inter-occurrence times. More precisely, N is a Poisson process with intensity ℓ if it is a renewal process with inter-occurrence time distribution

$$K_\ell(t) = 1 - e^{-\ell t}, \quad t \ge 0. \tag{5.10}$$

The density of $(T_1, \ldots, T_n)$, i.e. of the positions of first n points, is then

$$\begin{aligned} k_\ell(t_1)k_\ell(t_2-t_1)\cdots k_\ell(t_n-t_{n-1}) \\ = \ell e^{-\ell t_1}\cdot \ell e^{-\ell(t_2-t_1)}\cdots \ell e^{-\ell(t_n-t_{n-1})} \\ = \ell^n e^{-\ell t_n}, \quad 0<t_1<t_2<\ldots<t_n. \end{aligned} \tag{5.11}$$

Let now N be an ordinary renewal process, see Definition 3.8, with with inter-occurrence time distribution K_ℓ where ℓ belongs to some parameter space $\mathbb{L}$. Let ℓ be the outcome of a random variable Λ, defined on a complete probability space $(\mathbb{L}, \mathcal{L}, U)$ such that for each t the mapping $\ell \mapsto K_\ell(t)$ is $\mathcal{L}$-measurable. In the same way as for mixed Poisson processes we interpret the probability measure U as the structure or prior distribution. In order to simplify the presentation, we introduce the following rather restrictive assumption, due to Huang (1990).

Assumption 5.1. *Let U-a.s. the inter–occurrence time distribution $K_\ell(t)$ have a continuous density $k_\ell(t)$ such that $0 < k_\ell(t) < C$ for all $t > 0$, where $C < \infty$.*

Let

$$\hat\lambda_n(t,t_1,\ldots,t_n) \stackrel{\text{def}}{=} \frac{E[k_\Lambda(t_1)k_\Lambda(t_2-t_1)\cdots k_\Lambda(t_n-t_{n-1})k_\Lambda(t-t_n)]}{E[k_\Lambda(t_1)k_\Lambda(t_2-t_1)\cdots k_\Lambda(t_n-t_{n-1})\overline{K}_\Lambda(t-t_n)]} \tag{5.12}$$

for $0<t_1<\ldots<t_n<t$, where $\overline{K}_\Lambda(t) = 1-K_\Lambda(t)$. Notice that, under Assumption 5.1, the density $k_\ell(t)$ is $\mathcal{L}$-measurable, so the intensities in (5.12) are well-defined. Recall the definition of the $\boldsymbol{F}^N$-intensity given in Definition 3.7.

Definition 5.3. *Let U be a probability measure on a space $(\mathbb{L}, \mathcal{L})$ and let K_ℓ, $\ell \in \mathbb{L}$, be a family of $\mathcal{L}$-measurable distribution functions fulfilling Assumption 5.1. A point process N with $\boldsymbol{F}^N$-intensity*

$$\hat\lambda(t) = \hat\lambda_{N(t)}(t, T_1, \ldots, T_{N(t)}),$$

where $\hat\lambda_n(t,t_1,\ldots,t_n)$ is given be (5.12), is called a **mixed renewal process**.

The advantage of Definition 5.3 is that the $\boldsymbol{F}^N$-intensity is simply connected with K_ℓ. A definition in line with Definition 4.2 – although maybe more natural – would require measurability considerations of Π_ℓ which might be rather difficult in the renewal case.

Mixed renewal processes seem not to have been much considered in the literature. Segerdahl (1970, pp.164–165) did, however, propose that these processes might be used in actuarial models.

Example 5.2. Let K_ℓ be given by (5.10). Then $\mathbb{L} = \mathbb{R}_+ \overset{\text{def}}{=} [0, \infty)$ and it follows from (5.11) and (5.12) that

$$\hat{\lambda}_n(t, t_1, \dots, t_n) = \frac{E[\Lambda^n e^{-\Lambda t_n} \cdot \Lambda e^{-\Lambda\cdot(t-t_n)}]}{E[\Lambda^n e^{-\Lambda t_n} \cdot e^{-\Lambda\cdot(t-t_n)}]} = \frac{E[\Lambda^{n+1}e^{-\Lambda t}]}{E[\Lambda^n e^{-\Lambda t}]}$$

which implies, see (4.6) and Proposition 4.1, that N is a mixed Poisson process. □

A common choice of the inter-occurrence time distribution is the $\Gamma(\gamma, \beta)$-distribution, cf. Example 2.1. Then

$$k_\ell(t) = \frac{\beta^\gamma}{\Gamma(\gamma)} t^{\gamma-1} e^{-\beta t}, \quad t \geq 0,$$

and $\ell = (\gamma, \beta) \in \mathbb{L} = \mathbb{R}_+^2$. Notice that Assumption 5.1 is **not** fulfilled for $0 < \gamma < 1$ which may indicate that the assumption really is restrictive. In connection with renewal processes, the case $\gamma \geq 1$ might be 'most common', but the case $0 < \gamma < 1$ is by no means without interest. In fact, Yannaros (1988) has shown that a renewal process is a Cox process if and only if $0 < \gamma \leq 1$. Of course, in the case $\gamma = 1$ the process is a Poisson process.

Example 5.3. Let K_ℓ be an Erlang distribution of order 2, i.e. a $\Gamma(2, \ell)$-distribution. Then

$$k_\ell(t) = \ell^2 t e^{-\ell t} \quad \text{and} \quad \overline{K}_\ell(t) = (1 + \ell t)e^{-\ell t}, \quad t \geq 0,$$

and $\ell \in \mathbb{L} = \mathbb{R}_+$. Recall from Example 2.1 that K_ℓ is the distribution of the sum of 2 independent and exponentially distributed random variables.

Let a structure distribution U be given, and let N be the corresponding mixed renewal process. Let further $\widetilde{N}$ be MPP(U). Thus N consists of every second point of $\widetilde{N}$, or, more precisely, $T_n = \widetilde{T}_{2n}$. It follows from (5.12) that

$$\begin{aligned} \hat{\lambda}_n(t, t_1, \dots, t_n) &= \frac{E[\Lambda^{2n+2} \cdot (t - t_n)e^{-\Lambda t}]}{E[\Lambda^{2n}(1 + \Lambda \cdot (t - t_n))e^{-\Lambda t}]} \\ &= \frac{(t - t_n)E[\Lambda^{2n+2}e^{-\Lambda t}]}{E[\Lambda^{2n}e^{-\Lambda t}] + (t - t_n)E[\Lambda^{2n+1}e^{-\Lambda t}]} \\ &= \frac{(t - t_n)\kappa_{2n}(t)\kappa_{2n+1}(t)}{1 + (t - t_n)\kappa_{2n}(t)}. \end{aligned} \tag{5.13}$$

We will sketch an alternative derivation of (5.13), which does not rely on Definition 5.3. Suppose that at time t we know that $N(t) = n$ and that $T_n = t_n$. That is the same as to know that $\widetilde{T}_{2n} = t_n$ and that $\widetilde{T}_{2n+2} > t$. Under this condition we have

$$\hat{\lambda}_n(t, t_1, \ldots, t_n) = \begin{cases} 0, & \widetilde{T}_{2n+1} > t, \\ \tilde{\kappa}_{2n+1}(t), & \widetilde{T}_{2n+1} \leq t. \end{cases} \tag{5.14}$$

Recall from (4.8) that the conditional distribution U^* of Λ given $\widetilde{N}(t_n) = 2n$ is

$$U^*(x) = \frac{\int_{0-}^{x} \ell^{2n} e^{-t_n \ell} \, dU(\ell)}{\int_{0-}^{\infty} \ell^{2n} e^{-t_n \ell} \, dU(\ell)}.$$

Strictly speaking, we have here used the fact that $\widetilde{N}$ is a **strong Markov process**. This means that the arbitrary – but fixed – time s in Definition 3.1 may be replaced by an $\boldsymbol{F}^Y$**-stopping time** S, i.e. a random variable S such that $\{S \leq s\} \in \mathcal{F}_s^Y$ for each $s \geq 0$. A good reference on these questions is Davis (1993).

We have

$$\begin{aligned} &P\{\widetilde{T}_{2n+1} > t \mid \widetilde{N}(t_n) = 2n, \ \widetilde{T}_{2n+2} > t\} \\ &= \frac{P\{\widetilde{T}_{2n+1} > t \mid \widetilde{N}(t_n) = 2n\}}{P\{\widetilde{T}_{2n+2} > t \mid \widetilde{N}(t_n) = 2n\}} \\ &= \frac{P\{\widetilde{N}(t) - \widetilde{N}(t_n) = 0 \mid \widetilde{N}(t_n) = 2n\}}{P\{\widetilde{N}(t) - \widetilde{N}(t_n) \leq 1 \mid \widetilde{N}(t_n) = 2n\}} \\ &= \frac{E\left[e^{-(t-t_n)\Lambda} \mid \widetilde{N}(t_n) = 2n\right]}{E\left[(1 + (t - t_n)\Lambda)e^{-(t-t_n)\Lambda} \mid \widetilde{N}(t_n) = 2n\right]} \\ &= \frac{\int_{0-}^{\infty} e^{-(t-t_n)\ell} \ell^{2n} e^{-t_n \ell} \, dU(\ell)}{\int_{0-}^{\infty} (1 + (t - t_n)\ell) e^{-(t-t_n)\ell} \ell^{2n} e^{-t_n \ell} \, dU(\ell)} = \frac{1}{1 + (t - t_n)\tilde{\kappa}_{2n}(t)}. \end{aligned}$$

Using (5.14) we get

$$\begin{aligned} &\hat{\lambda}_n(t, t_1, \ldots, t_n) \\ &= \frac{1}{1 + (t - t_n)\tilde{\kappa}_{2n}(t)} \cdot 0 + \left(1 - \frac{1}{1 + (t - t_n)\tilde{\kappa}_{2n}(t)}\right) \cdot \tilde{\kappa}_{2n+1}(t) \\ &= \frac{(t - t_n)\kappa_{2n}(t)\kappa_{2n+1}(t)}{1 + (t - t_n)\kappa_{2n}(t)}, \end{aligned}$$

which is (5.13).

Let $K_\ell^{n*}(t)$ denote the distribution of T_n, i.e. K_ℓ^{n*} is the nth convolution of K_ℓ with itself. Thus K_ℓ^{n*} is an Erlang distribution of order

$2n$ and its density k_ℓ^{n*} is given by

$$k_\ell^{n*}(t) = \frac{\ell^{2n}}{(2n-1)!}\, t^{2n-1} e^{-\ell t}, \quad t \geq 0.$$

We have, cf. the end of Chapter 3,

$$E[N(t) \mid \Lambda = \ell] = \int_0^t \sum_{n=1}^\infty k_\ell^{n*}(s)\, ds$$

$$= \int_0^t \sum_{n=1}^\infty \frac{\ell^{2n}}{(2n-1)!}\, s^{2n-1} e^{-\ell s}\, ds = \int_0^t \ell \frac{e^{\ell s} - e^{-\ell s}}{2}\, e^{-\ell s}\, ds$$

$$= \int_0^t \ell \frac{1 - e^{-2\ell s}}{2}\, ds = \frac{\ell t}{2} - \frac{1 - e^{-2\ell t}}{4}. \tag{5.15}$$

The derivation of (5.15) is taken from Thorin (1970, p.33) where it is further shown that

$$\mathrm{Var}[N(t) \mid \Lambda = \ell] = \frac{\ell t}{4} + \frac{1}{16} - \frac{\ell t}{2}\, e^{-2\ell t} - \frac{1}{16}\, e^{-4\ell t}. \tag{5.16}$$

Denote, as before, the mean, variance and Laplace transform of Λ by μ_Λ, σ_Λ^2, and $\hat{u}(\cdot)$ respectively and recall that $\hat{u}'(v) = -E[\Lambda e^{-\Lambda v}]$.

It follows from (5.15) that

$$E[N(t)] = \frac{t\mu_\Lambda}{2} - \frac{1}{4} + \frac{\hat{u}(2t)}{4}. \tag{5.17}$$

The computation of $\mathrm{Var}[N(t)]$ is similar but more complicated.

We have $\mathrm{Var}[N(t)] = E[\mathrm{Var}[N(t) \mid \Lambda]] + \mathrm{Var}[E[N(t) \mid \Lambda]]$ where, cf. (5.16)

$$\begin{aligned} &E[\mathrm{Var}[N(t) \mid \Lambda]] \\ &\quad = E\left[\frac{\Lambda t}{4} + \frac{1}{16} - \frac{\Lambda t}{2}\, e^{-2\Lambda t} - \frac{1}{16}\, e^{-4\Lambda t}\right] \\ &\quad = \frac{t\mu_\Lambda}{4} + \frac{1}{16} + \frac{t\hat{u}'(2t)}{2} - \frac{\hat{u}(4t)}{16} \end{aligned}$$

and, cf. (5.15),

$$\begin{aligned} &\mathrm{Var}[E[N(t) \mid \Lambda]] \\ &\quad = \mathrm{Var}\left[\frac{\Lambda t}{2} - \frac{1 - e^{-2\Lambda t}}{4}\right] = \mathrm{Var}\left[\frac{\Lambda t}{2} + \frac{e^{-2\Lambda t}}{4}\right] \\ &\quad = \mathrm{Var}\left[\frac{\Lambda t}{2}\right] + \mathrm{Var}\left[\frac{e^{-2\Lambda t}}{4}\right] + 2\,\mathrm{Cov}\left[\frac{\Lambda t}{2}, \frac{e^{-2\Lambda t}}{4}\right] \\ &\quad = \frac{t^2\sigma_\Lambda^2}{4} + \left(\frac{\hat{u}(4t)}{16} - \frac{\hat{u}(2t)^2}{16}\right) + 2\left(-\frac{t\hat{u}'(2t)}{8} - \frac{t\mu_\Lambda \hat{u}(2t)}{8}\right). \end{aligned}$$

Putting these formulae together we get

$$\operatorname{Var}[N(t)] = \frac{t\mu_\Lambda}{4} + \frac{t^2\sigma_\Lambda^2}{4} + \frac{1}{16} + \frac{4t\hat{u}'(2t) - \hat{u}(2t)^2 - 4t\mu_\Lambda\hat{u}(2t)}{16}. \quad (5.18)$$

A first approximation, at least for large t, of $N(t)$ is $\widetilde{N}(t)/2$. It follows from Proposition 2.1 (i) and (ii) that

$$E\left(\frac{\widetilde{N}(t)}{2}\right) = \frac{t\mu_\Lambda}{2} \quad \text{and} \quad \operatorname{Var}\left(\frac{\widetilde{N}(t)}{2}\right) = \frac{t\mu_\Lambda}{4} + \frac{t^2\sigma_\Lambda^2}{4},$$

which agree with the leading terms in (5.17) and (5.18), cf. Segerdahl (1970, p.165). □

A renewal process with an inter-occurrence time distribution K fulfilling Assumption 5.1 is a birth process if and only if it is a Poisson process. This follows since then, see (3.6), we must have

$$r(t - t_n) = \kappa_n(t), \quad \text{for all } t_n,$$

which is possible if and only if $r(\cdot)$ is a constant. It is really enough to assume that K has a density, since the exponential distribution is the only distribution without memory, i.e. the only distribution such that

$$P\{T_n > s + t \mid T_n > s\} = P\{T_n > t\}, \quad s, t \geq 0.$$

Notice, however, that the trivial simple Markov point process $N(t) = \lfloor t \rfloor$, mentioned in the beginning of Chapter 4, is a renewal process with

$$K(t) = \begin{cases} 0, & t < 1, \\ 1, & t \geq 1. \end{cases}$$

The following proposition, due to Huang (1990), extends the above result to the mixed renewal case.

Proposition 5.2. *Let N be a mixed renewal process for which Assumption 5.1 holds. The following statements are equivalent:*

(i) *N is a mixed Poisson process.*

(ii) *N is birth process.*

Proof. (i) ⇒ (ii) is obvious, so we consider (ii) ⇒ (i).

We will first prove that

$$
\begin{aligned}
0 &= E[(k_\Lambda(t) - k_\Lambda(0)\overline{K}_\Lambda(t))^2] \\
&= E[k_\Lambda(t)^2] + E[k_\Lambda(0)^2\overline{K}_\Lambda(t)^2] - 2E[k_\Lambda(0)k_\Lambda(t)\overline{K}_\Lambda(t)]. \qquad (5.19)
\end{aligned}
$$

Let s, t, $\varepsilon > 0$. We consider the following four situations:

$$
\begin{array}{llll}
N(2t) = 2 & t_1 = t & t_2 = 2t - \varepsilon & \\
N(2t) = 2 & t_1 = \varepsilon & t_2 = t & \\
N(s+t) = 3 & t_1 = \varepsilon & t_2 = s & t_3 = s + t - \varepsilon \\
N(s+t) = 3 & t_1 = \varepsilon & t_2 = 2\varepsilon & t_3 = s + \varepsilon.
\end{array}
$$

The reason for the ε above, which shall be looked upon as small, is to make the chosen positions allowed.

Consider $N(2t) = 2$. We get

$$
\kappa_2(2t) = \frac{E[k_\Lambda(t)k_\Lambda(t-\varepsilon)k_\Lambda(\varepsilon)]}{E[k_\Lambda(t)k_\Lambda(t-\varepsilon)\overline{K}_\Lambda(\varepsilon)]} = \frac{E[k_\Lambda(\varepsilon)k_\Lambda(t-\varepsilon)k_\Lambda(t)]}{E[k_\Lambda(\varepsilon)k_\Lambda(t-\varepsilon)\overline{K}_\Lambda(t)]}
$$

and thus

$$
E[k_\Lambda(t)k_\Lambda(t-\varepsilon)\overline{K}_\Lambda(\varepsilon)] = E[k_\Lambda(\varepsilon)k_\Lambda(t-\varepsilon)\overline{K}_\Lambda(t)]
$$

or, by dominated convergence as $\varepsilon \to 0$,

$$
E[k_\Lambda(t)^2] = E[k_\Lambda(0)k_\Lambda(t)\overline{K}_\Lambda(t)]. \qquad (5.20)
$$

Similarly, for $N(s+t) = 3$, we get

$$
\begin{aligned}
\kappa_3(s+t) &= \frac{E[k_\Lambda(\varepsilon)k_\Lambda(s-\varepsilon)k_\Lambda(t-\varepsilon)k_\Lambda(\varepsilon)]}{E[k_\Lambda(\varepsilon)k_\Lambda(s-\varepsilon)k_\Lambda(t-\varepsilon)\overline{K}_\Lambda(\varepsilon)]} \\
&= \frac{E[k_\Lambda(\varepsilon)k_\Lambda(\varepsilon)k_\Lambda(s-\varepsilon)k_\Lambda(t-\varepsilon)]}{E[k_\Lambda(\varepsilon)k_\Lambda(\varepsilon)k_\Lambda(s-\varepsilon)\overline{K}_\Lambda(t-\varepsilon)]}
\end{aligned}
$$

and thus

$$
E[k_\Lambda(\varepsilon)k_\Lambda(s-\varepsilon)k_\Lambda(t-\varepsilon)\overline{K}_\Lambda(\varepsilon)] = E[k_\Lambda(\varepsilon)^2k_\Lambda(s-\varepsilon)\overline{K}_\Lambda(t-\varepsilon)]
$$

or, as $\varepsilon \to 0$,

$$
E[k_\Lambda(0)k_\Lambda(s)k_\Lambda(t)] = E[k_\Lambda(0)^2k_\Lambda(s)\overline{K}_\Lambda(t)].
$$

Integration with respect of s from t to ∞ yields

$$E[k_\Lambda(0)\overline{K}_\Lambda(t)k_\Lambda(t)] = E[k_\Lambda(0)^2\overline{K}_\Lambda(t)^2]. \tag{5.21}$$

From (5.20) and (5.21) formula (5.19) follow. Thus

$$\overline{K}_\ell'(t) = -k_\ell(0)\overline{K}_\ell(t) \quad U\text{-a.s.} \tag{5.22}$$

for any fixed value of t. Since (5.22) then holds U-a.s. for any countable and dense set of t-values it does in fact, due to Assumption 5.1, hold for all values of t. Thus

$$\overline{K}_\ell(t) = C_\ell e^{-k_\ell(0)t}, \quad t \geq 0 \quad U\text{-a.s.},$$

where, since K_ℓ is a distribution, $C_\ell = 1$ and $k_\ell(0) > 0$ U-a.s. Thus, cf. Example 5.2, the proposition follows. ■

CHAPTER 6

Characterization of mixed Poisson processes

We have already seen certain results connected with characterization. One category of such results is characterization of an individual process, i.e. determination of distributions of point processes, within certain classes.

A simple point process is characterized by, see Theorem 3.1, the probabilities $P\{N\{A\} = 0\}$ for all $A \in \mathcal{I}_U$. A birth process is determined, within the class of birth processes, cf. (4.2), by $p_n(t)$ for all t and n. Here it is not enough to consider $p_0(t)$, but it is if we consider a mixed Poisson process within the class of mixed Poisson processes.

We can also consider characterization of classes of point processes.

Watanabe's theorem (Theorem 3.2) characterized the Poisson processes, within the class of simple point processes with continuous compensator, as those with a deterministic compensator.

Theorem 4.1 gave a characterization of the class of Pólya processes within the class of mixed Poisson processes. That characterization is not valid within the class of birth processes. To see that we consider a Pólya process with intensities $\kappa_n(t) = \frac{a+n}{b+t}$. Let $\kappa_n^{(\beta)}(t) = \beta \cdot \frac{a+n}{b+t}$ be the intensities for some birth process. That process is, for $\beta \neq 1$, not even a mixed Poisson process, since – see Lemma 4.1 –

$$\kappa_n^{(\beta)}(t) - \frac{d\log(\kappa_n^{(\beta)}(t))}{dt} = \beta \cdot \frac{a+n}{b+t} + \frac{1}{b+t}$$

$$\neq \beta \cdot \frac{a+n}{b+t} + \beta \cdot \frac{1}{b+t} = \kappa_{n+1}^{(\beta)}(t).$$

We have already seen that it is of utmost importance to specify the class within which the characterization holds. From now on, we will consider characterizations of the class of mixed Poisson processes.

6.1 Characterization within birth processes

The following theorem is due to Lundberg (1964, pp.72, 84–89)

Theorem 6.1. *Let N be a birth process with intensities $\kappa_n(t)$ and marginal distribution $p_n(t)$. The following statements are equivalent:*

(*i*) *N is a mixed Poisson process.*

(*ii*) *$\kappa_n(t)$ satisfy $\kappa_{n+1}(t) = \kappa_n(t) - \frac{\kappa_n'(t)}{\kappa_n(t)}$ for $n = 0, 1, \ldots$*

(*iii*) *$\kappa_n(t)$ and $p_n(t)$ satisfy $p_n(t) = \frac{t}{n}\kappa_{n-1}(t)\, p_{n-1}(t)$ for $n = 1, 2, \ldots$*

(*iv*) *$E[\kappa_{N(t)}(t) \mid N(s) = m] = \kappa_m(s)$ for $0 < s \leq t$ and $m = 0, 1, \ldots$*

(*v*) *$E[N(t) - N(s) \mid N(s) = m] = \kappa_m(s)(t - s)$ for $s \leq t$ and $m = 0, 1, \ldots$*

(*vi*) *$P\{N(s) = m \mid N(t) = n\} = \binom{n}{m}\left(\frac{s}{t}\right)^m \left(1 - \frac{s}{t}\right)^{n-m}$ for $s \leq t$ and $m \leq n$.*

Proof. The proof to be given is, more or less, Lundberg's original proof. The structure of the proof is indicated below.

$$
\begin{array}{ccccc}
\text{(v)} & \Rightarrow & \text{(iv)} & & \text{(v)} \\
 & & \Downarrow & & \Uparrow \\
 & & \text{(ii)} & \Rightarrow & \text{(i)} \\
 & & \Uparrow & & \Downarrow \\
\text{(vi)} & \Rightarrow & \text{(iii)} & & \text{(vi)}
\end{array}
$$

(ii) $\Rightarrow$ (i). (Lundberg 1964, p.72)
Put $\hat{u}(t) = p_0(t)$ which implies $\hat{u}(t) \geq 0$ for $t \geq 0$. It follows from (4.2) that

$$\kappa_0(t) = -\frac{\hat{u}^{(1)}(t)}{\hat{u}(t)},$$

and thus $-\hat{u}^{(1)}(t) \geq 0$ for $t \geq 0$ since $\kappa_0(t) \geq 0$. Assume now that

$$\kappa_k(t) = -\frac{\hat{u}^{(k+1)}(t)}{\hat{u}^{(k)}(t)} \tag{6.1}$$

and that

$$(-1)^{k+1}\hat{u}^{(k+1)}(t) \geq 0 \text{ for } t \geq 0. \tag{6.2}$$

for $k = 0, \ldots, n$. From (ii) we now get, by taking the logarithmic derivative,

$$\kappa_{n+1}(t) = -\frac{\hat{u}^{(n+1)}(t)}{\hat{u}^{(n)}(t)} - \frac{\hat{u}^{(n+2)}(t)}{\hat{u}^{(n+1)}(t)} + \frac{\hat{u}^{(n+1)}(t)}{\hat{u}^{(n)}(t)} = -\frac{\hat{u}^{(n+2)}(t)}{\hat{u}^{(n+1)}(t)}.$$

Using $(-1)^{n+1}\hat{u}^{(n+1)}(t) \geq 0$ and $\kappa_{n+1}(t) \geq 0$, it follows that (6.1) and (6.2) hold for all $k \geq 0$.Thus $\hat{u}$ is completely monotone. Since $\hat{u}(0) = p_0(0) = 1$ it follows that $\hat{u}(t)$ is a Laplace transform and thus, see (4.6), the intensities define a mixed Poisson process.

(iii) $\Rightarrow$ (ii). (Lundberg 1964, p.85)
Taking the logarithmic derivative in (iii) we get

$$\frac{p_n'(t)}{p_n(t)} = \frac{1}{t} + \frac{\kappa_{n-1}'(t)}{\kappa_{n-1}(t)} + \frac{p_{n-1}'(t)}{p_{n-1}(t)}. \tag{6.3}$$

From (4.2) and (iii) we get

$$\frac{p_k'(t)}{p_k(t)} = -\kappa_k(t) + \kappa_{k-1}(t)\frac{p_{k-1}(t)}{p_k(t)} = -\kappa_k(t) + \frac{k}{t}. \tag{6.4}$$

Note that (6.4) holds also for $k = 0$. Inserting (6.4), for $k = n$ and $n - 1$, in (6.3) we get (ii).

(iv) $\Rightarrow$ (ii). (Lundberg 1964, p.88)
By (iv) we have

$$\kappa_m(s) = \sum_{n=m}^{\infty} \kappa_n(t)\, p_{m,n}(s,t).$$

From (4.4) and (iv) we get

$$\kappa_m'(s) = \sum_{n=m}^{\infty} \kappa_n(t)\, \frac{\partial p_{m,n}(s,t)}{\partial s}$$

$$= \kappa_m(s) \cdot \left(\kappa_m(t)\, p_{m,m}(s,t) + \sum_{n=m+1}^{\infty} \kappa_n(t)[p_{m,n}(s,t) - p_{m+1,n}(s,t)]\right)$$

$$= \kappa_m(s)[\kappa_m(s) - \kappa_{m+1}(s)],$$

which is (ii).

(v) ⇒ (iv). (Lundberg 1964, pp.87–88)
Let m and s be fixed and let $e(t)$ be defined by

$$e(t) = \sum_{n=m}^{\infty} (n-m)\, p_{m,n}(s,t).$$

By (4.3) we get

$$e'(t) = \sum_{n=m}^{\infty} (n-m)\big(-\kappa_n(t)\, p_{m,n}(s,t) + \kappa_{n-1}(t)\, p_{m,n-1}(s,t)\big)$$

$$= \sum_{n=m}^{\infty} \kappa_n(t)\, p_{m,n}(s,t).$$

Since, by (v), $e'(t) = \kappa_m(s)$, (iv) holds.

(vi) ⇒ (iii). (Lundberg 1964, p.86)
Since

$$P\{N(s)=m,\ N(t)=n\} = P\{N(s)=m \mid N(t)=n\}p_n(t)$$

$$= p_{m,n}(s,t)\, p_m(s)$$

it follows from (vi) that (4.9) holds. By taking the logarithmic derivative in (4.9) with respect to t we get

$$\frac{\partial p_{m,n}(s,t)}{\partial t} = \left(-\frac{n}{t} + \frac{n-m}{t-s} + \frac{p_n'(t)}{p_n(t)}\right) p_{m,n}(s,t). \tag{6.5}$$

From (4.3) and (4.2) for $m < n$ it is seen that (6.5) reduces to

$$\kappa_{n-1}(t)\, \frac{p_{m,n-1}(s,t)}{p_{m,n}(s,t)} = -\frac{n}{t} + \frac{n-m}{t-s} + \kappa_{n-1}(t)\, \frac{p_{n-1}(t)}{p_n(t)}. \tag{6.6}$$

Using (4.9) we get

$$\frac{p_{m,n-1}(s,t)}{p_{m,n}(s,t)} = \frac{n-m}{n}\,\frac{t}{t-s}\,\frac{p_{n-1}(t)}{p_n(t)},$$

and (6.6) reduces to

$$\kappa_{n-1}(t)\left(\frac{n-m}{n}\frac{t}{t-s}-1\right)\frac{p_{n-1}(t)}{p_n(t)}=-\frac{n}{t}+\frac{n-m}{t-s}$$

or

$$\kappa_{n-1}(t)\,\frac{p_{n-1}(t)}{p_n(t)}=\frac{n}{t},$$

which is (iii). Recall that we assumed $m<n$, but since (iii) is valid only for $n\geq 1$ and does not contain m, that is enough.

(i) $\Rightarrow$ (v).
Let U^* be given by (4.8). Then

$$E[N(t)-N(s)\mid N(s)=m]=\int_0^\infty (t-s)\,\ell\, dU^*(\ell)=(t-s)\,\kappa_m(s).$$

(i) $\Rightarrow$ (vi).
This follows from (4.9). One may also argue along the following line. Since (vi) holds for the Poisson process, independently of its intensity α, it also holds when the intensity is randomly generated. ■

Theorem 6.1 (iv) is most remarkable. Let us assume that $\kappa_0(0)<\infty$. Since N is Markovian we have

$$E[\kappa_{N(t)}(t)\mid N(s)]=E[\kappa_{N(t)}(t)\mid \mathcal{F}_s^N].$$

Thus that part of Theorem 6.1 can be rewritten in the following way:

A birth process N with $\kappa_0(0)<\infty$ is a mixed Poisson process if and only if $\kappa_{N(t)}(t)$ is an $\boldsymbol{F}^N$-martingale.

This is, to our knowledge, the first martingale characterization within point process theory. It is almost 25 years older than Watanabe's theorem!

Let $\{T_n;\ n\geq 1\}$ denote the jump times of N, i.e. T_n is the time of the nth jump, which implies that $N(T_n)=n$. Pfeifer and Heller (1987) have given the following variation of Lundberg's martingale characterization:

A birth process N with $\kappa_0(0)<\infty$ is a mixed Poisson process if and only if $E[\kappa_1(T_1)\mid T_1\geq t]=\kappa_0(t)$ for almost all $t\geq 0$ and $\kappa_n(T_n)$, $n\geq 1$, is an $\boldsymbol{F}^T$-martingale.

Theorem 6.1 (vi) it equivalent to $N(s)$ being an independent s/t-thinning of $N(t)$, cf. Definition 2.4. For fixed s and $t\to\infty$ it follows that $N(s)$ can be obtained by independent p-thinning for every $p \in (0,1)$. According to Theorem 2.3 this implies that $N(s)$ is mixed Poisson distributed. Thus Theorem 6.1 can also be looked upon as the first characterization of mixed Poisson in terms of p-thinning.

Theorem 6.1 is not true if N is defined only on a finite time-interval $[0, \tilde{t}]$. This it not surprising, since in the proof of (ii) $\Rightarrow$ (i), i.e. where it was proved that $p_0(t)$ is a Laplace transform, it was essential that t took its values in the whole of $[0, \infty)$ and not only in $[0, \tilde{t}]$. Also the analogy above with p-thinning indicates that, since it is essential in Theorem 2.3 that $N(s)$ can be obtained by p-thinning for **every** $p \in (0,1)$.

Let $X_1, \ldots, X_{\tilde{n}}$ be independent random variables, which are uniformly distributed on $[0, \tilde{t}]$. The point process N, defined by

$$N(t) = \text{the number of } X_n\text{s such that } X_n \le t, \quad t \in [0, \tilde{t}],$$

is called a **sample process**. This means that the location of the points of N are determined by $X_1, \ldots, X_{\tilde{n}}$. Sample processes were considered by Elfving (1946a, b), who called them **binomial processes**. It is easy to see that N is a birth process with, see Elfving (1946a, pp.187–188),

$$p_n(t) = \binom{\tilde{n}}{n} \left(\frac{t}{\tilde{t}}\right)^n \left(1 - \frac{t}{\tilde{t}}\right)^{\tilde{n}-n} \quad \text{and} \quad \kappa_n(t) = \frac{\tilde{n} - n}{\tilde{t} - t},$$

for $0 \le n \le \tilde{n}$ and $0 \le t \le \tilde{t}$.

Certainly N is **not** a mixed Poisson process, so Theorem 6.1 (i) does not hold. We have

$$\kappa_n(t) - \frac{d \log(\kappa_n(t))}{dt} = \frac{\tilde{n} - n}{\tilde{t} - t} - \frac{1}{\tilde{t} - t} = \frac{\tilde{n} - (n+1)}{\tilde{t} - t} = \kappa_{n+1}(t),$$

i.e. Theorem 6.1 (ii) holds. Simple calculations show that also (iii) holds. From the construction of the process, it follows that

$$P\{N(s) = m, \, N(t) = n\}$$

$$= \frac{\tilde{n}!}{m!\,(n-m)!\,(\tilde{n}-n)!} \left(\frac{s}{\tilde{t}}\right)^m \left(\frac{t-s}{\tilde{t}}\right)^{n-m} \left(1 - \frac{t}{\tilde{t}}\right)^{\tilde{n}-n} \tag{6.7}$$

for $0 \le m \le n \le \tilde{n}$ and $0 \le s \le t \le \tilde{t}$. Using (6.7) we get

$$p_{m,n}(s,t) = \frac{P\{N(s) = m, \, N(t) = n\}}{p_m(s)}$$

$$= \binom{\tilde{n} - m}{n - m} \left(\frac{t-s}{\tilde{t}-s}\right)^{n-m} \left(1 - \frac{t-s}{\tilde{t}-s}\right)^{(\tilde{n}-m)-(n-m)},$$

and it follows that

$$E[N(t) - N(s) \mid N(s) = m] = (\tilde{n} - m)\frac{t - s}{\tilde{t} - s} = \kappa_m(s)(t - s),$$

i.e. (v) holds. From this we get

$$E[\kappa_{N(t)}(t) \mid N(s) = m] = E\left[\frac{\tilde{n} - N(t)}{\tilde{t} - t} \mid N(s) = m\right]$$

$$= \frac{\tilde{n} - m - \kappa_m(s)(t - s)}{\tilde{t} - t} = \frac{\kappa_m(s)(\tilde{t} - s - t + s)}{\tilde{t} - t} = \kappa_m(s),$$

i.e. (iv) holds. Using (6.7) again we get

$$P\{N(s) = m \mid N(t) = n\} = \frac{P\{N(s) = m,\, N(t) = n\}}{p_n(t)}$$

$$= \binom{n}{m}\left(\frac{s}{t}\right)^m \left(1 - \frac{s}{t}\right)^{n-m},$$

which is (vi).

Let us now generalize the definition of a sample process in such a way that we regard $\tilde{n}$ as the outcome of a discrete random variable $\tilde{N}$ with generating function $G_{\tilde{N}}(s)$. The corresponding point process N is then called a **mixed sample process**. It is rather obvious that N is a Markov process; a formal proof is given by Puri (1982, p.42). Since $N(t)$ is an independent $t/\tilde{t}$-thinning of $N(t)$, it follows – see the discussion after Definition 2.4 – that

$$G_{N(t)}(s) = G_{\tilde{N}}(1 - (t/\tilde{t})(1 - s)) = \sum_{n=1}^{\infty} \frac{1}{n!}\left(\frac{t}{\tilde{t}}\right)^n G_{\tilde{N}}^{(n)}\left(1 - \frac{t}{\tilde{t}}\right) s^n, \quad (6.8)$$

where $G_{\tilde{N}}^{(n)}$ denotes the nth derivative of $G_{\tilde{N}}$. Thus we get

$$p_n(t) = \frac{1}{n!}\left(\frac{t}{\tilde{t}}\right)^n G_{\tilde{N}}^{(n)}\left(1 - \frac{t}{\tilde{t}}\right),$$

cf. Puri (1982, p.42). By differentiation of $p_n(t)$ it follows, in the same way as (4.6) followed from (4.2), that N is a birth process with intensities

$$\kappa_n(t) = \frac{G_{\tilde{N}}^{(n+1)}\left(1 - \frac{t}{\tilde{t}}\right)}{\tilde{t} \cdot G_{\tilde{N}}^{(n)}\left(1 - \frac{t}{\tilde{t}}\right)},$$

as pointed out by Puri (1982, p.42).

Let N be a birth process defined on $[0, \tilde{t}]$ such that

$$P\{N(t) = m \mid N(\tilde{t}) = n\} = \binom{n}{m}\left(\frac{t}{\tilde{t}}\right)^m \left(1 - \frac{t}{\tilde{t}}\right)^{n-m},$$

for all $0 \leq t \leq \tilde{t}$ and all $0 \leq m \leq n$ whenever $P\{N(\tilde{t}) = n\} \geq 0$. Since a birth process is determined by its one-dimensional marginal distributions, cf. (4.2), it follows that N is a mixed sample process. This characterization is due to Elfving (1946b, p.67). We will consider related – but much more general – characterizations in section 6.3.

Assume now that $\widetilde{N}$ is MP$(\tilde{t}, U)$, see Definition 2.2, for some distribution U and let N be the corresponding mixed sample process. This implies, see Proposition 2.1 (vi), that $G_{\widetilde{N}}(s) = \hat{u}(\tilde{t}(1-s))$ and thus by (6.8)

$$G_{N(t)}(s) = \hat{u}\big(\tilde{t}(1 - 1 + (t/\tilde{t})(1-s))\big) = \hat{u}(t(1-s)),$$

i.e. $N(t)$ is MP(t, U). Thus N is a mixed Poisson process. Since $N(\tilde{t}) = \widetilde{N}$, it follows that N is a mixed Poisson process if and only if $\widetilde{N}$ is mixed Poisson distributed, cf. Deffner and Haeusler (1985). Although it is hardly any risk for misunderstandings, we do point out that – very strictly speaking – N is a non-stationary mixed Poisson process with

$$A(t) = \begin{cases} t, & t \leq \tilde{t}, \\ \tilde{t}, & t > \tilde{t}, \end{cases}$$

cf. Remark 5.1.

Consider now the case where N is a stationary birth process. Put, see Definition 3.3, $N_s(t) = N(s+t) - N(s)$. Since N and N_s have the same distribution, also N_s is a birth process, and we have

$$\begin{aligned}\kappa_n(t) &= E[\kappa_{N(s+t)}(s+t) \mid N_s(t) = n] \\ &= \sum_{m=0}^{\infty} P\{N(s+t) = n+m \mid N_s(t) = n\}\kappa_{n+m}(s+t).\end{aligned}$$

Since

$$\begin{aligned}P\{N(s+t) = n+m \mid N_s(t) = n\} &= P\{N(s) = m \mid N_s(t) = n\} \\ &= \frac{P\{N(s) = m,\, N_s(t) = n\}}{P\{N_s(t) = n\}} = \frac{p_m(s)\, p_{m,m+n}(s, s+t)}{p_n(t)},\end{aligned}$$

we get

$$p_n(t)\kappa_n(t) = \sum_{m=0}^{\infty} p_m(s)\, p_{m,n+m}(s, s+t)\, \kappa_{n+m}(s+t). \tag{6.9}$$

Notice that, for $t = 0$ and $n = 1$, (6.9) reduces to

$$\kappa_0(0) = \sum_{m=0}^{\infty} p_m(s)\kappa_m(s),$$

which means that $E[N\{ds\}] = \kappa_0(0)ds$.

As mentioned, a mixed Poisson process is stationary. McFadden (1965) has shown that mixed Poisson processes are the only stationary birth processes for which $p_0(t)$ is infinitely differentiable. McFadden's – somewhat heuristic – proof is based on **Palm probabilities**, and contains arguments which we have not quite managed to follow. In section 6.2 we will discuss Palm probabilities in some detail, and we will in Remark 6.2 return to this question. We will therefore give a different proof. Here we will only mention that a Palm probability is a probability obtained by conditioning upon the occurrence of a point at time 0. In the proof to be given, we will consider the Palm probability

$$p_{1,k}(0,t) \stackrel{\text{def}}{=} \lim_{h\to 0} p_{1,k}(h,t). \tag{6.10}$$

Lemma 6.1. *For a stationary birth process with $\kappa_0(0) < \infty$, we have*

$$p_{1,k}(0,t) = \frac{\kappa_{k-1}(t)\, p_{k-1}(t)}{\kappa_0(0)}, \quad \textit{for } k \geq 1.$$

Proof. Consider $p_{1,k}(h,t)$ for small values of h. In order to approximate $p_{0,k-1}(h,t)$ we recall from (4.4) that

$$p^b_{0,k-1}(s,t) \stackrel{\text{def}}{=} \frac{\partial p_{0,k-1}(s,t)}{\partial s} = \kappa_0(s)[p_{0,k-1}(s,t) - p_{1,k-1}(s,t)].$$

By stationarity we get

$$p_{k-1}(t-h) = p_0(h)\, p_{0,k-1}(h,t) + p_1(h)\, p_{1,k}(h,t) + o(h),$$

or

$$p_{1,k}(h,t) = \frac{p_{k-1}(t-h) - p_0(h)\, p_{0,k-1}(h,t) + o(h)}{p_1(h)}$$

$$= \frac{p_{k-1}(t) - hp'_{k-1}(t)}{h\kappa_0(0)} - \frac{[1 - h\kappa_0(0)][p_{k-1}(t) + hp^b_{0,k-1}(0,t)]}{h\kappa_0(0)} + o(1)$$

$$= \frac{-p'_{k-1}(t) + \kappa_0(0)\, p_{k-1}(t) - p^b_{0,k-1}(0,t)}{\kappa_0(0)} + o(1)$$

$$= \frac{-p'_{k-1}(t) + \kappa_0(0)\, p_{1,k-1}(0,t)}{\kappa_0(0)} + o(1)$$

$$= \frac{\kappa_{k-1}(t)\, p_{k-1}(t) - \kappa_{k-2}(t)\, p_{k-2}(t) + \kappa_0(0)\, p_{1,k-1}(0,t)}{\kappa_0(0)} + o(1).$$

Thus we have the following recursive equation

$$p_{1,k}(0,t) = p_{1,k-1}(0,t) + \frac{\kappa_{k-1}(t)\, p_{k-1}(t)}{\kappa_0(0)} - \frac{\kappa_{k-2}(t)\, p_{k-2}(t)}{\kappa_0(0)}$$

which leads to the desired result. ∎

Assumption 6.1. *Let N be a birth process. Assume that*

(*a*) $\kappa_n(t)$ *is twice differentiable;*

(*b*) $\kappa_0(0) < \infty$;

(*c*) *For each n and $T < \infty$ there exists a constant $C_{n,T} < \infty$ such that $p_{k,n+k}(s,t)\kappa_{n+k}(t) \le C_{n,T}$ for all $k \ge 2$ and all $s \le t \le T$.*

We will comment upon Assumption 6.1 after the proof of Theorem 6.2.

Theorem 6.2 (McFadden, 1965). *Let N be a birth process for which Assumption 6.1 holds. The following statements are equivalent:*

(*i*) *N is a mixed Poisson process.*

(*ii*) *N is stationary.*

Proof. (i) $\Rightarrow$ (ii) is already shown, so we consider (ii) $\Rightarrow$ (i).

Let s be small in (6.9), and to emphasize that we put $s = h$. Then we get, where the letters over the equalities below refer to Assumption 6.1 and where the last equality follows from Lemma 6.1,

$$\begin{aligned}
p_n(t)\kappa_n(t) &= \sum_{k=0}^{\infty} p_k(h)\, p_{k,n+k}(h, h+t)\kappa_{n+k}(h+t) \\
&\overset{(c)}{=} p_0(h)\, p_{0,n}(h, h+t)\kappa_n(h+t) \\
&\quad + p_1(h)\, p_{1,n+1}(h, h+t)\kappa_{n+1}(h+t) + o(h) \\
&\overset{(a)}{=} [1 - h\kappa_0(0)]\, p_{0,n}(h, h+t)[\kappa_n(t) + h\kappa_n'(t)] \\
&\quad + h\kappa_0(0)\, p_{1,n+1}(0,t)\kappa_{n+1}(t) + o(h) \\
&\overset{(b)}{=} p_{0,n}(h, h+t)\kappa_n(t) + hp_n(t)[\kappa_n'(t) - \kappa_0(0)\kappa_n(t)] \\
&\quad + hp_n(t)\kappa_n(t)\kappa_{n+1}(t) + o(h).
\end{aligned}$$

In order to approximate $p_{0,n}(h, h+t)$ we recall from (4.3) and (4.4) that

$$p^f_{0,n}(s,t) \overset{\text{def}}{=} \frac{\partial p_{0,n}(s,t)}{\partial t} = -\kappa_n(t)\, p_{0,n}(s,t) + \kappa_{n-1}(t)\, p_{0,n-1}(s,t)$$

and, cf. the proof of Lemma 6.1,

$$p^b_{0,n}(s,t) \overset{\text{def}}{=} \frac{\partial p_{0,n}(s,t)}{\partial s} = \kappa_0(s)[\, p_{0,n}(s,t) - p_{1,n}(s,t)].$$

Then we get

$$\begin{aligned}
p_{0,n}(h, h+t) &= p_n(t) + h[\, p^f_{0,n}(0,t) + p^b_{0,n}(0,t)] + o(h) \\
&= p_n(t) + h\big[-\kappa_n(t)\, p_n(t) + \kappa_{n-1}(t)\, p_{n-1}(t) \\
&\quad + \kappa_0(0)[\, p_n(t) - p_{1,n}(0,t)]\big] + o(h) \\
&= p_n(t) + h\big[-\kappa_n(t)\, p_n(t) + \kappa_{n-1}(t)\, p_{n-1}(t) \\
&\quad + \kappa_0(0)\, p_n(t) - \kappa_{n-1}(t)\, p_{n-1}(t)\big] + o(h) \\
&= p_n(t) + hp_n(t)[\kappa_0(0) - \kappa_n(t)] + o(h).
\end{aligned}$$

Thus

$$p_n(t)\kappa_n(t) = p_n(t)\kappa_n(t) + hp_n(t)[\kappa_0(0) - \kappa_n(t)]\kappa_n(t)$$

$$+ hp_n(t)[\kappa_n'(t) - \kappa_0(0)\kappa_n(t)] + hp_n(t)\kappa_n(t)\kappa_{n+1}(t) + o(h),$$

and we get

$$0 = [\kappa_0(0) - \kappa_n(t)] + \frac{\kappa_n'(t)}{\kappa_n(t)} - [\kappa_0(0) - \kappa_{n+1}(t)]$$

$$= -\kappa_n(t) + \frac{\kappa_n'(t)}{\kappa_n(t)} + \kappa_{n+1}(t)$$

or

$$\kappa_{n+1}(t) = \kappa_n(t) - \frac{\kappa_n'(t)}{\kappa_n(t)}.$$

Thus, see Theorem 6.1 (ii), N is a mixed Poisson process. ■

We will now briefly comment upon Assumption 6.1:

With our method we certainly have to assume that $\kappa_n(t)$ is differentiable. Assumption (a) is made to ascertain that $\kappa_n(t+h) = \kappa_n(t) + h\kappa_n'(t) + o(h)$. It follows from (4.2) that we equally well can assume that $p_n(t)$ is three times differentiable. Assumption (b) is necessary with our approximation of $p_n(h)$ and it simplifies the use of Palm probabilities, cf. Remark 6.1. Assumption (c) is just a regularity assumption made in order to ascertain that

$$\sum_{k=2}^{\infty} p_k(h)\, p_{k,n+k}(h, h+t)\kappa_{n+k}(h+t) \le C_{n,T} \sum_{k=2}^{\infty} p_k(h) = o(h).$$

Most probably (c) can be much relaxed or even omitted.

6.2 Characterization within stationary point processes

Recall the definition of a stationary point process given by Definition 3.3.

Let N be MPP(U), where U is the distribution of a non-negative random variable Λ, and let U_c be the distribution of $c\Lambda$. Then the thinning D_pN is MPP(U_p). We can now 'compensate' the thinning by a compression of time. More formally, we define the compression operator K_p by $K_pN(t) = N(t/p)$. It readily follows that the time compressed process K_pN is MPP($U_{1/p}$). Thus $K_pD_p\Pi_U = \Pi_U$ for any mixed Poisson process. This is in fact a characterizing property,

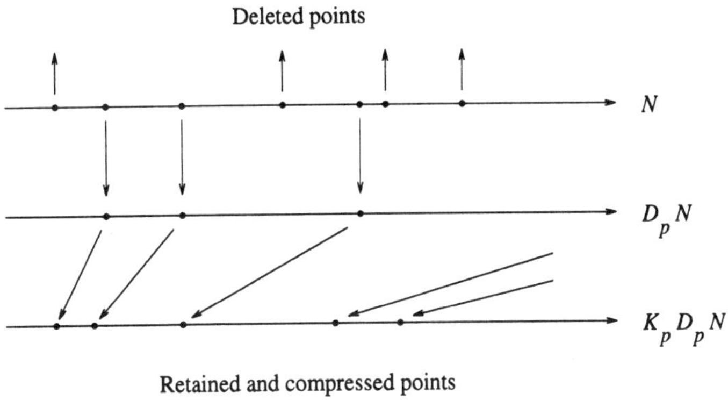

Figure 6.1. *Illustration of p-thinning and time compression.*

as shown by Nawrotzki (1962). The operator $K_p D_p$ is illustrated in Figure 6.1.

In Theorem 6.1 (vi) it was shown that – within birth processes – the mixed Poisson process is characterized by the conditional distribution of $N(s)$ given $N(t)$ for $s \leq t$. This characterization is a related to the fact that the location of jumps of N in $[0, t]$ given $N(t) = n$ is uniformly distributed. Formally, if T_k is the time of the kth jump and if $X_1, \ldots, X_n$ are independent and uniformly distributed on $[0, t]$, then

$$(T_1, \ldots, T_n) \text{ given } \{N(t) = n\} \stackrel{\mathrm{d}}{=} (X^{(1)}, \ldots, X^{(n)}), \tag{6.11}$$

$\stackrel{\mathrm{d}}{=}$ means 'equality in distribution' and $(X^{(1)}, \ldots, X^{(n)})$ denotes the ordered permutation of $(X_1, \ldots, X_n)$ where $X^{(k)} < X^{(k+1)}$.

Theorem 6.3 (Nawrotzki, 1962). *Let N be a stationary point process with distribution Π. The following statements are equivalent:*

(*i*) *N is a mixed Poisson process.*

(*ii*) *$K_p D_p \Pi = \Pi$ for some $p \in (0, 1)$.*

(*iii*) *The times of the jumps in $[0, t]$ given $N(t) = n$ are uniformly distributed for $n \geq 1$ and $t > 0$, as formally stated in (6.11).*

Proof. (i) ⇒ (ii) and (i) ⇒ (iii) are indicated or almost obvious.

(ii) ⇒ (i).

From $K_{p^{n+1}}D_{p^{n+1}}\Pi = K_{p^n}D_{p^n}K_pD_p\Pi = K_pD_p\Pi$ it follows that (ii) holds for p, p^2, $p^3 \ldots$. Put

$$p_n = p^n \quad \text{and} \quad N_n = K_{p_n}N.$$

Since $D_{p_n}N_n = N$ we have, trivially, that $D_{p_n}N_n \xrightarrow{\text{d}} N$ and thus it follows from Theorem 5.1 there exists a random measure Λ such that $p_nN_n \xrightarrow{\text{d}} \Lambda$. This implies that N is a Cox process with underlying random measure Λ. In particular, it follows that

$$p_nN_n(1) = p_nN(1/p_n) \xrightarrow{\text{d}} \Lambda(1) \text{ as } n \to \infty. \tag{6.12}$$

Since N is stationary, there exists a – possibly infinite – random variable $\overline{N}$, called the **individual intensity**, such that

$$N(s)/s \to \overline{N} \quad P\text{-a.s. as } s \to \infty. \tag{6.13}$$

It follows from (6.12) that $\overline{N} \stackrel{\text{d}}{=} \Lambda(1)$ and thus $P\{\overline{N} < \infty\} = 1$. Some further discussion of $\overline{N}$ is to be found in Chapter 3.

Further it follows from (6.13) that

$$\begin{aligned} p_nN_n(t) &= p_nN(t/p_n) \\ &= t \cdot \frac{p_n}{t}N(t/p_n) \to t \cdot \overline{N} \quad P\text{-a.s. as } n \to \infty. \end{aligned}$$

Thus $(\Lambda(t_1), \ldots, \Lambda(t_n)) \stackrel{\text{d}}{=} (t_1 \cdot \overline{N}, \ldots, t_n \cdot \overline{N})$ and (i) follows.

(iii) $\Rightarrow$ (i).

Since the uniform distribution is continuous, no more that one jump can occur at the same time, i.e. N is simple. Then, see Theorem 3.1, it is enough to show that

$$P\{N\{A\} = 0\} = \int_{0-}^{\infty} e^{\ell|A|}\,dU(\ell), \quad A \in \mathcal{I}_U, \tag{6.14}$$

where $|A|$ means the length of A, i.e. the sum of the lengths of the intervals in A. Recall from Chapter 3 that $\mathcal{I}_U$ denotes the set of finite unions of intervals in $[0, \infty)$.

Choose an $A \in \mathcal{I}_U$ and a $t > \sup\{x \mid x \in A\}$. Put $s = |A|$. Then

$$P\{N\{A\} = 0\} = \sum_{n=0}^{\infty} P\{N\{A\} = 0 \mid N(t) = n\}P\{N(t) = n\}$$

$$= \sum_{n=0}^{\infty} \left(\frac{t-s}{t}\right)^n P\{N(t) = n\},$$

and it is seen that $P\{N\{A\} = 0\} = P\{N(s) = 0\}$.

The random variable $N(s)$ can be regarded as obtained by independent p-thinning – see Definition 2.4 – from the random variable $N(t)$. Since a discrete random variable can be regarded as a point process defined on an one-point state space, we may apply Theorem 5.1 also to random variables. Choose $t_1, t_2, \ldots$ such that $\lim_{k\to\infty} t_k = \infty$. If we – with the obvious meaning – apply the thinning operator D_p to $N(t_k)$, we have $N(s) \stackrel{\mathrm{d}}{=} D_{s/t_k} N(t_k)$, and it follows from Theorem 5.1 that

$$\frac{s}{t_k} N(t_k) \stackrel{\mathrm{d}}{\to} \text{ some random variable } \Lambda_s \tag{6.15}$$

and that $N(s)$ is $\mathrm{MP}(U_s)$, where U_s is the distribution of Λ_s.

Since for any choice of A, the t_ks will sooner or later be large enough, we do not have to relate the choice of $t_1, t_2, \ldots$ to the chosen A. Thus the above reasoning holds for all A. Applying (6.15) to $s = 1$, it follows that $\Lambda_s \stackrel{\mathrm{d}}{=} s \cdot \Lambda_1$. Thus $N(|A|)$, for all $A \in \mathcal{I}_U$ is $\mathrm{MP}(|A|, U_1)$, and (6.14) follows. ■

In an actuarial model it is natural to regard N as a model for the times of the claims, while – from the point of view of the insurance company – it might be more interesting to describe the times when the claims are reported or when the costs are paid. In such cases there is a delay associated with each claim. Assume that these delays are described by a sequence $\{Y_k;\ k = 0, \pm 1, \pm 2, \ldots\}$ of independent random variables with common distribution F. This sequence is further independent of N. Naturally a claim which is reported after time 0 may have occurred before time 0, and therefore we consider N as defined on $\mathbb{R}$ and not, as usual, on $[0, \infty)$. Further we do not assume that $F(0-) = 0$, although it is natural within actuarial applications. If T_n is the time of the nth jump of N, normalized so that $N(0) = 0$ and $T_0 < 0 \leq T_1$, we denote by Π^F the distribution of the point process N^F with jumps at $T_k + Y_k$. We call N^F a **random translation** of N. Random translation is illustrated in Figure 6.2.

Assume that F is **non-lattice**, i.e. there exists no numbers c and d such that F is concentrated on c, $c \pm d$, $c \pm 2d, \ldots$. Consider the sequence of point processes $N_n = N^{F^{n*}}$ for $n = 1, 2, \ldots$, where F^{n*} denotes the nth convolution of F with itself. This means that N_n is

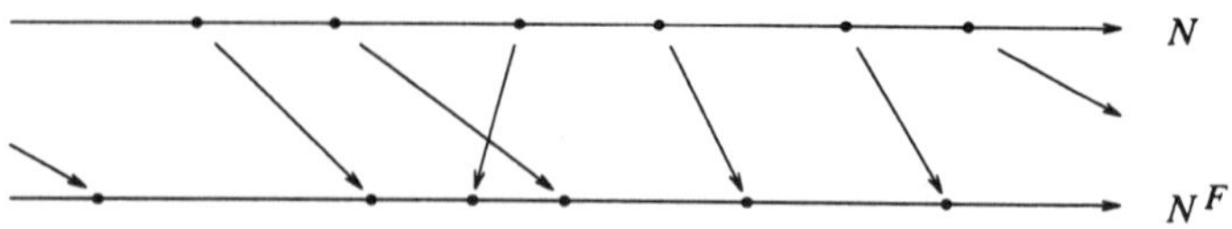

Figure 6.2. *Illustration of a randomly translated point process.*

obtained by n successive random translations. Let N be **ergodic** in the sense that the individual intensity $\overline{N} = \alpha$ Π-a.s. for some $\alpha < \infty$. For a high value of n, since F is non-lattice, most of the 'randomness' of N_n comes from the translations. Since the translations of each jump are independent of each other it ought to follow that N_n is approximately a Poisson process. This intuitive argument is 'doubtful', since a stringent version of it requires too restrictive regularity assumptions on N, cf. Daley and Vere-Jones (1988, p.303). Nevertheless, the following precise result holds:

If F is non-lattice and N is ergodic, then $N_n \xrightarrow{\text{d}}$ a Poisson process with intensity α.

Now we drop the assumption that N is ergodic. Assume that N has finite **intensity** α, i.e. $E[N(t)] = \alpha t$ where $\alpha < \infty$, which implies that $P\{\overline{N} < \infty\} = 1$. Let U be the distribution of $\overline{N}$. Then we have, see Matthes *et al.* (1978, p.467):

If F is non-lattice and N has finite intensity, then $N_n \xrightarrow{\text{d}}$ MPP(U).

If N is invariant under random translations, i.e. if $N^F \stackrel{\text{d}}{=} N$, it follows that $N_n \stackrel{\text{d}}{=} N$. Thus $N_n \xrightarrow{\text{d}} N$ holds trivially, and we get the following theorem.

Theorem 6.4. *Let N be a stationary point process defined on $\mathbb{R}$ and with finite intensity. The following statements are equivalent:*

(*i*) *N is a mixed Poisson process.*

(*ii*) *$\Pi^F = \Pi$ for some non-lattice distribution F.*

Theorem 6.4 and the corresponding convergence to Poisson and MPP have a rather long history going back to Maruyama (1955), Dobrushin (1956), Breiman (1963), Thedéen (1964), and Stone (1968).

Let Π be the distribution of a simple stationary point process N with finite intensity α. Here, as usual, N is defined on $[0,\infty)$. Now we will consider a characterization, due to Mecke (1976), in terms of Palm probabilities. As indicated in section 6.1, a Palm probability is a probability obtained by conditioning upon the occurrence of a point at time 0. For a stationary process the probability of the occurrence of a point at any fixed time t, $t > 0$, is zero. Very formally, since $\nu(0) = 0$ for all $\nu \in \mathcal{N}$, a point at time 0 is **impossible**. To cope with this purely notational problem we extend $\mathcal{N}$ to $\mathcal{N}_E$, where $\mathcal{N}_E$ is the set of all functions $\nu = \{\nu(t);\ t \geq 0\}$ such that either $\nu(\cdot) \in \mathcal{N}$ or $\nu(\cdot) - 1 \in \mathcal{N}$. The distribution Π may be extended to a measure on $(\mathcal{N}_E, \mathcal{B}(\mathcal{N}_E))$ simply by putting

$$\Pi\{B\} = \Pi\{B \cap \mathcal{N}\}, \quad B \in \mathcal{B}(\mathcal{N}_E).$$

This means that we want to consider probabilities of the form

$$\Pi\{B \mid N\{\{0\}\} = 1\}, \quad \text{for any } B \in \mathcal{B}(\mathcal{N}_E).$$

The problem is thus to give such probabilities a precise meaning for a general stationary point process. Intuitively we consider an event B and successively shift the process so that its 'points' fall at time 0. If this had been a statistical problem – and not the question of a probabilistic definition – we would probably have considered the proportion of times when the shifted point process belonged to B. Instead, we now consider the ratio of certain related intensities.

Consider a set $B \in \mathcal{B}(\mathcal{N}_E)$. Define the '$B$-thinned' process N^B by

$$N^B\{ds\} = 1_B(N_s)N\{ds\}, \quad \text{where } 1_B(N) = \begin{cases} 1 & \text{if } N \in B, \\ 0 & \text{if } N \notin B. \end{cases}$$

This means that N^B consists of those points in N for which the shifted point process N_s belongs to B. Obviously N^B is stationary.

Put $\mathcal{N}^0 = \{\nu \in \mathcal{N}_E;\ \nu\{\{0\}\} = 1\}$. Let $\alpha\{B\}$ be the intensity of N^B. It follows from Matthes et al. (1978, pp.309–311) that $\alpha\{\cdot\}$ is a measure, i.e. σ-additive, on $(\mathcal{N}^0,\ \mathcal{B}(\mathcal{N}^0))$.

Definition 6.1. *Let N be a simple stationary point process with distribution Π. The distribution Π^0, defined by*

$$\Pi^0\{B\} = \frac{\alpha\{B\}}{\alpha}, \quad B \in \mathcal{B}(\mathcal{N}^0),$$

is called the **Palm distribution**.

$\Pi^0\{B\}$ is the strict definition of '$\Pi\{B \mid N\{\{0\}\} = 1\}$'. The point process N^0 with distribution Π^0 is called the **Palm process**. For a precise interpretation of Π^0 as a conditional probability, cf. (6.10), we refer to Franken *et al.* (1981, pp.33, 38).

It is often convenient to withdraw the point at 0 and to consider the **reduced** Palm process $N^!$ with distribution $\Pi^!$. Formally $N^!$ is defined by

$$N^!\{A\} = \begin{cases} N^0\{A\} - 1 & \text{if } 0 \in A, \\ N^0\{A\} & \text{if } 0 \notin A \end{cases}$$

for any Borel set $A \subset [0, \infty)$, which implies that $\Pi^!$ is a measure on $(\mathcal{N}, \mathcal{B}(\mathcal{N}))$.

The process reduced Palm process $N^!$ is, in general, **not** stationary, but the inter-occurrence times $T_1^!, T_2^! - T_1^!, T_3^! - T_2^!, \ldots$ form a stationary sequence.

If N is a Poisson process we have $\Pi = \Pi^!$ which, in fact, is a characterization of the Poisson process. Intuitively this means that knowledge of a point at 0 has no influence on the distribution of the rest of the process, which is quite natural since the Poisson process is the only stationary point process with independent increments. In this case both the point process and the inter-occurrence times are stationary.

Let N be MPP(U), where U is the distribution of Λ and $\mu_\Lambda = \alpha < \infty$. Choose $A \in \mathcal{I}_U$ such that $0 \notin A$ and consider the set $B = \{\nu \in \mathcal{N};\ \nu\{A\} = 0\}$. Conditioned upon $\Lambda = \ell$, we have

$$E[N^B\{ds\} \mid \Lambda = \ell] =$$

$$E[1_B(N_s) \mid \Lambda = \ell] \cdot E[N\{ds\} \mid \Lambda = \ell] = e^{-\ell|A|} \cdot \ell\, ds.$$

Thus $\alpha\{B\} = \int_{0-}^{\infty} e^{-\ell|A|}\ell\, dU(\ell)$, from which it follows that

$$\Pi^!\{B\} = \int_0^\infty e^{-\ell|A|}\, dU^0(\ell), \tag{6.16}$$

where

$$dU^0(\ell) = \ell\, dU(\ell)/\alpha.$$

It follows from (6.16) and Theorem 3.1 that $N^!$ is MPP(U^0). As an example, we consider $B = \{\nu \in \mathcal{N};\ \nu(t) = n\}$. Then we have, cf.

(4.6),

$$p^!_n(t) \stackrel{\text{def}}{=} \Pi^!\{B\} = \int_0^\infty \frac{(\ell t)^n}{n!} e^{-\ell t}\, dU^0(\ell)$$

$$= \frac{\int_0^\infty \frac{\ell(\ell t)^n}{n!} e^{-\ell t}\, dU(\ell)}{\int_0^\infty \ell\, dU(\ell)} = \frac{\kappa_n(t)\, p_n(t)}{\kappa_0(0)}.$$

Thus, cf. Lemma 6.1, $p_{1,n+1}(0,t) = p^!_n(t)$. Further we have

$$\kappa^!_n(t) \stackrel{\text{def}}{=} \frac{\int_0^\infty \ell^{n+1} e^{-\ell t}\, dU^0(\ell)}{\int_0^\infty \ell^n e^{-\ell t}\, dU^0(\ell)} = \frac{\int_0^\infty \ell^{n+2} e^{-\ell t}\, dU(\ell)}{\int_0^\infty \ell^{n+1} e^{-\ell t}\, dU(\ell)} = \kappa_{n+1}(t),$$

which is quite natural and may serve as a motivation for the rather abstract Definition 6.1.

Theorem 6.5 (Mecke, 1976). *Let N be a simple stationary point process with finite intensity. The following statements are equivalent:*

(i) *N is a mixed Poisson process.*

(ii) *$N^!$ is stationary.*

Heuristic arguments for Theorem 6.5. A proof of the theorem would require Palm theory beyond the level of this book. We will, however, give some arguments, which indicate that Theorem 6.5 is natural.

Let us first assume that N is stationary and **mixing**. This is slightly stronger than to assume that N is ergodic, and means that the behaviour of the process on time intervals far away from each other is asymptotically independent. More precisely, mixing means that

$$\lim_{s\to\infty} \Pi\{A \cap T_s^{-1}B\} = \Pi\{A\}\Pi\{B\}, \quad A, B \in \mathcal{B}(\mathcal{N});$$

the shift operator T_s is defined in Chapter 3. It seems natural that

$$\lim_{s\to\infty} \Pi^!\{T_s^{-1}B\} = \Pi\{B\},$$

which, since $N^!$ is assumed to be stationary, implies $\Pi^! = \Pi$. As mentioned, this implies that N is a Poisson process.

Now we no longer assume that N is ergodic, and disregard the difference between ergodicity and mixing. Let $\mathcal{I}$ be the σ-algebra of invariant sets $B \in \mathcal{B}(\mathcal{N})$, cf. Chapter 3. Conditioned upon $\mathcal{I}$, the distribution $\Pi\{\cdot \mid \mathcal{I}\}$ is stationary and ergodic, or strictly speaking there exists such a version. Define the measure $\alpha^!\{\cdot\}$ by

$$\alpha^!\{B\} \stackrel{\text{def}}{=} \alpha \cdot \Pi^!\{B\}, \quad B \in \mathcal{B}(\mathcal{N}),$$

cf. Definition 6.1. Although not being a probability measure, $\alpha^!\{\cdot\}$ is stationary, and we may condition that measure upon $\mathcal{I}$. Conditioned upon $\mathcal{I}$, the reduced Palm probability is given by

$$\Pi^!\{B \mid \mathcal{I}\} = \frac{\alpha^!\{B \mid \mathcal{I}\}}{\bar{\nu}}, \quad B \in \mathcal{B}(\mathcal{N}),$$

where $\bar{\nu} = E[N(1) \mid \mathcal{I}]$ is the individual intensity. Since $\Pi^!\{\cdot \mid \mathcal{I}\}$ is stationary, it follows that $\Pi\{\cdot \mid \mathcal{I}\}$ is a Poisson distribution with intensity $\bar{\nu}$, i.e. Π is the distribution of a mixed Poisson process. ■

Remark 6.1. Mecke's result is slightly more general, since it includes the case of infinite intensity of the point process. His method is, essentially, to consider the Palm measure $\alpha\{\cdot\}$ instead of the Palm probability. □

Remark 6.2. In our discussion of McFadden's proof of Theorem 6.2, we mentioned that it contains arguments which we have not quite managed to follow. The method of his proof is to consider the reduced Palm process corresponding to $N^!$, and so on. Indirectly that means that he assumes that $N^!$ is stationary, which we now know implies that N is a mixed Poisson process. It must, however, be kept in mind that Mecke's characterization came more than ten years after McFadden's, and that it is possible – as we shall see in section 6.3 – to define Palm probabilities also in the non-stationary case. In an heuristic treatment, such as McFadden's, the difference is rather delicate. □

6.3 Characterization within general point processes

A point process N is called a (possibly non-stationary) mixed Poisson process, MPP(A, U), if it is a Cox process with $\Lambda(t) = \Lambda \cdot A(t)$, where Λ is a random variable with distribution U and A a function in $\mathcal{M}$, cf. Remark 5.1. Obviously we have $E[N(t)] = E[\Lambda]A(t)$. N is simple if and only if A is continuous. Note that A and U are not uniquely determined by N, since N determines only $\Lambda(\cdot)$ uniquely. By some suitable normalization, for example $A(1) = 1$, we can obtain uniqueness. This is certainly no problem, and we point it out only to avoid confusion.

The characterizations to be given will depend on whether $A(\infty) \stackrel{\text{def}}{=} \lim_{t\to\infty} A(t)$ is finite or not. Only in the cases $A(\infty) = 0$ or ∞ we will give characterizations of the mixed Poisson process. In those cases $N(\infty) = 0$ or ∞ P-a.s. The trivial case $N(\infty) = 0$ P-a.s. occurs when either $A(\infty) = 0$ or $\Lambda = 0$ P-a.s.

Assumption 6.1. $P\{N(\infty) = 0 \text{ or } \infty\} = 1$.

Assumption 6.1 holds for all stationary point processes.

We will start with an extension of Theorems 6.1 (vi) and 6.3 (iii), independently given by Kallenberg (1973) and Matthes *et al.* (1978, pp.99, 105). We will follow the terminology of Kallenberg. For each $t < \infty$, each $\mu \in \mathcal{M}$, and each discrete random variable $\widetilde{N}$ we may construct a point process N on $[0, t]$ in the following way:

(i) Let $X_1, X_2, \ldots$ be independent random variables on $[0, t]$ with distribution function $F_X(x) = \mu(x)/\mu(t)$ for $0 \le x \le t$. The X_ks are further independent of $\widetilde{N}$.

(ii) Let $N(t) = \widetilde{N}$ and let the times of the jumps be given by $X^{(1)}, \ldots, X^{(\widetilde{N})}$.

The point process defined above is called a **mixed sample process**. In the case $\mu(x) = \mu \cdot x$ this is the same as (6.11).

Let, for each t, ${}^tN$ denote the **restriction** of N on $[0, t]$, i.e. ${}^tN(s) = N(\min(s, t))$.

Theorem 6.6 (Kallenberg, 1973 and Matthes *et al.*, 1978). *Let N be a point process fulfilling Assumption 6.1 and let $t_1, t_2, \ldots$ be real numbers such that* $\lim_{n\to\infty} t_n = \infty$. *The following statements are equivalent:*

(i) N is a mixed Poisson process.

(ii) ${}^{t_k}N$ is a mixed sample process for each k.

Remark 6.3. Note that we have made no assumption on the relation between the μ and $\widetilde{N}$ corresponding to different t_ks. However, if ${}^{t_1}N$ and ${}^{t_2}N$, for $t_1 < t_2$, are mixed sample processes with μ_1, $\widetilde{N}_1$ and μ_2, $\widetilde{N}_2$ respectively, we must have $\widetilde{N}_1 \stackrel{\text{d}}{=} D_{\mu_2(t_1)/\mu_2(t_2)}\widetilde{N}_2$ and $\mu_1(s)/\mu_1(t_1) = \mu_2(s)/\mu_2(t_1)$ for $0 \le s \le t_1$. Thus, regarding μ_2 as fixed, we can always choose $\mu_1(s) = \mu_2(s)$ for $0 \le s \le t_1$.

Theorem 6.6 is not true without Assumption 6.1. If

$$P\{N(\infty) < \infty\} = 1$$

it is only implied by (ii) that N is a mixed sample process, i.e. $N(\infty)$ may have any distribution and not necessarily a mixed Poisson distribution.

In the case of simple point processes, Theorem 6.6 can be proved in a similar way as Theorem 6.3 (iii).

Some further aspects on Theorem 6.6 are found in Feigin (1979), Puri (1982), and Deffner and Haeusler (1985). □

Let us, for a moment, go back to the stationary case. A Poisson process has stationary and independent increments, which of course implies that

$$P\left\{\bigcap_{k=1}^{n}\left\{N\left(\frac{kT}{n}\right)-N\left(\frac{(k-1)T}{n}\right)\leq n_k\right\}\right\}$$
$$=\prod_{k=1}^{n}P\left\{N\left(\frac{kT}{n}\right)-N\left(\frac{(k-1)T}{n}\right)\leq n_k\right\}$$
$$=\prod_{k=1}^{n}P\left\{N\left(\frac{T}{n}\right)\leq n_k\right\}, \tag{6.17}$$

for all $n=1, 2,\ldots,$ all $n_1,\ldots, n_n$, and all T.

A (stationary) mixed Poisson process has **exchangeable** increments, which means that

$$P\left\{\bigcap_{k=1}^{n}\left\{N\left(\frac{kT}{n}\right)-N\left(\frac{(k-1)T}{n}\right)\leq n_k\right\}\right\}$$
$$=P\left\{\bigcap_{k=1}^{n}\left\{N\left(\frac{kT}{n}\right)-N\left(\frac{(k-1)T}{n}\right)\leq n_{i_k}\right\}\right\}, \tag{6.18}$$

for all $n = 1, 2,\ldots,$ all $n_1,\ldots, n_n$, all T, and all permutations $(i_1,\ldots, i_n)$ of $(1,\ldots, n)$.

We note that (6.17) really implies 'independent and identically distributed', and that (6.18) is an extension of that definition. The notion of exchangeability goes back to de Finetti (1931).

In the non-stationary case a Poisson process only has independent increments, and it obvious what we mean with that. It is less obvious how the concept of exchangeable increments shall be modified. We are, however, led to the following definition, cf. Kallenberg (1983, p.73).

Definition 6.2. *A point process N is said to be* **symmetrically distributed with respect to** $\mu \in \mathcal{M}$ *if the distribution of*

$$(N\{A_1\},\dots, N\{A_n\})$$

for all n and all disjoint sets $A_1,\dots, A_n \in \mathcal{I}_U$, only depends on $(\mu\{A_1\},\dots, \mu\{A_n\})$.

It is enough to require, cf. Kallenberg (1983, p.73), that the random variables $N\{A_1\},\dots,N\{A_n\}$ are exchangeable whenever $A_1,\dots, A_n$ are disjoint with $\mu\{A_1\} = \dots = \mu\{A_n\}$. Thus the notion of 'symmetric distributions' really is an extension of 'exchangeable increments'.

Let now N be simple and symmetrically distributed with respect to μ. In order to characterize N it is enough to consider $P\{N\{A\} = 0\}$. It follows from Definition 6.2 that those probabilities only depend on $\mu\{A\}$. The following theorem, due to Kallenberg (1973), extends results by de Finetti (1933, 1937) and Bühlmann (1960). For further extensions and related results, we refer to Kallenberg (1983, p.176).

Theorem 6.7 (Kallenberg, 1973). *Let N be a point process fulfilling Assumption 6.1 and μ a continuous function in $\mathcal{M}$ with $\mu(\infty) = \infty$. The following statements are equivalent:*

(*i*) N *is* MPP(μ, U).

(*ii*) $P\{N\{A\} = 0\} = \varphi(\mu\{A\})$ *for some function φ and all $A \in \mathcal{I}_U$.*

Theorem 6.7 can be fully understood without the notion of 'symmetric distributions', although some of its background got lost. If (i) holds, then (ii) holds with $\varphi = \hat{u}$. Recall that $\hat{u}$ is the Laplace transform of Λ.

A discrete time simple point process S is just an infinite sequence of random variables $X_1, X_2,\dots$ assuming the values 0 and 1 only. Let

$$S_n \stackrel{\text{def}}{=} \begin{cases} 0, & n = 0, \\ \sum_{k=1}^{n} X_k, & n \geq 1, \end{cases}$$

which corresponds to $N(t)$. In fact, the Pólya urn scheme, mentioned in Chapter 1, leads to a discrete time simple point process. The natural correspondence to the Poisson process is the **Bernoulli process**, where $X_1, X_2,\dots$ are independent and $P\{X_k = 1\} = p$. The process S is called a **mixed Bernoulli process** if

$$P\{X_1 = x_1,\dots, X_n = x_n\} = \int_{0-}^{1} p^k(1-p)^{n-k}\, dU(p), \quad k = \sum_{\nu=1}^{n} x_\nu,$$

for all n, all $x_1, \dots, x_n$ assuming the values 0 and 1, and some distribution U concentrated on $[0, 1]$.

The random variables $X_1, X_2, \dots$ are exchangeable, or – which is the same thing – S has exchangeable increments, if $(X_{i_1}, \dots, X_{i_n})$ have the same distribution for all n and all permutations.

The classical characterization due to de Finetti, see Feller (1971, p.228), is the following:

S has exchangeable increments if and only if it is a mixed Bernoulli process.

There is a fundamental difference between finite and infinite sequences of exchangeable random variables, and de Finetti's characterization is not true in the finite case, cf. Feller (1971, p.230). A simple counterexample is to consider $(X_1, \dots, X_n)$ for a fixed $n \geq 2$ and to pick out one of the X randomly. The chosen X_k is put equal to one and the others equal to zero. This example is the natural discrete time variant of a sample process.

We will now consider simple point processes N with finite mean $\alpha(t)$, i.e. $\alpha(t) \stackrel{\text{def}}{=} E[N(t)] < \infty$ for all $t < \infty$.

In the non-stationary case a Palm probability is a probability obtained by conditioning upon the occurrence of a point at any time s, which means that we want to consider probabilities of the form

$$\Pi\{B \mid N\{\{s\}\} = 1\}, \quad \text{for any } B \in \mathcal{B}(\mathcal{N}) \text{ and } s \geq 0.$$

We consider

$$1_B(N)N\{ds\}, \quad \text{where } 1_B(N) = \begin{cases} 1 & \text{if } N \in B, \\ 0 & \text{if } N \notin B, \end{cases}$$

which is a slight modification, $1_B(N)$ instead of $1_B(N_s)$, of the 'B-thinned' process used in the stationary case. Put, for fixed B,

$$\alpha\{B, ds\} = E[1_B(N)N\{ds\}],$$

which, since $1_B(N) \leq 1$, is absolutely continuous with respect to $\alpha\{ds\} \stackrel{\text{def}}{=} d\alpha(s)$. By the Radon–Nikodym theorem there exists a unique α-a.e. determined function $\Pi_s\{B\} : \mathbb{R}_+ \to [0, 1]$ such that

$$\int_A \Pi_s\{B\}\alpha\{ds\} = \int_{\mathcal{N}} 1_B(\nu)\alpha\{A\}\Pi\{d\nu\}, \text{ for all } A \in \mathcal{B}(\mathbb{R}_+),$$

where $\mathcal{B}(\mathbb{R}_+)$ denotes the Borel sets on $[0, \infty)$.

The family $\{\Pi_s\{B\};\ s \geq 0,\ B \in \mathcal{B}(\mathcal{N})\}$ can be chosen so that Π_s is a probability measure on $\mathcal{B}(\mathcal{N})$ for all $s \geq 0$ and so that the function $s \to \Pi_s\{B\}$ is $\mathcal{B}(\mathbb{R}_+)$-measurable for all $B \in \mathcal{B}(\mathcal{N})\}$. We are now led to the following definition.

Definition 6.3. *Let N be a simple point process with distribution Π. The distribution Π_s, determined by the Radon–Nikodym derivative,*

$$\Pi_s\{B\} = \frac{\alpha\{B, ds\}}{\alpha\{ds\}}, \quad B \in \mathcal{B}(\mathcal{N}),\ s \geq 0,$$

is called the **Palm distribution.**

We will be satisfied with this short discussion on Palm probabilities, which is taken from Grandell (1976, pp.53–57).

As in the stationary case the point process, or – to be precise – the family of point processes, N_s^0 with distribution Π_s is called the Palm process. The reduced Palm process $N_s^!$ is defined by

$$N_s^!\{A\} = \begin{cases} N_s^0\{A\} - 1 & \text{if } s \in A, \\ N_s^0\{A\} & \text{if } s \notin A \end{cases} \quad \text{for all } A \in \mathcal{B}(\mathbb{R}_+).$$

If N is MPP(A, U), then $N_s^!$ is MPP(A, U^0) almost everywhere with respect to α, where, like in the stationary case,

$$dU^0(\ell) = \ell\, dU(\ell)/\alpha.$$

We have now the following correspondence to Theorem 6.5.

Theorem 6.8 (Kallenberg, 1973). *Let N be a simple point process with finite intensity fulfilling Assumption 6.1. The following statements are equivalent:*

(*i*) *N is a mixed Poisson process.*

(*ii*) *The distribution of $N_s^!$ is independent of s almost everywhere with respect to α.*

Like in Theorem 6.5 the assumption of finite mean is really not necessary, cf. Matthes *et al.* (1978, p.253). The method is then to use the Campbell measure, i.e. the extension of $\alpha\{d\nu, ds\}$ to a measure on $\mathcal{B}(\mathcal{N} \times \mathbb{R}_+)$, instead of the Palm probability. Theorems 6.5 and 6.8 were obtained independently, although Theorem 6.5 may be looked upon as a corollary to Theorem 6.8, cf. Matthes *et al.* (1978, p.323).

We conclude this section by considering a martingale characterization, due to Grigelionis (1996b). For that purpose we need the concept of a 'local martingale'. Before giving its definition, we need two other concepts.

Recall from Chapter 3 that a filtration $\boldsymbol{F} = (\mathcal{F}_t;\ t \geq 0)$ is a non-decreasing family of sub-σ-algebras of $\mathcal{F}$.

Definition 6.4. *A random variable* S, $\Omega \to [0,\infty]$, *is called an* $\boldsymbol{F}$-**stopping** *time if* $\{S \le t\} \in \mathcal{F}_t$ *for each* $t \ge 0$.

This means that, knowing the history up to time t, one can decide if $S \le t$ or not. Note that the outcome $S = \infty$ is allowed. If S is a stopping time, so is $t \wedge S \stackrel{\text{def}}{=} \min(t, S)$ for each t. The jump times of a simple point process N are an increasing sequence of $\boldsymbol{F}^N$-stopping times.

Definition 6.5. *An* $\boldsymbol{F}$*-martingale* $M = \{M(t);\ t \ge 0\}$ *is called* **uniformly integrable** *if*

$$\lim_{x\to\infty} \sup_{t\ge 0} \int_{\{|M(t)|>x\}} |M(t)|\, dP = 0.$$

Definition 6.6. *A real valued process* $M = \{M(t);\ t \ge 0\}$ *is called an* $\boldsymbol{F}$**-local martingale** *if there exists an increasing sequence*

$$\{S_n;\ n \ge 1\}$$

of $\boldsymbol{F}$*-stopping times such that* $P\{\lim_{n\to\infty} S_n = \infty\} = 1$ *and* $M(t \wedge S_n)$ *is a uniformly integrable* $\boldsymbol{F}$*-martingale for each* n.

The $\boldsymbol{F}$-stopping times in Definition 6.6 are often called **localizing times.**

We now extend the definition of the compensator of a simple point process N by replacing the $\boldsymbol{F}$-martingale requirement with the corresponding $\boldsymbol{F}$-local martingale requirement. Definition 3.6 then reads as follows:

Definition 6.7. *Let* N *be a simple point process and let*

$$\hat{\Lambda} = \{\hat{\Lambda}(t);\ t \ge 0\}$$

be a random process with P*-a.s. non-decreasing and continuous realizations. The process* $\hat{\Lambda}$ *is called the* $\boldsymbol{F}$**-compensator** *of* N *if* $\hat{\Lambda}(t)$ *is* $\mathcal{F}_t$*-measurable and if* $N(t) - \hat{\Lambda}(t)$ *is an* $\boldsymbol{F}$*-local martingale.*

The reason for Definition 6.7 is to be able to treat simple point processes without finite mean.

Example 6.1. Let N be MPP(U), i.e. we are back in the stationary case, and consider the filtration $\boldsymbol{F}$ with

$$\mathcal{F}_t = \sigma\{\Lambda,\ N(s);\ s \le t\}.$$

This means that Λ is known all the time, and thus N is, essentially, a Poisson process with intensity Λ, cf. the proof of Proposition 4.1. The natural candidate for the $\boldsymbol{F}$-compensator is then, cf. Theorem 3.2, $\Lambda \cdot t$. If U has finite mean α there is no problem and Definition 3.6 applies. However, $\Lambda \cdot t$ 'ought to be' regarded as the $\boldsymbol{F}$-compensator also when $\alpha = \infty$. The technical problem is that $N(t) - \Lambda \cdot t$ must fulfil Definition 3.5 (ii). The way to overcome this problem is to apply Definition 6.7 with, for instance, the localizing times $S_n = n/\Lambda$. □

Let N be a simple point process with jump times $\{T_n;\ n \geq 1\}$ and let $T_0 \stackrel{\text{def}}{=} 0$. The $\boldsymbol{F}^N$-compensator can be defined by, see for instance Liptser and Shiryayev (1978, p.245),

$$\hat{\Lambda}(0) = 0,$$
$$\hat{\Lambda}(t) = \hat{\Lambda}(T_n) + \int_0^{t-T_n} \frac{dF_n(s)}{1 - F_n(s-)}, \tag{6.19}$$

for $T_n < t \leq T_{n+1}$ and $n = 0, 1, \ldots$, where

$$F_n(s) \stackrel{\text{def}}{=} P\{T_{n+1} - T_n \leq s \mid T_0, \ldots, T_n\}, \quad n = 0, 1, \ldots \tag{6.20}$$

The $\boldsymbol{F}^N$-compensator is also after the 'local martingale' extension uniquely (up to an equivalence) determined by the point process, and, conversely, the point process is uniquely determined by its $\boldsymbol{F}^N$-compensator.

Notice that (6.19) covers cases where the compensator need not to have continuous realizations. As mentioned after Definition 3.6, this is an important restriction, although it is always fulfilled in our applications.

Let N be the trivial simple point process with one point at each integer, i.e. $N(t) = \lfloor t \rfloor$. Then

$$F_n(s) = \begin{cases} 0, & s < 1, \\ 1, & s \geq 1, \end{cases}$$

and it follows that its $\boldsymbol{F}^N$-compensator $\hat{\Lambda}$ is N itself. A slightly less trivial example is $D_p N$. Then $F_n(s) = 1 - (1-p)^{\lfloor s \rfloor}$ and it follows from (6.19) that

$$\hat{\Lambda}(t) = \hat{\Lambda}(T_n) + \sum_{k=1}^{\lfloor t - T_n \rfloor} \frac{p(1-p)^{k-1}}{(1-p)^{k-1}} = p \cdot \lfloor t \rfloor.$$

It is not possible in Definition 6.7 only to require that the $\boldsymbol{F}$-compensator $\hat{\Lambda}$ shall have non-decreasing realizations and that $\hat{\Lambda}(t)$ shall be $\mathcal{F}$-measurable, since then one may always choose the $\boldsymbol{F}$-compensator $\hat{\Lambda} = N$, which obviously makes no sense in general. In fact, the general requirement is that $\hat{A}$ is an $\boldsymbol{F}$-predictable random measure.

Let $X(t)$ be a random process defined on $(\Omega, \mathcal{F}, P)$. This means that $X(\omega, t)$ is a mapping from $\Omega \times [0, \infty)$ into $(-\infty, \infty)$ such that $X(\cdot, t)$ is $\mathcal{F}$-measurable for each fixed t. The **$\mathcal{P}(\boldsymbol{F})$-predictable σ-algebra** is the σ-algebra generated by the sets

$$\begin{aligned} &B \times \{0\}, \quad B \in \mathcal{F}_0, \\ &B \times (s, t], \quad B \in \mathcal{F}_s, \; s < t, \end{aligned}$$

or, equivalently, by all mappings $\Omega \times [0, \infty) \to (-\infty, \infty)$ which are $\mathcal{F}_t$-measurable in ω and left-continuous in t.

The process X is said to be **$\boldsymbol{F}$-predictable** if X, as a function of (ω, t), is $\mathcal{P}(\boldsymbol{F})$-measurable. If $X(t)$ is $\mathcal{F}_t$-measurable and has left-continuous realizations, then it is $\boldsymbol{F}$-predictable. Thus the requirement in Definition 6.7 that $\hat{A}$ has continuous realizations does imply $\boldsymbol{F}$-predictability.

The $\boldsymbol{F}^N$-compensator $\hat{A}$ of a simple point process N is allowed to depend only on the strict history of N, i.e. $\hat{A}(t)$ is $\mathcal{F}_{t-}^N$-measurable where $\mathcal{F}_{t-}^N \stackrel{\text{def}}{=} \sigma\{N(s);\; s < t\}$. Notice, however, that the realizations of $\hat{A}$ are right-continuous. $\hat{A}$ has continuous realizations if all conditional distributions F_n, given by (6.20), are P-a.s. continuous. A more formal way to express this is by the following theorem, see for instance Liptser and Shiryayev (1978, p.243):

The $\boldsymbol{F}^N$-compensator $\hat{A}$ of a simple point process N has continuous realizations if and only if the jump times of N are **totally inaccessible** *$\boldsymbol{F}^N$-stopping times.*

Roughly speaking this means that, although we follow the history of N, there does not exist a predictor which hits a future jump time with positive probability.

Let N be MPP(U) and let, cf. (4.6),

$$\kappa_n(s) = \begin{cases} 0, & U(0) = 1, \\ \dfrac{\int_{0-}^{\infty} \ell^{n+1} e^{-\ell s}\, dU(\ell)}{\int_{0-}^{\infty} \ell^n e^{-\ell s}\, dU(\ell)}, & U(0) < 1. \end{cases} \tag{6.21}$$

Proposition 6.1. *Let N be* MPP(U). *N has $\boldsymbol{F}^N$-compensator*

$$\hat{A}(t) = \int_0^t \kappa_{N(s)}(s)\, ds,$$

where $\kappa_n(t)$ is given by (6.21).

Proof. The proposition follows essentially from Proposition 4.1 and Definition 3.6, at least when $E[\Lambda] < \infty$. In spite of that we will give the full proof, which is due to Grigelionis (1996b).

When $U(0) = 1$ the proposition is trivially true, the constant 0 is a martingale, so we assume that $U(0) < 1$.

Denote, as usual, the marginal distributions and transition probabilities of N by

$$p_n(t) = P\{N(t) = n\}$$

and

$$p_{m,n}(s,t) = P\{N(t) = n \mid N(s) = m\}, \quad 0 \le s \le t,$$

respectively. As mentioned in section 5.3, N is in fact a strong Markov process, i.e. the arbitrary but fixed time s in Definition 3.1 may be replaced by a stopping time. Thus we have

$$F_n(s) = 1 - p_{n,n}(T_n, T_n + s),$$

and it follows from (6.20) and (4.9) that

$$\begin{aligned}\hat{\Lambda}(t) &= \hat{\Lambda}(T_n) + \int_0^{t-T_n} \frac{-\frac{d}{ds} p_{n,n}(T_n, T_n + s)}{p_{n,n}(T_n, T_n + s)}\, ds \\ &= \hat{\Lambda}(T_n) - \log(p_{n,n}(T_n, t)) \\ &= \hat{\Lambda}(T_n) - n \log\left(\frac{T_n}{t}\right) - \log\left(\frac{p_n(t)}{p_n(T_n)}\right),\end{aligned}$$

for $T_n < t < T_{n+1}$ and $n = 0, 1, \ldots$ Differentiation yields, cf. the proof of Proposition 4.4 (iv),

$$\hat{\lambda}(t) \stackrel{\text{def}}{=} \hat{\Lambda}'(t) = \frac{n}{t} - \frac{p_n'(t)}{p_n'(t)} = \kappa_n(t),$$

for $T_n < t \le T_{n+1}$ and $n = 0, 1, \ldots$ This proves the proposition. ■

Theorem 6.9 (Grigelionis, 1996b). *Let N be a simple point process, let U be the distribution of a non-negative random variable, and let $\kappa_n(t)$ be given by (6.21). The following statements are equivalent:*

(*i*) *N is* MPP(U).

(*ii*) *$N(t) - \int_0^t \kappa_{N(s)}(s)\, ds$ is an $\boldsymbol{F}^N$-local martingale.*

Proof. (i) ⇒ (ii) is just Proposition 6.1. (ii) ⇒ (i) follows since N is uniquely determined by its $\boldsymbol{F}^N$-compensator. ■

Proposition 6.1 permits us to apply various general results about point processes and compensators to mixed Poisson processes. We will here only give one such example and refer to Grigelionis (1996a, b) for several others. The case we will consider is a consequence of the following general theorem, see for instance Liptser and Shiryayev (1978, p.280).

Theorem 6.10. *Let N be a simple point process with continuous $\boldsymbol{F}$-compensator $\hat{\Lambda}$ such that*

(i) $N(t)$ *is* $\mathcal{F}_t$*-measurable for* $t \geq 0$*;*

(ii) $\hat{\Lambda}(\infty) = \infty$ P*-a.s.*

The randomly time-transformed point process

$$\widetilde{N}(t) = N(\hat{\Lambda}^{-1}(t)),$$

where $\hat{\Lambda}^{-1}$ *is the inverse of* $\hat{\Lambda}$*, is a standard Poisson process.*

Recall from Chapter 4 that the inverse $\hat{\Lambda}^{-1}$ of $\hat{\Lambda}$ is defined by

$$\hat{\Lambda}^{-1}(t) = \sup(s \mid \hat{\Lambda}(s) \leq t).$$

$\hat{\Lambda}^{-1}$ is always right-continuous. Since $\hat{\Lambda}$ is continuous, $\hat{\Lambda}^{-1}$ is (strictly) increasing and $\hat{\Lambda}(\hat{\Lambda}^{-1}(t)) = t$ for all $t \geq 0$. Thus it follows that

$$N(t) = \widetilde{N}(\hat{\Lambda}(t)), \quad t \geq 0.$$

This relation seems, however, not very useful unless $\hat{\Lambda}(t)$ is $\mathcal{F}_0$-measurable for all $t \geq 0$. In that case N is a Cox process, cf. Definition 5.1. The characterization of Cox processes as those simple point processes having $\mathcal{F}_0$-measurable $\boldsymbol{F}$-compensators is a generalization of Watanabe's theorem (Theorem 3.2) due to Grigelionis (1975, p.105).

Some discussions about Theorem 6.10 and its actuarial applications and about Grigelionis' characterization of Cox processes are to be found in Grandell (1991, 38–40).

Let N be MPP(U) for some distribution U with $U(0) = 0$. Theorem 6.10 is trivially true for $\mathcal{F}_t = \sigma\{\Lambda, N(s);\ s \leq t\}$, cf. Definition 4.3, in which case $\hat{\Lambda}(t) = \Lambda \cdot t$ is $\mathcal{F}_0$-measurable.

Theorem 6.11 (Grigelionis, 1996b). *Let N be* MPP(U) *with* $U(0) = 0$ *and let* $\hat{\Lambda}(t) = \int_0^t \kappa_{N(s)}(s)\,ds$*, where* $\kappa_n(t)$ *is given by (6.21).*

Then

$$N(t) = \widetilde{N}(\hat{\Lambda}(t)) \quad \textit{and} \quad \widetilde{N}(t) = N(\hat{\Lambda}^{-1}(t)), \quad t \geq 0,$$

where $\hat{\Lambda}^{-1}$ *is the inverse of* $\hat{\Lambda}$ *and* $\widetilde{N}$ *is a standard Poisson process.*

Proof. According to Proposition 6.1 $\hat{\Lambda}$ is the $\boldsymbol{F}^N$-compensator of N and Theorem 6.10 (i) is fulfilled by definition and it only remains to show that $\hat{\Lambda}(\infty) = \infty$ P-a.s. Recall from Proposition 13 (i) and (iii) that $\kappa_n(t)$ is non-decreasing in n and that $\int_0^\infty \kappa_0(s)\,ds = \infty$ when $U(0) = 0$. Thus it follows that

$$\hat{\Lambda}(\infty) = \int_0^\infty \kappa_{N(s)}(s)\,ds \geq \int_0^\infty \kappa_0(s)\,ds = \infty \quad P\text{-a.s.},$$

which was to be proved. ■

CHAPTER 7

Reliability properties of mixed Poisson processes

In reliability theory, and sometimes also in actuarial applications, it is useful to classify distributions according to properties of their **failure rates.** For some examples of actuarial applications we refer to Gerber (1979, p.135), Klüppelberg (1989), and Willmot and Lin (1994). Let, as usual, Λ be a non-negative random variable with distribution U. If U is continuous with density u, the failure rate $r(\ell)$ is defined by

$$r(\ell) \stackrel{\text{def}}{=} \frac{u(\ell)}{1 - U(\ell)} \quad \text{if } 1 - U(\ell) > 0. \tag{7.1}$$

For any discrete random variable ξ, its failure rate r_n is defined by

$$r_n \stackrel{\text{def}}{=} P\{\xi = n \mid \xi \geq n\} = \frac{P\{\xi = n\}}{P\{\xi \geq n\}} \quad \text{if } P\{\xi \geq n\} > 0.$$

The distributions can now be written in the form

$$1 - U(\ell) = e^{-\int_0^\ell r(y)\, dy} \tag{7.2}$$

and

$$P\{\xi \geq n\} = \prod_{k=0}^{n-1} (1 - r_k), \text{ for } n = 1, 2, \ldots$$

In the continuous case the definition of the failure rate and the representation of the distribution are standard and in the discrete case we have followed Barlow and Proschan (1965, p.18).

The simplest case is certainly when the failure rate is constant. In the continuous case $r(\ell) \equiv \beta$ implies that Λ is exponentially distributed. Similarly, $r_n \equiv p$ implies that ξ is geometrically distributed with

$$P\{\xi \geq n\} = (1-p)^n \quad \text{or} \quad P\{\xi = n\} = p(1-p)^n.$$

Let N be MPP(U). We say that N belongs to some reliability class if the random variable $N(t)$ belongs to that class for each $t > 0$.

The following proposition gives a first example of a 'reliability relation' between Λ and N.

Proposition 7.1. *Let N be* MPP(U), *where U is a continuous distribution of a non-negative random variable Λ, and let $t > 0$ be given. Λ has constant failure rate if and only if $N(t)$ has constant failure.*

Proof. Assume that Λ has constant failure rate $r(\ell) \equiv \beta$. Then it follows from Example 2.1, for $\gamma = 1$, that $N(t)$ is geometrically distributed with $p = \frac{\beta}{\beta+t}$. Conversely, if $N(t)$ has constant failure rate p, it follows from Theorem 2.1 (i) that $N(t)$ is MP(t, U), where U is an exponential distribution with $r(\ell) \equiv \frac{pt}{1-p}$. Thus the proposition follows. ■

We will now consider more general reliability classes. As we did when we discussed infinite divisibility, we will distinguish between the general definition and its discrete counterpart. These classes are related to the 'residual life time', i.e. to the conditional **survivor function**

$$\overline{U}(\ell \mid x) \stackrel{\text{def}}{=} P\{\Lambda - x > \ell \mid \Lambda > x\} = \frac{\overline{U}(x+\ell)}{\overline{U}(x)}, \tag{7.3}$$

where

$$\overline{U}(\ell) = 1 - U(\ell). \tag{7.4}$$

Any statement about $\overline{U}(\ell \mid x)$ will be understood to hold for all x and $\ell \geq 0$ such that $\overline{U}(x) > 0$. Further, for N, we define

$$\overline{P}_n(t) \stackrel{\text{def}}{=} P\{N(t) \geq n\}$$

and

$$r_n(t) \stackrel{\text{def}}{=} P\{N(t) = n \mid N(t) \geq n\} = \frac{p_n(t)}{\overline{P}_n(t)}.$$

In actuarial applications, the case of monotone failure rate is important, so we will treat this case in some detail. In order to simplify the reading, we let in the rest of this chapter the term 'increasing' mean 'non-decreasing' and 'decreasing' mean 'non-increasing'.

We will now give a definition of 'monotone failure rate' based on the survivor function $\overline{U}$, which does not require the existence of a density. The definition does therefore include the case where the failure rate does not exist. If, however, U has a density, then Definition 7.1 is equivalent to $r(\ell)$ being increasing (decreasing), cf. Barlow and Proschan (1975, p.54).

Definition 7.1. *We say that Λ or U has* **increasing failure rate (decreasing failure rate)**, *or is IFR (DFR), if*

$$\overline{U}(\ell \mid x) \textit{ is decreasing (increasing) in } x \textit{ for each } \ell \geq 0.$$

We say that $N(t)$ is (discrete) IFR (DFR) if

$$\frac{\overline{P}_{n+1}(t)}{\overline{P}_n(t)} \textit{ is decreasing (increasing) in } n = 0, 1, \ldots$$

Let Λ_1 and Λ_2 have distributions U_1 and U_2 respectively. Λ_1 is said to be **stochastically larger (smaller)** than Λ_2 if $\overline{U}_1(\ell) \geq (\leq)$ $\overline{U}_2(\ell)$. This implies, among other things, that $E[\Lambda_1] \geq (\leq)$ $E[\Lambda_2]$. Thus to say that Λ is IFR (DFR) means that the residual life time is stochastically decreasing (increasing) in the survived time.

Notice that

$$P\{N(t) \geq k + n \mid N(t) \geq n\} = \frac{\overline{P}_{n+k}(t)}{\overline{P}_n(t)}$$

$$= \frac{\overline{P}_{n+1}(t)}{\overline{P}_n(t)} \cdot \frac{P_{n+2}(t)}{\overline{P}_{n+1}(t)} \cdots \frac{\overline{P}_{n+k}(t)}{\overline{P}_{n+k-1}(t)},$$

i.e. the given definition is the natural discrete correspondence. In the discrete case the failure rate always exists and, since $\overline{P}_{n+1}(t) = \overline{P}_n(t) \cdot (1 - r_n(t))$, an equivalent definition is that $r_n(t)$ is increasing (decreasing) in n.

The following proposition is due to Vinogradov (1973).

Proposition 7.2. *Let N be* MPP(U), *where U is the distribution of a non-negative random variable Λ. Λ is IFR (DFR) if and only if N is IFR (DFR).*

Proof. The proof of 'Λ is IFR (DFR) $\Rightarrow$ N is IFR (DFR)' is taken from Vinogradov (1973) while the proof of 'N is IFR (DFR) $\Rightarrow$ Λ is IFR (DFR)' is essentially due to Block and Savits (1980).

Assume that Λ is IFR (DFR). Consider $\bar{u}_t(\ell) \stackrel{\text{def}}{=} e^{-\ell t}\overline{U}(\ell)$ and the 'normalized moments'

$$a_n(t) = \int_0^\infty \frac{\ell^n}{n!}\bar{u}_t(\ell)\,d\ell \quad \text{for } n = 0, 1, \ldots$$

and

$$a_{-1}(t) = \bar{u}_t(0) = 1 - U(0).$$

Notice that

$$\frac{\bar{u}_t(x+\ell)}{\bar{u}_t(x)} = e^{-\ell t}\overline{U}(\ell \mid x).$$

It follows from Karlin *et al.* (1961 pp.1025, 1026) and Barlow *et al.* (1963, p.383) that if U is IFR (DFR), then

$$\frac{a_{n+1}(t)}{a_n(t)} \text{ is decreasing (increasing) in } n = -1, 0, 1, \ldots$$

(In Karlin *et al.* (1961) it is proved that the above result holds in the IFR case, but the proofs holds also for U being DFR. We refer to Barlow and Proschan (1975, p.76) for the relation between IFR and total positivity.)

It follows from Proposition 2.1 (iii) that

$$\overline{P}_{n+1}(t) = t^{n+1}a_n(t) \quad \text{for } n = 0, 1, \ldots$$

and

$$\overline{P}_0(t) = 1 = \frac{a_{-1}(t)}{1 - U(0)}$$

or

$$\frac{\overline{P}_{n+1}(t)}{\overline{P}_n(t)} = t\frac{a_n(t)}{a_{n-1}(t)} \quad \text{for } n = 1, 2, \ldots$$

and

$$\frac{\overline{P}_1(t)}{\overline{P}_0(t)} = t\overline{U}(0)\frac{a_0(t)}{a_{-1}(t)}.$$

If $\overline{U}(0) = 1$ the desired result follows. $\overline{U}(0) < 1$ is no problem in the DFR case and impossible in the IFR case, cf. Barlow and Proschan (1975, p.77).

Assume that N is IFR (DFR). Choose x, y and ℓ such that $x < y$ and that x, y, $x + \ell$ and $y + \ell$ are points of continuity of U. Let $\lceil x \rceil$

be the smallest integer larger or equal to x, i.e.

$$\lceil x \rceil = -\lfloor -x \rfloor = \begin{cases} x, & \text{if } x \text{ is an integer,} \\ \lfloor x \rfloor + 1, & \text{if } x \text{ is not an integer.} \end{cases}$$

It follows from Proposition 4.2 that

$$P\{\tfrac{1}{t}N(t) \geq x\} = P\{N(t) \geq tx\}$$

$$= P\{N(t) \geq \lceil tx \rceil + 1\} = \overline{P}_{\lceil tx \rceil + 1}(t) \to U(x)$$

as $t \to \infty$. A minor technical complication is that $\lceil t(x+\ell) \rceil - \lceil t\ell \rceil$ may depend on ℓ. It is easy to see that

$$\lceil t(x+\ell) \rceil \leq \lceil tx \rceil + \lceil t\ell \rceil \leq \lceil t(x+\ell) \rceil + 1,$$

and that also

$$\overline{P}_{\lceil tx \rceil + 1}(t) \to U(x) \text{ as } t \to \infty.$$

Thus

$$0 \leq (\geq) \frac{\overline{P}_{\lceil tx \rceil + \lceil t\ell \rceil}(t)}{\overline{P}_{\lceil tx \rceil}(t)} - \frac{\overline{P}_{\lceil ty \rceil + \lceil t\ell \rceil}(t)}{\overline{P}_{\lceil ty \rceil}(t)}$$

$$\to \overline{U}(\ell \mid x) - \overline{U}(\ell \mid y).$$

Since the points of continuity of U form a dense set, it follows that Λ is IFR (DFR). ■

Remark 7.1. Proposition 7.1 is actually a result about mixed Poisson distributions. It is natural to ask if, in fact, Proposition 7.2 also holds for distributions. We will show that this is not the case, by constructing a mixed Poisson process, where $N(t)$ is IFR only for some values of t.

Let $\widetilde{N}$ be a Poisson process with intensity one, cf. Definition 3.4, and denote its failure rate by $\tilde{r}_n(t)$. It is known, see Barlow and Proschan (1965, p.18) or Example 7.1, that $\widetilde{N}$ is IFR. Let N be MPP(U) where U is a two-point distribution with

$$u_0 = P\{\Lambda = 0\}, \quad u_1 = P\{\Lambda = 1\} \quad \text{and} \quad u_0 + u_1 = 1,$$

i.e. N has, with probability u_0, no points and is, with probability u_1, a Poisson process. Then we have

$$r_0(t) = P\{N(t) = 0\} = u_0 + u_1 e^{-t}$$
$$r_n(t) = \tilde{r}_n(t) \text{ for } n = 1, 2, \dots$$

Thus $N(t)$ is IFR if and only if $r_1(t)/r_0(t) \geq 1$. We have

$$f(t) \stackrel{\text{def}}{=} \frac{r_1(t)}{r_0(t)} = \frac{te^{-t}}{(1 - e^{-t})(u_0 + u_1 e^{-t})}.$$

Obviously $\lim_{t\to\infty} f(t) = 0$, so $N(t)$ is not IFR for t sufficiently large. This is quite natural, since then $N(t)$ is – with high probability – either zero or large.

For small values of t, routine calculations yield

$$f(t) = 1 + (\tfrac{1}{2} - u_0) + o(t).$$

Thus, for $u_0 < 1/2$, $N(t)$ is IFR for t sufficiently small. □

Proposition 7.2 has been extended to further reliability classes by Vinogradov (1976) in the NBU case and by Block and Savits (1980) in the other cases defined below. An overview of those classes, including their discrete counterparts, is to be found in Esary *et al.* (1973). Therefore we will only discuss those classes shortly, with some emphasis on the discrete case.

Decreasing (increasing) mean residual life.

Λ is DMRL (IMRL) if $E[\Lambda - x \mid \Lambda > x] = \int_0^\infty \overline{U}(x + \ell)/\overline{U}(x)\, d\ell$ is decreasing (increasing) in x. Similarly we have

$$E[N(t) - n \mid N(t) \geq n] = \sum_{k=n}^{\infty} \frac{(k-n)\, p_k(t)}{\overline{P}_n(t)}$$

$$= \sum_{k=n+1}^{\infty} \frac{\overline{P}_k(t)}{\overline{P}_n(t)} = \sum_{k=n}^{\infty} \frac{\overline{P}_k(t)}{\overline{P}_n(t)} - 1.$$

We say that $N(t)$ is DMRL (IMRL) if $\sum_{k=n}^{\infty} \overline{P}_k(t)/\overline{P}_n(t)$ is decreasing (increasing) in $n = 0, 1, \dots$ □

Increasing (decreasing) failure rate average.

Λ is IFRA (DFRA) if $\overline{U}(\ell)^{1/\ell}$ is decreasing (increasing) in $\ell > 0$. This definition may seem artificial, but it is important for reliability systems. If U has a density, and thus a failure rate, we have

$$\overline{U}(\ell)^{1/\ell} = e^{-\frac{1}{\ell}\int_0^\ell r(y)\,dy},$$

and thus Λ is IFRA (DFRA) if $\frac{1}{\ell}\int_0^\ell r(y)\,dy$ is increasing (decreasing).

In the discrete case $N(t)$ is said to be IFRA (DFRA) if $\overline{P}_n(t)^{1/n}$ is decreasing (increasing) in $n = 1, 2, \ldots$. It seems as this definition is merely made by 'analogy' than by considerations in terms of the failure rate. □

New better (worse) than used.

Λ is NBU (NWU) if $\overline{U}(\ell) \geq (\leq)\ \overline{U}(\ell \mid x)$, or if $\overline{U}(x)\overline{U}(\ell) \geq (\leq)\ \overline{U}(x+\ell)$, for $x,\ \ell \geq 0$.

$N(t)$ is NBU (NWU) if $\overline{P}_n(t)\overline{P}_k(t) \geq (\leq)\ \overline{P}_{n+k}(t)$, for $k,\ n = 0, 1, \ldots$ □

New better (worse) than used in expectation.

Λ is NBUE (NWUE) if $E[\Lambda] \geq (\leq)\ E[\Lambda - x \mid \Lambda > x]$ for $x \geq 0$.

$N(t)$ is NBUE (NWUE) if $\overline{P}_n(t)\sum_{k=0}^{\infty}\overline{P}_k(t) \geq (\leq)\ \sum_{k=n}^{\infty}\overline{P}_k(t)$ for $n = 0, 1, \ldots$, cf. the DMRL (IMRL) case. □

The fact that the reliability classes are defined in terms of $\overline{U}$ and $\overline{P}$ may, in connection with mixed Poisson processes, be regarded as a drawback. Often these quantities are difficult to write in a nice form. When U is absolutely continuous it is generally more simple to formulate conditions in terms of its density u. This is maybe even more accentuated when N is considered. We will therefore consider classification in terms of u and $p_n(t)$, related to IFR (DFR).

Let U have density u, and put

$$I = (\inf\{y \mid u(y) > 0\}, \sup\{y \mid u(y) > 0\}).$$

I may be looked upon as the 'extended' support of U. Assume that, for each $\ell \geq 0$,

$$\frac{u(x+\ell)}{u(x)}$$

is decreasing (increasing and with $u(\ell) > 0$ for $\ell > 0$) in x for $x \in I$. This is equivalent to, cf. Barlow and Proschan (1975, p.76), $\log u(\ell)$ being concave (convex for $\ell > 0$). Notice that either concavity or convexity implies that u is continuous and positive on I. Further convexity implies that $I = (0, \infty)$.

Since, for $x \in I$,

$$r(x) = \frac{u(x)}{\int_x^\infty u(\ell)\, d\ell} = \left(\int_0^\infty \frac{u(x+\ell)}{u(x)}\, d\ell \right)^{-1},$$

it follows that u being logarithmically concave (convex) implies that U is IFR (DFR).

Similarly we consider the case where

$$\frac{p_{n+1}(t)}{p_n(t)} \text{ is decreasing (increasing) in } n = 0, 1, \ldots$$

Obviously this is equivalent to

$$\frac{p_n(t)}{p_{n-1}(t)} \geq (\leq) \frac{p_{n+1}(t)}{p_n(t)} \quad \Leftrightarrow \quad p_n^2(t) \geq (\leq)\, p_{n-1}(t)\, p_{n+1}(t)$$

$$\Leftrightarrow \quad \log p_n(t) \geq (\leq) \frac{\log p_{n-1}(t) + \log p_{n+1}(t)}{2},$$

for $n = 1, 2, \ldots$, i.e. with $\{p_n(t)\}$ being logarithmically concave (convex).

Also in the discrete case $\{p_n(t)\}$ being logarithmically concave (convex) implies that $N(t)$ is IFR (DFR), since

$$r_n(t) = 1 - \frac{\overline{P}_{n+1}(t)}{\overline{P}_n(t)} = \frac{p_n(t)}{\overline{P}_n(t)} = \left(\sum_{k=0}^{\infty} \frac{p_{n+k}(t)}{p_n(t)} \right)^{-1}.$$

The following proposition, as pointed out by Vinogradov (1976), is almost contained is Proposition 7.2.

Proposition 7.3. *Let N be* MPP(U), *where U is absolutely continuous with a density u being continuous on I. u is logarithmically concave (convex) if and only if $\{p_n(t)\}$ is logarithmically concave (convex) for all $t > 0$.*

Proof. The proof of

$$\begin{gathered} u \text{ being logarithmically concave (convex)} \\ \Downarrow \\ \{p_n(t)\} \text{ being logarithmically concave (convex)} \end{gathered}$$

is as the first part of the proof of Proposition 7.2 with $\bar{u}_t(\ell)$ replaced by $u_t(\ell) \stackrel{\text{def}}{=} e^{-\ell t}u(\ell)$. Notice that $a_{-1}(t)$ does not need to be taken into account in this case.

Assume that $\{p_n(t)\}$ is logarithmically concave (convex). We will now follow Vinogradov (1973) more closely than we did in the proof of Proposition 7.2.

Recall that

$$p_n(t) = (-1)^n \frac{t^n}{n!} \hat{u}^{(n)}(t).$$

We have, by the inversion formula for Laplace transforms,

$$u(\ell) = \lim_{n\to\infty} \left(\frac{n}{\ell}\right)^{n+1} \frac{(-1)^n}{n!} \hat{u}^{(n)}\left(\frac{n}{\ell}\right) = \lim_{n\to\infty} \left(\frac{n}{\ell}\right) p_n\left(\frac{n}{\ell}\right),$$

and thus $tp_{\lceil t\ell \rceil}(t) \to u(\ell)$ as $t \to \infty$, for all continuity points of u. (The only possible points of discontinuity are the endpoints of I, i.e. $\inf\{y \mid u(y) > 0\}$ and $\sup\{y \mid u(y) > 0\}$). Continuing as in the proof of Proposition 7.2, the desired result follows. ■

Proposition 7.3 has been extended to related classes by Vinogradov (1976) and Block and Savits (1980).

Example 7.1. Let N be a Poisson process with intensity α. Then, see Definitions 2.2 and 3.4,

$$p_n(t) = \frac{(\alpha t)^n}{n!} e^{-\alpha t},$$

and thus

$$\frac{p_{n+1}(t)}{p_n(t)} = \alpha \frac{n!}{(n+1)!} = \frac{\alpha}{n+1}, \quad \text{for } n = 0, 1, \ldots$$

Thus $\{p_n(t)\}$ is logarithmically concave and therefore, as mentioned in Remark 7.1, N is IFR. Furthermore, this example shows that the absolute continuity of U must be assumed in Proposition 7.3. □

Example 7.2. Let Λ be $\Gamma(\gamma, \beta)$, see Example 2.1. Then

$$\log u(\ell) = \text{const.} + (\gamma - 1)\log(\ell) - \beta\ell,$$

which implies that u is logarithmically concave (convex) for $\gamma \geq (\leq)\ 1$.

Thus the same holds for the Pólya process. Naturally this is easily shown directly, since

$$\frac{p_{n+1}(t)}{p_n(t)} = \frac{n+\gamma}{n+1}\frac{t}{\beta+t}, \quad \text{for } n = 0, 1, \ldots$$

□

Example 7.3. We will now consider the Delaporte distribution discussed in Example 2.2. Recall that this means that Λ is shifted Γ-distributed. Let U_α denote its distribution, and notice that $U_\alpha(\ell) = U_0(\ell - \alpha)$, where U_0 is the $\Gamma(\gamma, \beta)$-distribution. The corresponding mixed Poisson process N is then the sum of a Poisson process with intensity α and an independent Pólya process.

For $\gamma \geq 1$ it follows from Examples 7.1 and 7.2 that both the Poisson and the Pólya processes are IFR. Since the class IFR is closed under addition, see Barlow and Proschan (1975, p.104), it follows that also N is IFR. Thus also U_α is IFR. Further u_0 is logarithmically concave, which implies that also u_α is logarithmically concave, since the fact that $\log u_\alpha(\ell) = -\infty$ for $\ell < \alpha$ does not 'destroy' the concavity.

For $\gamma < 1$ the situation is quite different. Although u_0 is logarithmically convex, this does not hold for any u_α with $\alpha > 0$. This is not just a technicality, due to the requirement of positive density. In fact, since

$$E[\Lambda - \alpha \mid \Lambda > \alpha] = E[\Lambda] - \alpha < E[\Lambda],$$

U_α is not NWUE, which is the largest class related to DFR. □

We will end this chapter by considering unimodality, although it may be regarded as doubtful whether that really ought to be called a reliability property. However, logarithmic concavity implies unimodality, see Barlow and Proschan (1965, p.230).

Definition 7.2. *A random variable Λ with distribution U is called* **unimodal** *with mode λ_0 if $U(\ell)$ is convex in $[0, \lambda_0)$ and concave in (λ_0, ∞).*

According to this definition the mode is not necessarily unique, and the possible modal values may form an interval. The mode may be a point of discontinuity of U, but apart from this unimodality requires that U has a density u which is increasing in $[0, \lambda_0)$ and decreasing in (λ_0, ∞), cf. Feller (1971, p.158).

A discrete distribution with at least two points of increase is thus never unimodal. In spite of that, unimodality has its natural counterpart in the discrete case.

Definition 7.3. *A discrete random variable ξ is called (discretely) unimodal with mode n_0 if $P\{\xi = n\}$ is increasing in $\{0, 1, \dots, n_0\}$ and decreasing in $\{n_0, n_0 + 1, \dots\}$.*

The following proposition is due to Holgate (1970).

Proposition 7.4. *Let U be the distribution of a non-negative unimodal random variable Λ and N be* MPP(U). *Then N is unimodal.*

Proof. Since Λ is unimodal the same holds for Λt, and it is therefore enough to show that that $N(1)$ is unimodal. Put $p_n = p_n(1)$.

Let u_0 be the possible discrete mass at λ_0 and $u(\cdot)$ the density of U. Integration by parts yields

$$p_{n+1} - u_0 \frac{\lambda_0^{n+1}}{(n+1)!} e^{-\lambda_0} = \int_0^\infty \frac{\ell^{n+1}}{(n+1)!} e^{-\ell} u(\ell)\, d\ell$$

$$= \int_0^\infty e^{-\ell} \cdot \frac{\ell^{n+1}}{(n+1)!} u(\ell)\, d\ell$$

$$= \left[-e^{-\ell} \frac{\ell^{n+1}}{(n+1)!} u(\ell) \right]_0^\infty + \int_0^\infty e^{-\ell} \left(\frac{\ell^n}{n!} u(\ell)\, d\ell + \frac{\ell^{n+1}}{(n+1)!}\, du(\ell) \right)$$

$$= 0 + \left(p_n - u_0 \frac{\lambda_0^n}{n!} e^{-\lambda_0} \right) + \int_0^\infty \frac{\ell^{n+1}}{(n+1)!} e^{-\ell}\, du(\ell),$$

cf. Proposition 2.1 (iii).

Letting $\Delta p_n = p_{n+1} - p_n$, we get

$$\Delta p_n = u_0 \frac{\lambda_0^n}{n!} e^{-\lambda_0} \left(\frac{\lambda_0}{n+1} - 1 \right) + \int_0^\infty \frac{\ell^{n+1}}{(n+1)!} e^{-\ell}\, du(\ell),$$

and thus, for $n \geq 1$,

$$
\begin{aligned}
(n+1)\Delta p_n &= u_0 \frac{\lambda_0^n}{n!} e^{-\lambda_0} (\lambda_0 - (n+1)) \\
&\quad + \int_0^{\lambda_0} \ell \cdot \frac{\ell^n}{n!} e^{-\ell} \, du(\ell) - \int_{\lambda_0}^{\infty} \ell \cdot \frac{\ell^n}{n!} e^{-\ell} \, |du(\ell)| \\
&\leq u_0 \frac{\lambda_0^n}{n!} e^{-\lambda_0} (\lambda_0 - (n+1)) + \lambda_0 \int_0^{\infty} \frac{\ell^n}{n!} e^{-\ell} \, du(\ell) \\
&= u_0 \frac{\lambda_0^n}{n!} e^{-\lambda_0} (\lambda_0 - (n+1)) \\
&\quad + \lambda_0 \left(\Delta p_{n-1} - u_0 \frac{\lambda_0^{n-1}}{(n-1)!} e^{-\lambda_0} \left(\frac{\lambda_0}{n} - 1 \right) \right) \\
&= - u_0 \frac{\lambda_0^n}{n!} e^{-\lambda_0} + \lambda_0 \Delta p_{n-1} \leq \lambda_0 \Delta p_{n-1}. \qquad (7.5)
\end{aligned}
$$

If $\lambda_0 = 0$ or if $\Delta p_0 \leq 0$ it follows from (7.5) that $N(1)$ is unimodal with mode $n_0 = 0$. For $\lambda_0 = 0$ we have $\Delta p_0 = -u_0 - \int_0^{\infty} \ell e^{-\ell} \, |du(\ell)| < 0$, so the mode is furthermore unique.

Let now $\lambda_0 > 0$ and $\Delta p_0 > 0$. Since $\sum_{n=0}^{\infty} p_n = 1$ there exists an $n_0 < \infty$ such that $\Delta p_n > 0$ for $n \leq n_0 - 1$ and $\Delta p_{n_0} \leq 0$. It follows from (7.5) that $\Delta p_n \leq 0$ for all $n \geq n_0$, i.e. $N(1)$ is unimodal with mode n_0. ■

Also in the discrete case logarithmic concavity implies unimodality. This follows by the same arguments as in the last lines of the proof of Proposition 7.4 since

$$
\frac{p_{n+1}}{p_n} \leq 1 \quad \Leftrightarrow \quad \Delta p_n \leq 0.
$$

CHAPTER 8

Compound mixed Poisson distributions

8.1 Compound distributions

We will in this section consider general compound distributions, i.e. distributions where the 'compounding' variable N is not necessarily mixed Poisson distributed. More precisely, a compound distribution is based on the following independent objects:

(i) a discrete random variable N having distribution

$$p_k = P\{N = k\}$$

with generating function $G(s) = E[s^N]$;

(ii) a sequence $\{Z_k\}_1^\infty$ of non-negative, independent and identically distributed random variables having the common distribution function F with mean value μ, variance σ^2, and Laplace transform

$$\hat{f}(v) = \int_{0-}^{\infty} e^{-vz}\, dF(z).$$

The random variable

$$Y = \sum_{k=1}^{N} Z_k, \quad \left(\sum_{k=1}^{0} Z_k \stackrel{\text{def}}{=} 0 \right),$$

is said to be **compound distributed** or to be a **random sum**.

In actuarial applications it is natural to interpret N as the number of claims in some time interval and $\{Z_k\}$ as the costs of the claims. The total amount of claims paid by the company in that interval is then described by Y.

In the special case where N is Poisson distributed, Y is said to be compound Poisson distributed. The compound Poisson distribution with discrete Z_ks was considered in Chapter 1 in connection with infinite divisibility.

Let $F_Y(y) = P\{Y \le y\}$. Then we have

$$F_Y(y) = \sum_{k=0}^{\infty} p_k F^{k*}(y), \tag{8.1}$$

where F^{k*} denotes the kth convolution of F with itself and

$$F^{0*}(y) = \begin{cases} 0 & \text{if } y < 0, \\ 1 & \text{if } y \ge 0. \end{cases}$$

Expressed in Laplace transforms, (8.1) is equivalent to

$$\hat{f}_Y(v) \stackrel{\text{def}}{=} \int_{0-}^{\infty} e^{-vz}\, dF_Y(z) = \sum_{k=0}^{\infty} p_k \hat{f}^k(v) = G(\hat{f}(v)). \tag{8.2}$$

We have

$$E[Y] = E[E[Y \mid N]] = \mu E[N] \tag{8.3}$$

and

$$\begin{aligned} \operatorname{Var}[Y] &= E[\operatorname{Var}[Y \mid N]] + \operatorname{Var}[E[Y \mid N]] \\ &= \sigma^2 E[N] + \mu^2 \operatorname{Var}[N]. \end{aligned} \tag{8.4}$$

Let $\hat{h}(r) \stackrel{\text{def}}{=} \hat{f}(-r)$. Thus $\hat{h}(r)$ is the **moment generating function** of Z_k.

Definition 8.1. *F is called* **light-tailed**, *if there exists $r_\infty > 0$ such that $\hat{h}(r) \uparrow +\infty$ when $r \uparrow r_\infty$ (we allow for the possibility $r_\infty = +\infty$).*

The value r_∞ is called the **abscissa of convergence** of $\hat{h}(r)$.

Definition 8.1, and its implications, will be more closely discussed in Chapter 9, where risk models will be discussed in some detail. Here we only notice that Definition 8.1 implies, when $r_\infty < \infty$, that

$$\limsup_{z \to \infty} e^{(r_\infty + \varepsilon)z} \overline{F}(z) = \infty, \quad \text{for each } \varepsilon > 0, \tag{8.5}$$

where $\bar{F}(z) \stackrel{\text{def}}{=} 1 - F(z)$. Integration by parts leads to, cf. Feller (1971, p.435),

$$\bar{f}(v) \stackrel{\text{def}}{=} \int_0^\infty e^{-vz}\bar{F}(z)\,dz = \frac{1-\hat{f}(v)}{v}. \tag{8.6}$$

and

$$\bar{h}(r) \stackrel{\text{def}}{=} \int_0^\infty e^{rz}\bar{F}(z)\,dz = \frac{\hat{h}(r)-1}{r}. \tag{8.7}$$

Thus $\bar{h}(r)$ and $\hat{h}(r)$ have the same abscissa of convergence. If (8.5) did not hold, we would have

$$\bar{h}(r_\infty + \varepsilon/2) \le \sup_{z\ge 0} e^{(r_\infty+\varepsilon)z}\bar{F}(z)\cdot \int_0^\infty e^{-(\varepsilon/2)v}\,dv < \infty,$$

which contradicts the fact that r_∞ is the abscissa of convergence of $\hat{h}(r)$. Therefore (8.5) holds.

The assumption '$\hat{h}(r) \uparrow +\infty$ when $r \uparrow r_\infty$' in Definition 8.1 is convenient, since many of the results to be given will depend on a constant $\phi < 1$ and the solution κ of the equation $\hat{h}(\kappa) = \phi^{-1}$. By Definition 8.1 such a solution is guaranteed. The case $\hat{h}(r_\infty) < \infty$ and $\hat{h}(r) = \infty$ for $r > r_\infty$ is thus excluded. The example $F'(z) = \text{const.}\, z^{-2}e^{-z}$ for $z > 1$ shows that such cases do exist. In Grandell (1991, p.3) those cases were referred to as 'rather pathological'. We will, however, see that those cases – pathological or not – are not at all without interest. Notice that also the completely uninteresting case $F(0) = 1$, where $\hat{h}(r) \equiv 1$, is excluded although no case hardly may regarded as 'more light-tailed'.

Similarly N is called light-tailed if $G(s)$ has **radius of convergence** $s_\infty > 1$, i.e. $G(s) < \infty$ for $|s| < s_\infty$ and, when $s_\infty < \infty$, $G(s) = \infty$ for $s > s_\infty$.

Let us first consider – somewhat primitive – exponential bounds, based on Chebyshev arguments. For any $\kappa \in [0, r_\infty)$ we have

$$\hat{h}(\kappa) = \int_{0-}^\infty e^{\kappa x}\,dF(x) \ge \int_z^\infty e^{\kappa x}\,dF(x) \ge e^{\kappa z}\int_z^\infty dF(x) = e^{\kappa z}\bar{F}(z),$$

or

$$\bar{F}(z) \le \hat{h}(\kappa)e^{-\kappa z}. \tag{8.8}$$

Naturally we want to choose κ as large as possible, but in general $\kappa = r_\infty$ is too large, since then $\hat{h}(\kappa) = \infty$. One possibility, when

$r_\infty < \infty$, is to say that (8.8) holds for $\kappa = r_\infty - \varepsilon$, where $\varepsilon \in (0, r_\infty]$ is arbitrary.

Exactly the same argument can now be applied to $\overline{F}_Y(y) \stackrel{\text{def}}{=} 1 - F_Y(y)$. Then, for all non-negative κ such that $\hat{h}(\kappa) < s_\infty$, we get

$$\overline{F}_Y(y) \leq G(\hat{h}(\kappa))e^{-\kappa y}. \tag{8.9}$$

Example 8.1. We will consider the simplest possible example. Let $Z_k = 1$, i.e. $\hat{h}(r) = e^r$, which implies that $Y = N$ and that $r_\infty = \infty$, and let N be geometrically distributed with $p_k = p(1-p)^k$. Thus we have

$$\overline{F}_Y(n) = P\{N > n\} = (1-p)^{n+1}. \tag{8.10}$$

Now we consider (8.9). We have

$$G(s) = \sum_{k=0}^{\infty} p(1-p)^k s^k = \frac{p}{1-(1-p)s},$$

and thus $s_\infty = 1/(1-p)$. Consider κ as the solution of

$$e^\kappa = \frac{1}{1-p}.$$

Thus this κ is the 'right' exponent, but it cannot be used in (8.9) since $G(e^\kappa) = \infty$. □

There may be cases where we have some estimate of s_∞. Let us assume that we have some restrictions on $\{p_k\}$, which ascertain that $s_\infty \geq \phi^{-1}$ where $\phi < 1$. If $s_\infty > \phi^{-1} > 1$ we can choose κ as the solution of

$$\hat{h}(\kappa) = \phi^{-1}. \tag{8.11}$$

Then it follows from (8.9) that

$$\overline{F}_Y(y) \leq G(\phi^{-1})e^{-\kappa y}. \tag{8.12}$$

However, if ϕ^{-1} is a good estimate of s_∞, i.e. if ϕ^{-1} is close to s_∞, then $G(\phi^{-1})$ ought in general to be large. If ϕ^{-1} is a bad estimate of s_∞, then κ is probably too small, which for large values of y is even worse. Thus it seems as if (8.12) can hardly be good in other than exceptional cases.

Put $\bar{P}_k = P\{N \geq k\}$. Then we have, cf. (8.1),

$$\bar{F}_Y(y) = \sum_{k=0}^{\infty}(\bar{P}_k - \bar{P}_{k+1})\bar{F}^{k*}(y)$$

$$= \sum_{k=0}^{\infty} \bar{P}_{k+1}\big(F^{k*}(y) - F^{(k+1)*}(y)\big), \tag{8.13}$$

where $\bar{F}^{k*}(y) \stackrel{\text{def}}{=} 1 - F^{k*}(y)$. The following theorem, and its proof, is due to Willmot and Lin (1994, pp.745–746). An alternative proof, based on martingale methods, is given by Gerber (1994). Extensions of Theorem 8.1, yielding non-exponential as well as lower bounds, are found in Willmot (1994), Willmot and Lin (1997) and Lin (1996).

Theorem 8.1. *Suppose that there exists a $\phi \in (0,1)$ such that*

$$\bar{P}_{n+1} \leq \phi\bar{P}_n, \quad \textit{for } n = 1, 2, \ldots \tag{8.14}$$

Then

$$\bar{F}_Y(y) \leq \frac{1 - p_0}{\phi}\, e^{-\kappa y},$$

where κ is the solution of $\hat{h}(\kappa) = \phi^{-1}$.

Proof. Put

$$\bar{F}_Y^{(n)}(y) = \sum_{k=0}^{n} \bar{P}_{k+1}(F^{k*}(y) - F^{(k+1)*}(y)).$$

Since $F^{k*}(y) - F^{(k+1)*}(y) \geq 0$ it follows that

$$\bar{F}_Y^{(n)}(y) \to \bar{F}_Y(y) \text{ as } n \to \infty.$$

It follows from (8.8) that

$$\bar{F}_Y^{(0)}(y) = \bar{P}_1\bar{F}(y) \leq \bar{P}_1\hat{h}(\kappa)e^{-\kappa y} = \frac{1 - p_0}{\phi}\, e^{-\kappa y}.$$

Assume now that $\bar{F}_Y^{(n)}(y) \leq (1 - p_0)\phi^{-1}e^{-\kappa y}$.

We have

$$
\begin{aligned}
\overline{F}_Y^{(n+1)}(y) &= \sum_{k=0}^{n+1} \overline{P}_{k+1}(F^{k*}(y) - F^{(k+1)*}(y)) \\
&= \overline{P}_1\overline{F}(y) + \sum_{k=1}^{n+1} \overline{P}_{k+1}(F^{k*}(y) - F^{(k+1)*}(y)) \\
&\leq \overline{P}_1\overline{F}(y) + \phi\sum_{k=1}^{n+1} \overline{P}_{k}(F^{k*}(y) - F^{(k+1)*}(y)) \\
&= \overline{P}_1\overline{F}(y) + \phi\sum_{k=1}^{n+1} \overline{P}_{k}(F^{(k-1)*} - F^{k*}) * F(y) \\
&= \overline{P}_1\int_y^\infty dF(x) + \phi\int_{0-}^{y} \overline{F}_Y^{(n)}(y-x)\,dF(x) \\
&\leq \overline{P}_1\int_y^\infty dF(x) + \phi\int_{0-}^{y} \frac{1-p_0}{\phi}\, e^{-\kappa(y-x)}dF(x) \\
&\leq (1-p_0)\int_{0-}^{\infty} e^{-\kappa(y-x)}dF(x) = (1-p_0)e^{-\kappa y}\phi^{-1},
\end{aligned}
$$

and thus the theorem follows by induction. ■

Put

$$\overline{G}(s) = \sum_{k=0}^{\infty} \overline{P}_{k+1}s^k$$

and notice that $\overline{P}_{k+1} = P\{N > k\}$. We have, cf. Feller (1968, p.265) and (8.6),

$$\overline{G}(s) = \frac{1-G(s)}{1-s}. \tag{8.15}$$

Thus $\overline{G}$ and G have the same radius of convergence.

Under assumption (8.14) in Theorem 8.1 we have $\overline{P}_{k+1} \leq \phi^k\overline{P}_1$, and thus

$$\overline{G}(s) \leq \overline{P}_1\sum_{k=0}^{\infty}\phi^k s^k = \frac{\overline{P}_1}{1-\phi s}, \quad \text{for } \phi s < 1.$$

Thus $s_\infty \geq \phi^{-1}$. Since, cf. Willmot and Lin (1994, p.747),

$$G(\phi^{-1}) = \sum_{k=0}^{\infty} p_k\phi^{-k} \geq \sum_{k=1}^{\infty} p_k\phi^{-k} \geq \phi^{-1}\sum_{k=1}^{\infty} p_k = \frac{1-p_0}{\phi},$$

it follows that the bound in Theorem 8.1 is better than (8.12). On the other hand, (8.14) is a much harder condition than $s_\infty > 1$, which is simplest seen by considering a non-random sum.

Example 8.1 (continued). The natural and best choice of ϕ is $\phi = 1 - p$. Then we have $p_0 = p$ and $e^{-\kappa} = 1 - p$. Thus

$$\overline{F}_Y(y) \leq (1-p)^y,$$

which, see (8.10) and choose $y = (n+1) - 0$, is the best possible exponential bound holding for all real y. □

Example 8.2. Let Z_k be exponentially distributed and N, as in Example 8.1, be geometrically distributed. Then

$$\hat{f}(v) = \frac{1}{\mu}\int_0^\infty e^{-vz}e^{-z/\mu}dz = \frac{1}{1+\mu v}$$

and, see Example 8.1,

$$G(s) = \frac{p}{1-(1-p)s} = p + (1-p)\frac{ps}{1-(1-p)s}.$$

From (8.2) it follows that

$$\hat{f}_Y(v) = p + (1-p)\frac{p}{1+\mu v-(1-p)} = p + (1-p)\frac{1}{1+\mu v/p}$$

or that

$$\begin{aligned} F_Y(y) &= p + (1-p)\left(1 - e^{-py/\mu}\right) \\ &= 1 - (1-p)e^{-py/\mu}, \quad \text{for } y > 0. \end{aligned} \tag{8.16}$$

Thus $Y = 0$, with probability $p = P\{N = 0\}$, and exponentially distributed with mean μ/p, with probability $1 - p = P\{N > 0\}$. From (8.16) it follows that

$$\overline{F}_Y(y) = (1-p)e^{-py/\mu}.$$

As in Example 8.1, the best choice of ϕ is $\phi = 1-p$. Further we have $p_0 = p$ and

$$\frac{1}{1-\mu\kappa} = (1-p)^{-1}, \quad \text{i.e. } \kappa = p/\mu.$$

From Theorem 8.1 we now get

$$\overline{F}_Y(y) \leq \frac{1-p}{1-p}e^{-py/\mu} = e^{-py/\mu}.$$

We will return to this example. □

Remark 8.1. The inequality in Theorem 8.1 can also be written as

$$\bar{F}_Y(y) \le \hat{h}(\kappa)\bar{P}_1 e^{-\kappa y}.$$

The probability $\bar{P}_1$ comes from the fact that $Y = 0$ on $\{N = 0\}$. Thus the inequality could also have been written as

$$P\{Y > y \mid N > 0\} \le \hat{h}(\kappa)e^{-\kappa y}.$$

In our applications of Theorem 8.1 in section 8.2, we will put different restrictions on $\{p_k\}$. Assume that $\{p_k\}$ fulfils two sets of such restrictions, which yield two values of ϕ and consequently two values of κ, κ_1 and κ_2, such that $\kappa_1 \ge \kappa_2$. Then $e^{-\kappa_1 y} \le e^{-\kappa_2 y}$ while $\hat{h}(\kappa_1) \ge \hat{h}(\kappa_2)$. Since these inequalities are of primary interest for large values of y, we will regard the bound $\hat{h}(\kappa_1)\bar{P}_1 e^{-\kappa_1 y}$ as better than the bound $\hat{h}(\kappa_2)\bar{P}_1 e^{-\kappa_2 y}$, although that is not true for small values of y.

From this argument it follows that, in applications of Theorem 8.1, ϕ shall be chosen as small as possible. Without comments, this was understood in Examples 8.1 and 8.2, when the used ϕ was called the best choice. □

We will now consider some asymptotic relations between N and Y, or, more precisely, relations between the tails of the distributions of N and Y.

In order to formulate those relations, we need the concepts of slow and regular variation. Standard references for regular variation are, for instance, Feller (1971) and Bingham *et al.* (1987).

Definition 8.2. *A positive measurable function L is said to be* **slowly varying at infinity**, $L \in \mathrm{SV}$, *if*

$$L(x\ell) \sim L(\ell), \quad \textit{i.e. if} \lim_{\ell\to\infty} \frac{L(x\ell)}{L(\ell)} = 1,$$

for all $x > 0$.

Definition 8.3. *A function C is said to be* **regularly varying at infinity with exponent** ρ, $C \in \mathrm{RV}_\rho$, *if*

$$C(\ell) = \ell^\rho L(\ell)$$

for some slowly varying function L.

Lemma 8.1, known as Potter's theorem, see Bingham *et al.* (1987, p.25), will be useful.

Lemma 8.1. *For each $A > 1$ and $\delta > 0$ there exists an ℓ_0 such that*

$$A^{-1}\left(\frac{x}{\ell}\right)^{-\delta} \leq \frac{L(x)}{L(\ell)} \leq A\left(\frac{x}{\ell}\right)^{\delta} \quad \textit{for all } x \geq \ell \geq \ell_0.$$

The following – so called Abelian and Tauberian – theorem relates the behaviour of regularly varying functions at infinity, with the behaviour of their Laplace transforms at zero, see Feller (1971, p.446).

Theorem 8.2. *Let $c(\cdot)$ be a monotone function and let $L \in \mathrm{SV}$. Then, for $0 < \rho < \infty$,*

$$\hat{c}(v) \stackrel{\text{def}}{=} \int_0^\infty e^{-v\ell} c(\ell)\, d\ell \sim \frac{1}{v^\rho} L\left(\frac{1}{v}\right), \quad v \downarrow 0, \tag{8.17}$$

if and only if

$$c(\ell) \sim \frac{1}{\Gamma(\rho)} \ell^{\rho-1} L(\ell), \quad \ell \to \infty. \tag{8.18}$$

The implication (8.18) ⇒ (8.17) is called an Abelian theorem; the converse (8.17) ⇒ (8.18) a Tauberian theorem.

The following theorem is a discrete version of Theorem 8.2, see Feller (1971, p.447).

Theorem 8.3. *Let $\{a_n\}$ be a non-negative and monotone sequence such that*

$$A(s) \stackrel{\text{def}}{=} \sum_{n=0}^{\infty} a_n s^n$$

converges for $0 \leq s < 1$. Let $L(\cdot)$ be slowly varying. Then, for $0 < \rho < \infty$,

$$A(s) \sim \frac{1}{(1-s)^\rho} L\left(\frac{1}{1-s}\right), \quad s \uparrow 1, \tag{8.19}$$

if and only if

$$a_n \sim \frac{1}{\Gamma(\rho)} n^{\rho-1} L(n), \quad n \to \infty. \tag{8.20}$$

All regularly varying functions belong to the much larger class $\mathcal{L}$, defined below.

Definition 8.4. *A positive measurable function R belongs to the class $\mathcal{L}$, $R \in \mathcal{L}$, if*

$$R(x+\ell) \sim R(\ell), \quad \ell \to \infty,$$

for all $x > 0$.

Notice that $R \in \mathcal{L} \Leftrightarrow R \circ \log \in \mathrm{SV}$, where $R \circ \log(\ell) = R(\log(\ell))$.

Let us first consider the case where N has a negative binomial distribution. This is certainly an important case in relation to mixed Poisson distributions. From Example 2.1, we then have

$$p_k = \binom{\gamma + k - 1}{k} (1-\phi)^{\gamma} \phi^k, \quad \text{for } k = 0, 1, \dots,$$

where $0 < \phi < 1$ and $\gamma > 0$.

By **Stirling's formula**,

$$\Gamma(x) \sim \sqrt{2\pi}\, x^{x-\frac{1}{2}} e^{-x}, \quad x \to \infty,$$

we get

$$\binom{\gamma + k - 1}{k} = \frac{\Gamma(\gamma + k)}{\Gamma(\gamma) k!} \sim \frac{k^{\gamma-1}}{\Gamma(\gamma)}, \quad k \to \infty \tag{8.21}$$

and thus

$$p_k \sim (1-\phi)^{\gamma} \frac{k^{\gamma-1}}{\Gamma(\gamma)} \phi^k. \tag{8.22}$$

A natural generalization of (8.22) will be the assumption on N in Theorem 8.4 below, due to Embrechts *et al.* (1985b). Theorem 8.4 is the main result in this section concerning the light-tailed case, cf. Definition 8.1. Recall from section 6.2, F is called non-lattice if there exist no numbers c and d such that F is concentrated on c, $c \pm d$, $c \pm 2d, \dots$.

For reference reasons, we give first the following simple result as a lemma.

Lemma 8.2. *We have*

$$\int_{\ell}^{\infty} x^{\rho} e^{-\kappa x}\, dx \sim \frac{1}{\kappa} \ell^{\rho} e^{-\kappa \ell},$$

for $\kappa > 0$.

Proof. The lemma follows immediately by l'Hospital's rule. ∎

Theorem 8.4. *Let $L(\cdot)$ be slowly varying. Suppose that*

$$p_k \sim L(k)k^{\gamma-1}\phi^k, \quad k \to \infty,$$

$-\infty < \gamma < \infty$ *and* $0 < \phi < 1$*, and that F is light-tailed and non-lattice. Then*

$$\overline{F}_Y(y) \sim \frac{(\phi\hat{h}'(\kappa))^{-\gamma}}{\kappa} L(y)y^{\gamma-1}e^{-\kappa y}, \quad y \to \infty,$$

where κ is the solution of $\hat{h}(\kappa) = \phi^{-1}$.

An essential step in the proof of Theorem 8.4 is Proposition 8.1 due to Embrechts *et al.* (1984), which is a Blackwell type theorem for generalized renewal measures. Let $\widetilde{F}$ be a distribution with mean $\tilde{\mu} > 0$. In our applications $\widetilde{F}$ will be the distribution of a non-negative random variable, but this assumption seems not to simplify the proof. Let $\{a_n\}_{n=0}^{\infty}$ be a sequence of non-negative numbers. The generalized renewal measure G_a, related to $\widetilde{F}$ and a is defined by

$$G_a(y) \stackrel{\text{def}}{=} \sum_{n=0}^{\infty} a_n \widetilde{F}^{n*}(y). \tag{8.23}$$

Notice, for $a_n \equiv 1$, that G_1 is the 'usual' renewal measure, cf. (3.3). Blackwell's theorem, cf. (3.4), states, for $\widetilde{F}$ being non-lattice, that

$$\lim_{y\to\infty} G_1(y+h) - G_1(y) = \frac{h}{\tilde{\mu}} \quad \text{for every } h > 0. \tag{8.24}$$

Proposition 8.1. *Let $a(y)$ be a positive function such that $a \in \mathrm{RV}_\rho$ and let $\widetilde{F}$ be light-tailed and non-lattice. Put $a_n = a(n)$. Then*

$$G_a(y+h) - G_a(y) \sim h\tilde{\mu}^{-(\rho+1)}a(y) \quad \textit{for every } h > 0.$$

Moreover, the convergence is uniform in h on compact sets.

Sketch of the proof of Proposition 8.1. The proof, for which we refer to Embrechts *et al.* (1984), is rather technical. We will here only give the main steps of it when $\rho > -1$, which, for our applications, seems to be the most interesting case. In that case no assumption on the tail of $\widetilde{F}$ is required. The assumption of $\widetilde{F}$ being light-tailed implies the assumptions on the tail, required when $\rho \le -1$.

Step 1.
Let $B(y) \in \mathrm{RV}_\rho$, $\rho \geq 0$ be non-decreasing put $b_n = B(n) - B(n-1)$. Using Theorems 8.2 and 8.3 it follows that

$$G_b(y) \sim \tilde{\mu}^{-\rho} B(y).$$

Step 2.
Let $R(y) \in \mathcal{L}$ be non-decreasing with $R(0) = 0$. Then

$$G_1 * R(y+h) - G_1 * R(y) \sim \frac{h}{\tilde{\mu}} R(y).$$

In order to show this, we choose $h > 0$ and $y_0 \in (0, y)$, and write

$$G_1 * R(y+h) - G_1 * R(y) = I_1(y) + I_2(y) + I_3(y),$$

where

$$I_1(y) = \int_0^{y-y_0} (G_1(y+h-z) - G_1(y-z))\, dR(z),$$

$$I_2(y) = \int_{y-y_0}^{y} (G_1(y+h-z) - G_1(y-z))\, dR(z)$$

and

$$I_3(y) = \int_y^{y+h} G_1(y+h-z)\, dR(z).$$

Using

$$0 \leq I_3(y) \leq G_1(h)(R(y+h) - R(y))$$

and

$$0 \leq I_2(y) \leq G_1(h+y_0)(R(y) - R(y-y_0)),$$

it follows that $I_2(y) + I_3(y) = o(R(y))$, and thus it is enough to consider $I_1(y)$.

Using Blackwell's theorem, with some more care than we do here, we get for y_0 large enough

$$I_1(y) \sim \int_0^{y-y_0} \frac{h}{\tilde{\mu}}\, dR(z) = \frac{h}{\tilde{\mu}} R(y-y_0) \sim \frac{h}{\tilde{\mu}} R(y).$$

Step 3.
Since $a(y) \in \mathrm{RV}_\rho$, $\rho > -1$ we have $ya(y) \in \mathrm{RV}_\rho$, $\rho > 0$. Thus $ya(y)$ is asymptotically non-decreasing. By some estimation it is shown that it is no restriction to assume that $ya(y)$ is non-decreasing.

Step 4.
Define

$$\tilde{Q}(y) = \int_0^y z\, d\tilde{F}(z), \quad Q_a(y) = \int_0^y z\, dG_a(z), \quad \text{and} \quad G_b(y),$$

where $b_n = (n+1)a(n+1) - na(n)$, i.e. $B(y) = (y+1)a(y+1)$, cf. Step 1. Then

$$G_1 * \tilde{Q} * G_b(y) = Q_a(y),$$

which follows by taking Laplace transforms of both sides. Since $ya(y) \in \mathrm{RV}_{\rho+1}$, $\rho > -1$, it follows from Step 1 that

$$G_b(y) \sim \tilde{\mu}^{-(\rho+1)} ya(y),$$

and thus, since $\tilde{Q}(y) \to \tilde{\mu}$,

$$\tilde{Q} * G_b(y) \sim \tilde{\mu}^{-\rho} ya(y).$$

Using this and Step 2, with $R(y) = \tilde{Q} * G_b(y)$, we get

$$Q_a(y+h) - Q_a(y) \sim \frac{h}{\tilde{\mu}} \tilde{Q} * G_b(y) \sim h\mu^{-(\rho+1)} ya(y).$$

From the definition of Q_a it follows that

$$y(Q_a(y+h) - Q_a(y)) \le G_a(y+h) - G_a(y)$$
$$\le (y+h)(Q_a(y+h) - Q_a(y)),$$

and the first part of Proposition 8.1 is proved.

Step 5.
For the proof of uniform convergence, we refer to Embrechts *et al.* (1984, p.569). ∎

Proof of Theorem 8.4. Define the **associated distribution**, which in insurance mathematics is often called the **Esscher transform**, F_κ by

$$dF_\kappa(z) = \phi e^{\kappa z} dF(z), \quad z \ge 0,$$

where $\phi = 1/\hat{h}(\kappa)$ is the norming factor. Since, cf. Definition 8.1, $\kappa < r_\infty$ also F_κ is light-tailed.

Notice that

$$\mu_\kappa \stackrel{\text{def}}{=} \int_0^\infty z\,dF_\kappa(z) = \phi \int_0^\infty z e^{\kappa z} dF(z) = \phi \hat{h}'(\kappa).$$

It is easily verified that

$$dF_\kappa^{k*}(z) = \phi^k e^{\kappa z} dF^{k*}(z) \quad \text{or} \quad dF^{k*}(z) = \phi^{-k} e^{-\kappa z} dF_\kappa^{k*}(z),$$

for all $k = 0, 1, \ldots$. Thus we have

$$\overline{F}_Y(y) = \int_y^\infty e^{-\kappa z} \sum_{k=0}^\infty p_k \phi^{-k} dF_\kappa^{k*}(z), \quad y \geq 0.$$

Let

$$H(y) = \sum_{k=0}^\infty p_k \phi^{-k} F_\kappa^{k*}(y).$$

Let $a(y) = y^{\gamma-1}L(y)$ and $\widetilde{F} = F_\kappa$. Since F_κ is light-tailed it follows from Proposition 8.1 that

$$H(y+h) - H(y) \sim h\mu_\kappa^{-\gamma} y^{\gamma-1} L(y).$$

Choose $\varepsilon > 0$, $\delta > 0$ and $A > 1$. Then, by uniform convergence:

$$\exists y_\varepsilon : \forall y \geq y_\varepsilon \quad \text{and} \quad \forall h,\ 0 \leq h \leq 1, \text{ say}:$$

$$\begin{aligned}(1-\varepsilon)h\mu_\kappa^{-\gamma} y^{\gamma-1} L(y) &\leq H(y+h) - H(y) \\ &\leq (1+\varepsilon)h\mu_\kappa^{-\gamma} y^{\gamma-1} L(y),\end{aligned}$$

and, by Lemma 8.1,

$$\exists y_0 : \forall z \geq y \geq y_0 :$$

$$L(y)A^{-1}\left(\frac{z}{y}\right)^{-\delta} \leq L(z) \leq L(y)A\left(\frac{z}{y}\right)^{\delta}$$

Thus, for $y \geq \max(y_\varepsilon, y_0)$ and $0 < h \leq 1$, we have

$$\overline{F}_Y(y) = \int_y^\infty e^{-\kappa z} dH(z) = \sum_{k=0}^\infty \int_{y+kh}^{y+(k+1)h} e^{-\kappa z} dH(z)$$

$$\le \sum_{k=0}^{\infty} e^{-\kappa(y+kh)}(H(y+(k+1)h) - H(y+kh))$$

$$\le (1+\varepsilon)\mu_\kappa^{-\gamma} \sum_{k=0}^{\infty} e^{-\kappa(y+kh)}(y+kh)^{\gamma-1}L(y+kh)h$$

$$\le (1+\varepsilon)A\mu_\kappa^{-\gamma}y^{-\delta}L(y) \sum_{k=0}^{\infty} e^{-\kappa(y+kh)}(y+kh)^{\gamma-1-\delta}h$$

$$\to (1+\varepsilon)A\mu_\kappa^{-\gamma}y^{-\delta}L(y) \int_y^\infty e^{-\kappa z}z^{\gamma-1-\delta}\,dz, \quad \text{as } h \downarrow 0.$$

By Lemma 8.2 we thus get

$$\limsup_{y\to\infty} \frac{\bar{F}_Y(y)}{L(y)y^{\gamma-1}e^{-\kappa y}} \le (1+\varepsilon)A\frac{\mu_\kappa^{-\gamma}}{\kappa}.$$

A similar argument yields the converse inequality

$$\liminf_{y\to\infty} \frac{\bar{F}_Y(y)}{L(y)y^{\gamma-1}e^{-\kappa y}} \ge (1-\varepsilon)A^{-1}\frac{\mu_\kappa^{-\gamma}}{\kappa}.$$

Since ε and A may be arbitrarily chosen, we have

$$\lim_{y\to\infty} \frac{\bar{F}_Y(y)}{L(y)y^{\gamma-1}e^{-\kappa y}} = \frac{\mu_\kappa^{-\gamma}}{\kappa} = \frac{(\phi\hat{h}'(\kappa))^{-\gamma}}{\kappa},$$

and the theorem is proved. ■

Remark 8.2. The assumption in Theorem 8.4 of F being light-tailed is useful since it implies certain 'extra' assumptions, which are needed when $\gamma \le 0$. For these assumptions to hold it is, in fact, enough to require that $\hat{h}(\kappa) = \phi^{-1}$ has a solution and that $\hat{h}(r) < \infty$ for some $r > \kappa$. This minor generalization may be useful, see Remark 8.8 below.

For $\gamma > 0$, no such extra assumptions are required and then Theorem 8.4 holds if $\hat{h}(\kappa) = \phi^{-1}$ has a solution and if $\hat{h}'(\kappa) < \infty$. This covers the case where N has a negative binomial distribution.

The proof of Theorem 8.4 was based on a renewal theorem, Proposition 8.1, and therefore the assumption of F being non-lattice is natural. However, usually renewal theorems have discrete variants. Willmot (1989b) considered the case where F is discrete, and obtained asymptotic results, very similar to Theorem 8.4. We will shortly compare his results with Theorem 8.4.

Suppose that

$$p_k \sim Ck^{\gamma-1}\phi^k, \quad k \to \infty,$$

$C > 0$, $-\infty < \gamma < \infty$ and $0 < \phi < 1$. The appearance of the constant C is of course a restriction in comparison with the formulation in Theorem 8.4, but can probably be relaxed. Suppose further that F is discrete, and let

$$f_n \stackrel{\text{def}}{=} P\{Z_i = n\}.$$

Assume that the greatest common divisor, d, of the set of values of n with $f_n > 0$ is 1. If $d > 1$, we just consider Z_i/d instead. Since F is discrete, it is natural to consider the generating function

$$H(s) \stackrel{\text{def}}{=} E[s^{Z_i}].$$

Assume that $H(s)$ has radius of convergence exceeding τ, where $\tau > 1$ is the solution of $H(\tau) = \phi^{-1}$. Then, see Willmot (1989b, p.176),

$$P\{Y = n\} \sim (\phi\tau H'(\tau))^{-\gamma} Cn^{\gamma-1}\tau^{-n}, \quad n \to \infty, \tag{8.25}$$

and

$$\overline{F}_Y(n) \sim \frac{(\phi\tau H'(\tau))^{-\gamma}}{\tau - 1} Cn^{\gamma-1}\tau^{-n}, \quad n \to \infty. \tag{8.26}$$

□

Example 8.2 (continued). Recall that

$$\overline{F}_Y(y) = (1-p)e^{-py/\mu},$$

when Z_k is exponentially distributed and N is geometrically distributed. Since $p_k = p \cdot (1-p)^k$, Theorem 8.4 applies with

$$\phi = 1 - p, \quad \gamma = 1, \quad \text{and} \quad L(y) = p.$$

Further $\kappa = p/\mu$ and

$$\hat{h}'(r) = \frac{d}{dr}\left(\frac{1}{1-\mu r}\right) = \frac{\mu}{(1-\mu r)^2},$$

and thus $\hat{h}'(\kappa) = \mu/\phi^2$. From Theorem 8.4 we then get

$$\overline{F}_Y(y) \sim \frac{(\mu/\phi)^{-1}}{p/\mu} pe^{-py/\mu} = \phi e^{-py/\mu} = (1-p)e^{-py/\mu},$$

i.e. Theorem 8.4 yields the exact result. This is, of course, not surprising, and is merely to be regarded as an illustration of Theorem 8.4.

Consider now the case where F is discrete, which was mentioned in Remark 8.2. The discrete correspondence to the exponential distribution is the geometric distribution. Let, as above, N be geometrically distributed with generating function

$$G(s) = \frac{p}{1-(1-p)s}$$

and let Z_k be geometrically distributed with generating function

$$H(s) = \frac{q}{1-(1-q)s}.$$

Notice that p and q are free parameters, both taking values in $(0, 1)$. In this case also Y is a discrete random variable, and its generating function, H_Y, cf. (8.2), is given by

$$H_Y(s) = G(H(s)) = p + (1-p)\frac{q_Y}{1-(1-q_Y)s},$$

where

$$q_Y = \frac{pq}{1-q(1-p)}.$$

Thus

$$P\{Y = n\} = (1-p)q_Y(1-q_Y)^n, \quad n > 0,$$

and

$$\overline{F}_Y(n) = (1-p)(1-q_Y)^{n+1}, \quad n > 0.$$

Willmot's asymptotic expressions (8.25) and (8.26) apply with

$$\phi = 1-p, \quad \gamma = 1, \quad \text{and} \quad C = p.$$

From $H(\tau) = \phi^{-1} = (1-p)^{-1}$ we get

$$\tau = \frac{1-q(1-p)}{1-q} = \frac{1}{1-q_Y}.$$

Since

$$H'(s) = \frac{q(1-q)}{(1-(1-q)s)^2}$$

we get

$$H'(\tau) = \frac{1-q}{q(1-p)^2}.$$

Thus (8.25) and (8.26) reduce to

$$P\{Y = n\} \sim (1-p)q_Y(1-q_Y)^n$$

and

$$\overline{F}_Y(n) \sim (1-p)(1-q_Y)(1-q_Y)^n$$

respectively, i.e. the asymptotic expressions hold exactly for $n > 0$. As in the continuous case, this is merely to be regarded as an illustration.

□

Let us now consider the case '$\phi = 1$' in Theorem 8.4, which would mean that we assume that $p_k \sim L(k)k^{\gamma-1}$. Naturally this makes sense only if $\gamma \leq 0$. However, since we are discussing tail behaviour, we choose to formulate the properties of N in term of the tail of its distribution, cf. Proposition 8.5. More precisely we assume that

$$P\{N > n\} \sim \omega L(n)n^{\gamma}, \quad \omega \geq 0 \text{ and } n \to \infty, \tag{8.27}$$

for some function $L \in \text{SV}$. Notice that this L is not exactly the same as for p_k, but (8.27) will now be our formal assumption about N. For $\omega = 0$, (8.27) means that

$$\lim_{n\to\infty} \frac{P\{N > n\}}{L(n)n^{\gamma}} = 0.$$

Further we weaken the assumption that $\hat{h}(r) < \infty$ for some $r > 0$ to

$$\overline{F}(z) \sim \chi L(z)z^{\gamma}, \quad \chi \geq 0 \text{ and } y \to \infty. \tag{8.28}$$

Notice that $\hat{h}(r) < \infty$ for some $r > 0$ implies $\chi = 0$, and that the L appearing in (8.27) and (8.28) is the same.

We are looking for results of the form

$$\overline{F}_Y(y) \sim aL(y)y^{\gamma}, \quad y \to \infty. \tag{8.29}$$

For $\gamma < -1$ we put $\mu_N = E[N]$ and, as usual, $\mu = E[Z_k]$.

The relations between (8.27), (8.28) and (8.29) are almost completely settled by the following theorem, due to Stam (1973, pp.311, 320).

Theorem 8.5.

(*i*) *Let (8.28) hold with $\gamma < -1$. Then (8.29) holds if and only if (8.27) holds and*

$$a = \omega\mu^{-\gamma} + \chi\mu_N. \tag{8.30}$$

(*ii*) *Let (8.27) hold with $\gamma < -1$. Then (8.29) holds if and only if (8.28) and (8.30) hold.*

(*iii*) *Let $\mu < \infty$ and $-1 < \gamma \leq 0$. Then (8.29) holds if and only if (8.27) holds and*

$$a = \omega\mu^{-\gamma}.$$

(*iv*) *Let $\mu_N < \infty$ and $-1 < \gamma \leq 0$. Then (8.29) holds if and only if (8.28) holds and*

$$a = \chi\mu_N.$$

In our applications, we will only apply the 'if parts' of Theorem 8.5. In order to facilitate the reading and the applications, we formulate those parts as three corollaries.

Corollary 8.1. *Let $L(\cdot)$ be slowly varying. Suppose that*

$$P\{N > n\} \sim \omega L(n)n^{\gamma}, \quad \omega \geq 0 \text{ and } n \to \infty,$$

and that

$$\bar{F}(z) \sim \chi L(z)z^{\gamma}, \quad \chi \geq 0 \text{ and } z \to \infty,$$

$\gamma < -1$. *Then*

$$\bar{F}_Y(y) \sim (\omega\mu^{-\gamma} + \chi\mu_N)L(y)y^{\gamma}, \quad y \to \infty.$$

Heuristic arguments for Corollary 8.1. A proof of the corollary is difficult, and we will therefore only give some arguments, which at the best may indicate that Corollary 8.1 is natural. We do, however, begin with a mathematically strict result.

Let ξ_1 and ξ_2 be two independent random variables, such that

$$P\{\xi_k > x\} \sim a_k L(x)x^{\gamma}, \quad a_k \geq 0,\ \gamma < -1 \text{ and } x \to \infty.$$

Then, cf. Feller (1971, p.279),

$$P\{\xi_1 + \xi_2 > x\} \sim (a_1 + a_2)L(x)x^{\gamma}, \quad x \to \infty,$$

which implies

$$P\{\xi_1 + \xi_2 > x\} \sim P\{\max(\xi_1, \xi_2) > x\}, \quad x \to \infty,$$

since

$$P\{\max(\xi_1, \xi_2) > x\} = P\{\{\xi_1 > x\} \cup \{\xi_2 > x\}\}$$
$$= P\{\xi_1 > x\} + P\{\xi_2 > x\} - P\{\xi_1 > x\}P\{\xi_2 > x\}.$$

An intuitive interpretation is that the event $\{\xi_1 + \xi_2 > x\}$ is caused by one of the variables ξ_1 or ξ_2, and not by the two in 'cooperation'. Naturally this can be extended to an arbitrary, but fixed, number of variables.

We will now extend this argument to the random sum Y and the event $\{Y > y\}$, for large values of y, in the following way:

If N is 'moderate', then one of $Z_1, \ldots, Z_N$ must be large. If we replace N by μ_N, we get

$$\overline{F}_Y(y) \sim \mu_N \overline{F}(y), \quad y \to \infty.$$

In this kind of arguments, we do not bother about whether μ_N is an integer or not.

If N is large, then, due to the law of large numbers,

$$\overline{Z}_N \stackrel{\text{def}}{=} \frac{1}{N} \sum_{k=1}^{N} Z_k \approx \mu$$

and

$$\overline{F}_Y(y) = P\{N\overline{Z}_N > y\} \sim P\{N\mu > y\} = P\{N > y/\mu\}, \quad y \to \infty.$$

Our use of the law of large numbers is certainly far from convincing.

Anyhow, putting the two cases together, we get

$$\begin{aligned}\overline{F}_Y(y) &\sim P\{N > y/\mu\} + \mu_N \overline{F}(y) \\ &= \omega L(y/\mu)(y/\mu)^\gamma + \mu_N \chi L(y) y^\gamma \\ &= \omega L(y)(y/\mu)^\gamma + \chi \mu_N L(y) y^\gamma, \quad y \to \infty,\end{aligned}$$

which is the correct result. ■

Corollary 8.2. *Let $L(\cdot)$ be slowly varying. Suppose that*

$$P\{N > n\} \sim \omega L(n) n^{\gamma}, \quad \omega \geq 0 \text{ and } n \to \infty,$$

$-1 < \gamma \leq 0$, *and that* $\mu < \infty$. *Then*

$$\bar{F}_Y(y) \sim \omega \mu^{-\gamma} L(y) y^{\gamma}, \quad y \to \infty.$$

Proof. With the same kind of heuristic arguments as used for Corollary 8.1, we could just say that N has a thicker tail than the Z_ks, and therefore $\bar{F}_Y(y) \sim P\{N > y/\mu\}$.

For $-1 < \gamma$ Theorems 8.2 and 8.3 apply with $\rho = \gamma + 1$, and the proof is therefore rather simple, cf. Stam (1973, pp.320–321).

We restrict ourselves to $\omega > 0$, which is the interesting case and, which is no restriction, put $\omega = 1$. From (8.2), (8.6), and (8.15), we get

$$\bar{f}_Y(v) = \frac{1 - G(\hat{f}(v))}{v} = \frac{1 - \hat{f}(v)}{v} \bar{G}(\hat{f}(v)).$$

For $v \downarrow 0$ we have $\hat{f}(v) \uparrow 1$, and it follows from Theorem 8.3, in the Abelian direction, that

$$\bar{G}(\hat{f}(v)) \sim \frac{\Gamma(\gamma + 1)}{(1 - \hat{f}(v))^{\gamma+1}} L\left(\frac{1}{1 - \hat{f}(v)}\right).$$

Since $1 - \hat{f}(v) \sim \mu v$, for $v \downarrow 0$, we get

$$\bar{f}_Y(v) \sim \mu \bar{G}(\hat{f}(v)) \sim \mu \frac{\Gamma(\gamma + 1)}{(\mu v)^{\gamma+1}} L\left(\frac{1}{\mu v}\right)$$

$$\sim \frac{\Gamma(\gamma + 1)}{\mu^{\gamma}} \frac{1}{v^{\gamma+1}} L\left(\frac{1}{v}\right), \quad v \downarrow 0,$$

where

$$L\left(\frac{1}{1 - \hat{f}(v)}\right) \sim L\left(\frac{1}{\mu v}\right), \quad v \downarrow 0,$$

follows, for instance, from Lemma 8.1. We now get, by Theorem 8.2 in the Tauberian direction,

$$\bar{F}_Y(y) \sim \mu^{-\gamma} y^{\gamma} L(y), \quad y \to \infty,$$

which is the corollary. ■

Corollary 8.3. *Let $L(\cdot)$ be slowly varying. Suppose that $\mu_N < \infty$ and that*

$$\bar{F}(z) \sim \chi L(z) z^\gamma, \quad \chi \geq 0 \text{ and } z \to \infty,$$

$-1 < \gamma \leq 0$. *Then*

$$\bar{F}_Y(y) \sim \chi \mu_N L(y) y^\gamma, \quad y \to \infty.$$

Proof. This Corollary is proved in the same way as Corollary 8.2, but the proof is a little simpler, cf. Stam (1973, p.320). Let $\chi = 1$. We have

$$\bar{f}_Y(v) = \frac{1 - G(\hat{f}(v))}{v} = \bar{f}(v) \frac{1 - G(\hat{f}(v))}{1 - \hat{f}(v)} \sim \mu_N \bar{f}(v), \quad v \downarrow 0,$$

since

$$\lim_{v \downarrow 0} \frac{1 - G(\hat{f}(v))}{1 - \hat{f}(v)} = G'(1) = \mu_N.$$

The corollary now follows from Theorem 8.2. ∎

Let us now assume that N is light-tailed, which is quite natural in risk models. The case where F is light-tailed does then correspond to the case where very large claims 'do not occur', while the case where $\bar{F}$ is regularly varying is natural when very large claims must be taken into account. Certain popular models in that case – a typical example is the lognormal distribution – do, however, not have regularly varying tails. It turns out that the class $\mathcal{S}$, introduced by Chistyakov (1964), has the 'right' properties.

Definition 8.5. *A distribution F on $[0, \infty)$ belongs to the class $\mathcal{S}$ of* **subexponential distributions** *if*

$$\lim_{z \to \infty} \frac{\bar{F}^{2*}(z)}{\bar{F}(z)} = 2.$$

Subexponentiality is an extension of regular variation, since $\bar{F} \in \mathrm{RV}_\rho$ implies $\bar{F}^{2*}(z) \sim 2\bar{F}(z)$, cf. the arguments for Corollary 8.1. The rôle of $\mathcal{S}$ in risk models will be discussed in Chapter 9. For a survey of subexponential distributions we refer to the survey article Goldie and Klüppelberg (1997) and to the book Embrechts *et al.* (1997).

The following lemma summarizes some important properties of $\mathcal{S}$, see Athreya and Ney (1972, pp.148–150). Recall from Definition 8.4 that $\bar{F} \in \mathcal{L}$ means that $\bar{F}(z + x) \sim \bar{F}(z)$ as $x \to \infty$.

Lemma 8.3. *If $F \in \mathcal{S}$, then*

(*i*) $\overline{F} \in \mathcal{L}$;

(*ii*) $\overline{F^{k*}}(z) \sim k\overline{F}(z)$, *for* $k = 0, 1, 2, \ldots$;

(*iii*) *Given any* $\varepsilon > 0$ *there exists a* $D < \infty$ *such that*

$$\frac{\overline{F^{k*}}(z)}{\overline{F}(z)} \leq D(1+\varepsilon)^k \quad \textit{for all } k \textit{ and } z.$$

One reason to call $\mathcal{S}$ the class of subexponential distributions is that

$$\lim_{z\to\infty} e^{\kappa z}\overline{F}(z) = \infty, \qquad \text{for all } \kappa > 0 \text{ and all } F \in \mathcal{S}.$$

This 'subexponential' property does, in fact, hold for all distributions F with $\overline{F} \in \mathcal{L}$ since, cf. the comments after Definition 8.4, $\overline{F} \in \mathcal{L} \Leftrightarrow \overline{F} \circ \log \in \mathrm{SV}$. By Lemma 8.1 we get

$$\lim_{z\to\infty} e^{\kappa z}\overline{F}(z) = \lim_{x\to\infty} e^{\kappa \log(x)}\overline{F}(\log(x)) = \lim_{x\to\infty} x^{\kappa}\overline{F}(\log(x)) = \infty,$$

for all $\kappa > 0$.

We can now formulate a correspondence to Corollaries 17 and 19, which in the mixed Poisson case has been proposed by Klüppelberg (1994).

Proposition 8.2. *Suppose that $F \in \mathcal{S}$ and that N is light-tailed. Then*

$$\overline{F}_Y(y) \sim \mu_N \overline{F}(y), \quad y \to \infty.$$

Proof. The proof to be given is essentially taken from Athreya and Ney (1972, p.150).

We have, cf. (8.1),

$$\overline{F}_Y(y) = \sum_{k=0}^{\infty} p_k \overline{F^{k*}}(y).$$

Since N is light-tailed, Lemma 8.3 (iii) implies, for $\varepsilon < s_\infty - 1$, that

$$\frac{\overline{F}_Y(y)}{\overline{F}(y)} = \sum_{k=0}^{\infty} p_k \frac{\overline{F^{k*}}(y)}{\overline{F}(y)}$$

$$\leq \sum_{k=0}^{\infty} p_k D(1+\varepsilon)^k = D \cdot G((1+\varepsilon)) < \infty.$$

By dominated convergence it follows from Lemma 8.3 (ii) that

$$\lim_{y\to\infty} \frac{\overline{F}_Y(y)}{\overline{F}(y)} = \sum_{k=0}^{\infty} p_k k = \mu_N,$$

which was to be proved. ■

It may very well be the case that a light-tailed F is too 'kind' while an $F \in \mathcal{S}$ is too 'dangerous'. Then it may be tempting to try to find some 'intermediate' case with nice properties. A natural way to do this – and to retain the flavour of $\mathcal{S}$ – is to choose some $V \in \mathcal{S}$ and some $\beta > 0$ and to use

$$F(z) \stackrel{\text{def}}{=} \frac{1}{\hat{v}(\beta)} \int_{0-}^{z} e^{-\beta x}\, dV(x),$$

where $\hat{v}$ is the Laplace transform of V. Then we have

$$\hat{h}(r) = \frac{\hat{v}(r-\beta)}{\hat{v}(\beta)}.$$

Thus $r_\infty = \beta$, cf. Definition 8.1, but $\hat{h}(r_\infty) < \infty$, which implies that F is not light-tailed. The distribution F belongs to the class $\mathcal{T}(\beta)$, defined below.

Definition 8.6. *A distribution F on $[0,\infty)$ belongs to the class $\mathcal{T}(\beta)$, $\beta \geq 0$, if*

(*i*) $\hat{h}(\beta) < \infty$;

(*ii*) *the Esscher transform V, defined by*

$$dV(z) \stackrel{\text{def}}{=} \hat{h}(\beta)^{-1} e^{\beta z} dF(z), \quad z \geq 0,$$

belongs to $\mathcal{S}$.

The class $\mathcal{T}(\beta)$ has been investigated by Teugels (1975) and Embrechts and Goldie (1982).

Due to the close connection between $\mathcal{T}(\beta)$ and $\mathcal{S}$, we can apply Proposition 8.2, and see what we get.

Corollary 8.4. *Suppose that $F \in \mathcal{T}(\beta)$ and that N is light-tailed with $s_\infty > \hat{h}(\beta)$. Then*

$$\int_y^\infty e^{\beta z} dF_Y(z) \sim G'(\hat{h}(\beta)) \int_y^\infty e^{\beta z} dF(z) \quad y \to \infty.$$

Proof. The proof to be given is essentially taken from Embrechts and Goldie (1982, p.272).

Let $V \in \mathcal{S}$ be given by Definition 8.6. Using

$$dF^{k*}(z) = \hat{h}(\beta)^k e^{-\beta z} dV^{k*}(z)$$

we get

$$dF_Y(z) = \sum_{k=0}^\infty p_k dF^{k*}(z) = \sum_{k=0}^\infty p_k \hat{h}(\beta)^k e^{-\beta z} dV^{k*}(z).$$

Multiplication with $e^{\beta z}/G(\hat{h}(\beta))$ and integration yields

$$\overline{V}_Y(y) \stackrel{\text{def}}{=} \frac{1}{G(\hat{h}(\beta))} \int_y^\infty e^{\beta z} dF_Y(z) = \sum_{k=0}^\infty \frac{p_k \hat{h}(\beta)^k}{G(\hat{h}(\beta))} \overline{V}^{k*}(y).$$

Let $\tilde{N}$ be a discrete random variable with

$$P\{\tilde{N} = k\} = p_k \hat{h}(\beta)^k / G(\hat{h}(\beta)).$$

Then $G_{\tilde{N}}(s) = G(s\hat{h}(\beta))/G(\hat{h}(\beta))$ and thus

$$\mu_{\tilde{N}} = G'_{\tilde{N}}(1) = \hat{h}(\beta) G'(\hat{h}(\beta))/G(\hat{h}(\beta)).$$

It follows from Proposition 8.2 that

$$\overline{V}_Y(y) \sim \frac{\hat{h}(\beta) G'(\hat{h}(\beta))}{G(\hat{h}(\beta))} \overline{V}(y)$$

or

$$G(\hat{h}(\beta)) \overline{V}_Y(y) \sim G'(\hat{h}(\beta)) \cdot \hat{h}(\beta) \overline{V}(y),$$

which is the corollary. ■

Formally Corollary 8.4 is a generalization – and therefore not a corollary – of Proposition 8.2, since $\mathcal{T}(0) = \mathcal{S}$. However, it is not the generalization we want, since it compares β-integrated tails rather than the tails themselves. In order to be able to compare the tails, we will consider the class $\mathcal{S}(\beta)$, like $\mathcal{S}$, introduced by Chistyakov (1964).

Definition 8.7. *A distribution F on $[0,\infty)$ belongs to the class $\mathcal{S}(\beta)$, $\beta \geq 0$, if*

(*i*) $\hat{h}(\beta) < \infty$;

(*ii*) $\lim_{z\to\infty} \overline{F}^{2*}(z)/\overline{F}(z) = c, \quad o < c < \infty$;

(*iii*) $\lim_{z\to\infty} \overline{F}(z+y)/\overline{F}(z) = e^{-\beta y}$ *for all real* y.

The class $\mathcal{S}(\beta)$ has been investigated by Embrechts and Goldie (1982) in great detail. Chover *et al.* (1973) have shown that $c = 2\hat{h}(\beta)$. Embrechts (1983) has shown that GIGDs, see Example 2.3, with $\gamma < 0$ belong to $\mathcal{S}(\psi/2)$.

Obviously $\mathcal{S}(0) = \mathcal{T}(0) = \mathcal{S}$. The relation between $\mathcal{T}(\beta)$ and $\mathcal{S}(\beta)$, for $\beta > 0$, is, however, not obvious at all. Embrechts and Goldie (1982, pp.270–271) has shown that $\mathcal{S}(\beta)$ is **strictly** smaller that $\mathcal{T}(\beta)$. In several papers the two classes are assumed to coincide – although wrong, it seems to be a rather natural misunderstanding – and this is one reason why we treat these classes relatively detailed. Intuitively, one may say the $\mathcal{S}(\beta)$ retains even more of the flavour of $\mathcal{S}$ than $\mathcal{T}(\beta)$ does. The following lemma, see Embrechts and Goldie (1982, p.265), which generalizes Lemma 8.3, may be looked upon as a support of that intuition.

Lemma 8.4. *If $F \in \mathcal{S}(\beta)$, then*

(*i*) $\overline{F}^{k*}(z) \sim k\hat{h}(\beta)^{k-1}\overline{F}(z), \quad$ *for* $k = 0, 1, 2, \ldots$;

(*ii*) *Given any $\varepsilon > 0$ there exists a $D < \infty$ such that*

$$\frac{\overline{F}^{k*}(z)}{\overline{F}(z)} \leq D(\hat{h}(\beta) + \varepsilon)^k \quad \text{for all } k \text{ and } z.$$

Using Lemma 10, we can now formulate a much more satisfying generalization of Proposition 8.2.

Proposition 8.3. *Suppose that $F \in \mathcal{S}(\beta)$ and that N is light-tailed with $s_\infty > \hat{h}(\beta)$. Then*

$$\overline{F}_Y(y) \sim G'(\hat{h}(\beta))\overline{F}(y), \quad y \to \infty.$$

Proof. Due to Lemma 10, this proposition is proved almost word for word as Proposition 8.2. The only change to be noticed is that

$$\lim_{y\to\infty}\frac{\overline{F}_Y(y)}{\overline{F}(y)}=\sum_{k=0}^{\infty}p_k k\hat{h}(\beta)^{k-1}=G'(\hat{h}(\beta)).$$

■

Example 8.3. Let N be Po(α)-distributed and $F\in\mathcal{S}(\beta)$. Then $G(s)=e^{\alpha(s-1)}$, and thus $s_\infty=\infty$. Further $G'(s)=\alpha G(s)$. Put, cf. (8.2), $\hat{h}_Y(r)=G(\hat{h}(r))$.

Let V_Y and V be the the Esscher transforms of F_Y and F respectively, cf. Definition 8.6. From Corollary 8.4 (or easier from the proof of it) we get

$$\overline{V}_Y(y)\sim\alpha\hat{h}(\beta)\overline{V}(y),\quad y\to\infty,$$

which is due to Embrechts and Goldie (1982, p.272).

From Proposition 8.3 we get

$$\overline{F}_Y(y)\sim\alpha\hat{h}_Y(\beta)\overline{F}(y),\quad y\to\infty,$$

which is also due to Embrechts and Goldie (1982, p.272). □

8.2 Exponential bounds

Now we will go back to the mixed Poisson case. Let then N be MPP(U) and consider $N(t)$ for some fixed value of t. Thus $N(t)$ is MP(t,U). Thus, in spite of the notation, the results to be given are about distributions. Similarly we denote the corresponding compound mixed Poisson variable $Y(t)$. Naturally we may interpret the distribution of $Y(t)$ as an one-dimensional marginal distribution of a compound mixed Poisson process. Those processes will be discussed in Chapter 9. We use the notation

$$p_n(t)=P\{N(t)=n\},\ \overline{P}_n(t)=P\{N(t)\geq n\},\ G_{N(t)}(s)=E[s^{N(t)}]$$

$$F(y,t)=P\{Y(t)\leq y\},\qquad \overline{F}(y,t)=P\{Y(t)>y\}.$$

As shown by Willmot and Lin (1994), on which this section is based, it is natural to express conditions on Λ in terms of its reliability properties. Recall from Chapter 7 that a condition on Λ leads to the same, discrete version, condition on $N(t)$, i.e. in that direction the results hold for distributions.

Let u_∞ denote the abscissa of convergence of $\hat{u}(-v)$ and let, as in section 8.1, s_∞ denote the radius of convergence of $G_{N(t)}(s)$. It follows from Proposition 2.1 (vi) that

$$s_\infty = 1 + \frac{u_\infty}{t}. \tag{8.31}$$

In order to avoid unnecessary complications, we exclude the trivial case $U(0) = 1$. Notice that then $p_n(t)$ and $\overline{P}_n(t)$ are positive for all n and all $t > 0$.

For the rest of this section we assume that F is light-tailed.

Suppose now that Λ is IFR. For the meaning of IFR and other reliability classes we refer to Chapter 7.

Theorem 8.6 (Willmot and Lin, 1994 p.748). *Let Λ be IFR. Then*

$$\overline{F}(y,t) \le \frac{\overline{P}_1(t)^2}{\overline{P}_2(t)}\, e^{-\kappa y},$$

where κ is the solution of $\hat{h}(\kappa)\overline{P}_2(t) = \overline{P}_1(t)$.

Proof. It follows from Proposition 7.2 that $\overline{P}_{n+1}(t)/\overline{P}_n(t)$ is decreasing. Thus Theorem 8.1 applies with $\phi = \overline{P}_2(t)/\overline{P}_1(t)$. ■

A similar theorem as Theorem 8.6 holds when Λ is DFR.

Theorem 8.7 (Willmot and Lin, 1994 p.747). *Suppose that U is light-tailed and let $u_\infty > 0$ denote the abscissa of convergence of $\hat{u}(-v)$. Let Λ be DFR. Then*

$$\overline{F}(y,t) \le \left(1 + \frac{u_\infty}{t}\right) \overline{P}_1(t)\, e^{-\kappa y},$$

where κ is the solution of $\hat{h}(\kappa) - 1 = u_\infty/t$.

Proof. It follows from Proposition 7.2 that $\overline{P}_{n+1}(t)/\overline{P}_n(t)$ is increasing. Put

$$\phi = \lim_{n\to\infty} \overline{P}_{n+1}(t)/\overline{P}_n(t),$$

which exists since $\overline{P}_{n+1}(t)/\overline{P}_n(t)$ is increasing and bounded from above by 1. By (8.15) and D'Alembert's ratio test for convergence it follows that $\phi^{-1} = s_\infty$. The desired result now follows from Theorem 8.1 and (8.31). ■

Remark 8.3. From the proofs of Theorems 8.6 and 8.7, it is seen the choices of ϕ are the best possible for which Theorem 8.1 applies.

Further, see (8.2), the choice $\phi^{-1} = s_\infty$, in Theorem 8.7, implies that the used κ is the best possible in the sense of (8.5). Such general optimality cannot be expected to hold for Theorem 8.6. This may indicate that DFR is easier to handle than IFR. □

We will now consider other reliability properties, starting with Λ being IFRA. What one can expect – and what we will get – is a weaker version of Theorem 8.6. Before considering that version, we will introduce two auxiliary quantities $\overline{U}_t$ and MRL_t.

Put

$$\overline{U}_t(\ell) = e^{-t\ell}\overline{U}(\ell),$$

where, cf. (7.4), $\overline{U}(\ell) = 1 - U(\ell)$. Notice that $\overline{U}_0(\ell) = \overline{U}(\ell)$. Let Λ_t be a random variable with distribution U_t, and put

$$\mathrm{MRL}_t(x) \stackrel{\text{def}}{=} E[\Lambda_t - x \mid \Lambda_t > x] = \frac{\int_x^\infty \overline{U}_t(\ell)\,d\ell}{\overline{U}_t(x)}, \tag{8.32}$$

where MRL stands for 'mean residual life'.

By Proposition 2.1 (iii) and integration by parts we get, for $n = 1, 2, \ldots$,

$$\overline{P}_{n+1}(t) = t\int_0^\infty \frac{(\ell t)^n}{n!}\, e^{-\ell t}\overline{U}(\ell)\,d\ell$$

$$= -\,t\left[\frac{(\ell t)^n}{n!}\int_\ell^\infty \overline{U}_t(x)\,dx\right]_0^\infty + t^2\int_0^\infty \frac{(\ell t)^{(n-1)}}{(n-1)!}\int_\ell^\infty \overline{U}_t(x)\,dx\,d\ell$$

$$= t^2\int_0^\infty \mathrm{MRL}_t(\ell)\frac{(\ell t)^{(n-1)}}{(n-1)!}e^{-\ell t}\overline{U}(\ell)\,d\ell. \tag{8.33}$$

Thus MRL_t is related to a kind of 'local' version of ϕ, and to make that relation more visible we put

$$\phi_t(\ell) \stackrel{\text{def}}{=} t\cdot \mathrm{MRL}_t(\ell).$$

Theorem 8.8 (Willmot and Lin, 1994 p.750). *Let Λ be IFRA. Then*

$$\overline{F}(y,t) \le e^{-\kappa y},$$

where κ is the solution of $\hat{h}(\kappa)\overline{P}_1(t) = 1$.

Proof. Since $\overline{U}(\ell)^{1/\ell}$ is decreasing and since

$$\overline{U}_t(\ell)^{1/\ell} = e^{-t}\overline{U}(\ell)^{1/\ell},$$

it follows that also U_t is IFRA. IFRA implies NBUE, cf. Barlow and Proschan (1975, p.159), and thus

$$\phi_t(\ell) = t \cdot \text{MRL}_t(\ell) \le t \cdot E[\Lambda_t] = \overline{P}_1(t).$$

Thus Theorem 8.1 applies with $\phi = \overline{P}_1(t)$. ■

Remark 8.4. We will now compare Theorems 8.6 and 8.8, when Λ is IFR. Since this implies that $N(t)$ is IFR, we have, cf. Definition 7.1, $\overline{P}_1(t) \ge \overline{P}_2(t)/\overline{P}_1(t)$. Let $\kappa_{8.6}$ and $\kappa_{8.7}$ be the κs in the two theorems. We have

$$\hat{h}(\kappa_{8.6}) = \frac{\overline{P}_1(t)}{\overline{P}_2(t)} \ge \frac{1}{\overline{P}_1(t)} = \hat{h}(\kappa_{8.7}),$$

and thus since $\hat{h}$ is increasing we get $\kappa_{8.6} \ge \kappa_{8.7}$. □

The following two theorems are related to reliability properties, but not directly related to some of the usual classes. In the first of the theorems, we assume that U has density u. Recall from (7.1) that its failure rate $r(\ell)$ is given by $r(\ell) = u(\ell)/\overline{U}(\ell)$.

Theorem 8.9 (Willmot and Lin, 1994 p.751). *Suppose that U has a density and a failure rate $r(\ell)$ which satisfies $r(\ell) \ge r_{\min} > 0$. Then*

$$\overline{F}(y,t) \le \left(1 + \frac{r_{\min}}{t}\right) \overline{P}_1(t)\, e^{-\kappa y},$$

where κ is the solution of $\hat{h}(\kappa) - 1 = r_{\min}/t$.

Proof. We have, cf. (7.2),

$$\overline{U}_t(\ell) = e^{-t\ell}\overline{U}(\ell) = e^{-\int_0^\ell (t+r(y))\,dy}.$$

From this and (8.32) we then get

$$\phi_t(x) = t\int_x^\infty e^{-\int_x^\ell (t+r(y))\,dy}\,d\ell$$

$$\le t\int_x^\infty e^{-\int_x^\ell (t+r_{\min}))\,dy}\,d\ell = \frac{t}{t+r_{\min}}. \tag{8.34}$$

Thus Theorem 8.1 applies with $\phi = t/(t+r_{\min})$, as may be seen from (8.33). ■

Example 8.4. Let Λ be $\Gamma(\gamma, \beta)$, see Example 2.1. Then we have

$$r_{\min} = \begin{cases} \lim_{\ell\to\infty} r(\ell) = \beta & \text{for } \gamma \le 1, \\ r(0) = 0 & \text{for } \gamma > 1. \end{cases}$$

Thus, for $\gamma \le 1$, Theorem 8.9 applies with $r_{\min} = \beta$. In this case Λ is DFR, see Example 7.2, and also Theorem 8.7 applies. Since, see again Example 2.1, we have

$$\hat{u}(-v) = \left(1 - \frac{v}{\beta}\right)^{-\gamma},$$

it follows that $u_\infty = \beta$. Thus Theorems 8.7 and 8.9 give the same bound. □

Remark 8.5. The fact, seen in Example 8.4, that Theorems 8.7 and 8.9 give the same bound when Λ is DFR is not dependent on the assumption of a Γ-distribution, but holds for all distributions fulfilling the assumptions in Theorem 8.9. In order to see that, we assume that U is DFR with failure rate $r(\ell)$. Since $r(\ell)$ is non-negative and decreasing, $r_{\min} \stackrel{\text{def}}{=} \lim_{\ell\to\infty} r(\ell)$ exists. Like in (8.6) we have

$$\bar{u}(v) \stackrel{\text{def}}{=} \int_0^\infty e^{-v\ell}\overline{U}(\ell)\,d\ell = \frac{1 - \hat{u}(v)}{v},$$

which implies that $\bar{u}$ and $\hat{u}$ have the same abscissa of convergence. Obviously u_∞ does only depend on the tail of $\overline{U}$ or, more precisely, for each ℓ_0 also

$$\int_{\ell_0}^\infty e^{-v\ell}\overline{U}(\ell)\,d\ell = \overline{U}(\ell_0)e^{-v\ell_0}\int_{\ell_0}^\infty e^{-\int_{\ell_0}^{\ell}(v+r(y))\,dy}\,d\ell$$

has the abscissa of convergence $-u_\infty$. Comparing with the exponential distribution, it follows that $r_{\min} \le u_\infty \le r(\ell_0)$. Since ℓ_0 can be arbitrarily chosen, $r_{\min} = u_\infty$ follows, and thus Theorems 8.7 and 8.9 give the same bound when Λ is DFR.

In the IFR case Theorem 8.6 gives a better bound than Theorem 8.9. Intuitively this is rather natural, since both Theorems 8.6 and 8.9 rely on Theorem 8.1, and the ϕ used in Theorem 8.6 is the best possible for which Theorem 8.1 applies, cf. Remark 8.3. We will, however, show that Theorem 8.8 is strictly better than Theorem 8.9, unless Λ is exponentially distributed. Recall from Remark 8.4 that Theorem 8.6 gives a better bound than Theorem 8.8 in the IFR case.

Suppose that U is IFR with failure rate $r(\ell)$. Really it is enough, as will be seen from the proof, to assume that U is IFRA. From (8.33) we get

$$\begin{aligned}\overline{P}_1(t) &= t\int_0^\infty e^{-\ell t}\overline{U}(\ell)\,d\ell \\ &= t\int_0^\infty e^{-\ell t}e^{-\int_0^\ell r(y)\,dy}\,d\ell \\ &\le t\int_0^\infty e^{-\ell t}e^{-\ell r_{\min}}\,d\ell = \frac{t}{t+r_{\min}}. \end{aligned} \tag{8.35}$$

Thus we have a smaller ϕ in Theorem 8.8 than in Theorem 8.9, unless we have equality in (8.35). Equality in (8.35) holds if and only if

$$\int_0^\ell r(y)\,dy = \ell r_{\min},$$

i.e. if and only if Λ is exponentially distributed. □

The next theorem is in terms of the residual life time of Λ. Recall from (7.3) the definition of the corresponding conditional survivor function

$$\overline{U}(\ell \mid x) \stackrel{\text{def}}{=} P\{\Lambda - x > \ell \mid \Lambda > x\} = \frac{\overline{U}(x+\ell)}{\overline{U}(x)},$$

for all x and $\ell \ge 0$ such that $\overline{U}(x) > 0$.

Similarly we define the conditional distribution by

$$dU(\ell \mid x) \stackrel{\text{def}}{=} \frac{dU(x+\ell)}{\overline{U}(x)}$$

and the conditional Laplace transforms

$$\hat{u}_x(v) \stackrel{\text{def}}{=} E[e^{-v(\Lambda - x)} \mid \Lambda > x] = \int_0^\infty e^{-v\ell}\,dU(\ell \mid x).$$

and

$$\bar{u}_x(v) \stackrel{\text{def}}{=} \int_0^\infty e^{-v\ell}\overline{U}(\ell \mid x)\,d\ell.$$

Put, cf. (8.32),

$$\text{MRL}(x) \stackrel{\text{def}}{=} E[\Lambda - x \mid \Lambda > x].$$

Theorem 8.10 (Willmot and Lin, 1994 p.751). *Suppose that* U *has the mean residual life time* $\mathrm{MRL}(x)$ *which satisfies* $\mathrm{MRL}(x) \le \mathrm{MRL}_{\max} < \infty$. *Then*

$$\overline{F}(y,t) \le \left(1 - e^{-t\,\mathrm{MRL}_{\max}}\right)^{-1} \overline{P}_1(t)\, e^{-\kappa y},$$

where κ *is the solution of* $\hat{h}(\kappa) = \left(1 - e^{-t\,\mathrm{MRL}_{\max}}\right)^{-1}$.

Proof. We have, cf. (8.32) and (8.6),

$$\phi_t(x) = t \cdot \mathrm{MRL}_t(x) = t \cdot \frac{\int_0^\infty e^{-(x+\ell)t}\overline{U}(x+\ell)\,d\ell}{e^{-xt}\overline{U}(x)}$$

$$= t \cdot \bar{u}_x(t) = 1 - \hat{u}_x(t) \le 1 - e^{-t\,\mathrm{MRL}(x)} \le 1 - e^{-t\,\mathrm{MRL}_{\max}},$$

where the first inequality follows by Jensen's inequality, see Feller (1971, pp.153–154).

Thus Theorem 8.1 applies with $\phi = 1 - e^{-t\,\mathrm{MRL}_{\max}}$, as may be seen from (8.33). ■

If Λ is DMRL or NBUE, then Theorem 8.10 applies with $\mathrm{MRL}_{\max} = \mu_\Lambda$. Similarly, if Λ is IMRL, then Theorem 8.10 applies with $\mathrm{MRL}_{\max} = \lim_{x\to\infty} \mathrm{MRL}(x)$.

Remark 8.6. Let U have failure rate $r(\ell) \ge r_{\min} > 0$, so that Theorem 8.9 applies. Since

$$\mathrm{MRL}(x) = \frac{\int_0^\infty \overline{U}(x+\ell)\,d\ell}{\overline{U}(x)} = \int_0^\infty e^{-\int_x^{x+\ell} r(y)\,dy}\,d\ell$$

$$\le \int_0^\infty e^{-\ell r_{\min}}\,d\ell = \frac{1}{r_{\min}},$$

Theorem 8.10 applies with $\mathrm{MRL}_{\max} = 1/r_{\min}$. Using

$$e^{-t/r_{\min}} = \frac{1}{e^{t/r_{\min}}} \le \frac{1}{1 + t/r_{\min}}$$

we get

$$1 - e^{-t/r_{\min}} \ge 1 - \frac{1}{1 + t/r_{\min}} = \frac{t}{t + r_{\min}},$$

i.e. Theorem 8.9 gives a better bound than Theorem 8.10. For further comparisons we refer to Willmot and Lin (1994, pp.752–753). □

8.3 Asymptotic behaviour

We will now consider some relations between the behaviour of $U(\ell)$ for large values of ℓ and the corresponding distributions of $N(t)$ and $Y(t)$, for fixed values of t when N is a mixed Poisson process.

An early and simple result in this direction is the following proposition due to Grandell (1970).

Proposition 8.4. *Suppose that*

$$\overline{U}(\ell) \sim L(\ell)\ell^{\gamma}, \quad \ell \to \infty,$$

for $-1 < \gamma < 0$. *Then*

$$P\{N(t) > n\} \sim L(n)(n/t)^{\gamma}, \quad n \to \infty.$$

Proof. From Proposition 2.1 (vi) it follows that $G_{N(t)}(s) = \hat{u}(t(1-s))$, and thus, cf. (8.15) and (8.6),

$$\bar{G}_{N(t)}(s) = t\bar{u}(t(1-s)),$$

where, as before,

$$\bar{G}_{N(t)}(s) = \sum_{k=0}^{\infty} P\{N(t) > n\}s^k \text{ and } \bar{u}(v) = \int_0^{\infty} e^{-v\ell}\overline{U}(\ell)\, d\ell.$$

It follows from Theorem 8.2 in the Abelian direction, with $\rho = \gamma + 1$, that

$$\bar{u}(v) \sim \frac{\Gamma(\gamma+1)}{v^{\gamma+1}} L\left(\frac{1}{v}\right), \quad v \to 0.$$

Thus

$$\begin{aligned}\bar{G}_{N(t)}(s) &\sim \frac{\Gamma(\gamma+1)}{t^{\gamma}(1-s)^{\gamma+1}} L\left(\frac{1}{t(1-s)}\right) \\ &\sim \frac{\Gamma(\gamma+1)}{t^{\gamma}} \frac{1}{(1-s)^{\gamma+1}} L\left(\frac{1}{1-s}\right), \quad s \uparrow 1.\end{aligned}$$

The proposition now follows from Theorem 8.3 in the Tauberian direction. ■

Then the following extensive generalization, which is due to Willmot (1990), holds.

Theorem 8.11. *Suppose that U is absolutely continuous with density u, and that*

$$u(\ell) \sim L(\ell)\ell^{\gamma-1}e^{-\beta\ell}, \quad \ell \to \infty, \tag{8.36}$$

for $\beta \geq 0$ and $-\infty < \gamma < \infty$ (with $\gamma < 0$ if $\beta = 0$). Then

$$p_n(t) \sim \frac{L(n)}{(\beta+t)^\gamma}\left(\frac{t}{\beta+t}\right)^n n^{\gamma-1}, \quad n \to \infty. \tag{8.37}$$

Proof. We will first show that it is no restriction to assume equality in (8.36). Define $C(\ell)$ by

$$u(\ell) = C(\ell)\ell^{\gamma-1}e^{-\beta\ell}.$$

From (8.36) it follows that $C(\ell) \sim L(\ell)$. Thus, if (8.37) holds for L replaced by C, it also holds as it stands.

Assume now that $u(\ell) = L(\ell)\ell^{\gamma-1}e^{-\beta\ell}$, and let

$$f_n(\ell) \stackrel{\text{def}}{=} \frac{(\beta+t)^{\gamma+n}}{\Gamma(\gamma+n)}\ell^{\gamma+n-1}e^{-(\beta+t)\ell},$$

i.e. $f_n(\ell)$ is, for $n > -\gamma$, the density of the $\Gamma(\gamma+n, \beta+t)$-distribution. Then we get, cf. (8.21),

$$\begin{aligned} p_n(t) &= \int_0^\infty \frac{(\ell t)^n}{n!} e^{-\ell t} L(\ell)\ell^{\gamma-1}e^{-\beta\ell}\, d\ell \\ &= \frac{1}{(\beta+t)^\gamma}\left(\frac{t}{\beta+t}\right)^n \frac{\Gamma(\gamma+n)}{n!}\int_0^\infty L(\ell)f_n(\ell)\, d\ell. \\ &\sim \frac{1}{(\beta+t)^\gamma}\left(\frac{t}{\beta+t}\right)^n n^{\gamma-1}\int_0^\infty L(\ell)f_n(\ell)\, d\ell, \quad n \to \infty. \end{aligned}$$

Thus the theorem is proved, if we show that

$$\int_0^\infty L(\ell)f_n(\ell)\, d\ell \sim L(n)\int_0^\infty f_n(\ell)\, d\ell = L(n). \tag{8.38}$$

In Willmot (1990) this is proved by applying an Abelian theorem due to Vuilleumier (1963), see Bingham *et al.* (1987, p.200). We will, however, use a more probabilistic argument.

Let X_n be a random variable with density f_n. We always assume that $n > -\gamma$. Define $Y_n \stackrel{\text{def}}{=} X_n/n$. It follows from Example 2.1 that

$$E[Y_n] = \frac{\gamma+n}{n(\beta+t)} \qquad \mathrm{Var}[Y_n] = \frac{\gamma+n}{n^2(\beta+t)^2},$$

and thus $Y_n \xrightarrow{\mathrm{P}} 1/(\beta+t)$ for $n \to \infty$, where $\xrightarrow{\mathrm{P}}$ means 'convergence in probability'. Thus, to prove (8.38) is the same as to prove

$$\lim_{n\to\infty} E\left[\frac{L(nY_n)}{L(n)}\right] = 1. \tag{8.39}$$

Since $\lim_{n\to\infty} L(ny)/L(n) = 1$ for all $y \geq 0$, it 'only' remains to find a dominator.

It follows from Lemma 8.1 that for each $A > 1$ and $\delta > 0$ there exists an x_0 such that

$$\frac{L(ny)}{L(n)} \leq A\max(y^{\delta}, y^{-\delta}) \tag{8.40}$$

for all $n \geq x_0$ and $y \geq x_0/n$. Now we act as if $L(x) = 0$ for $x < x_0$. This is no restriction, since – by the first arguments in this proof – $L(x)$ may be changed for small values of x and since no such x-values are involved in (8.40).

We have, for $\eta = \pm\delta$ and for n large enough,

$$E[Y_n^{\eta}] = n^{-\eta}E[X_n^{\eta}]$$

$$= n^{-\eta}\int_0^{\infty} \frac{(\beta+t)^{\gamma+n}}{\Gamma(\gamma+n)} \ell^{\gamma+\eta+n-1}e^{-(\beta+t)\ell}\, d\ell$$

$$= n^{-\eta}(\beta+t)^{-\eta}\frac{\Gamma(\gamma+\eta+n)}{\Gamma(\gamma+n)},$$

and thus, cf. (8.21),

$$E[Y_n^{\eta}] \sim n^{-\eta}(\beta+t)^{-\eta}n^{\eta} = (\beta+t)^{-\eta}, \quad n \to \infty,$$

and (8.39) follows by dominated convergence. The above moment estimate had also to be done by Willmot (1990), and thus we have proved Vuilleumier's Abelian theorem in our very special case. ■

Many distributions fulfil (8.36) as Γ, Pareto, generalized inverse Gaussian and certain stable distributions, cf. Willmot (1990). The original formulation of Theorem 8.11, given by Willmot (1990), also allow for discrete mixtures such that

$$P\{\Lambda = n\} \sim L(n)n^{\gamma-1}e^{-\beta n}, \quad n \to \infty.$$

Although the proof in 'principle' is the same, it becomes technically much more intricate.

Using Proposition 2.1 (iii) we get the following corollary of Theorem 8.11

Corollary 8.5. *Suppose that*

$$\overline{U}(\ell) \sim L(\ell)\ell^{\gamma}e^{-\beta\ell}, \quad \ell \to \infty,$$

for $\beta \geq 0$ *and* $-\infty < \gamma < \infty$ *(with* $\gamma < -1$ *if* $\beta = 0$*). Then*

$$P\{N(t) > n\} \sim \frac{L(n)\,t}{(\beta+t)^{\gamma+1}}\left(\frac{t}{\beta+t}\right)^n n^{\gamma}, \quad n \to \infty. \tag{8.41}$$

Comparing Corollary 8.5 with Proposition 8.4 we see that, very formally, Corollary 8.5 is not a strict generalization, since for $\beta = 0$ they apply to different values of γ. Notice, however, that for $\beta = 0$, formula (8.41) is reduced to

$$P\{N(t) > n\} \sim \frac{L(n)}{t^{\gamma}}n^{\gamma},$$

as in Proposition 8.4.

The assumptions on U have been formulated in terms of asymptotic properties of either $u(\cdot)$ or $\overline{U}(\cdot)$. The relation between those properties is given in the following proposition, which – although certainly known – we have not seen explicitly stated for $\beta > 0$.

Proposition 8.5. *Suppose that* U *is absolutely continuous with density* u*, and that*

$$u(\ell) \sim L(\ell)\ell^{\gamma-1}e^{-\beta\ell},$$

for $\beta \geq 0$ *and* $-\infty < \gamma < \infty$ *(with* $\gamma < 0$ *if* $\beta = 0$*). Then*

$$\overline{U}(\ell) \sim \begin{cases} \frac{L(\ell)\ell^{\gamma}}{-\gamma} & \text{if } \beta = 0, \\ \frac{L(\ell)\ell^{\gamma-1}}{\beta}e^{-\beta\ell} & \text{if } \beta > 0. \end{cases}$$

Proof. If $\beta = 0$ the result is, for instance, proved by Feller (1971, p.281), and therefore we consider the case $\beta > 0$.

Assume that $u(\ell) = L(\ell)\ell^{\gamma-1}e^{-\beta\ell}$, which is no restriction, cf. the proof of Theorem 8.11. Then we have

$$\overline{U}(\ell) = \frac{L(\ell)\ell^{\gamma-1}}{\beta}\, e^{-\beta\ell} \cdot \beta \int_\ell^\infty \frac{L(x)x^{\gamma-1}}{L(\ell)\ell^{\gamma-1}}\, e^{-\beta(x-\ell)}\, dx.$$

Choose $A > 1$ and $\delta > 0$. Then we get, by Lemmata 8.1 and 8.2,

$$\beta \int_\ell^\infty \frac{L(x)x^{\gamma-1}}{L(\ell)\ell^{\gamma-1}}\, e^{-\beta(x-\ell)}\, dx$$

$$\left\{ \begin{matrix} \leq A \\ \geq A^{-1} \end{matrix} \right\} \cdot \beta \int_\ell^\infty \left(\frac{x}{\ell}\right)^{\gamma-1\{\pm\}\delta} e^{-\beta(x-\ell)}\, dx \sim \left\{ \begin{matrix} A \\ A^{-1} \end{matrix} \right\},$$

for ℓ larger than some ℓ_0. Since A can be chosen arbitrarily close to 1, the proposition follows. ∎

Consider now $Y(t)$. The following theorem follows from Theorems 8.11 and 8.4.

Theorem 8.12. *Suppose that U is absolutely continuous with density u, that*

$$u(\ell) \sim L(\ell)\ell^{\gamma-1}e^{-\beta\ell},$$

for $\beta > 0$, and that F is light-tailed and non-lattice. Then

$$\overline{F}(y,t) \sim \frac{L(y)}{(\hat{h}'(\kappa)\,t)^\gamma \kappa} y^{\gamma-1}e^{-\kappa y}, \quad y \to \infty, \tag{8.42}$$

where $\kappa > 0$ satisfies $\hat{h}(\kappa) - 1 = \beta/t$.

Example 8.5. We will consider the simplest case. Let Λ be exponentially distributed with mean $1/\beta$. It follows from Example 2.1 that

$$p_n(t) = \frac{\beta}{\beta+t}\left(\frac{t}{\beta+t}\right)^n$$

and

$$P\{N(t) > n\} = \left(\frac{t}{\beta+t}\right)^{n+1}.$$

Thus it is seen, for $L(\ell) = \beta$ and $\gamma = 1$ and for $L(\ell) = 1$ and $\gamma = 0$ respectively, that (8.37) and (8.41) hold exactly.

Let now F be exponentially distributed with mean μ. It follows from (8.48) that

$$\overline{F}(y,t) = \frac{t}{\beta+t}\, e^{-\frac{\beta y}{\mu(\beta+t)}}. \tag{8.43}$$

Thus, for $L(\ell) = \beta$ and $\gamma = 1$, (8.42) is reduced to

$$\overline{F}(y,t) \sim \frac{\beta}{\hat{h}'(\kappa)\, t\kappa} e^{-\kappa y}. \tag{8.44}$$

Further, see Example 7.2, κ is the solution of $\frac{\mu r}{1-\mu r} = \frac{\beta}{t}$, which yields $\kappa = \frac{\beta}{\mu(\beta+t)}$. Since $\hat{h}'(r) = \frac{\mu}{(1-\mu r)^2}$ we get $\hat{h}'(\kappa) = \mu \Big/ \left(\frac{t}{\beta+t}\right)^2$. Putting this into (8.44) we get (8.43), i.e. (8.44) holds exactly. □

Remark 8.7. Theorem 8.12 is related to **saddlepoint**, or **Esscher, approximations** of $1 - F(y,t)$ as $y \to \infty$ and t is fixed; the idea goes back to Esscher (1932). Those approximations are, however, not identical with (8.42), and we will shortly consider the difference.

Let Λ be $\Gamma(\beta,\beta)$, and recall from Example 2.1 that this implies that $\mu_\Lambda = 1$ and $\sigma_\Lambda^2 = 1/\beta$. In this case (8.42) is reduced to

$$\overline{F}(y,t) \sim \frac{1}{(\hat{h}'(\kappa)\, t)^\beta \kappa} \frac{\beta^\beta}{\Gamma(\beta)} y^{\beta-1} e^{-\kappa y}. \tag{8.45}$$

The simplest saddlepoint approximations of $1 - F(y,t)$ is in this case, see Embrechts *et al.* (1985a, p.627–628),

$$\overline{F}(y,t) \sim \frac{\left(\frac{y}{\hat{h}'(\kappa_y)\, t}\right)^\beta - \left(\frac{\beta}{\beta+t}\right)^\beta}{\kappa_y y} \frac{\beta^\beta e^{-\beta}}{\Gamma(\beta)} e^{-\kappa_y y}, \tag{8.46}$$

where κ_y satisfies

$$\frac{\hat{h}'(\kappa_y)\, t}{1 - (\hat{h}(\kappa_y) \quad 1)\, t/\beta} = y. \tag{8.47}$$

The approximation (8.46) is called the 'first order short form approximation' and holds for $\beta > 1$ and under certain conditions on $\hat{h}(\,\cdot\,)$.

The most striking difference between (8.45) and (8.46) is certainly that κ_y depends on y. It follows from (8.47) that $\kappa_y \to \kappa$ as $y \to \infty$. Thus

$$\hat{h}(\kappa_y) \approx \hat{h}(\kappa) + (\kappa_y - \kappa)\hat{h}'(\kappa) = \frac{\beta}{t} + (\kappa_y - \kappa)\hat{h}'(\kappa)$$

for large values of y. Putting this into (8.47) we get

$$y \approx \frac{\beta}{\kappa - \kappa_y} \quad \text{or} \quad -\kappa_y y \approx -\kappa y + \beta.$$

Thus (8.45) and (8.46) agree asymptotically, which of course was to be expected.

For a survey of different numerical and approximate methods to evaluate $\overline{F}(y,t)$ as well as for numerical comparisons we refer to Bohman and Esscher (1963 and 1964). □

The following theorem may be looked upon as corresponding to Theorem 8.12 for $\beta = 0$.

Theorem 8.13. *Suppose that*

$$\overline{U}(\ell) \sim \omega L(\ell)\ell^\gamma, \quad \omega \geq 0 \text{ and } \ell \to \infty,$$

and that

$$\overline{F}(z) \sim \chi L(z) z^\gamma, \quad \chi \geq 0 \text{ and } z \to \infty,$$

for some $\gamma < -1$. *Then*

$$\overline{F}(y,t) \sim (\omega(\mu t)^{-\gamma} + \chi\mu_\Lambda t)L(y)y^\gamma, \quad y \to \infty.$$

Proof. Consider first $\omega > 0$. Then it follows from Corollary 8.5 that $\overline{U}(\ell) \sim \omega L(\ell)\ell^\gamma$ implies

$$P\{N(t) > n\} \sim \omega L(n)(n/t)^\gamma, \quad n \to \infty,$$

and the theorem follows from Corollary 8.1.

Consider now $\omega = 0$, i.e. assume that $\overline{U}(\ell) = o(\ell^\gamma L(\ell))$. By properties of the $\Gamma(n+1,t)$-distribution, and since $n! > n^n e^{-n}$, we get, for $a < 1$,

$$P\{N(t) > n\} = \int_0^\infty \frac{t^{n+1}\ell^n}{n!} e^{-\ell t}\overline{U}(\ell)\, d\ell$$

$$\leq \int_0^{an/t} \frac{t^{n+1}\ell^n}{n!} e^{-\ell t}\, d\ell + \int_{an/t}^\infty \frac{t^{n+1}\ell^n}{n!} e^{-\ell t}\, d\ell \cdot \overline{U}(an/t)$$

$$\le \frac{an}{t}\frac{t(an)^n}{n!}\,e^{-an} + \overline{U}(an/t) \le \frac{(an)^{n+1}e^{-an}}{n^n e^{-n}} + \overline{U}(an/t)$$
$$\le na^n e^n + \overline{U}(an/t).$$

Choose $a < 1/e$. Then

$$\lim_{n\to\infty} \frac{P\{N(t) > n\}}{n^\gamma L(n)} \le \lim_{n\to\infty} \frac{n(ae)^n + \overline{U}(an/t)}{n^\gamma L(n)}$$

$$= (a/t)^\gamma \lim_{n\to\infty} \frac{\overline{U}(an/t)}{(an/t)^\gamma L(an/t)} = 0,$$

i.e. also in this case

$$P\{N(t) > n\} \sim \omega L(n)(n/t)^\gamma, \quad n \to \infty,$$

and the theorem follows as above. ■

Recall the definition of the classes $\mathcal{S}$, Definition 8.5, and $\mathcal{S}(\beta)$, Definition 8.7. The following two theorems are merely 'translations' of Propositions 8.2 and 8.3.

Theorem 8.14. *Suppose that $F \in \mathcal{S}$ and that U is light-tailed. Then*

$$\overline{F}(y,t) \sim \mu_A t\overline{F}(y), \quad y \to \infty.$$

Proof. It follows from (8.31) that $u_\infty > 0$ implies $s_\infty > 1$, so $N(t)$ is light-tailed, and the theorem follows from Proposition 8.2. ■

Theorem 8.15. *Suppose that $F \in \mathcal{S}(\beta)$ and that U is light-tailed with $u_\infty > t(\hat{h}(\beta) - 1)$. Then*

$$\overline{F}(y,t) \sim |\hat{u}'(-t(\hat{h}(\beta) - 1))| \cdot t\overline{F}(y), \quad y \to \infty.$$

Proof. It follows from (8.31) that $u_\infty > t(\hat{h}(\beta)-1)$ implies $s_\infty > \hat{h}(\beta)$. From Proposition 2.1 (vi) we get

$$G_{N(t)}(s) = \hat{u}(t(1-s))$$

and

$$G'_{N(t)}(s) = -t\hat{u}'(t(1-s)) = |\hat{u}'(t(1-s))|t.$$

Thus the theorem follows from Proposition 8.3. ■

Theorem 8.14 is a strict generalization of Theorem 8.13 when U is light-tailed. The really nice thing with Theorem 8.13 is that F and U contribute $\overline{F}(y,t)$ in a rather symmetric way. It is therefore tempting to consider F light-tailed and $U \in \mathcal{S}$. However, $\mathcal{S}$ is defined in terms of convolution properties, while there seems not to be any natural convolution of U with itself involved. Therefore such an approach seems to be fruitless. In spite of that, the following proposition – due to Schmidli (1996) – holds.

Proposition 8.6. *Suppose that* $U \in \mathcal{S}$. *Then* $F_{N(t)} \in \mathcal{S}$, *where*

$$F_{N(t)}(x) \stackrel{\text{def}}{=} P\{N(t) \le x\}.$$

Proof. The main idea of the proof is to rely on the following simple relation between convolutions of structure distributions and convolutions of mixed Poisson distributions:

Let $F_{N(t)}$ *be* $\text{MP}(t,U)$. *Then* $F^{2*}_{N(t)}$ *is* $\text{MP}(t,U^{2*})$.

This relation follows immediately from Proposition 2.1 (vi) and was used in the proof of Proposition 2.5.

From the definition of $\mathcal{S}$, Definition 8.5, and the remarks after Lemma 8.3 it follows that for any $\varepsilon \in (0,1)$ there exists an $\ell_0 < \infty$ such that

$$2(1-\varepsilon)\overline{U}(\ell) \le \overline{U^{2*}}(\ell) \le 2(1+\varepsilon)\overline{U}(\ell) \quad \text{and} \quad e^{\ell}\,\overline{U}(\ell) \ge 1, \quad \forall \ell \ge \ell_0.$$

It follows from Proposition 2.1 (iii) that

$$\frac{\overline{F}^{2*}_{N(t)}(n)}{\overline{F}_{N(t)}(n)} = \frac{\int_0^\infty \frac{(\ell t)^n}{n!}\, e^{-\ell t}\overline{U^{2*}}(\ell)\,d\ell}{\int_0^\infty \frac{(\ell t)^n}{n!}\, e^{-\ell t}\overline{U}(\ell)\,d\ell}$$

$$= \frac{\int_0^{\ell_0} \frac{(\ell t)^n}{n!}\, e^{-\ell t}\overline{U^{2*}}(\ell)\,d\ell}{\int_0^\infty \frac{(\ell t)^n}{n!}\, e^{-\ell t}\overline{U}(\ell)\,d\ell} + \frac{\int_{\ell_0}^\infty \frac{(\ell t)^n}{n!}\, e^{-\ell t}\overline{U^{2*}}(\ell)\,d\ell}{\int_0^\infty \frac{(\ell t)^n}{n!}\, e^{-\ell t}\overline{U}(\ell)\,d\ell}.$$

Let $C \stackrel{\text{def}}{=} \inf_{0\le \ell<\infty} e^{\ell}\,\overline{U}(\ell)$ and notice that

$$C \ge \min\left(\inf_{0\le\ell<\ell_0} e^{\ell}\,\overline{U}(\ell), 1\right) \ge \min\left(\overline{U}(\ell_0), 1\right) = \overline{U}(\ell_0) > 0.$$

We have

$$0 \le \frac{\int_0^{\ell_0} \frac{(\ell t)^n}{n!} e^{-\ell t}\overline{U}^{2*}(\ell)\,d\ell}{\int_0^\infty \frac{(\ell t)^n}{n!} e^{-\ell t}\overline{U}(\ell)\,d\ell} \le \frac{\int_0^{\ell_0} \frac{(\ell t)^n}{n!} e^{-\ell t}\,d\ell}{\int_0^\infty \frac{(\ell t)^n}{n!} e^{-\ell t}Ce^{-\ell}\,d\ell}$$

$$\le \frac{\ell_0 \frac{(\ell_0 t)^n}{n!}}{Ct^n(1+t)^{-(n+1)}} = \frac{(\ell_0(1+t))^{n+1}}{Cn!} \to 0, \quad n \to \infty.$$

Similarly as above we get

$$\frac{\int_{\ell_0}^\infty \frac{(\ell t)^n}{n!} e^{-\ell t}\overline{U}^{2*}(\ell)\,d\ell}{\int_0^\infty \frac{(\ell t)^n}{n!} e^{-\ell t}\overline{U}(\ell)\,d\ell} \ge \frac{2(1-\varepsilon)\int_{\ell_0}^\infty \frac{(\ell t)^n}{n!} e^{-\ell t}\overline{U}(\ell)\,d\ell}{\int_0^\infty \frac{(\ell t)^n}{n!} e^{-\ell t}\overline{U}(\ell)\,d\ell}$$

$$= 2(1-\varepsilon)\left(1 - \frac{\int_0^{\ell_0} \frac{(\ell t)^n}{n!} e^{-\ell t}\overline{U}(\ell)\,d\ell}{\int_0^\infty \frac{(\ell t)^n}{n!} e^{-\ell t}\overline{U}(\ell)\,d\ell}\right)$$

$$\ge 2(1-\varepsilon)\left(1 - \frac{\int_0^{\ell_0} \frac{(\ell t)^n}{n!} e^{-\ell t}\,d\ell}{\int_0^\infty \frac{(\ell t)^n}{n!} e^{-\ell t}Ce^{-\ell}\,d\ell}\right) \to 2(1-\varepsilon), \quad n \to \infty.$$

Furthermore we have

$$\frac{\int_{\ell_0}^\infty \frac{(\ell t)^n}{n!} e^{-\ell t}\overline{U}^{2*}(\ell)\,d\ell}{\int_0^\infty \frac{(\ell t)^n}{n!} e^{-\ell t}\overline{U}(\ell)\,d\ell} \le \frac{2(1+\varepsilon)\int_{\ell_0}^\infty \frac{(\ell t)^n}{n!} e^{-\ell t}\overline{U}(\ell)\,d\ell}{\int_0^\infty \frac{(\ell t)^n}{n!} e^{-\ell t}\overline{U}(\ell)\,d\ell} \le 2(1+\varepsilon).$$

Those three estimates yield

$$2(1-\varepsilon) \le \liminf_{n\to\infty} \frac{\overline{F}^{2*}_{N(t)}(n)}{\overline{F}_{N(t)}(n)} \le \limsup_{n\to\infty} \frac{\overline{F}^{2*}_{N(t)}(n)}{\overline{F}_{N(t)}(n)} \le 2(1+\varepsilon).$$

Since $\varepsilon \in (0,1)$ may be arbitrarily chosen it follows that

$$\lim_{n\to\infty} \frac{\overline{F}^{2*}_{N(t)}(n)}{\overline{F}_{N(t)}(n)} = 2,$$

which completes the proof. ■

Recall from Proposition 8.4 and Corollary 8.5 that if $\overline{U} \in \mathrm{RV}_\gamma$, i.e. if $\overline{U}(\ell) \sim L(\ell)\ell^\gamma$, for $\gamma < 0$, cf. Definition 8.3 and the remarks after Definition 8.5, then

$$\overline{F}_{N(t)}(n) \sim L(n)(n/t)^\gamma \sim L(n/t)(n/t)^\gamma \sim \overline{U}(n/t), \quad n \to \infty.$$

A natural question is whether this holds for all $U \in \mathcal{S}$, which would be much more informative than Proposition 8.6. The proof of Proposition 8.6 gives no hints if this is the case. Formula (2.31) indicates that it might be the case if Λ is lognormally distributed.

It is tempting to try to modify the proofs of Proposition 8.4 or Theorem 8.11.

The proof of Proposition 8.4 relies on Abelian and Tauberian theorems, Theorems 8.2 and 8.3, and is thus highly dependent on the assumption of regular variation.

The main idea of the proof of Theorem 8.11 might be more promising to try to modify. Let

$$f_n(\ell) \stackrel{\text{def}}{=} \frac{t^{n+1}}{n!}\ell^n e^{-t\ell},$$

i.e. $f_n(\ell)$ is, for $n \geq 0$, the density of the $\Gamma(n+1,t)$-distribution. Let $Y_n = X_n/n$, where X_n is a random variable with density f_n, and notice that

$$E[Y_n] = \frac{n+1}{nt}, \quad \mathrm{Var}[Y_n] = \frac{n+1}{n^2t^2} \quad \text{and} \quad Y_n \xrightarrow{\text{P}} 1/t, \quad n \to \infty.$$

Then we have

$$\overline{F}_{N(t)}(n) = \int_0^\infty \overline{U}(\ell)f_n(\ell)\,d\ell = E[\overline{U}(nY_n)].$$

In order to follow the proof of Theorem 8.11 further, we could assume that $\lim_{n\to\infty} \overline{U}(ny)/\overline{U}(n/t)$ exists for all $y > 0$. However, this is the same as to assume regular variation, cf. Feller (1971, p.275).

Naturally from the discussion above one cannot draw the conclusion that regular variation is necessary. It is an open question for which distributions U the relation $\overline{F}_{N(t)}(n) \sim \overline{U}(n/t)$ holds.

Remark 8.8. Theorems 8.12, 8.13, and 8.14 are all formulated in such a way that, for some conditions on $U(\ell)$ and $F(z)$, we get a result about $F(y,t)$, which holds for all t. In Theorem 8.15 the conditions on $U(\ell)$ and $F(z)$ involves t and another way to express $u_\infty > t(\hat{h}(\beta)-1)$ is to say that Theorem 8.15 holds for $t < u_\infty/(\hat{h}(\beta)-1)$, i.e. for small values of t. A natural question is then: What happens for large values of t? In order to throw some light upon that question we consider the case

$$u(\ell) \sim L(\ell)\ell^{\gamma-1}e^{-\beta_u\ell}, \quad \text{for } \beta_u > 0,$$

where obviously $u_\infty = \beta_u$. Assume further that $F \in \mathcal{S}(\beta)$ and that F is non-lattice.

According to Theorem 8.15 we then have

$$\overline{F}(y,t) \sim |\hat{u}'(-t(\hat{h}(\beta)-1))| \cdot t\overline{F}(y), \quad \text{for } t < \frac{\beta_u}{\hat{h}(\beta)-1}.$$

Now consider Theorem 8.12. With our formulation, it does not apply. However, as mentioned in Remark 8.2, it is enough to require that $\hat{h}(\kappa) - 1 = \beta_u/t$ has a solution $\kappa < \beta$ or, which is the same, that $\hat{h}(\beta) - 1 > \beta_u/t$. According to Theorem 8.15 we then have

$$\overline{F}(y,t) \sim \frac{L(y)}{(\hat{h}'(\kappa)\,t)^{\gamma}\kappa} y^{\gamma-1}e^{-\kappa y}, \quad \text{for } t > \frac{\beta_u}{\hat{h}(\beta)-1},$$

with κ as above. □

8.4 Recursive evaluation

We restrict ourselves to the case where F is discrete, a case mentioned in Remark 8.2 and in the continuation of Example 8.2 in section 8.1. Let, as before,

$$p_m = P\{N = m\}, \quad G(s) = E[s^N]$$

and

$$f_k = P\{Z_i = k\}, \quad H(s) = E[s^{Z_i}].$$

In this case also Y is discrete, and let

$$g_k = P\{Y = k\}, \quad H_Y(s) = E[s^Y].$$

By using the g_k instead of the logical but clumsy notation $f_{Y,k}$ we follow the notation which seems to have become conventional. Recall further that

$$g_k = \sum_{m=0}^{\infty} p_m f_k^{m*} \quad \text{and} \quad H_Y(s) = G(H(s)),$$

where $f_k^{m*} = P\{Z_1 + \ldots + Z_m = k\}$, or

$$f_k^{(m+1)*} = \sum_{j=0}^{k} f_j f_{k-j}^{m*}. \tag{8.48}$$

Since the Z_is are identically distributed, it follows that

$$E[Z_i \mid Z_1 + \ldots Z_{m+1} = k] = \sum_{j=0}^{k} j \frac{f_j f_{k-j}^{m*}}{f_k^{(m+1)*}} = \frac{k}{m+1},$$

or

$$\frac{f_k^{(m+1)*}}{m+1} = \sum_{j=0}^{k} \frac{j}{k} f_j f_{k-j}^{m*}. \tag{8.49}$$

Let us first consider the case where the Panjer assumption (2.7) holds, i.e. where

$$p_{m+1} = \left(a + \frac{b}{m+1}\right) p_m, \quad m = 0, 1, \ldots \tag{8.50}$$

Theorem 8.16 (Panjer, 1981). *Assume that (8.50) holds. Then*

$$g_0 = G(f_0) \quad \textit{and} \quad g_k = \sum_{j=0}^{k} \left(a + \frac{bj}{k}\right) f_j g_{k-j}, \quad k = 1, 2, \ldots$$

Proof. We have

$$g_0 = \sum_{m=0}^{\infty} p_m f_0^{m*} = \sum_{m=0}^{\infty} p_m f_0^{m} = G(f_0)$$

and

$$\begin{aligned} g_k &= \sum_{m=0}^{\infty} p_m f_k^{m*} = \sum_{m=1}^{\infty} p_m f_k^{m*} = \sum_{m=0}^{\infty} p_{m+1} f_k^{(m+1)*} \\ &= \sum_{m=0}^{\infty} \left(a + \frac{b}{m+1}\right) p_m f_k^{(m+1)*} \\ &= \sum_{m=0}^{\infty} p_m \sum_{j=0}^{k} \left(a + \frac{bj}{k}\right) f_j f_{k-j}^{m*} = \sum_{j=0}^{k} \left(a + \frac{bj}{k}\right) f_j g_{k-j}, \end{aligned}$$

for $k = 1, 2, \ldots$. ■

If $f_0 > 0$ we have to solve the recursion given above for g_k. To avoid that we may write the recursion on the form

$$g_k = \frac{1}{1 - af_0} \sum_{j=1}^{k} \left(a + \frac{bj}{k} \right) f_j g_{k-j}.$$

Since $f_0 = 0$ is a natural – or at least an innocent – assumption we prefer the shorter form used in Theorem 8.16.

The recursion given in Theorem 8.16 is the **Panjer recursion**, mentioned in Chapter 2. Recall that the Panjer recursion only applies to the Poisson, the negative binomial and the binomial distribution, as was shown by Sundt and Jewell (1981). They (Sundt and Jewell 1981, p.33) further extended the Panjer assumption to

$$p_{m+1} = \left(a + \frac{b}{m+1} \right) p_m, \quad m = 1, 2, \ldots, \tag{8.51}$$

i.e. they let p_0 be (relatively) free. By this extension the assumption also includes the logarithmic distribution, see the continuation of Example 2.1.

Schröter (1990, p.165) extended the Panjer assumption, in a different way, to

$$p_{m+1} = \left(a + \frac{b}{m+1} \right) p_m + \frac{c}{m+1} p_{m-1}, \quad m = 0, 1, \ldots, \tag{8.52}$$

where $p_{-1} = 0$.

Sundt (1992, p.78) derived a recursion under the assumption

$$p_{m+1} = \sum_{i=1}^{K} \left(a_i + \frac{b_i}{m+1} \right) p_{m+1-i}, \quad m = M, M+1, \ldots, \tag{8.53}$$

where $p_m = 0$ for $m < 0$. In our applications it is enough to consider $K = 2$, $M = 1$ and $a_2 = 0$, in which case we let $a_1 = a$, $b_1 = b$ and $b_2 = c$. Then (8.53) reduces to

$$p_{m+1} = \left(a + \frac{b}{m+1} \right) p_m + \frac{c}{m+1} p_{m-1}, \quad m = 1, 2, \ldots, \tag{8.54}$$

and in that case Sundt's recursion is given by

$$g_k = \sum_{j=0}^{k} \left(\left(a + \frac{bj}{k} \right) f_j + \frac{cj}{2k} f_j^{2*} \right) g_{k-j}$$

$$+\big(p_1 - (a+b)p_0\big) f_k, \quad k = 1, 2, \ldots \tag{8.55}$$

It is seen that (8.54) contains (8.50), (8.51) and (8.52) as special cases. Recursion (8.55) does in those cases reduce to the original recursions.

For the rest of this section we consider the mixed Poisson case when the structure distribution U belongs to the Willmot class of absolutely continuous distributions:

(i) $U(\ell_0) = 0$ and $U(\ell_1) = 1$ for $0 \le \ell_0 < \ell_1 \le \infty$;

(ii) $$\frac{d}{d\ell} \log u(\ell) = \frac{\eta(\ell)}{\theta(\ell)} = \frac{\sum_{n=0}^{w} \eta_n \ell^n}{\sum_{n=0}^{w} \theta_n \ell^n} \quad \text{for } \ell_0 \le \ell \le \ell_1.$$

A distribution satisfying (i) and (ii) above is said to belong to that class $W(w)$. It is tacitly understood that at least one of η_w and θ_w is different from zero. It is, however, convenient to allow $\eta(\cdot)$ and $\theta(\cdot)$ to have common factors, cf. Willmot (1993, p.116). Thus $U \in W(w)$ merely means that U has a representation according to (ii).

This class – despite its usefulness in connection with recursive evaluation, as shown in Chapter 2 – has the nice property of forming a family of conjugate prior distributions relative to the Poisson distribution. Recall from Example 4.1 that this means that the posterior distribution U^*, given by (4.8), also belongs to the Willmot class. More precisely, we have the following proposition which is due to Hesselager (1996, pp.57–58).

Proposition 8.7. *Let $U \in W(w)$ and let U^* be the posterior distribution when $N(s) = m$ is observed.*

(1) *If $\theta_0 \neq 0$, then $U^* \in W(w+1)$ with*

$$\begin{aligned} \eta_n^* &= \eta_{n-1} + m\theta_n - s\theta_{n-1}, \\ \theta_n^* &= \theta_{n-1}, \end{aligned} \qquad n = 1, \ldots, w+1$$

with $\theta_{-1} = \eta_{-1} = \theta_{w+1} = 0$.

(2) *If $\theta_0 = 0$, then $U^* \in W(w)$ with*

$$\begin{aligned} \eta_n^* &= \eta_n + m\theta_{n+1} - s\theta_n, \\ \theta_n^* &= \theta_n, \end{aligned} \qquad n = 1, \ldots, w$$

with $\theta_{w+1} = 0$.

Proof. The density u^* of the posterior distribution U^* is given by

$$u^*(\ell) = \text{const. } \ell^m e^{-s\ell} \cdot u(\ell),$$

where the constant depends on m and s, but not on ℓ. Thus we get

$$\frac{d}{d\ell} \log u^*(\ell) = \frac{d}{d\ell} (\log \text{const.} + m \log \ell - s\ell + \log u(\ell))$$

$$= \frac{\sum_{n=0}^{w} \eta_n \ell^n}{\sum_{n=0}^{w} \theta_n \ell^n} + \frac{m - s\ell}{\ell}$$

for $\ell_0 \le \ell \le \ell_1$. Simple calculations lead to the desired results. Notice, for $\theta_w = 0$, that

$$\begin{aligned} \eta^*_{w+1} &= \eta_w \neq 0, \quad \text{in case (1)}, \\ \eta^*_w &= \eta_w \neq 0, \quad \text{in case (2)}, \end{aligned}$$

so that U^* really belongs to the asserted classes. ■

For simplicity of notation, we restrict ourselves to $t = 1$ and use the notation $p_m = p_m(1)$ and $N = N(1)$.

Let $U \in W(1)$. It follows from (2.20) that

$$(\theta_1 - \eta_1)(m + 1)p_{m+1} = \big(\theta_1 m + (\eta_0 - \theta_0 + \theta_1)\big)p_m + \theta_0 p_{m-1}$$

$$+ \pi_{\ell_0,m} u(\ell_0)\theta(\ell_0) - \pi_{\ell_1,m} u(\ell_1)\theta(\ell_1), \quad m = 1, 2, \ldots, \tag{8.56}$$

where $\pi_{\ell,m}$ is the Poisson probability, cf. Definition 2.1, and $p_{-1} = 0$. Thus (8.54) applies with

$$a = \frac{\theta_1}{\theta_1 - \eta_1}, \quad b = \frac{\eta_0 - \theta_0}{\theta_1 - \eta_1}, \quad c = \frac{\theta_0}{\theta_1 - \eta_1}$$

provided $\theta_1 \neq \eta_1$ and

$$\pi_{\ell_0,m} u(\ell_0)\theta(\ell_0) = \pi_{\ell_1,m} u(\ell_1)\theta(\ell_1), \quad m = 1, 2, \ldots. \tag{8.57}$$

Requirement (8.57) is fulfilled if either

$$u(\ell_0)\theta(\ell_0) = u(\ell_1)\theta(\ell_1) = 0$$

or if

$$\ell_0 = 0 \quad \text{and} \quad u(\ell_1)\theta(\ell_1) = 0.$$

In those cases (8.55) can be written as

$$(\theta_1 - \eta_1) g_k = \sum_{j=0}^{k} \left(\left(\theta_1 + \frac{(\eta_0 - \theta_0) j}{k} \right) f_j + \frac{\theta_0 j}{2k} f_j^{2*} \right) g_{k-j}$$

$$+ \begin{cases} 0, & \text{if } u(\ell_0)\theta(\ell_0) = u(\ell_1)\theta(\ell_1) = 0, \\ u(0)\theta_0 f_k, & \text{if } \ell_0 = 0 \text{ and } u(\ell_1)\theta(\ell_1) = 0, \end{cases} \tag{8.58}$$

for $k = 1, 2, \ldots$

As in Theorem 8.16 $g_0 = G(f_0) = \hat{u}(1 - f_0)$. If $f_0 = 0$ the initial condition reduces to $g_0 = p_0$. If $f_0 > 0$ we may consider a model with

$$\tilde{f}_0 = 0$$
$$\tilde{f}_k = f_k/(1 - f_0), \quad k \geq 0$$
$$\tilde{\Lambda} = (1 - f_0)\Lambda$$

or, which is the same,

$$\widetilde{U}(\ell) = U(\ell/(1 - f_0))$$
$$\widetilde{H}(s) = (H(s) - f_0)/(1 - f_0)$$
$$\widetilde{G}(s) = G(1 - (1 - f_0)(1 - s)).$$

Then we have

$$\widetilde{H}_Y(s) = \widetilde{G}(\widetilde{H}(s))$$

$$= G\left(1 - (1 - f_0)\left(1 - \frac{H(s) - f_0}{1 - f_0}\right)\right) = G(H(s)) = H_Y(s).$$

For $U \in W(w)$ we further have

$$\frac{d}{d\ell} \log \tilde{u}(\ell) = \frac{\widetilde{U}''(\ell)}{\widetilde{U}'(\ell)} = \frac{U''(\ell/(1 - f_0))}{(1 - f_0) U'(\ell/(1 - f_0))}.$$

Thus $\widetilde{U} \in W(w)$ with

$$\tilde{\ell}_0 = (1 - f_0)\ell_0, \quad \tilde{\ell}_1 = (1 - f_0)\ell_1$$

and

$$\tilde{\eta}_n = \eta_n/(1-f_0)^n \qquad\qquad \tilde{\eta}_n = (1-f_0)^{w-n}\eta_n$$
$$\tilde{\theta}_n = \theta_n/(1-f_0)^{n-1} \quad \text{or} \quad \tilde{\theta}_n = (1-f_0)^{w+1-n}\theta_n.$$

As will be seen in the examples below, many but not all cases are covered by (8.58).

Let $\gamma_{\ell,k}$ denote the compound Poisson probability, i.e.

$$\gamma_{\ell,k} = \sum_{m=0}^{\infty} \pi_{\ell,m} f_k^{m*}.$$

Theorem 8.17 (Willmot, 1993). *Let $U \in W(1)$ and assume that $\eta_1 \neq \theta_1$. Then $g_0 = G(f_0)$ and*

$$(\theta_1 - \eta_1)g_k = \sum_{j=0}^{k}\left(\left(\theta_1 + \frac{(\eta_0-\theta_0)j}{k}\right)f_j + \frac{\theta_0 j}{2k}f_j^{2*}\right)g_{k-j}$$

$$+ b_k(\ell_0) - b_k(\ell_1), \quad k = 1,2,\ldots, \tag{8.59}$$

where

$$b_k(\ell) = \begin{cases} \theta_0 u(0) f_k, & \ell = 0, \\ \ell^{-1}\theta(\ell)u(\ell)\gamma_{\ell,k}, & 0 < \ell < \infty, \\ 0, & \ell = \infty. \end{cases}$$

Proof. The original proof, given by Willmot (1993, pp.127–128), is based on the differential equation for $G(s)$ used in the proof of (2.20). The proof to be given is essentially due to Schröter (1990, pp.164–165).

From (8.56) we get

$$(\theta_1 - \eta_1)g_k = (\theta_1 - \eta_1)\sum_{m=0}^{\infty} p_{m+1} f_k^{(m+1)*}$$

$$= \sum_{m=0}^{\infty}\left(\theta_1 + \frac{\eta_0 - \theta_0}{m+1}\right)p_m f_k^{(m+1)*} \tag{$*$}$$

$$+ \sum_{m=1}^{\infty} \frac{\theta_0}{m+1} p_{m-1} f_k^{(m+1)*} \tag{$**$}$$

$$+\sum_{m=0}^{\infty}\frac{\pi_{\ell_0,m}u(\ell_0)\theta(\ell_0)-\pi_{\ell_1,m}u(\ell_1)\theta(\ell_1)}{m+1}f_k^{(m+1)*}. \qquad (***)$$

(*) From the proof of Theorem 8.16

$$\sum_{m=0}^{\infty}\left(\theta_1+\frac{\eta_0-\theta_0}{m+1}\right)p_m f_k^{(m+1)*}$$

$$=\sum_{j=0}^{k}\left(\eta_0+\frac{(\eta_0-\theta_0)j}{k}\right)f_j g_{k-j}$$

follows.

(**) We will need the following generalization of (8.49):

$$E[Z_1+Z_2 \mid Z_1+\dots Z_{m+1}=k]=\sum_{j=0}^{k} j\frac{f_j^{2*}f_{k-j}^{(m-1)*}}{f_k^{(m+1)*}}=\frac{2k}{m+1},$$

or

$$\frac{f_k^{(m+1)*}}{m+1}=\sum_{j=0}^{k}\frac{j}{2k}f_j^{2*}f_{k-j}^{(m-1)*}.$$

Using this we get

$$\sum_{m=1}^{\infty}\frac{\theta_0}{m+1}p_{m-1}f_k^{(m+1)*}=\sum_{m=1}^{\infty}\theta_0 p_{m-1}\sum_{j=0}^{k}\frac{j}{2k}f_j^{2*}f_{k-j}^{(m-1)*}$$

$$=\sum_{j=0}^{k}\frac{\theta_0 j}{2k}f_j^{2*}f_{k-j}^{(m-1)*}.$$

(***) Let $b_k(\ell)$ be defined as in Theorem 8.17 and let $\tilde{b}_k(\ell)$ be defined by

$$\tilde{b}_k(\ell)\stackrel{\text{def}}{=}\sum_{m=0}^{\infty}\frac{\pi_{\ell,m}u(\ell)\theta(\ell)}{m+1}f_k^{(m+1)*}.$$

Since $\pi_{0,0}=1$ we get $\tilde{b}_k(0)=u(0)\theta(0)f_k=b_k(0)$. Further $u(\infty)=0$, so $\tilde{b}_k(\infty)=b_k(\infty)=0$.

For $0<\ell<\infty$ and $k\neq 0$ we have

$$\sum_{m=0}^{\infty} \frac{1}{m+1} \pi_{\ell,m} f_k^{(m+1)*} = \sum_{m=1}^{\infty} \frac{1}{m+1} \frac{\ell^m}{m!} e^{-\ell} f_k^{(m+1)*}$$

$$= \frac{1}{\ell} \sum_{m=1}^{\infty} \frac{\ell^{m+1}}{(m+1)!} e^{-\ell} f_k^{(m+1)*} = \ell^{-1} \gamma_{\ell,k}$$

and $\tilde{b}_k(\ell) = b_k(\ell)$ follows. ■

Example 8.6. Recall from Example 2.2 that N is Delaporte distributed if Λ is shifted Γ-distributed with

$$u(\ell) = \frac{\beta^\gamma}{\Gamma(\gamma)} (\ell - \alpha)^{\gamma-1} e^{-\beta(\ell-\alpha)}, \quad \ell \geq \alpha,$$

and that (i) and (ii) hold with

$$\begin{array}{lcll} \ell_0 = \alpha & \text{and} & \eta_0 = \gamma - 1 + \alpha\beta & \eta_1 = -\beta \\ \ell_1 = \infty & & \theta_0 = -\alpha & \theta_1 = 1. \end{array}$$

Since $\theta(\alpha) = 0$ it follows that (8.54) reduces to

$$(1+\beta) g_k = \sum_{j=0}^{k} \left(\left(1 + \frac{(\gamma - 1 + \alpha\beta + \alpha) j}{k} \right) f_j \quad \frac{\alpha j}{2k} f_j^{2*} \right) g_{k-j}$$

for $k = 1, 2, \ldots$

This is the recursion of Schröter (1990), but it differs from the recursion derived by Willmot and Sundt (1989a) which was based on the infinite divisibility of the Delaporte distribution. According to Schröter (1990, p.166) the given recursion is more effective.

For $\alpha = 0$, i.e. for Λ being Γ-distributed and N negative binomial, the recursion reduces to the Panjer recursion:

$$(1+\beta) g_k = \sum_{j=0}^{k} \left(1 + \frac{(\gamma - 1) j}{k} \right) f_j g_{k-j}, \quad k = 1, 2, \ldots \tag{8.60}$$

□

Example 8.7. Consider now the special beta–Poisson distribution, which was touched upon in Example 2.4, with

$$u(\ell) = \frac{b(\ell_1 - \ell)^{b-1}}{\ell_1^b}, \quad 0 < \ell < \ell_1,$$

where $b > 0$. Recall that (i) and (ii) hold with

$$\begin{array}{lcll} \ell_0 = 0 & \text{and} & \eta_0 = 1 - b & \eta_1 = 0 \\ \ell_1 < \infty & & \theta_0 = \ell_1 & \theta_1 = -1. \end{array}$$

Since

$$u(\ell)\theta(\ell) = \frac{b(\ell_1 - \ell)^b}{\ell_1^b} = \begin{cases} b, & \ell = 0, \\ 0, & \ell = \ell_1, \end{cases}$$

it follows from (8.59) that

$$g_k = \sum_{j=0}^{k} \left(\left(1 + \frac{(b - 1 + \ell_1)j}{k} \right) f_j - \frac{\ell_1 j}{2k} f_j^{2*} \right) g_{k-j} - b f_k$$

for $k = 1, 2, \ldots$ □

Example 8.8. Let N be truncated normal–Poisson distributed. Recall from Example 2.5 that

$$U(\ell) = P\{X \le \ell \mid X \ge 0\} = \frac{\Phi\left(\frac{\ell - \mu_G}{\sigma_G}\right) - \Phi\left(-\frac{\mu_G}{\sigma_G}\right)}{1 - \Phi\left(-\frac{\mu_G}{\sigma_G}\right)},$$

where X is a normally distributed random variable with mean μ_G and variance σ_G^2, and that

$$\begin{array}{lcll} \ell_0 = 0 & \text{and} & \eta_0 = \mu_G & \eta_1 = -1 \\ \ell_1 = \infty & & \theta_0 = \sigma_G^2 & \theta_1 = 0. \end{array}$$

Thus (8.59) becomes

$$g_k = \sum_{j=0}^{k} \left(\frac{(\mu_G - \sigma_G^2)j}{k} f_j - \frac{\sigma_G^2 j}{2k} f_j^{2*} \right) g_{k-j} + \sigma_G^2 u(0) f_k \qquad (8.61)$$

for $k = 1, 2, \ldots$

Let us consider the case where N is Gauss–Poisson distributed, cf. Definition 5.2. Recall from Chapter 5 that

$$G(s) = \exp\{-\mu_\Lambda(1-s) + \tfrac{1}{2}\sigma_\Lambda^2(1-s)^2\}$$

and that

$$\begin{aligned} p_1 &= (\mu_\Lambda - \sigma_\Lambda^2)\, p_0, \\ (m+1)\, p_{m+1} &= (\mu_\Lambda - \sigma_\Lambda^2)\, p_m + \sigma_\Lambda^2 p_{m-1}, \quad m = 1, 2, \dots, \end{aligned}$$

where $\mu_\Lambda \geq \sigma_\Lambda^2$. Since the Gauss–Poisson distribution is not a mixed Poisson distribution we cannot apply Theorem 8.17. The Schröter recursion does, however, apply with

$$a = 0, \quad b = \mu_\Lambda - \sigma_\Lambda^2, \quad c = \sigma_\Lambda^2.$$

Thus (8.55) yields the recursion

$$\begin{aligned} g_0 &= \exp\{-\mu_\Lambda(1-f_0) + \tfrac{1}{2}\sigma_\Lambda^2(1-f_0)^2\} \\ g_k &= \sum_{j=0}^{k} \left(\frac{(\mu_\Lambda - \sigma_\Lambda^2)j}{k} f_j + \frac{\sigma_\Lambda^2 j}{2k} f_j^{2*} \right) g_{k-j}, \quad k = 1, 2, \dots \end{aligned}$$

Let us return to the truncated normal–Poisson distribution and consider

$$U(\ell) = P\{X \leq \ell \mid X \geq \ell_0\}, \quad \ell_0 > 0.$$

It is easily seen from Theorem 8.17 that the recursion (8.61) holds with $\sigma_G^2 u(0) f_k$ replaced with $b_k(\ell_0)$. □

Except for the very last case in the example above, all recursions do follow from Sundt's recursion (8.55). Although the last case may look somewhat artificial, it illustrates a general method of generating models by conditioning. Let Λ with distribution $U \in W(w)$ be given, and consider $\tilde{\Lambda}$ with distribution

$$\tilde{U}(\ell) \stackrel{\text{def}}{=} P\{\Lambda \leq \ell \mid \tilde{\ell}_0 \leq \Lambda \leq \tilde{\ell}_1\} = \frac{U(\ell) - U(\tilde{\ell}_0)}{U(\tilde{\ell}_1) - U(\tilde{\ell}_0)}$$

for $\ell_0 \leq \tilde{\ell}_0 \leq \ell \leq \tilde{\ell}_1 \leq \ell_1$. Since

$$\frac{d}{d\ell}\log \tilde{u}(\ell) = \frac{d}{d\ell}\log\left(\frac{u(\ell)}{U(\tilde{\ell}_1) - U(\tilde{\ell}_0)}\right) = \frac{d}{d\ell}\log u(\ell)$$

for $\ell_0 \leq \tilde{\ell}_0 \leq \ell \leq \tilde{\ell}_1 \leq \ell_1$ it follows that $U \in W(w)$ with

$$\begin{matrix} \tilde{\ell}_0 & & \tilde{\eta}_n = \eta_n \\ & \text{and} & \\ \tilde{\ell}_1 & & \tilde{\theta}_n = \theta_n. \end{matrix}$$

Thus, for $w = 1$, it follows from Theorem 8.17 that (8.59) holds for $\widetilde{U}$ with $b_k(\ell_0) - b_k(\ell_1)$ replaced by $b_k(\tilde{\ell}_0) - b_k(\tilde{\ell}_1)$.

Example 8.9. Let X be a $\Gamma(\alpha, \beta)$, cf. Example 2.1, and consider Λ with distribution

$$U(\ell) = P\{X \leq \ell \mid \ell_0 \leq X \leq \ell_1\}$$

$$= \int_{\ell_0}^{\ell} \frac{\beta^\gamma}{\Gamma(\gamma)} \ell^{\gamma-1} e^{-\beta\ell} \Big/ \int_{\ell_0}^{\ell_1} \frac{\beta^\gamma}{\Gamma(\gamma)} x^{\gamma-1} e^{-\beta x}\, dx, \quad \ell_0 \leq \ell \leq \ell_1.$$

The corresponding mixed Poisson distributed variable N is said to be **truncated Γ–Poisson** distributed.

It follows immediately from the Panjer recursion (8.60), and the remarks above, that

$$(1+\beta)g_k = \sum_{j=0}^{k}\left(1 + \frac{(\gamma-1)j}{k}\right) f_j g_{k-j} + b_k(\ell_0) - b_k(\ell_1)$$

for $k = 1, 2, \ldots$ For $\ell_0 = 0$ and $\ell_1 < \infty$ this is the tail-truncated Γ-distribution, mentioned in the end of Chapter 2. As noticed by Kemp (1968, p.199) the parameter γ may take any real value. For $\beta = 0$, and γ replaced by a we get

$$u(\ell) = \frac{a\ell^{a-1}}{\ell_1^a}, \quad 0 \leq \ell \leq \ell_1,$$

which is a special beta distribution, although not the same as considered in Example 8.7. Then the recursion reduces to

$$g_k = \sum_{j=0}^{k}\left(1 + \frac{(a-1)j}{k}\right) f_j g_{k-j} - b_k(\ell_1), \quad k = 1, 2, \ldots$$

For $\ell_0 > 0$ and $\ell_1 = \infty$ the truncated Γ–Poisson distribution may be regarded as an alternative to the Delaporte distribution, cf. Willmot (1993, p.125). □

Let now $U \in W(w)$ for $w > 1$. Of course we cannot apply Theorem 8.17 any more. Recall from Example 2.4 that the (general) beta distribution (2.27) belongs to $W(2)$. In that case

$$m(m+1)\,p_{m+1}(1)$$
$$= m(m-1+a+b+\ell_1)\,p_m(1) - \ell_1(m+a-1)\,p_{m-1}(1),$$

for $m = 1, 2, \ldots$, where $a > 0$ and $b > 0$. It seems natural to ask whether the above recursion might fulfil (8.53) for $K = 2$. If that would be the case, there must exist a_2 and b_2 such that

$$\frac{m+a-1}{m(m+1)} = -a_2 - \frac{b_2}{m+1}.$$

The above equality implies that $a_2 = 0$, by letting $m \to \infty$, and that $a = 1$ and $b_2 = -1$. However, for $a = 1$ we are back in the special beta distribution considered in Example 8.7 which belongs to $W(1)$. Thus we cannot rely on (8.53) anymore.

A natural idea is to generalize (8.53), so that it may be applied to $W(k)$. Wang and Sobrero (1994) have, in fact, derived a recursion under such a generalization. However, their recursion is rather complicated. Hesselager (1996, pp.55–56), on the other hand, has derived a recursion which, loosely speaking, is an extension of Theorem 8.17. This recursion is easier to implement and, according to Hesselager (1996), more effective. We will here only discuss Hesselager's recursion.

Define the auxiliary functions

$$g_k^{(n)} = \int_{\ell_0}^{\ell_1} \ell^n \gamma_{\ell,k} u(\ell)\, d\ell, \quad n = 0, \ldots, w, \tag{8.62}$$

and notice that $g_k = g_k^{(0)}$. Recall that $\gamma_{\ell,k} = \sum_{m=0}^{\infty} \pi_{\ell,m} f_k^{m*}$.

Theorem 8.18 (Hesselager, 1996). *Let $U \in W(w)$ and assume that $\eta_w \neq \theta_w$. Then*

$$g_0^{(n)} = \int_{\ell_0}^{\ell_1} \ell^n e^{-\ell(1-f_0)} u(\ell)\, d\ell, \quad n = 0, \ldots, w, \tag{8.63}$$

and, for $k = 1, 2, \ldots$,

$$g_k^{(n)} = \sum_{j=1}^{k} \frac{j}{k} f_j g_{k-j}^{(n+1)}, \quad n = 0, \ldots, w-1, \tag{8.64}$$

$$(\theta_w - \eta_w) g_k^{(w)} = \sum_{n=0}^{w-1} \big((n+1)\theta_{n+1} - \theta_n + \eta_n)\big) g_k^{(n)}$$

$$+ \sum_{n=0}^{w} \theta_n \sum_{j=0}^{k} f_j g_{k-j}^{(n)} + \theta(\ell_0) u(\ell_0) \gamma_{\ell_0,k} - \theta(\ell_1) u(\ell_1) \gamma_{\ell_1,k}. \tag{8.65}$$

Proof. The Panjer recursion, applied to $\gamma_{\ell,k}$, yields

$$\gamma_{\ell,0} = e^{-\ell(1-f_0)} \tag{8.66}$$

$$\gamma_{\ell,k} = \sum_{j=1}^{k} \frac{\ell j}{k} f_j \gamma_{\ell,k-j}, \quad k = 1, 2, \ldots. \tag{8.67}$$

The initial conditions (8.63) follow from (8.62) and (8.66). Multiplication of (8.67) with $\ell^n u(\ell)$ and integration yield (8.64). Rewrite $u'(\ell)/u(\ell) = \eta(\ell)/\theta(\ell)$ as

$$u(\ell) \sum_{n=0}^{w} \eta_n \ell^n = u'(\ell) \sum_{n=0}^{w} \theta_n \ell^n. \tag{8.68}$$

Multiplication of (8.68) with $\gamma_{\ell,k}$ and integration yield

$$\sum_{n=0}^{w} \eta_n g_k^{(n)} = \sum_{n=0}^{w} \theta_n \sum_{m=0}^{\infty} f_k^{m*} \int_{\ell_0}^{\ell_1} \ell^n \pi_{\ell,m} u'(\ell)\, d\ell. \tag{8.69}$$

Since $\frac{d}{d\ell}\pi_{\ell,m} = \pi_{\ell,m-1} - \pi_{\ell,m}$ with $\pi_{\ell,-1} = 0$, integration by parts leads to

$$\int_{\ell_0}^{\ell_1} \ell^n \pi_{\ell,m} u'(\ell)\, d\ell = \ell_1^n \pi_{\ell_1,m} u(\ell_1) - \ell_0^n \pi_{\ell_0,m} u(\ell_0)$$

$$- \int_{\ell_0}^{\ell_1} \big(n\ell^{n-1}\pi_{\ell,m} + \ell^n(\pi_{\ell,m-1} - \pi_{\ell,m})\big) u(\ell)\, d\ell.$$

Therefore, by (8.62) and (8.69),

$$\sum_{n=0}^{w} \eta_n g_k^{(n)} = \theta(\ell_1)u(\ell_1)\gamma_{\ell_1,k} - \theta(\ell_0)u(\ell_0)\gamma_{\ell_0,k}$$

$$-\sum_{n=1}^{w} n\theta_n g_k^{(n-1)} - \sum_{n=0}^{w} \theta_n \sum_{j=0}^{k} f_j g_{k-j}^{(n)} + \sum_{n=0}^{w} \theta_n g_k^{(n)}.$$

By collecting the terms involving $g_k^{(n)}$, except for those which vanish if $f_0 = 0$, (8.65) follows. ∎

It may be noticed that

$$\theta(\ell)u(\ell)\gamma_{\ell,k} = \begin{cases} \ell b_k(\ell), & \ell < \infty, \\ 0, & \ell = \infty, \end{cases} \qquad k = 1, 2, \ldots ,$$

where $b_k(\ell)$ was defined in Theorem 8.17. For $0 \leq \ell < \infty$ this is obvious. For $\ell = \infty$ we get, cf. Hesselager (1996, p.57), from (8.62)

$$\int_{\ell_0}^{\ell_1} \theta(\ell)\gamma_{\ell,k}u(\ell)\, d\ell = \sum_{n=0}^{w} \theta_n g_k^{(n)} < \infty,$$

which implies $\theta(\infty)u(\infty)\gamma_{\infty,k} = 0$.

Notice that the proof Theorem 8.18 is not based on (2.20). Let $f_1 = 1$ which implies that $g_k = p_k$ and that

$$g_k^{(n)} = \int_{\ell_0}^{\ell_1} \ell^n \pi_{\ell,k}u(\ell)\, d\ell = \frac{(k+n)!}{k!} p_{k+n} = (k+1)^{\overline{n}} p_{k+n}.$$

Recursion (8.65), with k replaced by m then reduces to

$$(\theta_w - \eta_w)(m+1)^{\overline{w}} p_{m+w}$$

$$= \sum_{n=0}^{w-1} \big((n+1)\theta_{n+1} - \theta_n + \eta_n\big)(m+1)^{\overline{n}} p_{m+n}$$

$$+ \sum_{n=0}^{w} \theta_n m^{\overline{n}} p_{m+n-1} + \theta(\ell_0)u(\ell_0)\pi_{\ell_0,m} - \theta(\ell_1)u(\ell_1)\pi_{\ell_1,m}$$

or

$$\sum_{n=0}^{w} \big((n+1)\theta_{n+1} - \theta_n + \eta_n\big)(m+1)^{\overline{n}} p_{m+n} + \sum_{n=-1}^{w} \theta_{n+1} m^{\overline{n+1}} p_{m+n}$$

$$= -\theta(\ell_1)u(\ell_1)\pi_{\ell_1,m} - \theta(\ell_0)u(\ell_0)\pi_{\ell_0,m},$$

which is (2.20). Hence Hesselager's proof of Theorem 8.18 may be considered as an alternative and somewhat shorter proof of (2.20).

Remark 8.9. For $U \in W(1)$ the recursion in Theorem 8.18 reduces to

$$g_k = \sum_{j=1}^{k} \frac{j}{k} f_j g_{k-j}^{(1)}$$

$$(\theta_w - \eta_w) g_k^{(1)} = (\theta_1 - \theta_0 + \eta_0) g_k$$

$$+ \theta_0 \sum_{j=0}^{k} f_j g_{k-j} + \theta_1 \sum_{j=0}^{k} f_j g_{k-j}^{(1)} + \ell_0 b_k(\ell_0) - \ell_1 b_k(\ell_1),$$

provided that $\ell_1 b_k(\ell_1)$ is interpreted as 0 for $\ell_1 = \infty$, with initial conditions

$$g_0 = \hat{u}(1 - f_0) \quad \text{and} \quad g_0^{(1)} = -\hat{u}'(1 - f_0).$$

Thus the Hesselager recursion, unless $\theta_1 = 0$, requires one recursion more than the Willmot recursion. Especially when U is a Γ-distribution, and the Willmot recursion reduces to the Panjer recursion, the Hesselager recursion uses two recursions. □

Among the examples discussed in Chapter 2, the beta distribution, mentioned above, and the generalized inverse Gaussian distribution, cf. Example 2.3, belong to $W(2)$. These two cases are considered by Hesselager (1996). For the GIGD case we refer to Hesselager (1996).

Example 8.10. Recall from Example 2.4 that N is beta–Poisson distributed if Λ is beta distributed with

$$u(\ell) = u(\ell) = \frac{\ell^{a-1}(\ell_1 - \ell)^{b-1}}{B(a,b)\ell_1^{a+b-1}}, \quad 0 < \ell < \ell_1,$$

where $a > 0$ and $b > 0$, and that (i) and (ii) hold with

$$\begin{matrix} \ell_0 = 0 \\ \ell_1 < \infty \end{matrix} \quad \text{and} \quad \begin{matrix} \eta_0 = (a-1)\ell_1 & \eta_1 = -(a+b-2) & \eta_2 = 0 \\ \theta_0 = 0 & \theta_1 = \ell_1 & \theta_2 = -1. \end{matrix}$$

Since $\theta(0) = \theta(\ell_1) = 0$ the Hesselager recursion is given by

$$g_k = \sum_{j=1}^{k} \frac{j}{k} f_j g_{k-j}^{(1)} \qquad g_k^{(1)} = \sum_{j=1}^{k} \frac{j}{k} f_j g_{k-j}^{(2)}$$

$$g_k^{(2)} = -a\ell_1 g_k + (a+b+\ell_1)g_k^{(1)} - \ell_1 \sum_{j=0}^{k} f_j g_{k-j}^{(1)} + \sum_{j=0}^{k} f_j g_{k-j}^{(2)}.$$

The recursion is obviously easy to implement.

Consider now, as in Example 8.7, $a = 1$. This special choice does not simplify the recursion and the recursion given in Example 8.6 is more tractable. Notice that we have

$$\frac{\eta(\ell)}{\theta(\ell)} = \frac{-(b-1)\ell}{\ell_1\ell - \ell^2},$$

i.e. $\eta(\ell)/\theta(\ell)$ has the common factor ℓ.

Let now $b = 1$, which yields

$$\frac{\eta(\ell)}{\theta(\ell)} = \frac{(a-1)\ell_1 - (a-1)\ell}{\ell_1\ell - \ell^2},$$

i.e. $\eta(\ell)/\theta(\ell)$ has the common factor $\ell - \ell_1$. Although this choice does not simplify the recursion, it may be an alternative to the recursion given in Example 8.9, since it does not contain $b_k(\ell_1)$.

In any case, these two special beta distributions illustrate a general method to avoid the 'non-homogeneous' part of a recursion by multiplying both $\eta(\ell)$ and $\theta(\ell)$ with the factors $\ell - \ell_0$ and $\ell - \ell_1$. This is the reason why it is convenient to allow $\eta(\cdot)$ and $\theta(\cdot)$ to have common factors. □

Compound mixed Poisson distributions appear naturally in an extension of the $M/G/1$ queue where the customers arrive in groups. A typical example may be a small self-service restaurant. Assume that the groups of customers arrive according to a Poisson process with intensity α and that the number of customers in each group are described by independent random variables $Z_1, Z_2, \ldots$ with, as before, $P\{Z_k = j\} = f_j$. Thus the arrival process is a compound Poisson process. The service times are described by independent and identically distributed random variables with distribution U. The arrival process and the service times are also independent. Notice that we are back in the 'usual' $M/G/1$ queue if $f_1 = 1$.

Let $X(t)$ be the number of customers in the system at time t. The problem in the analysis is that $X(\cdot)$ is not a Markov process unless U is an exponential distribution. The usual way to overcome this problem is to consider $X(t)$ at those times when customers leave the system. Let $T_1, T_2, \dots$ denote the departure times and put

$$\widetilde{X}_n = X(T_n), \quad n = 1, 2, \dots$$

The probability that k groups of customers arrive during one service period is

$$p_k = p_k(\alpha) = \int_{0-}^{\infty} \frac{(\ell\alpha)^k}{k!} e^{-\ell\alpha} \, dU(\ell), \quad k = 0, 1, \dots,$$

and thus the number of customers arriving during one service period is compound mixed Poisson distributed.

The embedded process $\{\widetilde{X}_n\}_{n=1}^{\infty}$ is a Markov chain with transition matrix

$$\boldsymbol{P} = \begin{pmatrix} \tilde{g}_0 & \tilde{g}_1 & \tilde{g}_2 & \tilde{g}_3 & \cdots \\ g_0 & g_1 & g_2 & g_3 & \cdots \\ 0 & g_0 & g_1 & g_2 & \cdots \\ 0 & 0 & g_0 & g_1 & \cdots \\ \vdots & \vdots & \vdots & \vdots & \end{pmatrix}$$

where

$$\tilde{g}_k = \sum_{j=1}^{k+1} f_j g_{k+1-j}, \quad k = 0, 1, \dots$$

Only the first row in the matrix requires explanation. If the system becomes empty when a customer leaves, nothing happens until a group of, let us say, j customers arrives. The system then behaves as if it had been j customers in it at a departure time. Hence the first row in $\boldsymbol{P}$ is a mixture of the other rows according to the probabilities f_j. A slightly different way to argue is to consider a customer who arrived to an empty system. Immediately after his departure the customers in system consists of the rest of his group and those customers arriving during his service period.

We will not discuss the analysis of the queueing system; for such a discussion we refer to Willmot (1986). Further queueing applications are to be found in Willmot (1990) and Willmot and Lin (1994).

CHAPTER 9

The risk business

The usual model of a risk business is based on the following independent objects:

(i) a point process N;

(ii) a sequence $\{Z_k\}_1^\infty$ of independent and identically distributed random variables, having the common distribution function F, with mean value μ, and variance σ^2.

Here N describes the times and $\{Z_k\}$ the costs of the claims. The total amount of claims paid by the company in the interval $(0, t]$ is then described by the **claim process** $Y = \{Y(t);\ t \geq 0\}$, defined by

$$Y(t) = \sum_{k=1}^{N(t)} Z_k, \qquad \left(\sum_{k=1}^{0} Z_k \stackrel{\text{def}}{=} 0 \right).$$

The notation is illustrated in Figure 9.1.

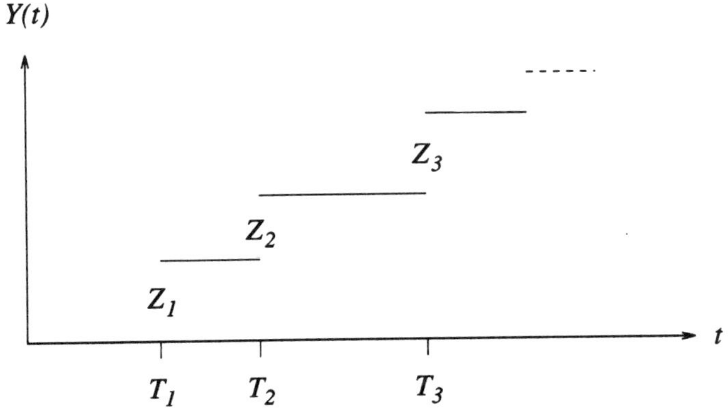

Figure 9.1. *Illustration of notation in the claim process.*

For fixed values of t, the random variable $Y(t)$ is a random sum. We have, see (8.3),

$$E[Y(t)] = \mu E[N(t)]$$

and, see (8.4),

$$\text{Var}[Y(t)] = \sigma^2 E[N(t)] + \mu^2 \text{Var}[N(t)].$$

We will consider the cases where N is a mixed Poisson process and where it is an Ammeter process, AP(U). Recall from Example 5.1 that an AP(U) is a Cox process with intensity process

$$\lambda(t) = L_k \quad \text{for } k\Delta \le t < (k+1)\Delta,$$

where $\Delta > 0$ is a fixed value and $\{L_k\}$ be a sequence of non-negative and independent random variables with common distribution U. Put $E[L_k] = \mu_L$ and $\text{Var}[L_k] = \sigma_L^2$. The intensity process in the Ammeter case is illustrated in Figure 9.2.

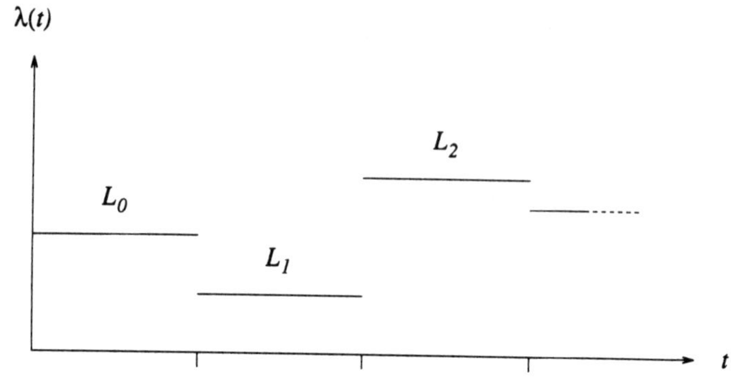

Figure 9.2. *Illustration of the intensity process in the Ammeter case.*

In this chapter we have chosen to use actuarial terminology, since the questions to be considered are mainly of risk theoretical interest. In the mixed Poisson case the claim process would, with the terminology used in Chapter 8, be called a **compound mixed Poisson process.**

9.1 The claim process

9.1.1 The mixed Poisson case

It follows from Proposition 2.1 that

$$E[Y(t)] = \mu\mu_\Lambda t \tag{9.1}$$

and

$$\mathrm{Var}[Y(t)] = (\sigma^2 + \mu^2)\mu_\Lambda t + \mu^2\sigma_\Lambda^2 t^2.$$

For large values of t we thus get

$$\mathrm{Var}[Y(t)] \sim \begin{cases} \mu^2\sigma_\Lambda^2 t^2 & \text{for } \mu \neq 0, \\ \sigma^2\mu_\Lambda t & \text{for } \mu = 0, \end{cases} \tag{9.2}$$

which indicates that there is a fundamental difference in the asymptotic behaviour whether μ is zero or not.

We will now consider the asymptotic distribution of $Y(t)$, for large values of t, in the sense of the 'central limit theorem'. This means that we, for suitable constants α_t and β_t, consider the distribution of

$$\frac{Y(t) - \alpha_t}{\beta_t}, \quad t \to \infty.$$

In most cases $\alpha_t \sim E[Y(t)]$ and $\beta_t^2 \sim \mathrm{Var}[Y(t)]$ are natural choices.

For $\mu \neq 0$ it follows from (9.1) and (9.2) that the natural choices are $\alpha_t = \mu\mu_\Lambda t$ and $\beta_t = t$ or $\mu\sigma_\Lambda t$. Since α_t/β_t is constant, we do choose $\alpha_t = 0$ and $\beta_t = t$, which is in agreement with Lundberg (1964, p.112).

Proposition 9.1. *Let N be* MPP(U). *Then*

$$\lim_{t\to\infty} \frac{Y(t)}{t} = \mu\Lambda \quad P\text{-a.s.}$$

Proof. On $\{\Lambda = 0\}$ we have $Y(t) = 0$ P-a.s. for all t, so the proposition holds. On $\{\Lambda > 0\}$ we get from Proposition 4.2 and the strong law of large numbers that

$$\frac{Y(t)}{t} = \frac{N(t)}{t} \cdot \frac{Y(t)}{N(t)} \to \Lambda \cdot \mu, \quad P\text{-a.s. as } t \to \infty.$$

■

It is natural to regard Proposition 9.1 more as a 'law of large numbers' than as a 'central limit theorem'. The intuitive explanation is that the randomness in $Y(t)$ due to $\{Z_k\}$ and the Poisson variation is, asymptotically, completely dominated by the randomness in Λ.

Let $F(y,t) = P\{Y(t) \le y\}$. Then we have

$$F(y,t) = \int_{0-}^{\infty} \sum_{k=0}^{\infty} \frac{(\ell t)^k}{k!} e^{-\ell t} F^{k*}(y)\, dU(\ell), \tag{9.3}$$

where F^{k*} denotes the kth convolution of F with itself.

Proposition 9.1 then implies, for $\mu > 0$, that

$$\lim_{t\to\infty} F(yt,t) = U(y/\mu),$$

for y/μ being a point of continuity of U.

Example 9.1. We will consider a simple example, due to Lundberg (1964, p.116), where $F(y,t)$ is easy to compute. Let both Λ and Z_k be exponentially distributed, i.e.

$$F(z) = 1 - e^{-z/\mu} \quad \text{for } z \ge 0$$

and

$$U(\ell) = 1 - e^{-\ell/\mu_\Lambda} \quad \text{for } \ell \ge 0.$$

In this case we have, cf. Example 2.1,

$$p_k(t) = \frac{1}{1+\mu_\Lambda t}\left(\frac{\mu_\Lambda t}{1+\mu_\Lambda t}\right)^k.$$

From Example 8.2, with $p = 1/(1+\mu_\Lambda)$, we get

$$F(y,t) = 1 - \frac{\mu_\Lambda t}{1+\mu_\Lambda t}\, e^{-\frac{y}{\mu(1+\mu_\Lambda t)}},$$

which implies

$$F(yt,t) = 1 - \frac{\mu_\Lambda t}{1+\mu_\Lambda t}\, e^{-\frac{ty}{\mu(1+\mu_\Lambda t)}}. \tag{9.4}$$

It is easily seen that $F(yt,t) \to U(y/\mu) = 1 - e^{-y/(\mu\mu_\Lambda)}$ as $t \to \infty$.

In order to illustrate Proposition 9.1 further, let ξ_t be an indicator variable, independent of Λ, defined by

$$\xi_t = \begin{cases} 0 & \text{if } N(t) = 0, \\ 1 & \text{if } N(t) > 0. \end{cases}$$

Simple calculations yield

$$\frac{Y(t)}{t} \stackrel{\mathrm{d}}{=} \Lambda \cdot \mu \cdot \frac{1 + \mu_\Lambda t}{\mu_\Lambda t} \cdot \xi_t.$$

Since

$$\frac{1 + \mu_\Lambda t}{\mu_\Lambda t} \cdot \xi_t \to 1 \cdot \begin{cases} 0 & \text{if } \Lambda = 0, \\ 1 & \text{if } \Lambda > 0, \end{cases} \qquad P\text{-a.s. as } t \to \infty,$$

Proposition 9.1 follows. □

One way to include the randomness in $Y(t)$ due to $\{Z_k\}$ and the Poisson variation is to consider an asymptotic expansion of $F(y, t)$, see Lundberg (1964, pp.117–124). Another way is to apply 'saddlepoint approximations', i.e. approximations of $1 - F(y, t)$ as $y \to \infty$ and t is fixed. Those approximations were briefly discussed in Remark 8.7.

Consider now the case $\mu = 0$. Note that Proposition 9.1 formally holds also in this case, but that it only says that $\lim_{t\to\infty} \frac{Y(t)}{t} \to 0$. It follows from (9.1) and (9.2) that the natural choice of normalizing constants are $\alpha_t = 0$ and $\beta_t = \sqrt{t}$, cf. Lundberg (1964, p.114).

Proposition 9.2. *Let N be* MPP(U) *and let W be a normally distributed random variable with mean zero and variance one. Then, for $\mu = 0$,*

$$\frac{Y(t)}{\sqrt{t}} \stackrel{\mathrm{d}}{\to} \sqrt{\Lambda} \cdot \sigma W \text{ as } t \to \infty,$$

where Λ and W are independent.

Proof. We will only give some arguments, in line with the proof of Proposition 9.1, in order to indicate that Proposition 9.2 is natural. For a real proof we refer to Lundberg (1964, pp.113–114).

On $\{\Lambda = 0\}$ we have $Y(t) = 0$ P-a.s. for all t, so the proposition holds. On $\{\Lambda > 0\}$ we get from Proposition 4.2 and the central limit theorem for random sums that

$$\frac{Y(t)}{\sqrt{t}} = \sqrt{\frac{N(t)}{t}} \cdot \frac{Y(t)}{\sqrt{N(t)}} \xrightarrow{\mathrm{d}} \sqrt{\Lambda} \cdot \sigma W, \quad t \to \infty.$$

The independence between Λ and W seems, at least intuitively, quite natural. ■

We will now consider a generalization where the distribution F of the costs of the claim may depend on time. This means that if a claim occur at time s, then its distribution is F_s independently of all other quantities in the claim process. From a practical point of view, it is natural to think of the time dependence as due to economical factors, as interest and inflation, or to seasonal variation.

This generalization is easy to handle as long as we consider the modified risk process at a fixed time t, but, nevertheless, it is mathematically rather delicate.

Theorem 9.1 (Jung, 1963). *Let $S_n(z_1, \dots, z_n)$, for each n, be a symmetric, i.e. permutation invariant, function of $z_1, \dots, z_n$. Then, for every fixed t,*

$$S_{N(t)}(Z_1, \dots, Z_{N(t)}) \stackrel{\mathrm{d}}{=} S_{N(t)}(\widetilde{Z}_1, \dots, \widetilde{Z}_{N(t)}),$$

where $\widetilde{Z}_1, \widetilde{Z}_2, \dots$ are independent and identically distributed random variables with distribution function

$$\widetilde{F}_t(z) = \frac{1}{t} \int_0^t F_s(z)\, ds$$

and independent of N.

Proof. Assume that $N(t) = n$, and let $J_1, \dots, J_n$ be a random permutation of the numbers $1, \dots, n$. It follows from (6.11) that $T_{J_1}, \dots, T_{J_n}$ are independent and uniformly distributed on $[0, t]$. Thus

$$P\{Z_{J_1} \le z_1, \dots, Z_{J_n} \le z_n\} = E\left[\prod_{k=1}^{n} F_{T_{J_k}}(z_k)\right]$$

$$= \prod_{k=1}^{n} E\left[F_{T_{J_k}}(z_k)\right] = \prod_{k=1}^{n} \frac{1}{t} \int_0^t F_s(z_k)\, ds = \prod_{k=1}^{n} \widetilde{F}_t(z_k),$$

i.e. $Z_{J_1}, \dots, Z_{J_n}$ are independent and with distribution $\widetilde{F}_t$. Since S_n is symmetric, the theorem follows. ■

Example 9.2. Time-dependent distributions of the costs of the claims occur naturally when inflation and interest are included in the model. Assume that both the forces of inflation and interest are constant. Let δ denote the net interest, i.e.

$$\delta = \text{interest} - \text{inflation}.$$

The discounted value of $Y(t)$ is then given by

$$Y(t) = \sum_{k=1}^{N(t)} Z_k e^{-\delta T_k}.$$

Thus it is natural to consider $F_s(z) = F(ze^{\delta s})$. In this case we get, cf. Willmot (1989a, pp.5–6),

$$\widetilde{F}_t(z) = \frac{1}{t}\int_0^t F(ze^{\delta s})\,ds$$

$$= \begin{bmatrix} y = ze^{\delta s} \\ dy = ze^{\delta s}\delta\,ds = \delta y\,ds \end{bmatrix} = \int_z^{ze^{\delta t}} \frac{F(y)}{\delta t y}\,dy.$$

Thus $\widetilde{F}_t$ is absolutely continuous, and we have

$$\widetilde{F}_t'(z) = e^{\delta t}\frac{F(ze^{\delta t})}{\delta t z e^{\delta t}} - \frac{F(z)}{\delta t z} = \frac{F(ze^{\delta t}) - F(z)}{\delta t z}.$$

□

Jung (1963) gave some examples of relevant choices of S_n, but we will restrict ourselves to

$$S_n(z_1, \dots, z_n) = \sum_{k=1}^{n} z_k.$$

With this choice it follows, for each fixed value of t, that the distribution of $Y(t)$ is given by (9.3) with F replaced by $\widetilde{F}_t$, cf. also Philipson (1965, p.221). From a mathematical point of view, this is a result about compound mixed Poisson distributions, since we can only consider one value of t at a time. On the other hand, the formulation of the modification requires that we consider the process. It may further be observed that the sequence $\{Z_k\}$, in the modified claim process, is in general neither independent, identically distributed, nor independent of N.

9.1.2 The Ammeter case

A fundamental property in the analysis of Ammeter processes is that $\{Y_k\}$, defined by

$$Y_k \stackrel{\text{def}}{=} Y(k\Delta) - Y((k-1)\Delta),$$

is a sequence of independent and identically distributed random variables. Thus we have

$$Y(t) = \sum_{k=1}^{\lfloor t/\Delta \rfloor} Y_k + (Y(t) - Y(\lfloor t/\Delta \rfloor \Delta)), \tag{9.5}$$

where $\lfloor \cdot \rfloor$ means 'integer part'.

From (9.1) we get

$$E[Y_k] = \mu\mu_L\Delta \quad \text{and} \quad \text{Var}[Y_k] = (\sigma^2 + \mu^2)\mu_L\Delta + \mu^2\sigma_L^2\Delta^2,$$

which, together with (9.5), leads to

$$E[Y(t)] = \mu\mu_L t \quad \text{and} \quad \text{Var}[Y(t)] \sim \big((\sigma^2 + \mu^2)\mu_L + \mu^2\sigma_L^2\Delta\big)t.$$

Proposition 9.3. *Let N be* AP(U). *Then*

$$\frac{Y(t) - \mu\mu_L t}{\sqrt{t}} \xrightarrow{\text{d}} \sqrt{(\sigma^2 + \mu^2)\mu_L + \mu^2\sigma_L^2\Delta} \cdot W, \quad t \to \infty,$$

where W is a normally distributed random variable with mean zero and variance one.

Proof. The proposition follows directly from the central limit theorem for $t = k\Delta$, $k \to \infty$. This restriction on t does not influence the limit distribution, so the proposition holds. ■

Thus we see that the asymptotic behaviour is quite different from the mixed Poisson case. Here the contribution of $\{Z_k\}$, the Poisson variation, and the intensity variation to randomness in $Y(t)$ is of the same order. For a discussion along these lines for Cox processes, we refer to Grandell (1971, pp.233–236).

9.2 Ruin probabilities

The **risk process**, $X = \{X(t);\ t \geq 0\}$, is defined by

$$X(t) = ct - Y(t),$$

where c is a positive real constant. We will here only treat the case with only non-negative risk sums, i.e. we assume that $F(0-) = 0$. Notice that, for technical reasons, 'zero risk sums' are allowed.

The **ruin probability** $\Psi(u)$ of a company facing the risk process X and having initial capital u is defined by

$$\Psi(u) = P\{u + X(t) < 0 \text{ for some } t > 0\}.$$

The notation is illustrated in Figure 9.3. Notice that the Z_k denote the claim size, while the jumps in the claim process are negative.

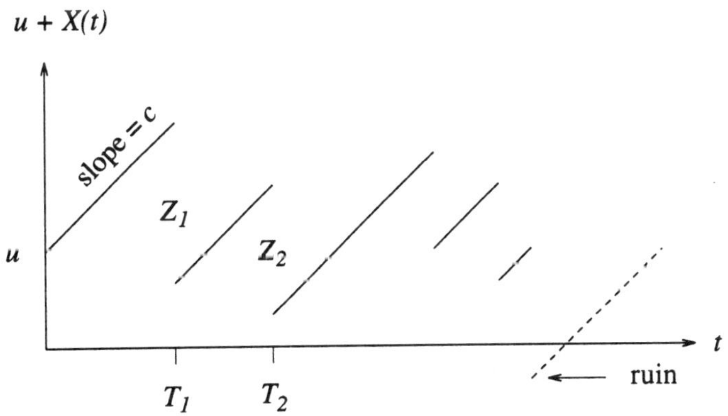

Figure 9.3. *Illustration of notation in the risk process.*

Although a reader is assumed to be acquainted with classical ruin theory, we will state some basic facts in the case where N is a Poisson process with intensity α. The claim process is in this case a **compound Poisson process.**

The expected profit of the risk business over the interval $(0, t]$ is

$$E[X(t)] = ct - E[Y(t)] = (c - \alpha\mu)\,t.$$

The relative **safety loading** ρ is defined by

$$\rho = \frac{c - \alpha\mu}{\alpha\mu}.$$

The risk process X is said to have **positive** safety loading if $\rho > 0$.

It is sometimes convenient to use the non-ruin probability

$$\Phi(u) \stackrel{\text{def}}{=} 1 - \Psi(u).$$

Notice that it follows from the definition that $\Psi(u) = 1$ for $u < 0$. The notation Φ is conventional; unfortunately Φ is also the conventional notation for the standard normal distribution. In this chapter Φ is used only in the first sense.

Following Feller (1971, p.183) we shall now derive a differential equation for Φ by a 'renewal argument'. Let T_1 be the epoch of the first claim. Then we have $X(T_1) = cT_1 - Z_1$. Since the Poisson process is a renewal process and since ruin cannot occur in $(0, T_1)$ we have

$$\Phi(u) = E[\Phi(u + cT_1 - Z_1)] = \int_0^\infty \alpha e^{-\alpha s} \int_{0-}^{u+cs} \Phi(u + cs - z)\, dF(z) ds.$$

The change of variables $x = u + cs$ leads to

$$\Phi(u) = \frac{\alpha}{c} e^{\alpha u/c} \int_u^\infty e^{-\alpha x/c} \int_{0-}^{x} \Phi(x - z)\, dF(z) dx.$$

Consequently Φ is differentiable and differentiation leads to

$$\Phi'(u) = \frac{\alpha}{c}\Phi(u) - \frac{\alpha}{c}\int_{0-}^{u} \Phi(u - z)\, dF(z). \tag{9.6}$$

Put, cf. Feller (1971, pp.469–470),

$$\bar{\phi}(v) \stackrel{\text{def}}{=} \int_0^\infty e^{-vz}\Phi(z)\, dz \quad \text{and} \quad \hat{\phi}(v) \stackrel{\text{def}}{=} \int_{0-}^\infty e^{-vz}\, d\Phi(z)$$

and notice that

$$\bar{\phi}(v) = \frac{1}{v}\int_0^\infty e^{-vz}\Phi'(z)\, dz + \frac{1}{v}\,\Phi(0) = \frac{1}{v}\hat{\phi}(v).$$

From (9.6) we get

$$v(\bar{\phi}(v) - \Phi(0)) = \frac{\alpha}{c}\bar{\phi}(v)(1 - \hat{f}(v))$$

or

$$\bar{\phi}(v) = \frac{\Phi(0)}{1 - \frac{\alpha}{c}\frac{1-\hat{f}(v)}{v}} \cdot \frac{1}{v} = \frac{\Phi(0)}{1 - \frac{\alpha}{c}\bar{f}(v)} \cdot \frac{1}{v}, \tag{9.7}$$

where, as before, $\hat{f}(v) = \int_{0-}^{\infty} e^{-vz}\, dF(z)$, $\overline{F}(z) = 1 - F(z)$, and $\bar{f}(v) \stackrel{\text{def}}{=} \int_0^\infty e^{-vz}\overline{F}(z)\, dz$. It follows from the law of large numbers that $\lim_{t\to\infty} X(t)/t = c - \alpha\mu$ with probability one. In the case of positive safety loading, $c > \alpha\mu$, there exists a random variable T, i.e. a function of N and $\{Z_k\}$, such that $X(t) > 0$ for all $t > T$. Since only finitely many claims can occur before T it follows that $\inf_{t>0} X(t)$ is finite with probability one and thus $\Phi(u) \to 1$. Applying Theorem 8.2, with $L(\ell) = 1$, it follows that $v\bar{\phi}(v) \to 1$ as $v \to 0$. Thus we get from (9.7)

$$1 = \frac{\Phi(0)}{1 - \frac{\alpha\mu}{c}} \quad \text{or} \quad \Phi(0) = 1 - \frac{\alpha\mu}{c},$$

and (9.7) itself can be written as

$$\hat{\phi}(v) = \frac{1 - \frac{\alpha\mu}{c}}{1 - \frac{\alpha}{c}\bar{f}(v)}. \tag{9.8}$$

Let F_I and $\hat{f}_I$ be defined by

$$F_I(z) \stackrel{\text{def}}{=} \frac{1}{\mu}\int_0^z \overline{F}(x)\, dx, \tag{9.9}$$

and

$$\hat{f}_I(v) \stackrel{\text{def}}{=} \int_0^\infty e^{-vz}\, dF_I(z) = \frac{\bar{f}(v)}{\mu}.$$

Now it follows from (9.8) that

$$\hat{\phi}(v) = \frac{1 - \frac{\alpha\mu}{c}}{1 - \frac{\alpha\mu}{c}\hat{f}_I(v)} = \left(1 - \frac{\alpha\mu}{c}\right)\sum_{n=0}^{\infty}\left(\frac{\alpha\mu}{c}\right)^n \hat{f}_I^n(v). \tag{9.10}$$

Thus we get

$$\Psi(u) = \left(1 - \frac{\alpha\mu}{c}\right)\sum_{n=0}^{\infty}\left(\frac{\alpha\mu}{c}\right)^n \overline{F}_I^{n*}(u), \tag{9.11}$$

which, in fact, is the **Pollaczek–Khinchine formula.**

The Pollaczek–Khinchine formula has a natural probabilistic interpretation, in terms of 'ascending ladder points' of a random walk. For a discussion of this interpretation we refer to Asmussen (1997). The Pollaczek–Khinchine formula applies to much more general situations than above, but in general the distribution $\overline{F}_I$ then has to be replaced by a more complicated distribution. This happens, for instance, when N is a renewal process, a case which we will consider in section 9.2.2. In actuarial literature, it is often referred to as **Beekman's convolution formula.**

The tail behaviour of the claim distribution F is of utmost importance for the ruin probability. Put

$$h(r) \stackrel{\text{def}}{=} \int_{0-}^{\infty} (e^{rz} - 1)\, dF(z) = \hat{h}(r) - 1,$$

where $\hat{h}$ is the moment generating function of Z_k used in Chapter 8. The change of notation is done in order to get nicer formulae.

Definition 9.1 below is just a reformulation of Definition 8.1.

Definition 9.1. *We talk about* **small claims**, *or say that F is* **light-tailed**, *if there exists $r_\infty > 0$ such that $h(r) \uparrow +\infty$ when $r \uparrow r_\infty$ (we allow for the possibility $r_\infty = +\infty$).*

The important part of Definition 9.1 is that $h(r) < \infty$ for some $r > 0$. This means that the tail of F decreases at least exponentially fast, and thus for instance the lognormal and the Pareto distributions are excluded.

The following basic results go back to the pioneering works Lundberg (1926) and Cramér (1930):

$$\Psi(u) = \frac{\alpha\mu}{c} e^{-\frac{(c-\alpha\mu)u}{c\mu}} = \frac{1}{1+\rho} e^{-\frac{\rho u}{\mu(1+\rho)}} \tag{9.12}$$

when the claim costs are exponentially distributed with mean μ;

the **Cramér–Lundberg approximation**

$$\lim_{u\to\infty} e^{Ru}\,\Psi(u) = \frac{\rho\mu}{h'(R) - c/\alpha}, \tag{9.13}$$

where the **Lundberg exponent** R is the positive solution of $h(r) = cr/\alpha$;

the **Lundberg inequality**

$$\Psi(u) \leq e^{-Ru}. \tag{9.14}$$

If $\rho \leq 0$, then $\Psi(u) = 1$.

Modern standard proofs of (9.12)–(9.14) are, for instance, found in Grandell (1991).

We will show that those results also simply follows from Chapter 8, since (9.11) gives the ruin probability as a compound distribution where the compounding variable N is geometrically distributed with $p = (c - \alpha\mu)/c$.

When F is an exponential distribution we have, see (8.16),

$$\Psi(u) = (1-p)e^{-pu/\mu},$$

and since

$$\frac{pu}{\mu} = \frac{(c-\alpha\mu)u}{c\mu}$$

(9.12) follows.

Apply Theorem 8.4 with $\phi = 1 - p$, $\gamma = 1$, and $L(y) = p$ and put $\hat{h}_I(r) = \hat{f}_I(-r)$. Since κ is the solution of

$$\hat{h}_I(r) = \phi^{-1} \;\Leftrightarrow\; \frac{1}{\mu}\frac{h(r)}{r} = \frac{c}{\alpha\mu} \;\Leftrightarrow\; h(r) = \frac{cr}{\alpha} \tag{9.15}$$

it follows that $\kappa = R$. We have

$$\hat{h}'_I(r) = \frac{rh'(r) - h(r)}{\mu r^2},$$

and thus

$$\hat{h}'_I(R) = \frac{h'(R) - c/\alpha}{\mu R}.$$

Further it follows that

$$\Psi(u) \sim \frac{(\phi\hat{h}'_I(R))^{-1}}{R}pe^{-Ru} = \frac{\mu}{\frac{\alpha\mu}{c}(h'(R) - c/\alpha)}\frac{c-\alpha\mu}{c}e^{-Ru},$$

which is (9.13).

From Theorem 8.1, with the same ϕ and $\hat{h}_I$ as above, (9.14) immediately follows.

Let us now consider the case where F is a **Pareto distribution** with $\delta > 1$, i.e.

$$\overline{F}(z) = \left(\frac{z}{a}\right)^{-\delta} \quad \text{for } z \geq a > 0. \tag{9.16}$$

Then we have $\mu = \delta a/(\delta - 1)$ and

$$\overline{F}_I(z) = \frac{\delta - 1}{\delta a} \int_z^{\infty} \left(\frac{x}{a}\right)^{-\delta} dx = \frac{1}{\delta}\left(\frac{z}{a}\right)^{-(\delta-1)} \quad \text{for } z \geq a > 0.$$

We can now apply Corollary 8.1 with $\gamma = -\delta$, $L(z) = 1$, $\omega = 0$, $\chi = a^{-(\delta-1)}/\delta$, and $\mu_N = \alpha\mu/(c - \alpha\mu) = 1/\rho$. Then we get

$$\Psi(u) \sim \frac{1}{\delta\rho}\left(\frac{u}{a}\right)^{-(\delta-1)}, \quad u \to \infty. \tag{9.17}$$

A slight generalization of the Pareto case is where

$$\overline{F}(z) \sim z^{-\delta} L(z), \quad z \to \infty. \tag{9.18}$$

Distributions fulfilling (9.18) are sometimes said to be of the **Pareto type**. We will only consider $\delta > 1$ so that $\mu < \infty$. Since, cf. Proposition 8.5,

$$\overline{F}_I(z) \sim \frac{1}{\mu(\delta - 1)} z^{-(\delta-1)} L(z).$$

Exactly as in the Pareto case it follows from Corollary 8.1, with $\chi^{-1} = \mu(\delta - 1)$, that

$$\Psi(u) \sim \frac{1}{\rho\mu(\delta - 1)} u^{-(\delta-1)} L(u), \quad u \to \infty, \tag{9.19}$$

which is due to von Bahr (1975). We can write (9.19) on the form

$$\Psi(u) \sim \frac{1}{\rho}\overline{F}_I(u), \quad u \to \infty. \tag{9.20}$$

Thorin and Wikstad (1977) showed that (9.20) also holds when F is a lognormal distribution.

The Pareto and lognormal distributions have both been used as models for the costs of the claims in 'dangerous' situations. It is, however, not a coincidence that (9.20) holds in both cases. Recall from Definition 8.5 that $F_I \in \mathcal{S}$ if

$$\overline{F}_I^{2*}(z) \sim 2\overline{F}_I(z), \quad z \to \infty.$$

Exactly as above it follows from Proposition 8.2, with

$$\mu_N = \frac{\alpha\mu}{c - \alpha\mu} = \frac{1}{\rho}$$

that (9.20) holds for all $F_I \in \mathcal{S}$. Furthermore, it is shown by Embrechts and Veraverbeke (1982, p.70), that (9.20) does hold **exactly** when $F_I \in \mathcal{S}$. Due to this, the following definition is natural, although not quite satisfying.

Definition 9.2. *We talk about* **large claims** *if* $F_I \in \mathcal{S}$.

The Cramér–Lundberg approximation (9.13) is known to be numerically very accurate already for reasonably small values of u, see for instance Grandell (1991, pp.20–22). The same can in general not be said about the large claim approximation (9.20). In order to throw some light upon that question we consider $\alpha = 1$ and

$$\overline{F}(z) = (1 + 2z)^{-3/2}, \quad z \geq 0.$$

This distribution, which is obtained by a simple translation of a Pareto distribution, has been considered by Thorin and Wikstad (1973), from which the exact values of the ruin probability is taken. The approximation (9.20) is in this case reduced to

$$\Psi(u) \sim \frac{1}{\rho}(1 + 2u)^{-1/2}, \quad u \to \infty.$$

In Table 9.1 $\Psi(u)$ and $\rho^{-1}(1 + 2u)^{-1/2}$ are compared for some values of u and ρ.

It is naturally difficult to draw any general conclusions from Table 9.1 other than that large claim approximations should be used with care in numerical applications. Omey and Willekens (1987) have studied the asymptotic precision of these approximations in connection with random sums. Their results are not quite easy to apply and indicate that the question is far from easy.

It would, of course, have been much nicer if 'large claims' had been expressed more directly as a property of F itself. We will return to this question in section 9.2.2, but now we will consider the 'intermediate' case where $F \in \mathcal{S}(\beta)$, $\beta > 0$, see Definition 8.7.

Table 9.1. *Values of $\Psi(u)$ and its large claim approximation (9.20) in the case $\overline{F}(z) = (1+2z)^{-3/2}$.*

ρ	u	$\Psi(u)$	$\rho^{-1}(1+2u)^{-1/2}$
5%	100	0.65777	1.41069
5%	1000	0.36209	0.44710
5%	10000	0.13710	0.14142
10%	100	0.47654	0.70535
10%	1000	0.20754	0.22355
10%	10000	0.07010	0.07071
15%	100	0.36842	0.47023
15%	1000	0.14349	0.14903
15%	10000	0.04695	0.04714

Proposition 9.4. *If $F \in \mathcal{S}(\beta)$, $\beta > 0$, then*

(*i*) $F_I \in \mathcal{S}(\beta)$;

(*ii*) $\overline{F}_I(z) \sim (\mu\beta)^{-1}\overline{F}(z), \quad z \to \infty.$

Proof. The proof to be given is essentially taken from Embrechts and Veraverbeke (1982, p.71).

Using Definition 8.7 (iii), $\lim_{z\to\infty} \overline{F}(z+y)/\overline{F}(z) = e^{-\beta y}$, we get

$$\lim_{z\to\infty} \frac{\overline{F}_I(z)}{\overline{F}(z)} = \lim_{z\to\infty} \frac{1}{\mu} \int_z^\infty \frac{\overline{F}(y)}{\overline{F}(z)}\, dy$$

$$= \lim_{z\to\infty} \frac{1}{\mu} \int_0^\infty \frac{\overline{F}(z+y)}{\overline{F}(z)}\, dy = \frac{1}{\mu} \int_0^\infty e^{-\beta y}\, dy = \frac{1}{\mu\beta},$$

provided we find a domination. In order to do that we will use a standard trick in connection with $\mathcal{S}(\beta)$. Put

$$L(x) \stackrel{\text{def}}{=} x^\beta \overline{F}(\log x),$$

which implies

$$\lim_{x\to\infty} \frac{L(vx)}{L(x)} = v^\beta \lim_{x\to\infty} \frac{\overline{F}(\log v + \log x)}{\overline{F}(\log x)} = x^\beta e^{-\beta \log v} = 1,$$

i.e. $L(x)$ is slowly varying. It follows from Lemma 8.1 that for each $A > 1$ and $\delta > 0$ there exists an x_0 such that

$$\frac{L(vx)}{L(x)} \leq Av^{\delta} \quad \text{for all } x \geq x_0 \text{ and } v \geq 1.$$

Thus

$$\frac{\bar{F}(z+y)}{\bar{F}(z)} = e^{-\beta y}\frac{L(e^{z+y})}{L(e^z)} \leq Ae^{-(\beta-\delta)y}, \quad z \geq e^{x_0} \text{ and } y \geq 0,$$

which is integrable for $\delta < \beta$, and (ii) follows by dominated convergence.

From this, (i) follows by the closure property of $\mathcal{S}(\beta)$, see Embrechts and Goldie (1982, p.265). ■

The closure property of $\mathcal{S}(\beta)$, mentioned in the proof above, means that if $F \in \mathcal{S}(\beta)$ and if V and F are **tail-equivalent**, i.e. if

$$\bar{V}(z) \sim \text{const. } \bar{F}(z), \quad z \to \infty,$$

for some const. $\in (0,\infty)$, then also $V \in \mathcal{S}(\beta)$. This holds also for $\beta = 0$, and shows that the property $F \in \mathcal{S}$ or $F \in \mathcal{S}(\beta)$ really is a tail-property of F.

Notice that $\beta > 0$ was essential in the proof of Proposition 9.4. In fact, the result is not true for $\beta = 0$, cf. Proposition 8.5.

Proposition 9.5, due to Embrechts and Veraverbeke (1982, p.71), may be looked upon as a correspondence to (9.13) and (9.20) when $F \in \mathcal{S}(\beta)$.

Proposition 9.5. *Suppose that $F \in \mathcal{S}(\beta)$ and that $h(\beta) < c\beta/\alpha$. Then*

$$\Psi(u) \sim \frac{\frac{\alpha}{c\beta}\left(1-\frac{\alpha\mu}{c}\right)}{\left(1-\frac{\alpha h(\beta)}{c\beta}\right)^2}\bar{F}(u), \quad u \to \infty.$$

Proof. Like above, we rely on (9.11). Apply Proposition 8.3, with

$$G(s) = \frac{1-\frac{\alpha\mu}{c}}{1-\frac{\alpha\mu s}{c}}.$$

Then we have $s_\infty = c/(\alpha\mu)$ and

$$G'(s) = \frac{\frac{\alpha\mu}{c}\left(1-\frac{\alpha\mu}{c}\right)}{\left(1-\frac{\alpha\mu s}{c}\right)^2}.$$

Provided $s_\infty > \hat{h}_I(\beta)$, it follows from Proposition 8.3 that

$$\Psi(u) \sim G'(\hat{h}_I(\beta))\overline{F}_I(u). \tag{9.21}$$

Since, see (9.15), $\hat{h}_I(r) = h(r)/(\mu r)$ it follows that

$$s_\infty > \hat{h}_I(\beta) \Leftrightarrow \frac{c}{\alpha\mu} > \frac{h(\beta)}{\mu\beta} \Leftrightarrow h(\beta) < \frac{c\beta}{\alpha}.$$

Further

$$G'(\hat{h}_I(\beta)) = G'\left(\frac{h(\beta)}{\mu\beta}\right) = \frac{\frac{\alpha\mu}{c}\left(1 - \frac{\alpha\mu}{c}\right)}{\left(1 - \frac{\alpha h(\beta)}{c\beta}\right)^2}.$$

The proposition now follows from (9.21) and Proposition 9.4. ■

9.2.1 The mixed Poisson case

From a purely mathematical point of view, ruin calculations are very easily performed. The simple idea is to first condition upon the outcome of Λ, and then to weight over ruin probabilities computed in the Poisson case. This approach is discussed in Bühlmann (1972), Grandell and Peiram (1973), and Eade (1983). In spite of that, misunderstandings, which have been clarified by Goovaerts (1982), can be found in the literature.

Let $\Psi(u, \ell)$ be the ruin probability when N is a Poisson process with intensity ℓ. Note that $\Psi(u, \ell) = 1$ for $c - \ell\mu \leq 0$, i.e. for $\ell \geq c/\mu$. Thus we have

$$\Psi(u) = \int_{0-}^{c/\mu} \Psi(u, \ell)\, dU(\ell) + 1 - U(c/\mu). \tag{9.22}$$

Example 9.1 (continued). Consider the case where both Λ and Z_k are exponentially distributed. Then it follows from (9.12) and (9.22) that

$$\Psi(u) = \frac{\mu}{c}\frac{1}{\mu_\Lambda}\int_0^{c/\mu} \ell e^{-\frac{(c-\ell\mu)u}{c\mu}} e^{-\ell/\mu_\Lambda}\, d\ell + e^{-c/(\mu\mu_\Lambda)}$$

$$= \frac{\mu_\Lambda}{(u\mu_\Lambda - c)^2}\left((u^2\mu_\Lambda - uc - c\mu)e^{-c/(\mu\mu_\Lambda)} + \mu c e^{-u/\mu}\right)$$

$$= \frac{\left(u^2 - (u+\mu)\mu(1+\rho)\right)e^{-(1+\rho)} + \mu^2(1+\rho)e^{-u/\mu}}{(u - \mu(1+\rho))^2}, \tag{9.23}$$

where $\rho = (c - \mu\mu_\Lambda)/(\mu\mu_\Lambda)$ is the safety loading. It is seen that (9.12) and (9.23) are very different. Tables of the ruin probability are given by Grandell and Peiram (1973). □

Assume now that there exists an $\ell_1 < c/\mu$ such that $U(\ell_1) = 1$, which may be interpreted as if the risk business is almost surely 'profitable'. It is natural to let ℓ_1 be the **right extremity** of U, i.e. $\ell_1 = \inf\{\ell \mid U(\ell) = 1\}$. A typical example of distributions of this kind are the beta distributions. As mentioned in Example 2.4, Quinkert (1957) studied the beta–Poisson distribution as a model for the number of claims in a time-interval. It follows from (9.11) that

$$\Psi(u) = \sum_{n=0}^{\infty} p_n \overline{F}_I^{n*}(u), \tag{9.24}$$

where

$$p_n = \int_{0-}^{\ell_1} \left(1 - \frac{\ell\mu}{c}\right)\left(\frac{\ell\mu}{c}\right)^n dU(\ell). \tag{9.25}$$

Similarly as in the Poisson case, we can apply results from Chapter 8, since (9.24) gives the ruin probability as a compound distribution. Here the compounding variable N is mixed geometrically distributed. Put, like in Chapter 8,

$$\overline{P}_n = P\{N \geq n\}, \quad G(s) = E[s^N] \text{ and } \overline{G}(s) = \sum_{n=0}^{\infty} \overline{P}_{n+1} s^n$$

and recall that $\overline{G}$ and G have the same radius of convergence s_∞. The following lemma is essentially due to Willmot and Lin (1994, pp.749–750).

Lemma 9.1. *Let ℓ_1 be the right extremity of U and suppose that $\ell_1 < c/\mu$. Then $s_\infty = c/(\ell_1\mu) > 1$, where s_∞ is the radius of convergence of G.*

Proof. We have

$$\overline{P}_{n+1} = \int_0^{\ell_1} \left(\frac{\ell\mu}{c}\right)^{n+1} dU(\ell) \leq \left(\frac{\ell_1\mu}{c}\right)^{n+1}$$

and therefore $\overline{G}$ has radius of convergence $s_\infty \geq c/(\ell_1\mu)$.

For each $\varepsilon > 0$ we have $U(\ell_1 - \varepsilon) < 1$, since ℓ_1 is the right extremity of U. Therefore

$$\overline{P}_{n+1} \geq \int_{\ell_1-\varepsilon}^{\ell_1} \left(\frac{\ell\mu}{c}\right)^{n+1} dU(\ell) > (1 - U(\ell_1 - \varepsilon)) \left(\frac{(\ell_1 - \varepsilon)\mu}{c}\right)^{n+1},$$

from which it follows that $s_\infty \leq c/((\ell_1 - \varepsilon)\mu)$.

Since ε can be arbitrarily chosen, and since $\overline{G}$ and G have the same radius of convergence, the lemma is proved. ■

Consider now the Lundberg inequality. A trivial Lundberg inequality follows from

$$\Psi(u) = \int_{0-}^{\ell_1} \Psi(u,\ell)\, dU(\ell) \leq \Psi(u,\ell_1) \leq e^{-R_1 u},$$

where R_1 is the positive solution of $h(r) = cr/\ell_1$. This inequality is certainly very crude.

The following theorem, which is due to Willmot and Lin (1994, p.755), gives a more refined inequality.

Theorem 9.2. *Let ℓ_1 be the right extremity of U. Suppose that $\ell_1 < c/\mu$ and that F is light-tailed. Then*

$$\Psi(u) \leq \frac{\mu_\Lambda}{\ell_1}\, e^{-R_1 u},$$

where R_1 is the positive solution of $h(r) = cr/\ell_1$.

Proof. It follows by Schwarz's inequality that

$$\overline{P}_n^2 = E[(\Lambda\mu/c)^n]^2 \leq E\left[(\Lambda\mu/c)^{n-1}\right] E\left[(\Lambda\mu/c)^{n+1}\right] = \overline{P}_{n-1}\overline{P}_{n+1}.$$

Thus $\overline{P}_{n+1}/\overline{P}_n \geq \overline{P}_n/\overline{P}_{n-1}$, i.e. N is DFR, see Definition 7.1. This also follows from general DFR properties, cf. Barlow and Proschan (1975, p.103).

By the same arguments as in the proof of Theorem 8.7, it follows that we can apply Theorem 8.1 with $\phi^{-1} = s_\infty = c/(\ell_1\mu)$. Recall, cf. (9.15), that $\hat{h}_I(r) = h(r)/(\mu r)$. Thus $\hat{h}(\kappa) = \phi^{-1}$ implies that $\kappa = R_1$. Further

$$\frac{1-p_0}{\phi} = \frac{\mu_\Lambda \mu/c}{\ell_1\mu/c} = \frac{\mu_\Lambda}{\ell_1},$$

and the theorem is proved. ■

If the distribution of Λ is concentrated on $\{0, \ell_1\}$, we have, since $\Psi(u, 0) = 0$ and $\mu_\Lambda = P\{\Lambda = \ell_1\}\ell_1$,

$$\Psi(u) = \frac{\mu_\Lambda}{\ell_1}\Psi(u, \ell_1),$$

which indicates that we cannot expect any essential improvements of Theorem 9.2. This is not surprising, cf. Remark 8.9.

Another question is, if in general exponential bounds are natural in the mixed Poisson case. In order to throw some light upon that question, we let Λ be uniformly distributed on $[0, \ell_1]$ and Z_k exponentially distributed. Then, cf. (9.12), $R = \frac{c - \ell_1\mu}{c\mu}$ and we have

$$\begin{aligned}\Psi(u) &= \frac{\mu}{c}\frac{1}{\ell_1}\int_0^{\ell_1} \ell e^{-\frac{(c-\ell\mu)u}{c\mu}}\, d\ell \\ &= \frac{\mu c}{u^2\ell_1}\left(e^{-\frac{u}{\mu}} + \left(\frac{u\ell_1}{c} - 1\right)e^{-R_1 u}\right) \\ &\sim \frac{\mu}{u}e^{-R_1 u}, \quad u \to \infty, \qquad (9.26)\end{aligned}$$

which indicates that exponential bounds cannot be expected to work as well as in the Poisson case. This may not be understood as if the mixed Poisson case was less dangerous than the Poisson case, since R_1 corresponds to the worst possible outcome of Λ.

The above arguments also indicate that one cannot expect to get a Cramér–Lundberg approximation in this case. Naturally it is tempting to try to apply Theorem 8.4 in a similar way as was done in the Poisson case. However, that approach seems not to work since in Theorem 8.4 rather precise assumptions on p_n are required. Those assumptions seem not to be fulfilled when U is continuous in a neighbourhood of ℓ_1. Suppose we approximate U with

$$U_\varepsilon(\ell) = \begin{cases} U(\ell) & \text{for } 0 \le \ell < \ell_1 - \varepsilon, \\ U(\ell_1 - \varepsilon) & \text{for } \ell_1 - \varepsilon \le \ell < \ell_1. \end{cases}$$

Thus U_ε has the masspoint $1 - U(\ell_1 - \varepsilon)$ at ℓ_1. Let $p_n^{(\varepsilon)}$ be given by (9.25), but with U replaced by U_ε. Then

$$p_n^{(\varepsilon)} \sim (1 - U(\ell_1 - \varepsilon))\left(1 - \frac{\ell_1\mu}{c}\right)\left(\frac{\ell_1\mu}{c}\right)^n,$$

so the assumptions in Theorem 8.4 are fulfilled for any fixed ε. It seems, however, not possible to get anything reasonable by letting $\varepsilon \to 0$, unless U has a masspoint at ℓ_1.

The following analogy to the Cramér–Lundberg approximation, see Theorem 9.3 below, was proposed by Asmussen (1996). The crucial point in the proof of it is the following 'extension' of the Poisson case Cramér–Lundberg approximation:

$$\lim_{u\to\infty} e^{R\left(\ell_1 - \frac{d}{u}\right)u} \Psi\left(u, \ell_1 - \tfrac{d}{u}\right) = C(\ell_1) \tag{9.27}$$

where $R(\ell)$ is the positive solution of $h(r) = cr/\ell$, d a positive constant, and $C(\ell) = \frac{c-\ell\mu}{\ell h'(R(\ell))-c}$, cf. (9.13).

A natural way to ascertain (9.27) is to verify that

$$\lim_{u\to\infty} e^{R(\ell)u} \Psi(u, \ell) = C(\ell) \quad \text{uniformly in } \ell \in [\ell_1 - \delta, \ell_1], \tag{9.28}$$

for some $\delta \in (0, \ell_1)$, or that

$$e^{R(\ell)u} \Psi(u, \ell) \to C(\ell) \text{ monotonically as } u \to \infty \tag{9.29}$$

for $\ell \in [\ell_1 - \delta, \ell_1]$. By Dini's theorem, cf. any textbook on real analysis, (9.29) implies (9.28).

Although (9.27) ought be true, since ℓ_1 is quite an innocent value, we have not managed to prove it. In order to 'overcome' this difficulty we **assume** that (9.27) holds. We will return to this assumption in Remark 9.1; a reader who strongly dislike this kind of solution of a mathematical difficulty is advised to have a quick look at Remark 9.1 before deciding whether it is worth reading Theorem 9.3.

Theorem 9.3. *Let ℓ_1 be the right extremity of U and suppose that $\ell_1 < c/\mu$. Assume that U has a density u such that*

$$u(\ell) \sim B \cdot (\ell_1 - \ell)^{b-1}, \quad \ell \uparrow \ell_1, \tag{9.30}$$

for some $b > 0$ and $B > 0$. Suppose that F is light-tailed and that (9.27) holds. Then

$$\Psi(u) \sim \frac{BC_1\Gamma(b)}{(D_1u)^b} e^{-R_1u}, \quad u \to \infty,$$

where R_1 is the positive solution of $h(r) = cr/\ell_1$,

$$C_1 = \frac{c - \ell_1\mu}{\ell_1 h'(R_1) - c} \quad \textit{and} \quad D_1 = \frac{cR_1C_1}{\ell_1(c - \ell_1\mu)}.$$

Proof. Let $R(\ell)$ be the positive solution of $h(r) = cr/\ell$ and notice that $R(\ell_1) = R_1$. Using $h(R(\ell)) = cR(\ell)/\ell$ we get

$$R'(\ell_1) = -\frac{cR_1}{\ell_1^2 h'(R_1) - \ell_1 c}.$$

Notice that $R'(\ell_1) < 0$ and that $D_1 = -R'(\ell_1) > 0$. Thus

$$R(\ell) \downarrow R_1 \quad \text{and} \quad R_1 - R(\ell) \sim -D_1(\ell_1 - \ell), \quad \ell \uparrow \ell_1.$$

Let, cf. (9.13),

$$C(\ell) = \frac{c - \ell\mu}{\ell h'(R(\ell)) - c} \quad \text{and} \quad C_1 = C(\ell_1).$$

For each $A > 1$ there exists a $\delta \in (0, \ell_1)$ such that

$$u(\ell_1 - \ell) \le AB\ell^{b-1}, \quad \forall \ell \in (0, \delta),$$

$$R_1 - R(\ell_1 - \ell) \le -A^{-1}D_1\ell, \quad \forall \ell \in (0, \delta).$$

We have

$$\lim_{u\to\infty} u^b e^{R_1 u} \Psi(u) = \lim_{u\to\infty} \int_0^{\ell_1} u^b e^{R_1 u} \Psi(u, \ell) u(\ell)\, d\ell$$

$$= \lim_{u\to\infty} \int_0^{\ell_1} u^b e^{R_1 u} \Psi(u, \ell_1 - \ell) u(\ell_1 - \ell)\, d\ell$$

$$= \lim_{u\to\infty} \int_0^{\delta} u^b e^{R_1 u} \Psi(u, \ell_1 - \ell) u(\ell_1 - \ell)\, d\ell$$

$$+ \lim_{u\to\infty} \int_\delta^{\ell_1} u^b e^{R_1 u} \Psi(u, \ell_1 - \ell) u(\ell_1 - \ell)\, d\ell.$$

Since $R(\ell) \downarrow R_1$ as $\ell \uparrow \ell_1$ we get by (9.14)

$$\lim_{u\to\infty} \int_\delta^{\ell_1} u^b e^{R_1 u} \Psi(u, \ell_1 - \ell) u(\ell_1 - \ell)\, d\ell \le \lim_{u\to\infty} u^b e^{R_1 u} \Psi(u, \ell_1 - \delta)$$

$$\leq \lim_{u\to\infty} u^b e^{R_1 u} e^{-R(\ell_1-\delta)u} = 0,$$

and thus

$$\lim_{u\to\infty} u^b e^{R_1 u}\,\Psi(u) = \lim_{u\to\infty} \int_0^\delta u^b e^{R_1 u}\,\Psi(u,\ell_1-\ell)u(\ell_1-\ell)\,d\ell.$$

By the transformation

$$\begin{array}{ll} v = D_1 u\ell & dv = D_1 u\,d\ell \\ \ell = \frac{v}{D_1 u} & d\ell = \frac{dv}{D_1 u} \end{array}$$

we get

$$\int_0^\delta u^b e^{R_1 u}\,\Psi(u,\ell_1-\ell)u(\ell_1-\ell)\,d\ell$$

$$= \frac{1}{D_1}\int_0^{D_1 u\delta} u^{b-1} e^{R_1 u}\,\Psi\left(u,\ell_1-\tfrac{v}{D_1 u}\right)u\left(\ell_1-\tfrac{v}{D_1 u}\right)dv$$

$$= \frac{1}{D_1}\int_0^\infty I_{D_1 u\delta}(v)u^{b-1}e^{R_1 u}\Psi\left(u,\ell_1-\tfrac{v}{D_1 u}\right)u\left(\ell_1-\tfrac{v}{D_1 u}\right)dv,$$

where

$$I_{D_1 u\delta}(v) = \begin{cases} 1, & v \leq D_1 u\delta, \\ 0, & v > D_1 u\delta. \end{cases}$$

Since

$$I_{D_1 u\delta}(v)u^{b-1}e^{R_1 u}\Psi\left(u,\ell_1-\tfrac{v}{D_1 u}\right)u\left(\ell_1-\tfrac{v}{D_1 u}\right)$$

$$\leq I_{D_1 u\delta}(v)u^{b-1}e^{R_1 u}e^{-R\left(\ell_1-\frac{v}{D_1 u}\right)u}AB\left(\tfrac{v}{D_1 u}\right)^{b-1}$$

$$\leq \text{const. } I_{D_1 u\delta}(v)e^{\left(R_1-R\left(\ell_1-\frac{v}{D_1 u}\right)\right)u}v^{b-1}$$

$$\leq \text{const. } I_{D_1 u\delta}(v)e^{-A^{-1}v}v^{b-1} \leq \text{const. } e^{-A^{-1}v}v^{b-1},$$

which is integrable, it follows by dominated convergence that

$$\lim_{u\to\infty} u^b e^{R_1 u}\,\Psi(u)$$

$$= \frac{1}{D_1}\int_0^\infty \lim_{u\to\infty} I_{D_1 u\delta}(v)u^{b-1}e^{R_1 u}\Psi\left(u,\ell_1-\tfrac{v}{D_1 u}\right)u\left(\ell_1-\tfrac{v}{D_1 u}\right)dv$$

$$= \frac{B}{D_1} \int_0^\infty \lim_{u\to\infty} u^{b-1} e^{R_1 u} \Psi\Big(u, \ell_1 - \tfrac{v}{D_1 u}\Big) \Big(\tfrac{v}{D_1 u}\Big)^{b-1} dv$$

$$= \frac{B}{D_1^b} \int_0^\infty \lim_{u\to\infty} e^{R_1 u} \Psi\Big(u, \ell_1 - \tfrac{v}{D_1 u}\Big) v^{b-1}\, dv.$$

We have

$$\lim_{u\to\infty} e^{R_1 u} \Psi\Big(u, \ell_1 - \tfrac{v}{D_1 u}\Big)$$

$$= \lim_{u\to\infty} e^{\left(R_1 - R\left(\ell_1 - \frac{v}{D_1 u}\right)\right)u} e^{R\left(\ell_1 - \frac{v}{D_1 u}\right)u} \Psi\Big(u, \ell_1 - \tfrac{v}{D_1 u}\Big)$$

$$= e^{-v} \lim_{u\to\infty} e^{R\left(\ell_1 - \frac{v}{D_1 u}\right)u} \Psi\Big(u, \ell_1 - \tfrac{v}{D_1 u}\Big) = e^{-v} C_1,$$

where the last equality follows by the assumption (9.27). Thus we have

$$\lim_{u\to\infty} u^b e^{R_1 u} \Psi(u) = \frac{BC_1}{D_1^b} \int_0^\infty v^{b-1} e^{-v}\, dv = \frac{BC_1 \Gamma(b)}{D_1^b},$$

which is the desired result. ■

Example 9.3. Let Z_k is exponentially distributed. Then, cf. (9.12),

$$R(\ell) = \frac{c - \ell\mu}{c\mu} \quad \text{and} \quad C_1 = \frac{\ell_1 \mu}{c},$$

and (9.27) is trivially fulfilled. Since $D_1 = 1/c$ we get

$$\Psi(u) \sim \frac{B\ell_1 \mu c^{b-1} \Gamma(b)}{u^b} e^{-\frac{(c-\alpha\mu)u}{c\mu}}, \quad u \to \infty.$$

For $b = 1$ and $B = 1/\ell_1$, which is the case when Λ is uniformly distributed, we get (9.26). □

We will consider two cases, which are interesting in themselves, where (9.28), and thus also (9.27), does hold. In order to simplify notation, we suppress the dependence on ℓ. Let

$$\pi(r) \stackrel{\text{def}}{=} h(r) - cr/\ell, \quad r < r_\infty;$$

then (9.13) implies that

$$\Psi(u) \sim \frac{c - \ell\mu}{\ell\pi'(R)} e^{-Ru}, \quad u \to \infty,$$

where R is the positive solution of the Lundberg equation $\pi(r) = 0$. In the cases to be considered, $r_\infty < \infty$ while $\pi(r)$ is simple and has an obvious analytical continuation to the complex plane $\mathbb{C}$ excluding r_∞ and possible other the poles.

Example 9.4. Let F be a mixture of m exponential distributions with means $\mu_1, \ldots, \mu_m$, i.e.

$$F(z) = \sum_{k=1}^{m} p_k \left(1 - e^{-z/\mu_k}\right) \quad \text{or} \quad f(z) = \sum_{k=1}^{m} \frac{p_k}{\mu_k} e^{-z/\mu_k}, \quad z \geq 0,$$

where $p_k > 0$, $\sum_{k=1}^{m} p_k = 1$, and $\mu_1 > \cdots > \mu_m > 0$. In this case we have

$$\pi(s) = \sum_{k=1}^{m} \frac{p_k\, \mu_k s}{1 - \mu_k s} - \frac{cs}{\ell}, \quad s \in \mathbb{C}.$$

Obviously $\pi(s)$ is a rational function with poles in $1/\mu_1, \ldots, 1/\mu_m$. The (extended) Lundberg equation $\pi(s) = 0$ is an algebraic equation of degree $m+1$ with roots $0, R_1, \ldots, R_m$. Let $R_1 = R < 1/\mu_1 = r_\infty$. Since $\pi(r + i0)$ must have one root between each pole it follows that all roots are real and $R_k > R$ for $k = 2, \ldots, m$.

It follows from Cramér (1955, p.82) that

$$\Psi(u) = \sum_{k=1}^{m} \frac{c - \ell\mu}{\ell\pi'(R_k)} e^{-R_k u} \tag{9.31}$$

and that $\frac{c-\ell\mu}{\ell\pi'(R_k)} > 0$ for all k. Thus (9.29) holds, which implies (9.27). □

Example 9.5. Let F be $\Gamma(\gamma, \beta)$, i.e.

$$f(z) = \frac{\beta^\gamma}{\Gamma(\gamma)} z^{\gamma-1} e^{-\beta z}, \quad z \geq 0.$$

It follows from Example 2.1 that

$$\pi(s) = \left(1 - \frac{s}{\beta}\right)^{-\gamma} - 1 - \frac{cs}{\ell}, \quad s \in \mathbb{C}.$$

Since $\pi(R) = 0$ we get

$$\pi'(R) = \frac{\gamma}{\beta}\left(1 - \frac{R}{\beta}\right)^{-\gamma-1} - \frac{c}{\ell} = \frac{(\gamma+1)cR/\beta - (c - \gamma\ell/\beta)}{(1 - R/\beta)\ell},$$

and thus

$$\Psi(u) \sim g(R)\, e^{-Ru}, \quad u \to \infty,$$

where

$$g(s) \stackrel{\text{def}}{=} \frac{(c-\gamma\ell/\beta)(1-s/\beta)}{(\gamma+1)cs/\beta-(c-\gamma\ell/\beta)}, \quad s \in \mathbb{C}.$$

Thorin (1973) studied this case for $\gamma < 1$, Dickson and Gray (1984) for γ being an integer, and finally Thorin (1986) in full generality. Let $R_1, \ldots, R_m$ be those roots of $\pi(s) = 0$ with positive real parts. We have (Thorin 1986)

$$\Psi(u) = \sum_{k=1}^{m} g(R_k)\, e^{-R_k u}$$

$$+ \frac{\sin(\pi\gamma)}{\pi} \int_{\beta}^{\infty} \frac{\frac{c-\gamma\ell/\beta}{\ell}(\frac{x}{\beta}-1)^{\gamma} e^{-xu}\, dx}{\{(1+\frac{cx}{\ell})(\frac{x}{\beta}-1)^{\gamma}-\cos(\pi\gamma)\}^2+\sin^2(\pi\gamma)} \tag{9.32}$$

where $R_1 = R < \beta = r_\infty$ and

$$m = \begin{cases} \gamma, & \text{when } \gamma \text{ is an even integer,} \\ 2k+1, & \text{when } 2k < \gamma < 2k+2 \text{ for } k = 0,1,\ldots \end{cases}$$

When γ is not an even integer, R is the only real root and the remaining roots are located in complex-conjugate pairs. When γ is an even integer, R_γ is real while the other roots are located as above. All roots are simple, all $\operatorname{Re} R_k > R$ for $k > 1$, and any two roots in different complex-conjugate pairs have different real parts. Notice that $\sin(\pi\gamma) = 0$ when γ is an integer; in that case $\pi(s)$ is a rational function which explains for the similarity between (9.31) and (9.32).

Since $e^{(R-x)u} \downarrow 0$ for all $x \geq \beta$ as $u \to \infty$ if follows by monotone convergence that the required uniformity holds for the integral part of (9.32), and thus that part causes no problems. Similarily the term corresponding to $k = \gamma$ in the sum causes no problems when γ is an even integer. Let R_k, $k > 1$, be a complex root. We have

$$\left|g(R_k)\, e^{-R_k u}\right| = |g(R_k)|\, e^{-\operatorname{Re} R_k u}$$

which implies that $e^{Ru}\left|g(R_k)\, e^{-R_k u}\right| \downarrow 0$ as $u \to \infty$. Thus the required uniformity holds and (9.27) follows. □

Remark 9.1. Example 9.3 shows that (9.27) does hold for exponentially distributed claims. Examples 9.4 and 9.5 give some support for our belief that (9.27) ought to hold, or at least ought to hold very generally. In particular the Γ-distribution is a 'popular' model in connection with small claims, and this is the main reason why we have chosen to formulate Theorem 9.3 as we have done and not only for exponentially distributed claims.

In this context the structure distribution U may be more interesting than the claim distribution. We believe that the assumption (9.30) is (close to) the 'right' one. For the beta distribution, see (2.27), that assumption holds with $B = 1/(B(a,b)\ell_1^b)$.

A safe conclusion is that Theorem 9.3 is far from its final form regarding the claim distribution. In spite of our 'defense' of the formulation, it is, of course, **most unsatisfactory**. The remaining part of this remark is mainly intended for those readers who are tempted to try to reformulate Theorem 9.3 into a decent theorem, thereby using an author's privilege to regard his shortcommings as 'proposals for research'.

Under the following assumption, which certainly is far from necessary, (9.27) does hold.

Assumption 9.1. *Suppose that $\ell < c/\mu$, that F is light-tailed, that $\overline{F}(z) > 0$ for all z, and that one of the following conditions*

(a) $R(\ell)E[Z_k - z \mid Z_k > z] > \frac{c-\ell\mu}{c}, \quad z \geq 0,$

(b) $R(\ell)E[Z_k - z \mid Z_k > z] = \frac{c-\ell\mu}{c}, \quad z \geq 0,$

(c) $R(\ell)E[Z_k - z \mid Z_k > z] < \frac{c-\ell\mu}{c}, \quad z \geq 0,$

holds.

Lemma 9.2. *Suppose, for some $\delta \in (0, \ell_1)$, that Assumption 9.1 holds for all $\ell \in [\ell_1 - \delta, \ell_1]$. Then (9.29) is valid.*

Proof. It follows from 'the standard proof' of (9.13) that $e^{R(\ell)u}\Psi(u,\ell)$ satisfies the proper renewal equation, cf. Grandell (1991, p.6),

$$e^{R(\ell)u}\Psi(u,\ell) = \frac{\ell\mu}{c}e^{R(\ell)u}\overline{F}_I(u) + \frac{\ell}{c}\int_0^u e^{R(\ell)(u-z)}\Psi(u-z,\ell)e^{R(\ell)z}\overline{F}(z)\,dz.$$

The equation is also a simple consequence of (9.11) provided it is noticed that $\overline{F}_I^{0*}(u) = 0$ for $u \geq 0$ and that

$$\overline{F}_I^{n*}(u) = 1 - F_I^{(n-1)*} * F_I(u) = \overline{F}_I(u) + \overline{F}_I^{(n-1)*} * F_I(u).$$

It follows from Weiner (1965, 1977) that

$$e^{R(\ell)u}\Psi(u,\ell) \text{ is } \begin{array}{c}\text{increasing}\\ \text{constant}\\ \text{decreasing}\end{array} \text{ if } \begin{array}{c}(a)\\ (b)\\ (c)\end{array} \text{ is valid,}$$

i.e. the lemma holds. ■

Obviously Assumption 9.1 (b) is fulfilled for the exponential distribution. It is fairly easy to show that Assumption 9.1 does not hold for the mixture of exponential distributions, treated in Example 9.4, where $e^{R(\ell)u}\Psi(u,\ell)$ is decreasing. Neither does Assumption 9.1 hold for the Γ-distribution.

Assumption 9.1 is easy to interpret. If, for instance, (a) holds, then $z = 0$ implies $R(\ell) > \frac{c-\ell\mu}{c\mu}$ if $F(0) = 0$. This means that F – from the point of ruin – is less dangerous than the exponential distribution. Since the tail behaviour of F is of utmost importance for the ruin probability it seems natural to expect $E[Z_k - z \mid Z_k > z]$ to be smaller than μ for large values of z. It may, however, not be too much smaller if (a) is to be valid. This seems to be a delicate balancing act, and we have in fact not managed to find any distribution, except the exponential one, where Assumption 9.1 holds.

Although much is known about ruin probabilities in the Poisson case, there is generally no reason to consider varying values of ℓ. A feasible way to obtain less restrictive conditions for (9.29) than Assumption 9.1 may therefore be to try to generalize Lamma 9.2, although we have have no ideas how that could be done. □

Let us now leave the case of small claims. For large claims we have the following theorem.

Theorem 9.4. *Let ℓ_1 be the right extremity of U. Suppose that $\ell_1 < c/\mu$ and that $F_I \in \mathcal{S}$. Then*

$$\Psi(u) \sim E\left[\frac{1}{\rho(\Lambda)}\right]\overline{F}_I(u), \quad u \to \infty,$$

where

$$\rho(\ell) = \frac{c - \ell\mu}{\ell\mu}.$$

Proof. It follows from Lemma 9.1 that $s_\infty > 1$, i.e. that N is light-tailed. From (9.24) and Proposition 8.2 we get

$$\Psi(u) \sim \mu_N \overline{F}_I(u), \quad u \to \infty,$$

and since

$$\mu_N = \int_0^{\ell_1} \frac{\ell\mu}{c - \ell\mu}\, dU(\ell) = \int_0^{\ell_1} \frac{1}{\rho(\ell)}\, dU(\ell)$$

the theorem follows. ■

In the case where $F \in \mathcal{S}(\beta)$ we have the following theorem.

Theorem 9.5. *Let ℓ_1 be the right extremity of U. Suppose that $F \in \mathcal{S}(\beta)$ and that $h(\beta) < c\beta/\ell_1$. Then*

$$\Psi(u) \sim E\left[\frac{\frac{\Lambda}{c\beta}\left(1 - \frac{\Lambda\mu}{c}\right)}{\left(1 - \frac{\Lambda h(\beta)}{c\beta}\right)^2}\right] \overline{F}(u), \quad u \to \infty.$$

Proof. The theorem follows, with the obvious changes, in the same way as Proposition 9.5. ■

9.2.2 The Ammeter case

We will first consider the ruin problem in the Ammeter case by using relatively recent methods, and after that briefly comment on Ammeter's original approach. For a more detailed treatment of the Ammeter case, we refer to Grandell (1995). Let, cf. (9.5),

$$X_k \stackrel{\text{def}}{=} X(k\Delta) = \sum_{j=1}^{k} (c\Delta - Y_j),$$

and

$$\Psi_\Delta(u) = P\{u + X_k < 0 \text{ for some } k > 0\}. \tag{9.33}$$

This notion of ruin is illustrated in Figure 9.4, for the same realization of the risk process as used in Figure 9.3.

Thus $\Psi_\Delta(u)$ is the probability for the risk process to be ruined at some time of the form $t = k\Delta$, which is the probability studied by Ammeter. The analysis can be given in terms of either random walks or risk processes generated by a renewal process. We will choose the latter approach.

The first treatment of the ruin problem when the occurrence of the claims is described by a renewal process is due to Sparre Andersen

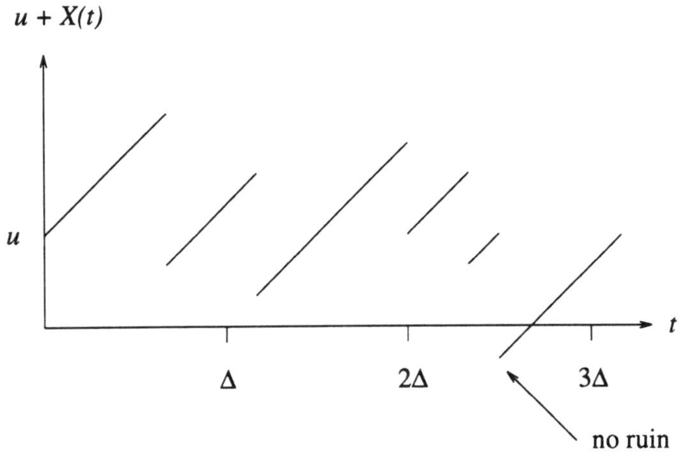

Figure 9.4. *Illustration of a risk process, inspected at equidistant times.*

(1957). In a series of papers Thorin has carried through a systematic study, see for instance Thorin (1982).

Let $\widetilde{N}$ be a point process and let T_k denote the time of the kth claim. Recall from Definition 3.8 that $\widetilde{N}$ is called an ordinary **renewal process** (with inter-occurrence time distribution K) if the variables T_1, $T_2 - T_1$, $T_3 - T_2, \ldots$ are independent and have a common distribution K with mean $1/\tilde{\alpha}$. Let $K(0) = 0$ which is equivalent to $\widetilde{N}$ being simple. Let $\hat{k}$ denote the Laplace transform, i.e. $\hat{k}(v) = \int_0^\infty e^{-vs}\, dK(s)$. When K is an exponential distribution, we are back in the Poisson case.

Let $\widetilde{X}$ be the risk process, where the position of the claims are described by the ordinary renewal process $\widetilde{N}$, defined above. The distribution of the costs of the claims is denoted by $\widetilde{F}$ where $\widetilde{F}(0-) = 0$. Thus ruin can only occur at claim epochs. To avoid some technical complications we assume that $\widetilde{F}$ contains an absolutely continuous component. Furthermore $\tilde{\mu}$, $\tilde{h}(r)$ and so on have their natural interpretation with respect to $\widetilde{F}$. The safety loading $\tilde{\rho}$ is defined by

$$\tilde{\rho} \stackrel{\text{def}}{=} \frac{c - \tilde{\alpha}\tilde{\mu}}{\tilde{\alpha}\tilde{\mu}}.$$

Consider first the case of small claims. The Lundberg exponent R is then the positive solution of

$$(\tilde{h}(r) + 1)\hat{k}(cr) = 1.$$

The following generalizations of (9.12–14) hold, see Grandell (1991):

$$\Psi(u) = (1 - \tilde{\mu}R)e^{-Ru} \tag{9.34}$$

when the claim costs are exponentially distributed with mean $\tilde{\mu}$;

the Cramér–Lundberg approximation

$$\lim_{u\to\infty} e^{Ru}\,\Psi(u) = \widetilde{C}, \tag{9.35}$$

where $\widetilde{C}$ in general is not at all as explicit as in the Poisson case, cf. Thorin (1974, p.94);

the Lundberg inequality

$$\Psi(u) \le e^{-Ru}. \tag{9.36}$$

Consider now the large claim case, i.e. the case where $\widetilde{F}_I \in \mathcal{S}$. Then, cf. Embrechts and Veraverbeke (1982, p.65),

$$\widetilde{\Psi}(u) \sim \frac{1}{\tilde{\rho}}\,\overline{\widetilde{F}}_I(u), \quad u \to \infty. \tag{9.37}$$

Notice that the asymptotic behaviour of the ruin probability in the large claim case only depends on the underlying renewal process through $\tilde{\alpha}$.

In the intermediate case, i.e. for $\widetilde{F}_I \in \mathcal{S}(\beta)$, $\beta > 0$, there also exist certain results in the renewal case, see Embrechts and Veraverbeke (1982, p.67). Those results are, however, not as satisfying as in the Poisson case. We will therefore restrict ourselves to small and large claims.

Now we go back to the Ammeter case. The sequence $\{X_k\}$ can be looked upon as the risk business in a renewal model with

$$K(s) = \begin{cases} 0 & \text{if } s < \Delta \\ 1 & \text{if } s \ge \Delta \end{cases} \quad \text{or} \quad \hat{k}(v) = e^{-v\Delta} \tag{9.38}$$

and, see (9.3),

$$\widetilde{F}(z) = F(z, \Delta). \tag{9.39}$$

This implies that $\tilde{\alpha} = 1/\Delta$ and $\tilde{\mu} = \mu\mu_L\Delta$. The safety loading ρ is defined by

$$\rho \stackrel{\text{def}}{=} \tilde{\rho} = \frac{c - \mu\mu_L}{\mu\mu_L}. \tag{9.40}$$

Consider now the small claim case. Assume further that $\hat{u}(-v)$ fulfils the same condition as h, i.e. essentially that the tail of U decreases at least exponentially fast. Under these assumptions $\widetilde{F}$ is light-tailed, and we have

$$\tilde{h}(r) + 1 = \int_{0-}^{\infty} \sum_{k=0}^{\infty} \frac{(\ell\Delta)^k}{k!} e^{-\ell\Delta}(h(r)+1)^k \, dU(\ell)$$

$$= \int_{0-}^{\infty} e^{\ell\Delta h(r)} \, dU(\ell) = \hat{u}(-\Delta h(r)).$$

Since $(\tilde{h}(r) + 1)\hat{k}(cr) = \hat{u}(-\Delta h(r))e^{-cr\Delta}$ it follows that R is the positive solution of

$$\hat{u}(-\Delta h(r))e^{-cr\Delta} = 1. \tag{9.41}$$

Example 9.6. Let U be $\Gamma(\gamma, \beta)$, and recall from Example 2.1 that then

$$\hat{u}(s) = \left(1 + \frac{s}{\beta}\right)^{-\gamma}.$$

Thus R is the positive solution of

$$h(r) = \frac{\beta}{\Delta}\left(1 - e^{-cr\Delta/\gamma}\right), \tag{9.42}$$

which is (9.35) in Ammeter (1948, p.196). This is the case treated by Ammeter. His derivation was, however, completely different. □

Example 9.7. Now we simplify Example 9.6 further and consider the Ammeter case corresponding to Example 9.1, i.e. the case

$$F(z) = 1 - e^{-z/\mu} \quad \text{for } z \geq 0$$

and

$$U(\ell) = 1 - e^{-\ell/\mu_L} \quad \text{for } \ell \geq 0.$$

Then

$$h(r) = \frac{\mu r}{1 - \mu r} \quad \text{and} \quad \hat{u}(s) = \frac{1}{1 + \mu_L s},$$

which reduces (9.42) to

$$\mu \mu_L \Delta r = (1 - \mu r)\left(1 - e^{-cr\Delta}\right),$$

Further if follows from (9.4) that

$$\widetilde{F}(z) = 1 - \frac{\mu_L \Delta}{1 + \mu_L \Delta} e^{-\frac{z}{\mu(1+\mu_L \Delta)}},$$

i.e. 'almost' an exponential distribution. The idea is now to modify K so that the corresponding claim distribution becomes exponential and to use (9.34).

Let Y be a random variable with distribution $\widetilde{F}$. Let

$$\check{p} \stackrel{\text{def}}{=} \widetilde{F}(0) = \frac{1}{1 + \mu_L \Delta}, \qquad \check{\mu} = \mu(1 + \mu_L \Delta)$$

and

$$\check{F}(y) \stackrel{\text{def}}{=} P\{Y \le y \mid Y > 0\} = \frac{\widetilde{F}(y) - \widetilde{F}(0)}{1 - \widetilde{F}(0)} = 1 - e^{-y/\check{\mu}}.$$

Let $\{I_j\}_1^\infty$ denote the successive values of those k where $Y_k > 0$. Then we have, since ruin cannot occur for a value of k with $Y_k = 0$,

$$\Psi_\Delta(u) = P\{u + X_{I_j} < 0 \text{ for some } j > 0\}.$$

The sequence $\{\check{T}_j\}$, where $\check{T}_j = I_j \Delta$, defines a renewal process with discrete inter-occurrence time distribution $\check{K}$ given by $P\{\check{T}_1 = k\Delta\} = \check{p}^{k-1}(1 - \check{p})$ for $k = 1, 2, \ldots$ Thus we can equally well consider a renewal model defined by $\check{K}$ and $\check{F}$. Since $\check{F}$ is exponential, it follows from (9.34) that

$$\Psi_\Delta(u) = (1 - \mu(1 + \mu_L \Delta)R)e^{-Ru}. \tag{9.43}$$

□

Although certainly $\Psi_\Delta(u)$ may be interesting from many points of view, the mathematically most natural probability ought to be $\Psi(u)$.

Assume that $u + X(t_0) < 0$ for some $t_0 \in (k_0\Delta, (k_0+1)\Delta]$. Since $F(0) = 0$ it follows that

$$u + X((k_0+1)\Delta) \le u + X(t_0) + c\Delta < c\Delta$$

or

$$u - c\Delta + X((k_0+1)\Delta) < 0.$$

Thus we have

$$\Psi_\Delta(u) \le \Psi(u) \le \Psi_\Delta(u - c\Delta) \quad \text{for } u \ge c\Delta, \tag{9.44}$$

where the first inequality is trivial.

The following Lundberg inequality is due to Embrechts *et al.* (1993).

Theorem 9.6. $\Psi(u) \le Ce^{-Ru}$, *for some* $C < \infty$.

Proof. Embrechts *et al.* (1993) consider a much more general situation, and their proof is based on the 'martingale approach', due to Gerber (1973). In this case the theorem is an immediate consequence of (9.36) and (9.44):

$$\Psi(u) \le \Psi_\Delta(u - c\Delta) \le e^{-R(u-c\Delta)} = e^{cR\Delta}e^{-Ru}.$$

Since $e^{cR\Delta}e^{-Ru} \ge 1$ for $u < c\Delta$, the theorem holds for all values of u. ■

From the proof of Theorem 9.6 it is seen that we can choose $C = e^{cR\Delta}$. Although we do not at all claim that this is the smallest possible value of C we must in general, since, cf. (9.36), $\Psi_\Delta(u) \le e^{-Ru}$, accept that $C > 1$. For a further discussion on this matter we refer to Björk and Grandell (1988), Grandell (1991), and Embrechts *et al.* (1993). It is shown by Schmidli (1997) that the Cramér–Lundberg approximation also holds in this case, i.e. that

$$\lim_{u\to\infty} e^{Ru}\Psi(u) = \widetilde{C}_\Delta,$$

but no explicit form of constant $\widetilde{C}_\Delta$ is known. However, it follows from (9.44) that

$$C_\Delta \le \widetilde{C}_\Delta \le e^{cR\Delta}C_\Delta, \tag{9.45}$$

where, cf. (9.35), $C_\Delta \stackrel{\text{def}}{=} \lim_{u\to\infty} e^{Ru}\Psi_\Delta(u)$.

In order to apply the inequality (9.45), the following proposition may be used.

Proposition 9.6. *Assume that there exists a positive solution r_0 of*

$$\hat{u}(-r/\mu_L)e^{-(1+\rho)r} = 1. \tag{9.46}$$

Then we have

$$e^{cR\Delta} \le e^{(1+\rho)r_0}.$$

Proof. Put $R_0 = \mu\mu_L\Delta R$. Noticing that $cR\Delta = (1+\rho)R_0$ it follows from (9.41) that R_0 is the positive solution of

$$\hat{u}\left(-\Delta h\left(\frac{r}{\mu\mu_L\Delta}\right)\right)e^{-(1+\rho)r} = 1.$$

Since $\hat{u}(-v)$ is increasing in v and since $h(r) \ge r\mu$ it follows that

$$\hat{u}\left(-\Delta h\left(\frac{r}{\mu\mu_L\Delta}\right)\right) \ge \hat{u}(-r/\mu_L).$$

Thus we have

$$1 = \hat{u}\left(-\Delta h\left(\frac{R_0}{\mu\mu_L\Delta}\right)\right)e^{-(1+\rho)R_0} \ge \hat{u}(-R_0/\mu_L)e^{-(1+\rho)R_0},$$

which implies

$$\hat{u}(-R_0/\mu_L)e^{-(1+\rho)R_0} \le \hat{u}(-r_0/\mu_L)e^{-(1+\rho)r_0}.$$

Since $\hat{u}(-r/\mu_L)e^{-(1+\rho)r}$ is convex in r, it follows that $R_0 \le r_0$ and the proposition is proved. ∎

Example 9.7 (continued). The simplest non-trivial example is probably when L_k is exponentially distributed. For the moment, we do, however, not assume that F is an exponential distribution. Then $\hat{u}(v) = 1/(1+\mu_L v)$ and (9.46) reduces to

$$\frac{1}{1-r}e^{-(1+\rho)r} = 1.$$

Some values of r_0 and $e^{(1+\rho)r_0}$ are given in Table 9.2.

It is seen from (9.46), and illustrated in Table 9.2, that r_0 increases with increasing values of ρ. This is quite natural, since the higher safety loading the quicker the risk process 'recovers' after ruin.

Table 9.2. *Values of r_0 and $e^{(1+\rho)r_0}$ for exponentially distributed intensity.*

ρ	r_0	$e^{(1+\rho)r_0}$
5%	0.0937	1.1034
10%	0.1761	1.2138
15%	0.2490	1.3316
20%	0.3137	1.4571
25%	0.3714	1.5908
30%	0.4230	1.7330

□

Example 9.6 (continued). Let U be $\Gamma(\gamma, \beta)$ and Y and X be the corresponding Ammeter claim and risk processes. Then it follows from (2.17) that

$$Y(\Delta) \stackrel{\mathrm{d}}{=} \sum_{k=1}^{\bar{N}} Z_k^*,$$

where $\bar{N}$ is $\mathrm{Po}(\bar{\alpha})$ with $\bar{\alpha} = \gamma \log\left(\frac{\beta+\Delta}{\beta}\right)$ and where Z_k^* has distribution F^* with $h^*(r) \stackrel{\mathrm{def}}{=} \int_{0-}^{\infty} e^{rz}\, dF^*(z) - 1$ given by

$$h^*(r) = -\frac{\log\left(1 - \frac{\Delta(h(r)+1)}{\beta+\Delta}\right)}{\log\left(\frac{\beta+\Delta}{\beta}\right)} - 1 = -\frac{\log\left(\frac{\beta-\Delta h(r)}{\beta}\right)}{\log\left(\frac{\beta+\Delta}{\beta}\right)}.$$

Let N^* be a Poisson process with intensity

$$\alpha^* = \frac{\bar{\alpha}}{\Delta} = \frac{\gamma}{\Delta} \log\left(\frac{\beta+\Delta}{\beta}\right)$$

and let the claim cost of the kth claim be Z_k^*, as defined above. Let X^* be the corresponding risk process with the same c as X. Since N^* is a Poisson process, information about its ruin probability $\Psi^*(u)$ is given by (9.13) and (9.14). Although the processes X and X^* are different, the sequences $\{X(k\Delta)\}_{k=0}^{\infty}$ and $\{X^*(k\Delta)\}_{k=0}^{\infty}$ have the same distribution. Thus

$$\Psi_\Delta(u) = \Psi_\Delta^*(u),$$

which is the main observation in Ammeter's approach. Since we are 'close' to the misunderstanding, which we have mentioned, we will point out that Ammeter (1948, p.184) completely understood the relation between X and X^*.

Consider now $\Psi^*(u)$. It is easily seen that $h^*(r) = cr/\alpha^*$ reduces to (9.42), so we have the same R in $\Psi^*(u)$ and $\Psi_\Delta(u)$. This is 'as it shall be' and may be regarded as a check of the calculations. Further we have

$$h^{*\prime}(r) = \frac{\Delta h'(r)}{(\beta - \Delta h(r))\log\left(\frac{\beta+\Delta}{\beta}\right)},$$

and thus, cf. (9.42),

$$h^{*\prime}(R) = \frac{\Delta h'(R)}{\beta\log\left(\frac{\beta+\Delta}{\beta}\right)e^{-cR\Delta/\gamma}}.$$

Since $\mu^*\alpha^* = \mu\gamma/\beta = \mu\mu_L$, which follows from the two representations of $Y(\Delta)$ or by direct calculations, we get from (9.13)

$$C^* \stackrel{\text{def}}{=} \lim_{u\to\infty} e^{Ru}\,\Psi^*(u) = \frac{\rho\mu\mu_L e^{-cR\Delta/\gamma}}{\mu_L h'(R) - ce^{-cR\Delta/\gamma}}, \tag{9.47}$$

where $\rho = (c - \mu\mu_L)/(\mu\mu_L)$ is the safety loading. The next step is to relate C^* with

$$C^*_\Delta \stackrel{\text{def}}{=} \lim_{u\to\infty} e^{Ru}\,\Psi^*_\Delta(u).$$

It follows from (9.44) that

$$\begin{aligned}\Psi^*(u + c\Delta) &\le \Psi^*_\Delta(u) = \Psi_\Delta(u) \le \Psi(u)\\ &\le \Psi_\Delta(u - c\Delta) = \Psi^*_\Delta(u - c\Delta) \le \Psi^*(u - c\Delta)\end{aligned} \tag{9.48}$$

for $u \ge c\Delta$. From the above inequalities it follows, cf. (9.45), that

$$\begin{aligned} e^{-cR\Delta}C^* &\le C^*_\Delta \le \liminf_{u\to\infty} e^{Ru}\,\Psi(u)\\ &\le \limsup_{u\to\infty} e^{Ru}\,\Psi(u) \le e^{cR\Delta}C^*_\Delta \le e^{cR\Delta}C^*.\end{aligned} \tag{9.49}$$

Naturally we are better off if we can calculate $C_\Delta = C^*_\Delta$, but (9.49) gives bounds which only require C^*. We know only one case where C_Δ can be simply calculated, and will consider that case in the continuation of Example 9.7. For some further details about this example and Ammeter's derivation, we refer to Grandell (1995). □

The inequalities (9.48) and (9.49) in the example above are nice, since the bounds are given in terms of an 'associate' Poisson risk process X^*. This is a consequence of the infinite divisibility of the Γ-distribution. A systematic discussion of the case where U is ID is given by Grandell (1995).

Example 9.7 (continued). We will now consider the relation between C^*_Δ and C^* in some detail. From (9.43) it is seen that

$$C_\Delta = C^*_\Delta = 1 - \mu(1 + \mu_L \Delta)R.$$

Since $h'(r) = \mu/(1 - \mu r)^2$ it follows from (9.47) that

$$C^* = \frac{\rho\mu\mu_L(1-\mu R)C^*_\Delta}{\mu\mu_L - c(1-\mu R)C^*_\Delta} = \frac{\rho(1-\mu R)C^*_\Delta}{1-(1+\rho)(1-\mu R)C^*_\Delta}.$$

In Table 9.3 the different bounds in (9.49) are illustrated in the case $\mu = \mu_L = 1$ and $\rho = 10\%$.

Table 9.3. *Values of bounds in the case* $\rho = 10\%$.

Δ	$e^{-cR\Delta}C^*$	C^*_Δ	$e^{cR\Delta}C^*_\Delta$	$e^{cR\Delta}C^*$
1	0.8472	0.8792	0.9396	0.9676
10	0.7540	0.8370	0.9852	1.0446
100	0.7275	0.8254	0.9983	1.0642
1000	0.7244	0.8240	0.9998	1.0664

□

We will now consider large claims. As for small claims, we base the analysis on the renewal model, specified by (9.38), (9.39) and (9.40). The following lemma will be useful, although the assumption is given in terms of an 'unnatural' distribution.

Lemma 9.3. *Assume that* $\widetilde{F}_I \in \mathcal{S}$. *Then*

$$\Psi(u) \sim \frac{1}{\rho}\overline{\overline{F}}_I(u), \quad u \to \infty.$$

Proof. It follows from (9.37) that $\Psi_\Delta(u)$ satisfy the asymptotic relation in the lemma. From (9.44), which holds in general and not only for small claims, it follows that

$$1 \le \frac{\Psi(u)}{\Psi_\Delta(u)} \le \frac{\Psi_\Delta(u - c\Delta)}{\Psi_\Delta(u)} \quad \text{for } u \ge c\Delta.$$

From the closure property of $\mathcal{S}$, discussed after Proposition 9.4, and Lemma 8.3, it follows that

$$\Psi(u) \sim \Psi_\Delta(u), \quad u \to \infty, \tag{9.50}$$

and the lemma is proved. ∎

The relation (9.50) in the proof above is interesting in itself. Intuitively it means that ruin is caused by a claim so large that the risk process will remain negative until the next epoch of the form $k\Delta$.

Since $\widetilde{F}(z) = F(z, \Delta)$ is a mixed compound Poisson distribution, we can apply results given in section 8.3. Let us first consider the case where F is heavy-tailed and U is light-tailed. It turns out that we will need both that $F \in \mathcal{S}$ as well as that $F_I \in \mathcal{S}$. As mentioned, the relation between these two conditions is not at all trivial. In order to cope with these requirements we will rely on the class $\mathcal{S}^*$, introduced by Klüppelberg (1987).

Definition 9.3. *A distribution F on $[0, \infty)$ belongs to the class $\mathcal{S}^*$ if it has finite expectation μ and*

$$\lim_{x\to\infty} \int_0^x \frac{\overline{F}(x-y)\overline{F}(y)}{\overline{F}(x)}\, dy = 2\mu.$$

From Klüppelberg (1989) it follows that the lognormal and the Pareto distribution belong to $\mathcal{S}^*$. Further properties of the class $\mathcal{S}^*$ are to be found in the survey of subexponential distributions by Goldie and Klüppelberg (1997).

Theorem 9.7 below was proposed by Klüppelberg (1994).

Theorem 9.7. *Let $F \in \mathcal{S}^*$ and U be light-tailed. Then*

$$\Psi(u) \sim \frac{1}{\rho}\overline{F}_I(u), \quad u \to \infty.$$

Proof. Since $F \in \mathcal{S}^*$ implies (Klüppelberg 1988, p.135) $F \in \mathcal{S}$, it follows from Theorem 8.14 that

$$1 - \widetilde{F}(u) \sim \mu_L \Delta \overline{F}(u), \quad u \to \infty,$$

which, by l'Hospital's rule, implies

$$\lim_{u\to\infty} \frac{\overline{\widetilde{F}}_I(u)}{\overline{F}_I(u)} = \lim_{u\to\infty} \frac{\frac{1}{\mu\mu_L\Delta}\int_u^\infty \overline{\widetilde{F}}(x)\,dx}{\frac{1}{\mu}\int_u^\infty \overline{F}(x)\,dx}$$

$$= \frac{1}{\mu_L \Delta} \lim_{u\to\infty} \frac{\overline{\overline{F}}(u)}{\overline{F}(u)} = 1.$$

Since $F \in \mathcal{S}^*$ also implies (Klüppelberg 1988, p.135) $F_I \in \mathcal{S}$, it follows from the closure property of $\mathcal{S}$ that $\widetilde{F}_I \in \mathcal{S}$. Thus the theorem follows from Lemma 9.3. ∎

In the case covered by Theorem 9.7, there is no essential influence of the intensity. This observation is in agreement with Asmussen *et al.* (1994).

Example 9.8. Let F be a Pareto distribution with $\delta > 1$, which, cf. (9.16), implies that

$$\overline{F}_I(z) = \frac{1}{\delta}\left(\frac{z}{a}\right)^{-(\delta-1)} \quad \text{for } z \geq a > 0.$$

Thus the result in Theorem 9.7 agrees with (9.17), which illustrate the little influence of the intensity. □

Naturally $\widetilde{F}$ may have a thick tail, not only due to F, but also due to the tail behaviour of U. The discussion after Theorem 8.15 does, however, indicate that it is fruitless to relate U with $\mathcal{S}$. In order to find symmetric conditions on F and U, we restrict ourselves to regularly varying tails.

Assume, for some $\delta > 1$ and some slowly varying function L, that

$$\overline{U}(\ell) \sim \omega \ell^{-\delta} L(\ell) \quad \text{and} \quad \overline{F}(z) \sim \chi z^{-\delta} L(z) \tag{9.51}$$

where $\omega \geq 0$ and $\chi \geq 0$. If $\omega = 0$, then (9.51) means that $\overline{U}(\ell) = o(\ell^{-\delta} L(\ell))$ and similarly if $\chi = 0$.

Theorem 9.8. *Assume that (9.51) holds and that at least one of the constants ω or χ are strictly positive. Then*

$$\Psi(u) \sim \frac{\omega(\mu\Delta)^\delta + \chi\mu_L\Delta}{\rho\mu_L\mu(\delta-1)\Delta} u^{-(\delta-1)} L(u), \quad u \to \infty. \tag{9.52}$$

Proof. It follows from Theorem 8.13 that

$$\overline{\overline{F}}(u) \sim (\omega(\mu\Delta)^{-\gamma} + \chi\mu_L\Delta) L(u) u^\gamma, \quad u \to \infty.$$

Since $\tilde{\mu} = \mu\mu_L\Delta$ it follows, cf. Proposition 8.5, that

$$\bar{\bar{F}}_I(u) \sim \frac{\omega(\mu\Delta)^\delta + \chi\mu_L\Delta}{\mu_L\mu(\delta-1)\Delta} u^{-(\delta-1)}L(u), \quad u \to \infty.$$

Since subexponentiality is an extension of regular variation, it follows that $\widetilde{F}_I \in \mathcal{S}$. Thus the theorem follows from Lemma 9.3. ■

The case where both ω and χ are positive seems somewhat artificial, since it means that U and F have almost exactly the same tail behaviour. If we let one of those constants be zero, it is no restriction to let the other be one.

Let us first consider the case $\chi = 0$ and $\omega = 1$. This means that the heavy-tail behaviour of $\widetilde{F}$ is caused by the variation of the intensity. It does not imply that F is necessarily light-tailed, but it is less heavy-tailed than U. Then it follows from (9.52) that

$$\Psi(u) \sim \frac{(\mu\Delta)^{\delta-1}}{\rho\mu_L(\delta-1)} u^{-(\delta-1)}L(u), \quad u \to \infty.$$

In this case the intensity has an essential influence on the ruin probability.

Now consider the case $\chi = 1$ and $\omega = 0$. This means that F is heavy-tailed, and we can really talk about large claims. Then it follows from (9.52) that

$$\Psi(u) \sim \frac{1}{\rho\mu(\delta-1)} u^{-(\delta-1)}L(u), \quad u \to \infty. \tag{9.53}$$

Example 9.8 (continued). When F is a Pareto distribution with $\delta > 1$, we have $\mu = \delta a/(\delta-1)$. Thus (9.53) agrees with (9.17). This is a slightly more general observation than the observation that Theorem 9.7, in the Pareto case, agreed with (9.17), since $\omega = 0$ only means that U is less heavy-tailed that F and not necessarily light-tailed. □

We can apply Theorem 9.1 also to ruin problems when Δ is equal to the length of a seasonal variation in the distribution of the costs of the claims, i.e. when $F_{s+k\Delta}(z) = F_s(z)$ for all z, s and k. Natural lengths may be one day, one week, or one year and multiples of these units. If we are only interested in $\Psi_\Delta(u)$, it follows from Theorem 9.1 that we just have to replace F with $\frac{1}{\Delta}\int_0^\Delta F_s(z)\,ds$.

References and author index

Adell, J. A. and de la Cal, J (1993) On the uniform convergence of normalized Poisson mixtures to their mixing distribution. *Statist. Probab. Lett. 18*, 227–232. [22] *

Albrecht, P. (1981) Über einige Eigenschaften des gemischten Poissonprozesses. *Mitt. Ver. Schweiz. Vers. Math. 81*, 241–250. [64]

Albrecht, P. (1982) On Some Statistical Methods Connected with the Mixed Poisson Process. *Scand. Actuarial J.*, 1–14. [xi]

Albrecht, P. (1983) A note on a limiting behaviour of the occurrence times of a mixed Poisson process. *Adv. Appl. Prob. 15*, 460. [66]

Albrecht, P. (1984) Laplace Transforms, Mellin Transforms and Mixed Poisson Processes. *Scand. Actuarial J.*, 58–64. [45, 49]

Albrecht, P. (1985) Mixed Poisson processes. In *Encyclopedia of Statistical Sciences, Vol. 5.* Ed. by Kotz, S., Johnson, N. L. and Read, C. B., Wiley, New York, 556–559. [ix]

Ammeter, H. (1948) A generalization of the collective theory of risk in regard to fluctuating basic probabilities. *Skand. AktuarTidskr.*, 171–198. [xi, 11, 87, 239]

Andersen, E. Sparre (1957) On the collective theory of risk in the case of contagion between the claims. *Transactions XVth International Congress of Actuaries, II*, New York, 219–229. [236]

Asmussen, S., Fløe Henriksen, L. and Klüppelberg, C. (1994) Large claims approximations for risk processes in a Markovian environment. *Stochastic Proc. Applic. 54*, 29–43. [247]

Asmussen, S. (1996) Personal communications. [228]

Asmussen, S. (1997) *Ruin Probability.* World Scientific Publishers, Singapore. To appear. [218]

Athreya, K. B. and Ney, P. E. (1972) *Branching Processes.* Springer-Verlag, Berlin. [166, 167]

Aven, T. (1985) A Theorem for Determining the Compensator of a Counting process. *Scand. J. Statist. 12*, 69–72. [65]

* Pages on which references are cited are given in square brackets.

von Bahr, B. (1975) Asymptotic ruin probabilities when exponential moments do not exist. *Scand. Actuarial J.*, 6–10. [220]

Barlow, R., Marshall, A. and Proschan, F. (1963) Properties of probability distributions with monotone hazard rate. *Ann. Math. Statist. 34*, 375–389. [136]

Barlow, R. E. and Proschan, F. (1965) *Mathematical Theory of Reliability.* Wiley, New York. [133, 137, 142]

Barlow, R. E. and Proschan, F. (1975) *Statistical Theory of Reliability and Life Testing: Probability Models.* Holt, Rinehart and Winston, New York. [135, 136, 140, 142, 174, 226]

Bartholomew, D. J. (1969) Sufficient conditions for a mixture of exponential distributions to be a probability density function. *Ann. Math. Statist. 40*, 2183–2188. [31]

Bates G. E. and Neyman J. (1952a) Contributions to the theory of accident proneness. I. An optimistic model of the correlation between light and severe accidents. *Univ. Calif. Publ. Statistics 1*, 215–254. [8, 9]

Bates G. E. and Neyman J. (1952b) Contributions to the theory of accident proneness. II. True or false contagion. *Univ. Calif. Publ. Statistics 1*, 255–275. [8, 9]

Beall, G. and Rescia, R. R. (1953) A generalization of Neyman's contagious distributions. *Biometrics 9*, 354–386. [45]

Berljand, O. S., Nazarov, I. M. and Pressman, A. Ja. (1962) i^n erfc- или сложное гаусс–пуассоновское распределение. *Доклады Академии наук СССР 147*, 1005–1007. [47]

Bhattacharya, S. K. and Holla, M. S. (1965) On a Discrete Distribution with Special Reference to the Theory of accident Proneness. *J. Amer. Stat. Assoc. 60*, 1060–1066. [46]

Billingsley, P. (1968) *Convergence of Probability Measures.* Wiley, New York. [44]

Bingham, N. B., Goldie, C. M. and Teugels, J. L. (1987) *Regular variation.* Cambridge University Press, Cambridge. [152, 179]

Björk, T. and Grandell, J. (1988) Exponential inequalities for ruin probabilities in the Cox case. *Scand. Actuarial J.*, 77–111. [241]

Block, H. and Savits, T. (1980) Laplace transforms for classes of life distributions. *Ann. Prob. 8*, 465–474. [135, 138, 141]

Bohman, H. and Esscher. F. (1963) Studies in Risk Theory with Numerical Illustrations Concerning Distribution Functions and Stop Loss Premiums. Part I. *Skand. AktuarTidskr.*, 173–225. [184]

Bohman, H. and Esscher. F. (1964) Studies in Risk Theory with Numerical Illustrations Concerning Distribution Functions and Stop Loss Premiums. Part II. *Skand. AktuarTidskr.*, 1–40. [184]

Breiman, L. (1963) The Poisson tendency in traffic distribution. *Ann. Math. Statist. 34*, 308–311. [116]

Brillinger, D. R. (1978) A note on a representation for the Gauss–Poisson process. *Stochastic Proc. Applic. 6*, 135–137. [87, 91, 92]

Brémaud, P. (1981) *Point Processes and Queues. Martingale Dynamics.* Springer-Verlag, New York. [63]

Brown, S. and Holgate, P. (1971) Table of the Poisson log normal distribution. *Sankhyā B, 33*, 235–258. [48]

Bühlmann, H. (1960) Austauschbare stochastiche Variabeln und ihre Grenzwertsätze. *Univ. California Publ. Statist. 3*, 1–36. [123]

Bühlmann, H. (1970) *Mathematical Methods in Risk Theory.* Springer-Verlag, Berlin. [17, 22, 33]

Bühlmann, H. and Buzzi, R. (1971) On a transformation of the weighted compound Poisson process. *Astin Bulletin VI*, 42–46. [71]

Bühlmann, H. (1972) Ruinwahrscheinlichkeit bei erfahrungstarifiertem Portefeuille. *Mitt. Ver. Schweiz. Vers. Math. 72*, 211–224. [224]

Bulmer, M. G. (1974) On fitting the Poisson lognormal distribution to species–abundance data. *Biometrics 30*, 235–258. [48]

Chistyakov, V. P. (1964) Теорема о суммах независимых положительных случайных величн и ее приложения к ветвящимся случайным процессам. *Теор. Вероят. и Примен. IX*, 710–718. English translation: A theorem on sums of independent random variables and its applications to branching processes. *Theor. Prob. Appl. 9*, 640–648. [166, 169]

Chover, J., Ney, P. and Wainger, S. (1973) Functions of probability measures. *J. Analyse Math. 26*, 255–302. [170]

Cox, D. R. (1955) Some statistical methods connected with series of events. *J. R. Statist. Soc B, 17*, 129–164. [84]

Cramér, H. (1930) *On the Mathematical Theory of Risk.* Skandia Jubilee Volume, Stockholm. Reprinted in *Harald Cramér Collected works Vol. I.* (1994), Ed. by Martin-Löf, A., 601 – 678. Springer Verlag, Berlin. [218]

Cramér, H. (1955) *Collective Risk Theory.* Skandia Jubilee Volume, Stockholm. Reprinted in *Harald Cramér Collected works Vol. II.* (1994), Ed. by Martin-Löf, A., 1028–1115. Springer-Verlag, Berlin. [232]

Cramér, H. (1969) On streams of random events. *Skand. AktuarTidskr. Suppl.*, 13–23. [85]

Daley, D. J. and Vere-Jones, D. (1988) *An Introduction to the Theory of Point Processes.* Springer-Verlag, New York. [116]

Davis, M. H. A. (1993) *Markov Models and Optimization.* Chapman and Hall, London. [96]

Deffner, A. and Haeusler, E. (1985) A characterization of order statistic point processes that are mixed Poisson processes and mixed sample processes simultaneously. *J. Appl. Prob. 22*, 314 – 323. [108, 122]

Delaporte, P. (1960) Un problème de tarification de l'assurance accidents d'automobile examiné par la statistique mathématique. *Trans. of the XVIth International Congress of Actuaries II*, Brussels, 121–135. [34]

Dickson, D. C. M. and Gray, J. R. (1984) Exact solutions for ruin probability in the presence of an absorbing upper barrier. *Skand. AktuarTidskr.*, 143–158. [233]

Dobrushin, R. L. (1956) О законе Пуассона для распределения частиц в пространстве. *Укр. матем. журнал VIII*, 127–134. [116]

Dubourdieu, J. (1938) Remarques relatives a la théorie mathématique de l'assurance-accidents. *Bull. Trim. de l'Inst. des Actuaires Français 49*, 76 –126. [ix, 80–82]

Dubourdieu, J. (1952) *Théorie mathématique des assurances. Premier livre: Théorie mathématique du risque dans les assurances de répartition.* Gauthier-Villars, Paris. [82]

Eade, J. P. (1983) The ruin problem for mixed Poisson risk processes. *Scand. Actuarial J.*, 193–210. [224]

Eggenberger, F. (1924) Die Wahrscheinlichkeitsansteckung. *Mitt. Ver. Schweiz. Vers. Math. 24*, 31–144. [ix]

Elfving, G. (1946a) Contributions to the theory of integer-valued Markoff processes. *Skand. AktuarTidskr.*, 175–205. [106]

Elfving, G. (1946b) On compound binomial processes. *Försäkringsmatematiska studier tillägnade Filip Lundberg*, 48–78. [106, 108]

Embrechts, P. and Goldie, C. M. (1982) On convolution tails. *Stochastic Proc. Applic. 13*, 263–278. [168, 169–171, 223]

Embrechts, P. and Veraverbeke, N. (1982) Estimates for the probability of ruin with special emphasis on the possibility of large claims. *Insurance: Mathematics and Economics 1*, 55–72. [221–223, 238]

Embrechts, P. (1983) A property of the generalized inverse Gaussian distribution with some applications. *J. Appl. Prob. 20*, 537 – 544. [170]

Embrechts, P., Maejima, M. and Omey, E. (1984) A renewal theorem of Blackwell type. *Ann. Prob. 12*, 561–570. [155, 157]

Embrechts, P., Jensen, J. L., Maejima, M. and Teugels, J. L. (1985a) Approximations for compound Poisson and Pólya processes. *Adv. Appl. Prob. 17*, 623–637. [183]

Embrechts, P., Maejima, M. and Teugels, J. L. (1985b) Asymptotic behaviour of compound distributions. *Astin Bulletin 15*, 45–48. [154]

Embrechts, P., Grandell, J. and Schmidli, H. (1993) Finite-time Lundberg inequalities in the Cox case. *Scand. Actuarial J.*, 17–41. [241]

Embrechts, P., Klüppelberg, C. and Mikosch, T. (1997) *Modelling Extremal Events for Insurance and Finance.* Springer-Verlag, Berlin. [166]

Esary, J., Marshall, A. and Proschan, F. (1973) Shock models and wear processes. *Ann. Prob. 1*, 627–649. [138]

Esscher. F. (1932) On the probability function in the collective theory of risk. *Skand. AktuarTidskr.*, 175–195. [182]

Feigin, P. D. (1979) On the characterization of point processes with the order statistic property. *J. Appl. Prob. 16*, 297 – 304. [122]

Feller, W. (1943) On a general class of "contagious" distributions. *Ann. Math. Statist. 14*, 389–400. [7, 16]

Feller, W. (1968) *An Introduction to Probability Theory and its Applications. Vol I.* 3rd ed. Wiley, New York. [27, 150, 152]

Feller, W. (1971) *An Introduction to Probability Theory and its Applications. Vol II.* 2nd ed. Wiley, New York. [16, 23, 24, 30, 32, 41, 56, 72, 124, 143, 147, 153, 163, 177, 182, 188, 216]

de Finetti, B. (1931) Funzione caratteristica di un fenomeno aleatorio. *Mem. Reale Academia Naz. Lincei. Ser. 6, Vol. 4*, 251–299. [122]

de Finetti, B. (1933) Sulla legge di distribuzione dei valori in una successione di numeri aleatori equivalenti. *Atti Reale Academia Naz. Lincei. Rendiconti, Classe Fisiche, Matematiche e Naturali. 18*, 279–284. [123]

de Finetti, B. (1937) La prevision, ses lois logiques, ses sources subjectives. *Annales de l'Institut Henri Poincaré. 7*, 1 – 68. [123]

Franken, P., König, D., Arndt, U., and Schmidt, V. (1981) *Queues and Point Processes.* Akademie-Verlag, Berlin and Wiley, New York. [118]

Gerber, H. U. (1973) Martingales in risk theory. *Mitt. Ver. Schweiz. Vers. Math. 73*, 205–216. [241]

Gerber, H. U. (1979) *An Introduction to Mathematical Risk Theory.* S. S. Heubner Foundation monograph series 8, Philadelphia. [133]

Gerber, H. U. (1983) On the asymptotic Behaviour of the Mixed Poisson Process. *Scand. Actuarial J.*, 256. [66]

Gerber, H. U. (1994) Martingales and tail probabilities. *Astin Bulletin 24*, 145–146. [147]

Godambe, A. V. (1977) On representation of Poisson mixtures as Poisson sums and a characterization of the gamma distribution. *Math. Proc. Camb. Phil. Soc. 82*, 297–300. [28, 32]

Goel, P. K. (1982) On Implications of Credible Means Being Exact Bayesian. *Scand. Actuarial J.*, 41–46. [23]

Goldie, C. M. (1967) A class of infinitely divisible random variables. *Proc. Camb. Phil. Soc. 63*, 1141–1143. [30]

Goldie, C. M. and Klüppelberg, C. (1997) Subexponential distributions. In *A Practical Guide to Heavy Tails: Statistical Techniques for Analysing Heavy Tailed Distributions.* Ed. by Adler, R., Feldman, R. and Taqqu, M. S., Birkhäuser, Boston. To appear. [166, 246]

Good, I. J. (1953) The population frequencies of species and the estimation of population parameters. *Biometrika 40*, 237–260. [38]

Goovaerts, M. J. (1982) A Remark on Survival Probabilities for a Weighted Poisson Process. *Scand. Actuarial J.*, 211–215. [224]

Grandell, J. (1970) A note on the behaviour of the "tail" of the mixture-Poisson distribution. *Skand. AktuarTidskr.*, 76–77. [178]

Grandell, J. (1971) On stochastic processes generated by a stochastic intensity function. *Skand. AktuarTidskr.*, 204–240. [214]

Grandell, J. and Peiram, L. (1973) A note on the ruin problem for a class of stochastic processes with interchangeable increments. *Astin Bull. 7*, 81–89. [224, 225]

Grandell, J. (1976) *Doubly Stochastic Poisson Processes.* Lecture Notes in Math. 529, Springer-Verlag, Berlin. [22, 63–66, 77, 125]

Grandell, J. (1991) *Aspects of risk theory.* Springer-Verlag, New York. [25, 66, 86, 130, 147, 219, 221, 234, 238, 241]

Grandell, J. (1995) Some remarks on the Ammeter risk process. *Mitt. Ver. Schweiz. Vers. Math. 95*, 43–72. [236, 244]

Greenwood, M. and Yule, G. U. (1920) An inquiry into the nature of frequency distributions representative of multiple happenings with particular reference to the occurrence of multiple attack of disease or of repeated accidents. *J. R. Statist. Soc. 83*, 255–279. [3, 11]

Grigelionis, B. (1975) Случайные точечные процессы и мартингалы. *Liet. Matem. Rink. 15*, 101–114. English translation: Random point processes and martingales. *Lithuanian Math. J. 15*, 444–453. [130]

Grigelionis, B. (1996a) On the asymptotic behaviour of mixed Poisson processes. *Liet. Matem. Rink. 36.* In press. [130]

Grigelionis, B. (1996b) On mixed Poisson processes and martingales. *Scand. Actuarial J.* To appear 1998. [125, 128–130]

Gurland, J. (1958) A generalized class of contagious distributions. *Biometrics 14*, 229–249. [43, 45]

Haight, F. A. (1965) On the effect of removing persons with N or more accidents from an accident prone population. *Biometrika 52*, 298–300. [10]

Haight, F. A. (1967) *Handbook of the Poisson Distribution.* Wiley, New York. [ix, 49]

Hesselager, O. (1996) A Recursive Procedure for Calculation of Some Mixed Compound Poisson Distributions. *Scand. Actuarial J.*, 54–63. [192, 201, 203, 204]

Hofmann, M. (1955) Über zusammengesetzte Poisson- Prozesse und ihre Anwendungen in der Unfallversicherung. *Mitt. Ver. Schweiz. Vers. Math. 55*, 499–575. [72–74]

Holgate, P. (1970) The modality of some compound Poisson distributions. *Biometrika 57*, 666–667. [143]

Holla. M. S. (1967) On a Poisson-inverse Gaussian distribution. *Metrika 11*, 115–121. [41]

Huang, W.-J. (1990) On the characterization of point processes with the exchangeable and Markov properties. *Sankhyā A, 52*, 16 – 27. [94, 98]

Huff, B. (1978) Some comments on mixtures and infinite divisibility. *Siam J. Appl. Math. Vol. 35, No. 1*, 17–20. [28]

Johnson, N. L. (1957) Uniqueness of a result in the theory of accident proneness. *Biometrika 44*, 530–531. [22]

Johnson, N. L. (1967) Note on a uniqueness relation in certain accident proneness models. *J. Amer. Stat. Assoc. 62*, 288–289. [22]

Jung, J. (1963) A Theorem on Compound Poisson Processes with Time-dependent Change Variables. *Skand. AktuarTidskr.*, 95 – 98. [212, 213]

Jung, J. and Lundberg, O. (1969) Risk Processes Connected with the Compound Poisson Process. *Skand. AktuarTidskr. Suppl.*, 118 – 131. [ix]

Jørgensen, B. (1982) *Statistical properties of the Generalized Inverse Gaussian Distribution.* Lecture Notes in Stat. 9, Springer-Verlag, New York. [39, 40, 42]

Kallenberg, O. (1973) Characterization and convergence of random measures and point processes. *Z. Wahrschein. und verw. Geb. 27*, 9–21. [121, 123, 125]

Kallenberg, O. (1975) Limits of compound and thinned point processes. *J. Appl. Prob. 12*, 269–278. [86]

Kallenberg, O. (1983) *Random Measures.* 3rd ed. Akademie-Verlag, Berlin and Academic Press, New York. [16, 26, 28, 122, 123]

Karlin, S., Proschan, F and Barlow, R. E. (1961) Moment inequalities of Pólya frequency functions. *Pacific J. Math. 11*, 1023–1033. [136]

Kemp, C. D. and Kemp, A. W. (1965) Some properties of the 'Hermite' distribution. *Biometrika 52*, 381–394. [87–90]

Kemp, A. W. and Kemp, C. D. (1966) An alternative derivation of the Hermite distribution. *Biometrika 53*, 627–628. [87, 90]

Kemp, A. W. (1968) A Limited Risk cPp. *Skand. AktuarTidskr.*, 198–203. [49, 200]

Kemp, C. D. and Papageorgiou, H. (1982) Bivariate Hermite distributions. *Sankhyā A, 44*, 269–280. [91]

Klüppelberg, C. (1987) Subexponentielle Verteilungen und Charakterisierungen verwandter Klassen. Diss. University of Mannheim, Mannheim. [246]

Klüppelberg, C. (1988) Subexponential distributions and integrated tails. *J. Appl. Prob. 25*, 132–141. [246, 247]

Klüppelberg, C. (1989) Estimation of ruin probabilities by means of hazard rates. *Insurance: Mathematics and Economics 8*, 279–285. [133, 246]

Klüppelberg, C. (1994) Personal communications. [167, 246]

Kupper, J. (1962a) Wahrscheinlichkeitstheoretische Modelle in der Schadenversicherung, Teil I: Die Schadenzahl. *Blätter der deutschen Gesellschaft für Versicherungsmathematik 5*, 451–503. [47]

Kupper, J. (1962b) Wahrscheinlichkeitstheoretische Modelle in der Schadenversicherung, Teil II: Schadenhöhe und Totalschaden. *Blätter der deutschen Gesellschaft für Versicherungsmathematik 6*, 95–130. [47]

Lévy, P. (1937) Sur les exponentielles de polynômes et sur l'arithmétique des produits de lois de Poisson. *Ann. École Normale Supérieure, 3e série 54*, 231–292. Reprinted in *Œuvres de Paul Lévy Vol. III.* (1976), Ed. by Dugué, D., 356–417. Gauthier-Villars, Paris. [29]

Lin, X. (1996) Tail of compound distributions and excess time. *J. Appl. Prob. 33*, 184–195. [149]

Liptser, R. S. and Shiryayev, A. N. (1978) *Statistics of Random Processes II. Applications.* Springer-Verlag, New York. [127, 128]

Lukacs, E. (1960) *Characteristic functions.* Charles Griffin, London. [28]

Lundberg, F. (1926) *Försäkringsteknisk Riskutjämning.* F. Englunds boktryckeri A.B., Stockholm. [218]

Lundberg, O. (1964) *On random processes and their application to sickness and accident statistics.* 1st ed. 1940, Almqvist och Wiksell, Uppsala. [ix, x, 7, 22, 60–64, 66–68, 70, 76, 77, 79, 102–104, 209–211]

Maceda, E. C. (1948) On the compound and generalized Poisson distributions. *Ann. Math. Statist. 19*, 414–416. [27]

Maruyama, G. (1955) On the Poisson distribution derived from independent random walks. *Nat. Sci. Rep. Ochanomiza Univ. 6*, 1–6. [116]

Matthes, K., Kerstan, J., and Mecke, J. (1978) *Infinitely Divisible Point Processes.* Wiley, New York. [71, 116, 117, 121, 125]

McFadden, J. A. (1965) The mixed Poisson process. *Sankhyā A, 27*, 83–92. [82, 109, 110]

McNolty, F. (1964) Reliability density functions when the failure rate is randomly distributed. *Sankhyā A, 26*, 287–292. [44]

Mecke, J. (1968) Eine charakteristische Eigenschaft der doppelt stochastischen Poissonschen Prozesse. *Z. Wahrschein. und verw. Geb. 11*, 74–81. [26, 85]

Mecke, J. (1976) A characterization of mixed Poisson processes. *Rev. Roumaine math. pures appl. XXI*, 1355–1360. [117, 119]

Milne, R. K. and Westcott, M. (1972) Further results for Gauss–Poisson processes. *Adv. Appl. Prob. 4*, 151–176. [87, 88]

Milne, R. K. and Westcott, M. (1993) Generalized multivariate Hermite distributions and related point processes. *Ann. Inst. Statist. Math. 45*, 367–381. [93]

Nawrotzki, K. (1962) Ein Grenzwertsatz für homogene zufällige Punktfolgen. *Math. Nachr. 24*, 201–217. [113]

Newbold, E. M. (1926) *A contribution to the study of the human factor in the causation of accidents.* Industrial Health Research Board, Report No. 34, London [3, 11]

Newbold, E. M. (1927) Practical Applications of the Statistics of Repeated Events, Particularly to Industrial Accidents. *J. R. Statist. Soc. 90*, 487–547. [ix]

Newman, D. S. (1970) A new family of point processes which are characterized by their second moment properties. *J. Appl. Prob.* *7*, 338–358. [87]

Omey, E. and Willekens (1987) Second-order behaviour of distributions subordinate to a distribution with finite mean. *Commun. Statist. - Stochastic models 3*, 311–342. [221]

Ottestad, P. (1944) On certain compound frequency distributions. *Skand. AktuarTidskr.*, 32–42. [15]

Panjer, H. H. (1981) Recursive evaluation of a family of compound distributions. *Astin Bulletin 12*, 22–26. [20, 190]

Panjer, H. H. and Wang, S. (1993) On the stability of recursive formulas. *Astin Bulletin 23*, 227–258. [37, 42]

Patil, G. P. (1964) On certain compound Poisson and compound binomial distributions. *Sankhyā A*, *26*, 293–294. [47]

Pesonen, E. (1968) On definition of compound Poisson processes. *Skand. AktuarTidskr.*, 156–157. [84]

Philipson, C. (1960) The theory of confluent hypergeometric functions and its application to compound Poisson processes. *Skand. AktuarTidskr.*, 136–162. [49]

Philipson, C. (1965) A Generalized model for the Risk Process and its application to a tentative Evaluation of outstanding Liabilities. *Astin Bulletin III*, 215–238. [213]

Philipson, C. (1968) A review of the collective theory of risk. *Skand. AktuarTidskr.*, 45–68 and 117–133. [83, 87]

Pfeifer, D. and Heller, U. (1987) A martingale characterization of the mixed Poisson process. *J. Appl. Prob. 24*, 246–251. [105]

Pólya, G. (1930) Sur quelques points de la théorie des probabilités. *Ann. Inst. H. Poincaré I*, 117–161. [ix, 4, 6]

Prekopa, A. (1958) On secondary processes generated by a random point distribution of Poisson type. *Ann. Univ. Sci. Budapest Sect. Math. 1*, 153–170. [56]

Puri, P. S. and Goldie, C. M. (1979) Poisson mixtures and quasi-infinite divisibility of distributions. *J. Appl. Prob. 16*, 138–153. [24, 28, 29, 32, 33]

Puri, P. S. (1982) On the characterization of point processes with the order statistic property without the moment condition. *J. Appl. Prob. 19*, 39–51. [107, 122]

Quinkert, W. (1957) Die kollektive Risikotheorie unter Berücksichtigung schwankender Grundwahrscheinlichkeiten mit endlichen Schwankungsbereich. Diss. University of Cologne, Cologne. [43, 225]

Rubin, I. (1972) Regular point processes and their detection. *IEEE Trans. Inform. Theory 18*, 547–557. [10]

Ruohonen, M. (1988) On a model for the claim number process. *Astin Bulletin 18*, 57–68. [34]

Schmidli, H. (1996) Compound Sums and Subexponentiality. Tech. Report. University of Aarhus, Aarhus. [186]

Schmidli, H. (1997) An extension to the renewal theorem and an application to risk theory. *Ann. Appl. Prob. 7*. In press. [241]

Schröter, K. J. (1990) On a Family of Counting Distributions and Recursions for Related Compound Distributions. *Scand. Actuarial J.*, 161–175. [34, 191, 195, 197]

Segerdahl, C.-O. (1970) Stochastic Processes and Practical Working Models or Why is the Pólya process approach defective in modern practice and how to cope with its deficiencies? *Skand. AktuarTidskr.*, 146–166. [94, 98]

Shaban, S. A. (1988) Poisson-Lognormal Distributions. In *Lognormal distributions – theory and applications.* Ed. by Crow, E. L. and Shimizu, K., 195–210. Marcel Dekker, New York. [48, 49]

Shanbhag, D. N. and Westcott, M. (1977) A Note on Infinitely Divisible Point Processes. *J. R. Statist. Soc B, 39*, 331–332. [28]

Sichel, H. S. (1971) On a family of discrete distributions particular suited to represent long tailed frequency data. *Proceedings of the Third Symposium on Mathematical Statistics.* Ed. by Laubscher, N. F., CSIR, Pretoria, 51–97. [41]

Sichel, H. S. (1974) On a distribution representing sentence-length in written prose. *J. R. Statist. Soc. A 137*, 25–34. [41]

Sichel, H. S. (1975) On a distribution law for word frequencies. *J. Amer. Statist. Ass. 70*, 542–547. [42]

Stam, A. J. (1973) Regular variation of the tail of a subordinated probability distribution. *Adv. Appl. Prob. 5*, 308–327. [162, 165, 166]

Steutel, F. W. (1967) Note on the infinite divisibility of exponential mixtures. *Ann. Math. Statist. 38*, 1303–1305. [31]

Steutel, F. W. (1970) *Preservation of infinite divisibility under mixing and related topics.* Mathematical Centre Tracts 33, Mathematical Centre, Amsterdam. [28]

Stone, C. (1968) On a theorem of Dobrushin. *Ann. Math. Statist. 39*, 1391–1401. [116]

Sundt, B. and Jewell, W. (1981) Further Results on Recursive evaluation of Compound Distributions. *Astin Bulletin 12*, 27–39. [38, 191]

Sundt, B. (1992) On some extensions of Panjer's class of counting distributions. *Astin Bulletin 22*, 61–80. [191]

Szász, D. (1970) Once more on the Poisson process. *Studia Sci. Math. Hung. 5*, 441–444. [54]

Teugels, J. L. (1975) The class of subexponential distributions. *Ann. Prob. 31*, 1000–1011. [168]

Thedéen, T. (1964) A note on on the Poisson tendency in traffic distribution. *Ann. Math. Statist. 35*, 1823–1824. [116]

Thorin, O. (1970) Some remarks on the ruin problem in the case the epochs of claims form a renewal process. *Scand. Actuarial J.*, 81–99. [97]

Thorin, O. (1973) The ruin problem in case the tail of the claim distribution is completely monotone. *Skand. AktuarTidskr.*, 100–119. [233]

Thorin, O. and Wikstad, N. (1973) Numerical evaluation of ruin probabilities for a finite period. *Astin Bulletin VII*, 137–153. [221]

Thorin, O. (1974) On the asymptotic behavior of the ruin probability for an infinite period when the epochs of claims form a renewal process. *Scand. Actuarial J.*, 81–99. [238]

Thorin, O. and Wikstad, N. (1977) Calculation of ruin probabilities when the claim distribution is lognormal. *Astin Bulletin IX*, 231–246. [220]

Thorin, O. (1982) Probabilities of ruin. *Scand. Actuarial J.*, 65–102. [237]

Thorin, O. (1986) Ruin probabilities when the claim amounts are gamma distributed. Försäkringstekniska forskningsnämnden. Meddelande nr 69. [233]

Vinogradov, O. P. (1973) Определение стареющих функций распределения в терминах преобразования Лапласа. *Теор. Вероят. и Примен. XVIII*, 856–858. English translation: The definition of distribution functions with increasing hazard rate in terms of the Laplace transform. *Theor. Prob. Appl. 18*, 811–814. [135, 141]

Vinogradov, O. P. (1976) О применениях одной формулы обращения преобразования Лапласа. *Теор. Вероят. и Примен. XXI*, 857–860. English translation: On applications of an inversion formula for the Laplace transform. *Theor. Prob. Appl. 21*, 836–839. [138, 140, 141]

Vuilleumier, M. (1963) *On asymptotic behaviour of linear transformations of slowly varying sequences and of sequences of regular asymptotic behaviour.* Mathematics Research Center Tech. Reports 435, Madison, Wisconsin. [179]

Wang, S. and Panjer, H. H. (1993) Critical starting points for stable evaluation of mixed Poisson probabilities. *Insurance: Mathematics and Economics 13*, 287–297. [37]

Wang, S. and Sobrero, M. (1994) Further results on Hesselager's procedure for calculation of some compound distributions. *Astin Bulletin 24*, 161–166. [201]

Watanabe, S. (1964) On discontinuous additive functionals and Lévy measures of a Markov process. *Japan. J. Math. 34*, 53–70. [55]

Waymire, E. and Gupta, V. K. (1983) An analysis of the Pólya point process. *Adv. Appl. Prob. 15*, 39–53. [72]

Weiner, H. J. (1965) An integral equation in age dependent branching processes. *Ann. Math. Statist. 36*, 1569–1573. [235]

Weiner, H. J. (1977) Correction to "An integral equation in age dependent branching processes". *Ann. Prob. 5*, 159. [235]

Willmot, G. (1986) Mixed compound Poisson distributions. *Astin Bulletin 16, S*, S59–S79. [34, 49, 206]

Willmot, G. (1987) The Poisson-Inverse Gaussian Distribution as an Alternative to the Negative Binomial. *Scand. Actuarial J.*, 113 - 127. [41]

Willmot, G. (1989a) The total claims distribution under inflationary conditions. *Scand. Actuarial J.*, 1–12. [197, 213]

Willmot, G. (1989b) Limiting tail behaviour of some discrete compound distributions. *Insurance: Mathematics and Economics 8*, 175–185. [159, 160]

Willmot, G. and Sundt, B. (1989a) On Evaluation of the Delaport distribution and Related Distributions. *Scand. Actuarial J.*, 101–113. [34, 197]

Willmot, G. and Sundt, B. (1989b) On Posterior Probabilities and Moments in Mixed Poisson Processes. *Scand. Actuarial J.*, 139–146. [62]

Willmot, G. (1990) Asymptotic tail behaviour of Poisson mixtures with applications. *Adv. Appl. Prob. 22*, 147–159. [178–181, 206]

Willmot, G. (1993) On recursive evaluation of mixed Poisson probabilities and related quantities. *Scand. Actuarial J.*, 114–133. [34, 35, 37, 47, 49, 192, 195, 201]

Willmot, G. (1994) Refinements and distributional generalizations of Lundberg's inequality. *Insurance: Mathematics and Economics 15*, 49–63. [147, 149]

Willmot, G. and Lin, X. (1994) Lundberg Bounds on the Tails of Compound Distributions. *J. Appl. Prob. 31*, 743–756. [133, 149, 150, 171–174, 177, 206, 225, 226]

Willmot, G. and Lin, X. (1997) Simplified bounds on the tails of compound distributions. *J. Appl. Prob. 34*, 127–133. [149]

Yannaros, N. (1988) On Cox processes and gamma renewal processes. *J. Appl. Prob. 25*, 423–427. [95]

List of notation

Latin notation and abbreviations

AP(U) Ammeter process
$\mathcal{B}$ Borel algebra; σ-algebra
Bin binomial distribution
D_p p-thinning operator
DFR decreasing failure rate
DFRA decreasing failure rate average
DID discretely infinitely divisible
DMRL decreasing mean residual life
F distribution of Z_k
$\bar{F}$ $\bar{F}(z) = 1 - F(z)$; survivor function
F^{n*} nth convolution of F with itself
$\bar{F}^{n*}$ $\bar{F}^{n*}(z) = 1 - F^{n*}(z)$
f density of Z_k; $f(z) = F'(z)$
f_j $= P\{Z_k = j\}$
$\hat{f}$ Laplace transform of F or Z_k; $\hat{f}(v) = E[e^{-vZ_k}]$
$\mathcal{F}_t$ σ-algebra; $\mathcal{F}_s \subseteq \mathcal{F}_t,\ s \le t$
$\boldsymbol{F}$ filtration; $\boldsymbol{F} = (\mathcal{F}_t;\ t \ge 0)$
$\mathcal{F}_t^Y$ $= \sigma\{Y(s);\ s \le t\}$
$\boldsymbol{F}^Y$ natural filtration, $\boldsymbol{F}^Y = (\mathcal{F}_t^Y;\ t \ge 0)$
g_k $= P\{Y = y\}$ where Y is a random sum
G_N generating function of N; $G_N(s) = E[s^N]$
GIGD generalized inverse Gaussian distribution
$\hat{h}$ moment generating function of Z_k; $\hat{h}(r) = E[e^{rZ_k}]$
h $h(r) = E[e^{rZ_k} - 1] = \hat{h}(r) - 1$
$\bar{h}$ $\bar{h}(r) = \int_0^\infty e^{rz}\bar{F}(z)\,dz = h(r)/r$
$\mathcal{I}_U$ the set of finite unions of intervals in $[0, \infty)$
ID infinitely divisible
IFR increasing failure rate

IFRA	increasing failure rate average
IGD	inverse Gaussian distribution
IMRL	increasing mean residual life
K_p	compression operator; $K_pN(t) = N(t/p)$
$\mathcal{L}$	$R \in \mathcal{L}$ if $R(x+\ell) \sim R(\ell)$ as $\ell \to \infty$
MP	mixed Poisson distribution
MP(t, U)	N has random mean $t\Lambda$
MPP	mixed Poisson process
MPP(U)	$N(\cdot)$ has random intensity Λ
N	discrete random variable or point process
N^0	Palm process in the stationary case, 118
$N^!$	reduced Palm process in the stationary case, 118
N_s^0	Palm process in the general case, 125
$N_s^!$	reduced Palm process in the general case, 125
$\mathcal{N}$	the set of counting functions
$\mathcal{N}_S$	$= \{\nu \in \mathcal{N};\ \nu(t) - \nu(t-) = 0 \text{ or } 1\}$
NBU	new better than used
NBUE	new better than used in expectation
NWU	new worse than used
NWUE	new worse than used in expectation
$o(x)$	$f(x) = o(x)$ if $f(x)/x \to 0$ as $x \to 0$ or ∞
$\mathcal{P}$	the set of distributions of point processes
$p_n(t)$	$= P\{N = n\}$ if N is MP(t, U)
$p_n(t)$	$= P\{N(t) = n\}$ if N is a point process
$p_{m,n}(s,t)$	$= P\{N(t) = n \mid N(s) = m\}$
Po	Poisson distribution
$\mathbb{R}$	the real numbers; $\mathbb{R} = (-\infty, \infty)$
$\mathbb{R}_+$	the non-negative real numbers; $\mathbb{R}_+ = [0, \infty)$
$r(\cdot)$	failure rate, 133
RV$_\rho$	regularly varying functions, 152
$\mathcal{S}$	subexponential distributions, 166
$\mathcal{S}(\beta)$	170
$\mathcal{S}^*$	246
T_s	shift operator; $(T_s\nu)(t) = \nu(s+t) - \nu(s), \quad s \geq 0$
T_n	the time of the nth jump in a point process
$\mathcal{T}(\beta)$	169
SV	slowly varying functions, 152
U	structure distribution of Λ
$\overline{U}$	$\overline{U}(\ell) = 1 - U(\ell)$; survivor function
u	density of Λ; $u(\ell) = U'(\ell)$

u_k	$= P\{\Lambda = \ell_k\}$
$\hat{u}$	Laplace transform of U or Λ; $\hat{u}(v) = E[e^{-v\Lambda}]$
$\bar{u}$	$\bar{u}(v) = \int_0^\infty e^{-vz}\overline{F}(z)\,dz = (1 - \hat{u}(v))/v$
$W(w)$	the Willmot class of distributions, 192
Y	often random sum or compound point process
Z_n	often random variable associated to the nth jump

Greek notation and abbreviations

α	non-random intensity or mean; $\alpha = E[\Lambda]$
$\Gamma(\cdot)$	Γ-function, 17
$\Gamma(\gamma, \beta)$	Γ-distribution, 17
$\kappa_n(t)$	transition intensity
κ	exponent measuring the 'speed of decrease'
Λ	random intensity or mean with distribution U
Λ_B^*	best estimate of Λ, 20
Λ_L^*	best linear estimate of Λ, 22
$\hat{\Lambda}$	compensator, 55
μ	$= E[Z_k]$
μ_Λ	$= E[\Lambda]$
$\nu(\cdot)$	counting function, 50
Π	distribution of a point process
Π_α	distribution of a Poisson process
Π_U	distribution of a mixed Poisson process
Π^0	Palm distribution in the stationary case, 117,
Π_s	Palm distribution in the general case, 125
$\Pi^!$	reduced Palm distribution
σ^2	$= \mathrm{Var}[Z_k]$
σ_Λ^2	$= \mathrm{Var}[\Lambda]$
Φ	standard normal distribution, except in Chapter 9
Φ	non-ruin probability; $\Phi(u) = 1 - \Psi(u)$, in Chapter 9
Ψ	ruin probability

General notation

$\stackrel{\mathrm{d}}{=}$	equality in distribution
$\stackrel{\mathrm{def}}{=}$	equality by definition
$\stackrel{\mathrm{d}}{\to}$	convergence in distribution
$\sim$	asymptotic equivalence

$\circ$	composition of functions; $f \circ g(x) = f(g(x))$
$*$	convolution
$\lfloor x \rfloor$	integer part; the largest integer smaller or equal to x
$\lceil x \rceil$	the smallest integer larger or equal to x
$(m+1)^{\overline{n}}$	$= (m+n)!/m!$

Subject index

Recent Results in Cancer Research 126

R. Ackermann V. Diehl (Eds.)

Malignancies of the Genitourinary Tract

With 63 Figures and 93 Tables

Springer-Verlag
Berlin Heidelberg New York
London Paris Tokyo
Hong Kong Barcelona
Budapest

Professor Dr. Rolf Ackermann
Urologische Klinik, Heinrich-Heine-Universität
Moorenstrasse 5, W-4000 Düsseldorf, FRG

Professor Dr. Volker Diehl
Medizinische Klinik I, Universität Köln
Joseph-Stelzmann-Strasse 9, W-5000 Köln 41, FRG

ISBN 3-540-54433-X Springer-Verlag Berlin Heidelberg New York
ISBN 0-387-54433-X Springer-Verlag New York Berlin Heidelberg

Library of Congress Cataloging-in-Publication Data
Malignancies of the genitourinary tract/A. Ackermann, V. Diehl (eds.).
p. cm. Includes bibliographical references and index.
ISBN 3-540-54433-X (alk. paper). – ISBN 0-387-54433-X (alk. paper)
1. Genitourinary organs-Cancer. I. Ackermann, R. (Rolf), 1941–
II. Diehl, V. (Volker), 1938– .
[DNLM: 1. Urogenital Neoplasms-diagnosis. 2. Urogenital
Neoplasms-therapy. W1 RE106P v.126 / WJ 160 M251]
RC261.R35 vol. 126 [RC280.G4] 616.99'4606-dc20 DNLM/DLC
92-49944

Printed in Germany

Typesetting: Best-set Typesetter Ltd., Hong Kong
25/3130-5 4 3 2 1 0 – Printed on acid-free paper

Preface

Malignancies originating from the urogenital tract account for about 20% of all solid neoplasias. The various tumor types, including carcinoma of the prostate, transitional cell carcinoma, renal cell carcinoma, and testicular germ cell tumors, differ widely in terms of biological features, incidence and appearance, diagnostic work-up and management. It is not surprising that this complexity of urologic oncology with its related problems has increasingly attracted both experimental investigators and clinical experts. Only time will tell whether the achievements that today are considered as progress will turn out to be significant contributions towards the ultimate goal: the cure of the patients' life-threatening diseases with little or no side effects. However, many innovations and improvements have already taken place in the field of urologic oncology within recent years. The high pace of innovative development requires continuous, thorough, and critical updating of our kowledge so that persisting deficits can be defined: the basic requirement for posing new scientific questions. The following examples may serve to illustrate the recent developments and the resulting new problems and questions:

Recently, it has become evident that carcinoma of the prostate must be detected earlier to enable us to cure more of the affected patients in the future than currently. Since the patients do not present symptoms in the early stage of this disease, screening programs for populations at risk may be helpful in reducing the high rate at which advanced disease is still found, i.e., in about 80% of the patients. Several strategies for screening that take into consideration new labaratory tests such as analysis of the prostate-specific antigen have been advocated. However, a consensus has not yet been reached about the overall strategy. This is of particular relevance since radical prostatectomy has become a safe curative procedure with decreasing morbidity and sequelae due to our better understanding of the complex anatomy of the pelvis. Although this operation provides a cure in cases where the tumor is confined to the gland, its effectiveness is still being debated in locally advanced disease. Reevaluation of the results obtained within the last two decades and a critical analysis of the results of new approaches, e.g., adjuvant androgen deprivation, are needed urgently to better define the indications for radical prostatectomy.

Radical nephrectomy is considered the treatment of choice for nonmetastic renal cell carcinoma, regardless of the local extent of the disease. With the widespread use of sonography, increasingly small solid renal lesions are being discovered. This development has raised doubts whether removal of the whole kidney with the suprarenal gland is mandatory in these particular cases. Organ-sparing excision of small renal cell carcinomas has been used in selected cases. Based on this experience, it is now important to further clarify whether organ-preserving operative procedures are as safe as radical nephrectomy for small tumors.

Although metastatic renal cell carcinoma remains an unsolved therapeutic problem, new approaches using various cytokines, either alone or in combination, are currently under investigation as the result of the rapid progress in tumor immunology. In spite of some promising results, the reported observations are rather confusing and require further clarification.

Similar to the current situation in advanced renal cell carcinoma, the relevance of cisplatin-based adjuvant and neoadjuvant chemotherapy in locally advanced and/or metastatic transitional cell carcinoma remains to be determined. This is of particular importance for defining the limitations of radical cystectomy as a curative therapy. Early performance of radical cystectomy in multifocal, superficial, high-grade lesions is increasingly being advocated as new forms of continent urinary diversions are emerging, although its advantages remain to be proven. This aggressive approach must be evaluated critically and compared with other treatment strategies, e.g., radiochemotherapy or TUR with and without laser coagulation.

Furthermore, the efficacy of various forms of intravesical instillation therapy, e.g., BCG, mitomycin C, or doxorubicin, may influence the indications for radical cystectomy. More sensitive diagnostic tools, such as DNA-ploidy determination for a better definition of the malignant potential of a given tumor or immunocytology for earlier detection of local recurrence, may improve the current management of patients with transitional cell carcinoma.

In contrast to the tumor types mentioned above, efforts in germ cell tumor management are focused predominantly on the development of new treatment strategies that provide the same effectiveness as the traditional strategies but with less morbidity and fewer sequelae. In addition, new therapeutic approaches for patients with advanced disease and for those suffering from treatment failures are currently under investigation. It remains to be determmined which of the concepts should be followed. Close collaboration among experts in the various fields is clearly needed to come to conclusions and recommendations.

For this reason it was the intention of the "Gesellschaft zur Bekämpfung von Krebskrankheiten (GBK)" of the state of North Rhine-Westphalia to bring together some of the most eminent experts in the field of urologic oncology, people whose names are associated with the rapid developments in this area in recent years, in order to evaluate some of the problems we have just outlined in a comprehensive manner. This critical analysis of selected subjects should help readers to determine the relevance of some innovations which are still being vigorously debated. The editors would like to thank the authors for sharing their knowledge with a wider audience by contributing to this book.

Düsseldorf/Köln, autumn 1992

R. Ackermann
V. Diehl

Contents

*List of Contributors**

---

* The address of the principal author is given on the first page of each contribution.

[1] Page on which contribution begins.

Screening for Carcinoma of the Prostate: Epidemiological and Methodological Aspects

P. Iversen and S.T. Torp-Pedersen

Department of Urology, Herlev Hospital, 2730 Herlev, Denmark

Introduction

Cancer of the prostate constitutes a significant health problem. In Denmark prostatic cancer is the third leading cause of death from cancer in men, following lung and colorectal cancer [14,28]. The incidence and mortality have increased significantly over the last four decades [7]. In part, this trend may be explained by better diagnostic capabilities and improved registration, but data from other Nordic countries and international reports [41] support a true increase in incidence of the disease.

The prevalence of preclinical cancer of the prostate found at autopsy and transurethral resection (TURP) does not vary significantly around the world, although the incidence and mortality of clinically manifest cancer vary considerably [6,43]. An explanation for this has been sought in genetic and environmental factors, but, although some correlations have been demonstrated, no certain causal associations between prostatic cancer and suspected risk factors have so far been confirmed [43].

The percentage of cancers confined within the prostatic capsule at the time of diagnosis is increasing [7,27,39]. However, a substantial number of patients with a clinically localized tumor prove to have more advanced disease when surgical staging is performed [40].

The treatment of disseminated disease is palliative androgen deprivation [32], and prognosis in these patients is poor with a median survival of 2–2.5 years. Chemotherapy has so far been discouraging [35].

The best treatment of localized cancer is still an unsettled issue. Radiotherapy and radical prostatectomy are the two main modalities for curative therapy [5,40]. Nevertheless, because of the long natural history of the disease, many advocate the "wait and see" policy.

Recent Results in Cancer Research, Vol. 126

Rationale for Screening for Prostatic Cancer

We are faced with a malignant disease with substantial morbidity and mortality. We are not as yet able to identify and eliminate causal factors. Cure of advanced disease is not possible; cure of localized disease may be possible, but when diagnosed the majority of patients have cancer beyond the prostatic capsule.

Thus it seems logical to conclude that we should detect more cancers when they are still localized and potentially curable, i.e., advance the diagnosis and treatment in time. To do so, it is necessary to examine large populations of asymptomatic males.

Screening may be defined as the use of simple tests or examinations to differentiate those who probably have disease from those who probably do not [13]. While the idea of screening is simple, the implementation and validation of the concept is complex [2,11,17,23]. The difference between normal medical practice and exposing healthy individuals to screening procedures is obvious, and emphasizes the need for conclusive evidence that screening will result in a decrease in morbidity and mortality.

To some extent, screening is already in effect. The American Cancer Society advocates that all men over a certain age have a yearly digital rectal exam. Furthermore, the ongoing sophistication of tools for diagnosis of prostate cancer generates interest in earlier diagnosis – perhaps simply because it is feasible. Or, perhaps for monetary reasons. In the United States "prostate clinics" are emerging – clinics advertising early diagnosis with digital rectal exams, prostate ultrasound and serum prostate specific antigen (PSA). This uncontrolled screening activity has generated relevant critique since a beneficial effect so far is unproven.

Whether a screening program will influence morbidity and mortality from prostatic cancer depends upon:

1. Natural history of prostatic cancer
   a) The duration of the preclinical period
   b) The proportion of cancers detected by screening that would progress, if left untreated
2. The reliability of the screening test(s) in detecting preclinical cancer
3. The extent to which available treatment alters the natural history of prostatic cancer, i.e., the efficacy of therapy
4. Compliance in the population, i.e., the percentage of the population in question attending the program

The outcome of treatment has briefly been discussed, and although the need for an effective therapy is indisputable when initiating screening programs, this issue will not be discussed in more detail.

Epidemiology and Natural History

In 1985, 1293 new cases of prostatic cancer were diagnosed in Denmark and 820 males died from the disease [14,28]. Thus, approximately 63% (820/1293) with a diagnosis of prostatic cancer will die from the disease, assuming a stable age composition of the male population and only minor changes in incidence and mortality from year to year.

The still existing and often referred to myth that prostatic cancer is a malignancy that patients are likely to die with and not from is contradicted when considering Danish incidence and mortality rates in 1985 (Fig. 1) [14,28]. When comparing the incidence rate with the mortality rate in the age group 5 years older, a striking resemblance is observed. This indicates that once a prostate cancer is *diagnosed* in a patient, he is likely to die from it, if he escapes intercurrent mortality. This concurs with the results of a Swedish nationwide survey [3], investigating 44 300 cases of prostatic cancer. Median survival from diagnosis (all ages) was 3.1 years. Adjustment for expected mortality from other causes revealed continuous decrease in relative survival. A patient escaping death from other causes had only 33.7% chance of surviving his prostatic cancer 10 years following diagnosis, and only a 17% chance at 20 years. The study demonstrated that prostatic cancer kills, but it also confirmed the long natural history of the disease.

Johansson et al. [19] followed 223 patients with newly diagnosed prostatic cancer without distant metastases. Five-year corrected survival was 93.8%. However, only nine patients with poorly differentiated tumors were included in the study; of these six progressed and five died within the observation period.

Carcinoma of the prostate is unique among human cancers. Even though many patients endure substantial suffering and finally succumb to the disease, many more live with an *undiagnosed* cancer, completely free of morbidity and eventually die from other causes. The evidence for this is found in numerous autopsy studies [15,34], and it is reflected in the so-called incidental carcinoma found in men undergoing TURP for clinically benign disease [31].

In essence, the therapeutic dilemma in prostate cancer rests upon the discrepancy between the number of patients actually dying from cancer, and the many more with histologic but clinically silent cancer. Comparing an autopsy prevalence of 30% in men over 50 [15] with the Danish age-specific mortality, it can be calculated that each Danish man dying from prostatic cancer only represents 1/259 of those over the age of 50 who histologically have the disease. This ratio is often quoted, although it is basically irrelevant. More to the point, how many of the 259 will *eventually* die from prostatic cancer? Consider the cumulated mortality that can be estimated from Fig. 1. A 6% risk of dying from prostatic cancer from age 50 to 85 can be deduced, assuming no other cause for mortality. Unfortunately, we do not know if these 6% will originate from the 30% with "latent" cancer at age 50.

McNeal et al. advocate that latent carcinoma is not a specific subgroup of cancer, but merely a very early carcinoma with a slow growth rate. As volume increases, dedifferentiation and capacity to invade and metastasize develop [25]. If this is correct, then a long preclinical period must be assumed to exist, and the risk of a preclinical cancer "surfacing" clinically is only a matter of time. On the other hand it is clear that only a fraction of these early cancers have time enough to become clinical cancers. If we were able to detect all cancers in a group of 50-year-old males, we would not be able to select those that eventually would become a threat to the individual.

Thus, it is evident that we do not know enough about the natural history and pathogenesis of prostatic cancer.

Screening Tests

The ideal screening test is simple, safe, and may be performed by paramedical personnel. It is quick and well accepted by the subjects. The financial costs are low. Sensitivity and specificity are high [12].

In a clinical setting a high specificity is required, while a lower sensitivity may be acceptable. The reason for this is that sensitivity affects the *few* with disease, while specificity affects the *many* without disease. A low specificity would mean a high rate of false positives thus adding to the costs and exposing healthy individuals to the psychological impact and potential risks of subsequent diagnostic tests (biopsy etc.).

A sensitivity less than 100% theoretically results in preclinical cancers not being detected. However, the true sensitivity can not be calculated without knowing the prevalence of the disease in the population. If the prevalence at autopsy is used, a sensitivity of 100% would lead to detection of cancers that would never have required treatment. An estimate of the prevalence with relevance to screening should ideally include all preclinical cancers that would surface as clinical cancers, i.e., the cumulated incidence rate. Still, a high sensitivity calculated from this prevalence would not ensure that the cancers detected at screening would be identical to the ones emerging clinically years later.

Clinical Experience with Screening for Prostatic Cancer

From 1948 to 1969, 5856 healthy men underwent a total of 28407 annual rectal examinations at the Cancer Detection Centre at the University of Minnesota [16]. This screening revealed 75 subsequently confirmed carcinomas of the prostate, or one carcinoma for every 78 patients screened, but only one for every 379 rectal examinations. Twenty-two patients underwent radical prostatectomy. The author claimed 5- and 10-year survival rates similar to survival in the age-matched American male population.

In 1978, 1.3 million German males underwent rectal examination in a screening program. Suspicious findings on palpation were present in 1%, but only 0.1% proved to be malignant. The costs were reported to be 28 DM per screened patient [18]. The program was ambitious, but attendance rate was low and follow-up has been poor. The study does not provide any conclusions on the effect of screening.

Thompson et al. [36] did not find rectal examination useful as a screening tool, finding only 65 nodules, of which 17 proved to be cancers in 2005 men. Of nine who underwent pathological staging, six were stage C or D1. In a later report on more than 5000 patients screened by annual rectal examinations, a significantly larger portion of patients had clinically localized disease compared to the distribution of stages before the screening program [37]. However, almost 60% of those pathologically staged had tumor outside the prostate. Chodak et al. [9] found almost the same percentage of understaging in 36 cancers detected by screening 2131 males with rectal examination. In an earlier presentation of the same study [10], the calculated cost per patient was reported to be 39 dollars per examination, or 4135 dollars per detected cancer.

Varenhorst et al. [38] invited 1494 randomly selected males between the ages of 50 and 69 years to participate in a screening program. Of these, 1163 (78%) were screened by rectal palpation by both general physician and a urologist. Thirteen cancers were found (1.1%); only one patient had clinically advanced cancer, and ten underwent radical prostatectomy. One of these had lymph node metastases, and seven had positive margins and were treated with adjuvant radiotherapy.

In a study of 4843 men screened by serial rectal examinations, Mueller et al. [26] found an increasing percentage of clinically localized cancers at follow-up as compared to the first visit.

Logically, rectal examination can only detect palpable cancers, a fact underlined by the significant percentage of incidental carcinoma found at TURP for a clinically benign hyperplasia [31]. Hopes that markers like prostatic acid phosphatase or PSA would be of value in early diagnosis have not been fulfilled [33,42].

Employing transrectal ultrasound scanning (TRUS) in a screening scenario, the detection rate (Table 1) has been found to be twice that of rectal examination [20,29]. In a group of 784 self-referred men age 60 or more, Lee et al. [20] found cancer in 20 (2.6%) compared to only 1.3% detected with rectal palpation in the same group. A total of 64 underwent biopsy because of abnormalities detected by TRUS. It is significant that 17 of the tumors were less than 1.5 cm in diameter, and only 1/14 of those surgically staged were not confined to the prostate. Only 7/17 of the small tumors were detected by rectal palpation.

A detection rate of 2.6% is not impressive when compared to a prevalence of more than 30% as suggested by autopsy studies. However, if we consider the risk of dying from prostatic cancer within the next 20 years for a 60-year-old male, this is also close to 2.6% in the United States [4]. Whether TRUS is able to recognize the same cancers that would surface is not known, however; in the series by Lee et al. [20] it is, nevertheless noteworthy that although small confined tumors were detected, only one tumor less than 0.8 cm in diameter was found. From their autopsy studies, McNeal et al. [25] suggest that tumor volumes less than $0.5\,cm^2$ (0.8 cm diameter) are of no clinical significance.

Therefore, TRUS may prove useful as a screening tool. Its major drawbacks are time consumption and need of trained personnel. The cost per screened individual has been estimated at 166 dollars [20], or four times the cost of rectal examination.

Evaluation of Screening – Bias

No effect on mortality and morbidity from prostatic cancer as a result of screening has been demonstrated. Detection of more localized tumors and prolongation of survival from diagnosis do not in themselves prove screening to be beneficial. When evaluating screening programs, several factors must be taken into consideration [2,23].

Selection Bias. This refers to factors that determine population participation in screening. If these factors are in any possible way related to prevalence of the disease, then this will affect comparisons with nonscreened groups. For instance, the study by Lee et al. [20] was based on self-referred men. It may be argued that these men had some unrecorded reason to seek medical attention.

Diagnostic Bias. As discussed earlier, the risk of detecting prostatic cancers that would never surface clinically is not negligible. When such cancers are included in analysis, survival and stage distribution can not be compared to nonscreened groups with only clinical cancers. This applies to all the studies referred to.

Length Time Bias. This relates to the detection of cancers in the preclinical phase. The longer this phase, the greater the chance of the cancer being detected in screening before it surfaces clinically. In screening, this will favor the detection of slowly growing tumors, while more rapidly growing tumors may develop and emerge clinically between two screening programs. This bias applies to all the studies referred to.

Lead Time Bias. An important aim for every screening is to advance diagnosis in time; if this aim is achieved, the interval from diagnosis to death will logically be prolonged. So even though "survival" – calculated from diagnosis to death – is prolonged, death from prostatic cancer may not be postponed. Lead time is an important concept in screening and designates the interval from detection by screening to the time where clinical diagnosis would have been made. The lead time bias is exemplified in the Minnesota study [16] claiming improved survival.

Considering the situation where a screened population is compared to a control population, two more biases may be defined.

Treatment Bias. In order to prove a screening effect, the only difference between the two groups should be the screening itself. Any discrepancy in patient management in the two groups would skew the results. Therefore, treatment policy must be identical in the two groups, and a change of therapy in the groups must occur simultaneously. Even if this is ensured, it is still impossible to totally avoid treatment bias. Based upon the basic concept of screening, i.e., to advance diagnosis in time, the cancers detected in the screened population may be exposed to a treatment which will be considered insufficient and old-fashioned when the "corresponding" cancers in the control group are diagnosed X number of years later (lead time).

Surveillance Bias. When evaluating morbidity and cause of death, the two groups may not be comparable. Most likely, the screened group will be under a more scrutinous observation. Furthermore, the screened population has been alerted to the risk of prostatic cancer. Consequently, it is likely that morbidity and mortality from prostatic cancer will be more easily recognized in the screened population.

Risks and Adverse Effects of Screening

The psychological impact of screening in general has never been meticulously investigated. In a screening program for prostatic cancers *few* will gain directly from the program, while *many* healthy individuals will be reminded of the possibility of a malignant disease. Some will have to go through supplementary tests because of a false positive screening test. In others, detected cancers will eventually be classified as advanced and incurable, thus making these individuals live like patients earlier than they otherwise would. Whether or not early hormonal therapy improves prognosis in these patients, we do not know.

Even if it is assumed that a screening test would only select patients harboring prostatic cancers with the potential to progress and kill within a period of time, we would still treat a significant number of patients that would die from other causes. As an example it may be calculated from available survival statistics [21,28], that 26% of a population of Danish 65-year-old males will die from causes other than prostatic cancer in a period of 8 years. To some extent we would be able to predict which patients would die – from concurrent malignancy, heart-lung diseases etc., but still the dilemma persists; we would inevitably treat patients that would die before their prostatic cancer would ever have become a clinical reality.

As stated earlier, an effective treatment for early prostatic cancer is a prerequisite to screening for the disease. Because it will have to be implemented on healthy men, some of whom may never suffer from the disease, the treatment must be associated with low mortality and morbidity.

It has been suggested [8,24] that a screening scenario could lead to a situation where treatment killed more patients than it saved from disease. This polemic statement needs attention. Assume that a hypothetical screening test was able to detect all latent cancers, approximately 40% in a population of 100 000 65-year-old males. If they all underwent a radical prostatectomy associated with a realistic mortality of 1% [1,40], then 400 would die as a result of the treatment. When considering a mortality rate of 93/100 000 in the 65–69 age group (Fig. 1), the comparison is scaring. However, if we accept that the 40% with latent cancer include all those later to progress and die from the disease, then the comparison should be made with the cumulated mortality for a population of 100 000 untreated 65-year-old males. Corrected for other mortality [21], 1230 of the 100 000 would die from prostatic cancer within a 10-year period. Thus, it would take a detection rate of 40% and a treatment mortality of more than 3% before the number killed by treatment equaled those saved from disease, presuming all those treated were cured of their cancer. This, of course, is an extreme situation. Hopefully, we would never detect and treat 40%. Still, such a hypothesis exemplifies the risk of inflicting an iatrogenic burden on the population.

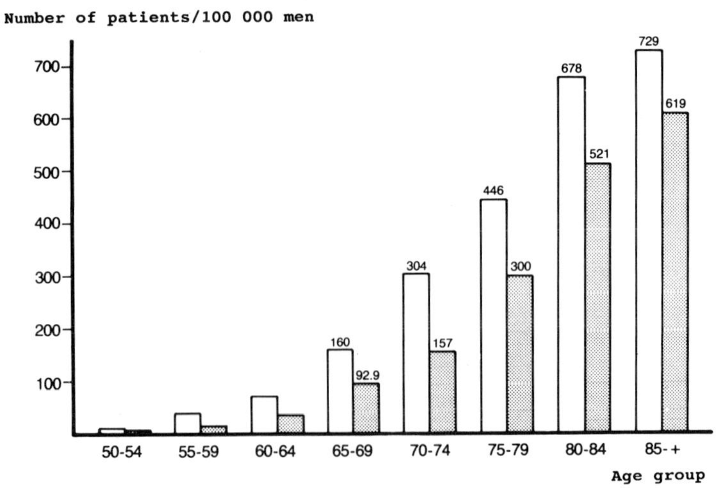

Fig. 1. Age-specific incidence (*open bars*) and mortality rates (*stippled bars*) for prostatic cancer in Denmark 1985

Advancing diagnosis and treatment in time makes it essential to compare gain by early therapy with loss by complications of that therapy. Because of the long natural history and the late onset of the disease, the gain in life span will be modest. In contrast, death at a younger age from complications to early treatment following screening may represent a loss of a substantial number of years. Thus although mortality from treatment may seem low, it may outweigh the gains of a screening program.

In this discussion, only mortality has been addressed, but also morbidity related to treatment, although more elusive to quantitate, also constitutes a problem. The balance between reduced quality of life and gained survival in the treated individual is always intricate – in a screening program for prostatic cancer it is even more troublesome.

Possible Designs for Future Screening Trials

Future studies must be designed to answer the following questions:

1. How much will screening advance the time of diagnosis?
2. Is an earlier diagnosis associated with an earlier stage of disease?
3. With present therapy, is the earlier diagnosis associated with improved survival?
4. Considering autopsy prevalence will screening overdiagnose – i.e., diagnose insignificant cancers?
5. Will screening underdiagnose – i.e., have an unacceptably low sensitivity?
6. At what age should screening begin?
7. How often should screening take place?

If we can answer these questions we will know the natural history of prostate cancer (prevalence, preclinical phase), the power of our screening tool(s) (sensitivity, specificity), and the power of our treatment (cure rate broken down by stage).

Any discussion about screening for prostate cancer is often summed up by remarks like "we need controlled prospective clinical trials with a screening group and a control group, which demonstrate reduced mortality in the screening group before screening for prostate cancer can be considered to be worthwhile."

We therefore consider it essential to focus our attention on such a trial. Before doing so, however, it is necessary to elaborate on basic epidemiological parameters, namely incidence and prevalence. In the following calculations and examples, Danish statistics from 1985 are used ([14,21,28]; Fig. 1).

Incidence and Prevalence Related to Screening

Incidence

In Fig. 2 we address the age-specific incidence, age-specific incidence rate, and their relation to population size. The *age-specific incidence rate* is the number of cancers diagnosed per year in 100 000 men of that age. When the age-specific incidence rate for a 65-year-old is 126, it means that a 65-year-old has a 0.126% risk of having prostatic cancer diagnosed while he is 65 years old. A year later, age 66, his risk has increased to 0.143% and continues to increase to 0.725% at age 87. The *age-specific incidence* (not rate) is the actual number of cancers for that age diagnosed in that year. If we consider 100 000 males aged 65, then, according to life table statistics [23], 69 714 will still be alive at age 74. At this age, the incidence rate is 389, but only 255 new cases from the 69 714 will be diagnosed. Thus, the increased risk of prostatic cancer with advancing age does not lead to a parallel rise in actual numbers because of the simultaneous decline in population at risk (Fig. 2).

How many cancers will be diagnosed if we follow 100 000 65-year-olds for 15 years? This can be calculated as (number of 65-year-olds × age-specific incidence rate for 65-year-olds) + (number of 66-year-olds × age-specific incidence rate for 66-year-olds) + . . . + (number of 79-year-olds × age-specific incidence rate for 79-year-olds) = 3226 (this number is corrected for already diagnosed cancers, i.e., a male with a diagnosed cancer will not be at risk for cancer in the years to follow). Assuming a preclinical phase of 15 years, all these cancers are present at age 65. In other words: Within the next 15 years, prostatic cancer will be diagnosed in 3.2% of the 65-year-olds.

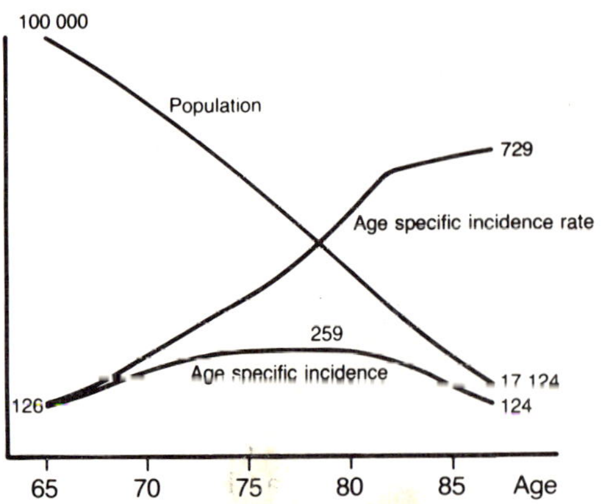

Fig. 2. The relationship between size of population, age-specific incidence, and age-specific incidence rate

Similarly, in 1.9% the cancer will surface within 10 years. This we will name the *cumulative incidence*.

If we assume, as indicated in the literature, an *autopsy prevalence* of 40% in 65-year-olds, then it can be calculated from the above that only 1 out of 12.5 cancers present in 65-year-olds would surface in 15 years, and only 1 out of 21 would surface in 10 years.

However, not all of these cancers which are biologically destined to surface within 10 years will do so, because not all 65-year-olds live 10 years. When we screen 100 000 65-year-olds we do not know which 69 714 will still be alive 10 years later. A 65-year-old may harbor a T4 cancer that will never surface because he will die of cardiovascular disease. Therefore, unavoidably we will overdiagnose cancer.

We can account for this overdiagnosis by cumulating the incidence assuming no mortality. Again considering 100 000 65-year-olds, 126 cancers will be diagnosed the 1st year leaving 99 874 at risk for cancer the 2nd year, when 143 cancers (99 874 × 143/100 000) will be diagnosed leaving 99 731 at risk the 3rd year and so forth. We end up with a *screening-related prevalence*: In 100 000 65-year-olds there are 4542 cancers biologically destined to surface within the next 15 years. However, only 3222 will actually be diagnosed in the absence of screening. In other words there is a screening-related prevalence of 4.5% and a cumulative incidence of 3.2% for cancer surfacing within the next 15 years. Looking ten years ahead the corresponding percentages are 2.3% and 1.9%, respectively.

Thus, the difference between screening-related prevalence and *cumulative incidence* represents the unavoidable overdiagnosis – unavoidable because it relates to other mortality and not to the screening test itself.

Hypothetical Setting for Screening

To illustrate the problems associated with the interpretion of screening studies and hopefully to resolve some of the misunderstandings concerning prevalence and incidence, we will create a screening scenario. It includes a hypothetical biological model for prostatic cancer, a theorctical screening test, and a theoretical treatment.

Biological Model for Prostatic Cancer

All prostatic cancers are alike, slow growing with a preclinical phase of 15 years. Thus, in this model we are not subject to length time bias. In 65-year-olds there is a high autopsy prevalence (40%) of cancer – the majority being very small and biologically destined to surface so late that the host will have died from other causes. The largest of these cancers (= the oldest) are biologically destined to surface soon. The smaller the cancer, the later it will be diagnosed. With a preclinical phase of 15 years, all the cancers that will be diagnosed in a group of 65-year-olds as we follow them for 15 years are actually present at year zero, i.e., when the individual patient was 65.

In the model, individuals with locally advanced and/or metastatic disease will eventually die from prostatic cancer and are considered "lost to disease". Thus, survival for this patient category is not included in the following calculations.

The Screening Test

We will use a hypothetical screening test with a sensitivity dependent upon the size of the tumor. If the cancer is above a certain size, all are detected by our test. If below a certain limit, none are detected. In our biological model, size of the cancer predicted when it would surface. Thus, we may describe the sensitivity as a function of lead time, i.e., sensitivity will be high for cancers that will surface within the first years following screening, and low for cancers destined to surface late.

Thus, we postulate a screening test with 100% sensitivity for cancers destined to surface in the year of screening, 90% for cancers which will surface 1 year later, 80% for 2 years later . . . , ending with a sensitivity of 10% for cancers destined to surface 9 years following screening.

Treatment

Our hypothetical therapy can cure cancer confined to the prostate. Individuals with more advanced cancer are lost to disease.

Furthermore, we assume that 20% of cancers are confined to the prostate at the time of clinical diagnosis (without screening). When cancers are detected earlier by screening, a higher percentage of localized cancer is presumed. Again, we relate this to lead time. We postulate *X*: 1 year lead time: 30% localized; 2 years lead time: 40% localized . . . 8 years lead time: 100% localized. Or, we could be even more optimistic *Y*: 1 year lead time: 40% localized; 2 years lead time: 60% localized . . . 4 years lead time: 100% localized. When screening is instituted cancers with different lead times will be detected and calculation of overall cure rate will have to take this into account.

Hypothetical Screening Studies

Using the above definitions and hypothetical setting we will now conduct three screening trials.

Study A

A population of 100000 65-years-olds undergo the screening test and all curable cancers are treated. Another 100000 comparable 65-year-olds constitute the control group and all curable cancers diagnosed clinically receive the same treatment as those in the screening group. All men in both groups are followed, and the yearly incidences are recorded (Figs. 3 and 4).

A total of 1066 cancers are detected with the screening test: 100% of the 126 that would have been diagnosed anyway that year (lead time: 0 years); 90% of the 143 that would have had a cancer diagnosed 1 year later, had they all stayed alive (lead time: 1 year); 80% of the 160 2 years later and so on.

One year following screening, 14 cancers that were not originally detected surface clinically in the screening group (because the test had a 90% sensitivity for cancers with a lead time of 1 year). Two years later 30 undetected cancers surface and so forth.

Ten years later the effect of screening is lost. This occurs when the yearly incidences in the two groups are identical (Fig. 3). In the cumulative graph (Fig. 4) this event is signalled when the two lines run parallel. The distance between the two lines reveals that the screening resulted in unnecessary detection of 121 cancers. Unnecessary, not because insignificant cancers were detected, but because of other mortality.

When the effect of screening is lost, further follow-up of the two groups is unnecessary, and only the 2065 cancers in the screening group (1066 detected at screening, 999 surfacing later) and the 1944 cancers in the control group are followed to register the difference in cure rate.

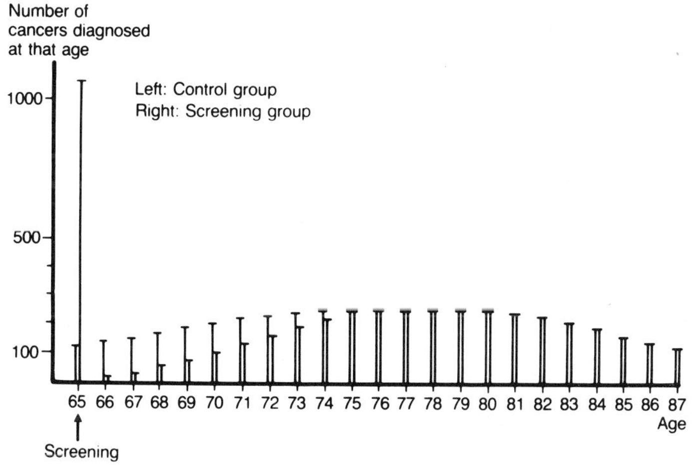

Fig. 3. Numbers of prostatic cancers detected/diagnosed annually in hypothetical screening study A at age 65

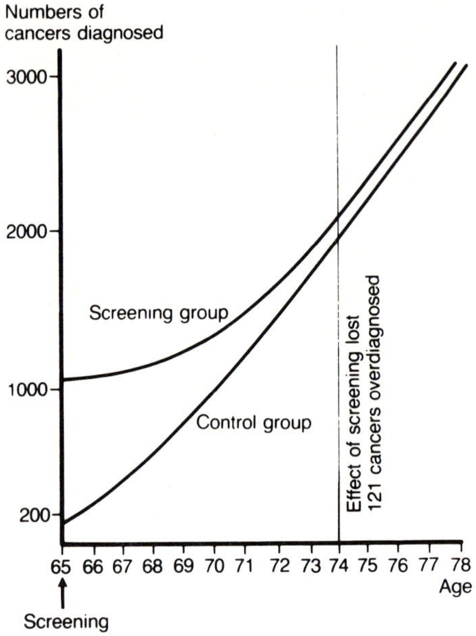

Fig. 4. Cumulated numbers of prostatic cancers detected/diagnosed in hypothetical screening study A at age 65

Tables 1–4 show the numbers of cured and dead (lost to disease) according to our hypothetical presumptions. Due to screening, mortality from prostatic cancer is reduced from 1.6% to 1.2% or 1.0% according to the two treatment postulates *X* and *Y* (see above).

This theoretical study included 200 000 males. How many individuals do we actually need, assuming that our presumptions are realistic, to reach statistically valid conclusions? Accepting a type 1 error of 5% and a type 2 error of 10%, 23 722 males in each group are necessary to demonstrate the difference in mortality as a result of postulate *X*. For postulate *Y*, 11 215 individuals in each group are needed.

Study B

Again, 100 000 males undergo screening at age 65, which is repeated at age 70. Control group and therapy are similar to study A.

The incidences for the screened group and the control group are graphically displayed in Figs. 5 and 6. Naturally, males with diagnosed prostatic cancer are not included in the second screening.

The second screening is instituted when the sensitivity of the first test has fallen to 50% (according to assumption). The effect of the first test is lost

Table 1. Cancers detected in one screening: study A

Lead time	No. of patients with lead time	Overdiagnosed due to screening (*n*)	Gained compared to controls (*n*)	Mortality[a] (%)		Cancer deaths[a]	
				X[b]	Y[b]	X[b]	Y[b]
0	126	0	(126)	80	80	101	101
1	129	4	125	70	60	88	75
2	128	7	121	60	40	73	48
3	132	12	120	50	20	60	24
4	130	15	115	40	0	46	0
5	122	18	104	30	0	31	0
6	109	20	89	20	0	18	0
7	90	19	71	10	0	7	0
8	65	16	49	0	0	0	0
9	35	10	25	0	0	0	0
	1066	121	945			424	248

[a] According to our biological model of prostatic cancer and our hypothetical treatment, patients with locally advanced and/or metastatic cancer are "lost" to disease and will eventually die. Without screening, 20% of cancers will be localized.
[b] X and Y refer to assumptions as to how lead time relates to percentage of localized cancer (see text).

Table 2. Cancers overlooked (not detected) in screening: study A

Number	Deaths[a] from prostatic cancer
999	799

[a] See footnote a in Table 1.

Table 3. Control group (100 000 65-year-olds followed through age 74): study A

Lead time	No. of patients	Mortality[a] (%)	Deaths[a] from prostatic cancer
0	1944	80	1555

[a] See footnote a in Table 1.

Table 4. Results of one screening: study A

	%
Screening group	
1066 cancers detected at screening	1.1
121 cancers overdiagnosed at screening	0.1
999 cancers overlooked at screening	1.0
According to X[b]: 1223 deaths[a] from prostatic cancer	1.2
Accroding to Y[b]: 1047 deaths[a] from prostatic cancer	1.0
Control group	
1944 cancers diagnosed	1.9
1555 deaths[a] from prostatic cancer	1.6

[a,b] See footnotes a and b in Table 1.

when the males are 75 years old and the effect of the second test when they are 80 years old.

Once again, as in study A, we overdiagnose – in this case 321 cancers. The mortality from prostatic cancer in the screening group is 1.9% or 1.5%, depending upon which mortality postulate is chosen, as compared to 2.6% in the control group (Tables 5–8).

Repeating the statistical requirements (errors type 1 and 2: 5% and 10%), we need 12 566 or 5480 individuals in each group according to postulates *X* and *Y*.

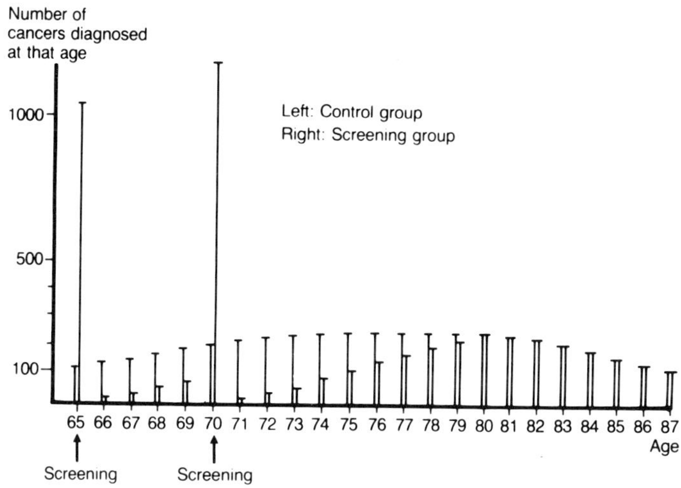

Fig. 5. Numbers of prostatic cancers detected/diagnosed annually in hypothetical screening study B at ages 65 and 70

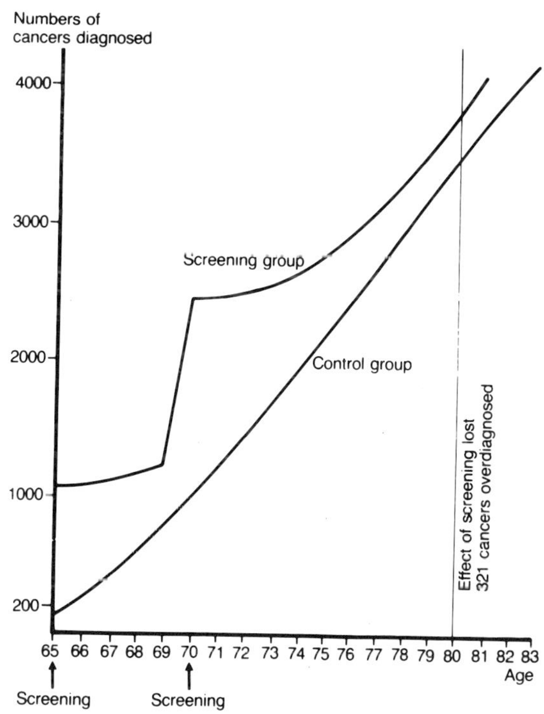

Fig. 6. Cumulated numbers of prostatic cancers detected/diagnosed in hypothetical screening study B at ages 65 and 70

Table 5. Cancers detected in two screenings: study B

Lead time	No. of patients with lead time[c]	Overdiagnosed due to screening[c] (*n*)	Gained compared to controls[c] (*n*)	Mortality[a] (%)		Cancer deaths[a]	
				X[b]	Y[b]	X[b]	Y[b]
0	230	0	(230)	80	80	184	184
1	252	6	246	70	60	172	148
2	269	18	251	60	40	151	100
3	286	29	257	50	20	129	51
4	291	40	251	40	0	100	0
5	282	51	231	30	0	69	0
6	246	55	191	20	0	38	0
7	199	52	147	10	0	15	0
8	145	44	101	0	0	0	0
9	79	28	51	0	0	0	0
	2279	323	1956			858	483

[a] According to our biological model of prostatic cancer and our hypothetical treatment, patients with locally advanced and/or metastatic cancer are "lost" to disease and will eventually die. Without screening, 20% of cancers will be localized.
[b] X and Y refer to assumptions as to how lead time relates to percentage of localized cancer (see text).
[c] Summing up cancers detected in each of the two screenings ranked by lead time.

Table 6. Cancers overlooked (not detected) in two screening: study B

Number	Deaths[a] from prostatic cancer
1268	1014

[a] See footnote a in Table 5.

Table 7. Control group (100 000 65-year-olds followed through age 79): study B

Lead time	No. of patients	Mortality[a] (%)	Deaths[a] from prostatic cancer
0	3226	80	2581

[a] See footnote a in Table 5.

Table 8. Results of two screening: study B

Screening group	%
2279 cancers diagnosed at screening	2.3
323 cancers overdiagnosed at screening	0.3
1268 cancers overlooked at screening	1.3
According to X[b]: 1878 deaths[a] from prostatic cancer	1.9
Accroding to Y[b]: 1497 deaths[a] from prostatic cancer	1.5
Control group	
3226 cancers diagnosed	3.2
2581 deaths[1] from prostatic cancer	2.6

[a,b] See footnotes a and b in Table 5.

Reflections on Theoretical Studies A and B

These theoretical screening scenarios demonstrate that the more often we screen, the greater the impact on mortality. This is hardly surprising, since we thereby increase the number of detected cancers with long lead time. The two hypothetical studies are indeed simplifications of real life, but do serve the purpose of disclosing the problem of overdiagnosing due to screening. It is important to observe that overdiagnosis in this theoretical setup is only caused by other mortality, since our screening test only detected cancers contained in the screening-related prevalence. In real life, our test may detect cancers contained in the autopsy prevalence, but "outside" the screening-related prevalence, i.e., cancers that would never surface clinically. In that situation, overdiagnosis may be much more significant.

The overwhelming, but required size of such studies must impress anyone with experience in clinical trials. Also, they exemplify the magnitude of decrease in mortality to be expected.

We will, by conducting trials like studies A and B, be able to answer most of the questions asked. However, we are left with major problems:

1. The natural history of prostatic cancer will only partly be revealed. By following yearly incidences in the screening group, we will be able to confirm whether or not we are able to detect cancers at an earlier point, and to what extent. By comparison with the control group, we will also have an average measure for the number of cancers detected at screening that would have surfaced later if left alone. But we will never know *which* of the cancers detected at screening would have surfaced and which would have remained silent.
2. Furthermore, within the design of studies A and B we are in fact determining the effect of screening *plus treatment*. In a situation where the best management of localized prostatic cancer is still a hot issue of debate, this is unfortunate. In the extreme situation where treatment does not improve survival, we will not be able to detect any difference in mortality between the two groups, even though the screening test served its purpose.
3. At first glance, the design of a controlled trial seems simple. For instance, we may randomly construct a screening group and a matching control group from a national registry (design I). However, only a certain percentage of the males offered screening will accept the offer. Thus, the population actually undergoing screening will be a selected group. How do we create a similarly selected control group? To avoid this, a population of males could be offered participation in a randomized study (design II). Once they had agreed to take part, the males would be allocated to either a screening group or a control group. However, in this situation the control group will be alerted to the risk of prostatic cancer and their threshold to seek medical attention may be lowered. Thus, the control group may not behave like the background population. Consequently, designs I and II are both subject to *selection bias*. In design I the bias affects the screening group, in design II the control group. In design I this could be dealt with by analysing on an "intention to treat/screen" basis, including in the analysis all males who were originally *offered* but did not necessarily undergo screening.

Surveillance bias (defined earlier) will represent a major hurdle in studies A and B. Defining cause of death in large populations will always be difficult. An individual with a previously diagnosed prostatic cancer will upon death easily be classified as dying from prostatic cancer. Hence, the unavoidable overdiagnosis due to screening (see earlier), may result in a falsely high prostatic cancer death rate in the screening group. In its extreme consequence, this phenomenon could camouflage a possible beneficial effect of screening.

Treatment bias (defined earlier) can only be totally eliminated if treatment for prostatic cancer (all stages) remains unaltered throughout the study period, typically 15–20 years. This is of course impossible.

Awed by the magnitude of all these shortcomings and practical problems associated with studies like A and B, we will propose a third model.

Study C

A total of 10 000 males are screened at age 65. They are not informed of the result of the test, even when a cancer is detected. Each detected cancer is T staged. No control group is included.

Each year a number of cancers in the group will be diagnosed clinically with an incidence similar to the control group in studies A and B.

For each cancer detected in the screening, we will eventually know (a) if it ever surfaced, (b) how much later it surfaced (lead time), and (c) the stage at clinical diagnosis compared to stage at detection in screening.

Furthermore, we can estimate the sensitivity as it relates to lead time (the sensitivity we postulated as 100%, 90%, 80% . . . in studies A and B).

This study is more an epidemiological than a screening study because no treatment is included, but it has the advantage of identifying, for each cancer, the effect of screening on lead time and stage. It requires no control group and it does not depend upon an effective therapeutic modality. Offering no treatment, on the other hand, it may be difficult to motivate sufficient numbers of men to participate, and a selection bias is possible. Further, individuals taking part in the study will be alerted to the existence of prostatic disease, lowering the threshold to seek medical attention.

Ethical Aspects of Studies A, B, and C

From a superficial point of view, it seems highly unethical to conduct study C. How can you diagnose a cancer, and not tell the patient? How is it possible to refrain from treating a potentially curable cancer?

We allow the cancers to behave as they would had there not been a screening. Once diagnosed clinically, however, they may be treated in the usual manner. It is still a matter of debate how to treat patients with localized cancer of the prostate. Furthermore, when we diagnose a prostatic cancer in a screening we do not know if it would ever have surfaced without screening. This can only be answered by a study like C.

Studies A and B may seem more ethically acceptable at first glance. Studies like these have been proposed by several authors [2,9–11]. But such studies are also associated with ethical considerations. If you detect cancer in an asymptomatic male, how do you inform him of his cancer when you do not know if he will ever suffer from, let alone die from prostatic cancer?

And along the same lines, would you urge that asymptomatic person to undergo treatment (with its mortality and morbidity associations)?

Conclusion

As stated earlier, the concept of screening is a simple one. We hope that this discussion has proved, however, that putting the concept into clinical practice is an entirely different matter. Although it is not dealt with in this discussion, the huge costs of a screening program are beyond dispute; in essence, however, the decision to pay that price is a political question. Our task is to clarify whether or not a screening program will benefit society. At present, the truth is we do not know. Therefore, we cannot recommend that the authorities initiate large scale programs. However, the screening concept cannot be disregarded. Further research is warranted.

In our opinion, studies like A and B entail such immense methodological problems that they will be impossible to interpret.

Therefore, albeit highly controversial, we find that a study like C must have high priority. It has its toll of ethical hurdles but comes closest to describing in detail the preclinical phase of prostatic cancer and will, with fewer patients (and less resources) more accurately describe the capability of present screening modalities. It also allows concurrent continued research to identify prostatic tumor predictors of biological potential and enables further clarification of the efficacy of therapeutic modalities. Study C could be carried out in a country where localized prostatic cancer is not routinely treated. If, ultimately, studies like A and B are necessary to verify an impact of screening on community mortality, study C will provide knowledge about the natural history of prostatic cancer which will help in the design of such studies.

Finally, the scene of medical capability is constantly shifting. If, in the near future, cure of advanced prostatic cancer in early stages with low-tumor burden becomes feasible with modern immuno- and/or chemotherapeutic modalities, then screening for prostatic cancer might become an even more intriguing subject.

References

1. Ackermann R, Frohmüller HGW (1983) Complications and morbidity following radical prostatectomy. World J Urol 1:62–67
2. Adami H-O (1988) Strategies for control of prostatic cancer mortality with some methodologic comments. Scand J Urol Nephrol Suppl 110:15–22
3. Adami H-O, Norlén BJ, Malker B, Meirik O (1986) Long-term survival in prostatic carcinoma, with special reference to age as a prognostic factor. A nation-wide study. Scand J Urol Nephrol 20:107–112

4. Annual Cancer Statistics Review including Cancer Trends: 1950–1985 (1988). National Cancer Institute, Bethesda
5. Bagshaw MA, Ray GR, Cox RS (1987) Selecting initial therapy for prostate cancer. Radiation therapy perspective. Cancer 60:521–525
6. Breslow N, Chan CW, Dhom G, Drury RAB, Franks LM, Gellei B, Lee YS, Lundberg S, Sparke B, Sternby NH, Tulinius H (1977) Latent carcinoma of prostate at autopsy in seven areas. Int J Cancer 20:680–688
7. Carstensen B, Storm HH (1987) Overlevelseskurver fra Cancerregisteret 1948–82. Danish Cancer Registry
8. Chodak GW (1988) Early detection of prostate cancer. Presented at the International Symposium on Endocrine Therapy, Monaco
9. Chodak GW, Keller P, Schoenberg H (1989) Assessment of screening for prostate cancer using the digital rectal examination. J Urol 141:1136–1138
10. Chodak GW, Keller P, Schoenberg H (1988) Routine screening for prostate cancer using the digital rectal examination. In: Schroeder FH et al. (eds) EORTC Genitourinary Group, monograph 5: progress and controversies in oncological urology II. Liss, New York, pp 87–95
11. Cochrane AL, Holland WW (1971) Validation of screening procedures. Br Med Bull 27:3–8
12. Cole P, Morrison AS (1978) Basic issues in cancer screening. In: Miller EB (ed) Screening in cancer. International Union Against Cancer, Geneva
13. Commission on Chronic Illness in the United States. Harvard University Press, Cambridge
14. Danish Cancer Society (1985) Danish Cancer Registry: cancer incidence in Denmark 1985
15. Franks LM (1954) Latent carcinoma of the prostate. J Pathol Bacteriol 68: 603–616
16. Gilbertsen VA (1971) Cancer of the prostate gland. Results of early diagnosis and therapy undertaken for cure of the disease. JAMA 215:81–84
17. Henderson M (1976) Validity of screening. Cancer 37:573–581
18. Jacobi GH (1987) Early detection of prostatic carcinoma: the German experience. Presented at: Progress and Controversies in Oncological Urology II, Amsterdam
19. Johansson J-E, Adami H-O, Andersson S-O, Bergström R, Krusemo UB, Kraaz W (1989) Natural history of localised prostatic cancer: A population-based study in 223 untreated patients. Lancet i:799–803
20. Lee F, Littrup PJ, Torp-Pedersen ST, Mettlin C, McHugh TA, Gray JM, Kumasaka GH, McLeary RD (1988) Prostate cancer: comparison of transrectal US and digital rectal examination for screening. Radiology 168:389–394
21. Life table for the years 1985–1986. (1988) Statistical Year Book
22. Love RR (1985) Principles of cancer screening. In: Stoll BA (ed) New horizons in oncology: screening and monitoring of cancer, vol 4. Wiley, New York
23. Love RR, Camilli AE (1981) The value of screening. Cancer 48:489–494
24. McClennan BL (1988) Transrectal US of the prostate: is the technology leading the science? Radiology 168:571–575
25. McNeal JE, Bostwick DG, Kindrachuk RA, Redwine EA, Freiha FS, Stamey TA (1986) Patterns of progression in prostate cancer. Lancet i:60–63
26. Mueller EJ, Crain TW, Thompson IM, Rodriguez FR (1988) An evaluation of serial digital rectal examination in screening for prostate cancer. J Urol 140:1445–1447
27. Murphy GP, Natarajan N, Pontes JE, Schmitz RL, Smart CR, Schmidt JD, Mettlin C (1982) The national survey of prostate cancer in the United States by the American College of Surgeons. J Urol 127:928–934

28. National Board of Health (1985) Causes of death in Denmark
29. Ragde H, Aldape HC, Blasko JC, Bagley CM (1988) Prostate cancer screening with high resolution transrectal ultrasound. Presented at the 3rd International Symposium on Transrectal Ultrasound of the Prostate, Kiel
30. Resnick M (1988) Background for screening – epidemiology and cost effectiveness. In: Schroeder FH et al. (eds) EORTC Genitourinary Group monograph 5: progress and controversies in oncological urology II. Liss, New York, pp 111–120
31. Sheldon CA, Williams RD, Fraley EE (1980) Incidental carcinoma of the prostate: a review of the literature and critical reappraisal of classification. J Urol 124:626–631
32. Sogani PC, Fair WR (1987) Treatment of advanced prostatic cancer. Urol Clin North Am 14:353–371
33. Stamey TA, Yang N, Hay AR, McNeal JE, Freiha FS, Redwine E (1987) Prostate-specific antigen as a serum marker for adenocarcinoma of the prostate. N Engl J Med 317:909–916
34. Starklint H (1950) Undersøgelser over latent og manifest prostatacancer. Munksgård, Copenhagen
35. Tannock IF (1985) Is there evidence that chemotherapy is of benefit to patients with carcinoma of the prostate? J Clin Oncol 3:1013–1021
36. Thompson IM, Ernst JJ, Gangai MP, Spence CR (1984) Adenocarcinoma of the prostate: results of routine urological screening. J Urol 132:690–692
37. Thompson IM, Rounder JB, Teague JL, Peek M, Spence CR (1987) Impact of routine screening for adenocarcinoma of the prostate on stage distribution. J Urol 137:424–426
38. Varenhorst E, Enlund A-L, Herder A, Malmquist E, Berglund K, Löfman O, Carlsson P, Pedersen K (1989) Prostatacancerscreening med rektalpalpation kan organiseres til rimlig kostnad. Läkartidningen 86:3475–3477
39. Veterans Administration Cooperative Urological Research Group (1967) Treatment and survival of patients with cancer of the prostate. Surg Gynecol Obstet 124:1011–1017
40. Walsh PC, Lepor H (1987) The role of radical prostatectomy in the management of prostatic cancer. Cancer 60:526–537
41. Waterhouse J, Muir C, Shanmugaratnam K, Powell J (1982) Cancer incidence in five continents, vol 4. IARC, Lyon (IARC scientific publication no 42)
42. Watson RA, Tang DB (1980) The predictive value of prostatic acid phosphatase as a screening test for prostate cancer. N Engl J Med 303:497–499
43. Zaridze DG, Boyle P (1987) Cancer of the prostate: epidemiology and aetiology. Br J Urol 59:493–502

Screening for Carcinoma of the Prostate in a Randomly Selected Population

E. Varenhorst, K.V. Pedersen, P. Carlsson, K. Berglund, and O. Löfman

Department of Surgery and Urology, Lasarettet, 601 82 Norrköping, Sweden

Introduction

Cancer of the prostate is an increasing health problem in many countries. Scandinavian countries have a relatively high incidence with the highest incidence rate in Sweden where carcinoma of the prostate accounts for 23% of all forms of cancer among men [7]. It is thus the chief form of cancer in Sweden and the most common cause of death from neoplasm in men [8]. At present, only early detection by screening in combination with effective treatment provides a means of reducing the mortality of prostatic cancer. In order to show that screening reduces mortality from prostatic cancer, however, it is essential to carry out a controlled study comparing a screened ant treated risk population with an unscreened one. This requires random selection of a high risk population and a long follow-up period.

The only established non-invasive method of diagnosing prostate cancer today is digital rectal examination. This is the simplest method, and any doctor can perform it. The accusation has been levelled, however, that it is subjective and difficult to record accurately; in addition, no study has yet been carried out as to whether the examination should be performed by a urologist or a general practitioner (GP). Recent studies assessing digital rectal examination and transrectal ultrasonography as screening modalities were conducted in men who had sought advise on their own initiative, and who may, therefore, have constituted a biased, preselected population [1,4]. Our study was designed to evaluate an organisation model, and to study the cost, acceptance, and psychological consequences of a screening programme in a randomly selected population.

Recent Results in Cancer Research, Vol. 126

Material and Methods

A total of 1494 men were randomly selected from a risk population of 9026 men aged between 50 and 69 years from the City of Norrköping, Sweden. This represented the total male population for this age group in this area.

An invitation was sent to the men selected. The digital rectal examinations were performed at primary Health Care Centres with the men standing and supported on their forearms with the knees flexed and pronated.

A senior urologist and a GP performed the examination independently and all findings were recorded. A firm nodular consistency was the main criterion of malignancy. Prostates with consistency only slightly diverging from the normal, such as slight induration, were classified as benign, and were not examined further. If one or both of the examiners suspected malignancy, the man was informed immediately, and fine needle aspiration biopsy, with three aspirates from each lobe (total six), was performed 1 week later by an experienced cytologist. The man was told the cytological diagnosis after yet another week.

Immediately after the examination we asked each man about discomfort during the examination and the amount of time the examination had taken up. After fine needle aspiration biopsy patients completed questionnaires about mental stress during the week before the procedure, physical discomfort, and the information they had been given. They completed another questionnaire on mental stress after they obtained the results of the biopsy. Travelling time and cost, and loss of income were recorded. The costs were calculated from the internal accounts of the health district administration and an estimate of travelling expenses.

Results

Of the 1494 men contacted 1163 (77.8%) came for examination. The highest acceptance rate was found in the age group 60–64 years (Table 1). In 45 cases (3.9%), malignancy was suspected and fine needle aspiration biopsy was carried out in all but one who declined. Carcinoma was confirmed in 13 of these 44 cases (29.5%).

The incidence of cancer increased with increasing age, the highest incidence being 2.9% in the oldest group. The overall incidence of carcinoma was 1.1% (Table 1). Using the TNM classification [3] 1 patient belonged to clinical category T1, 11 to T2, and 1 to T3. As regards the grade of malignancy, seven patients were classed G1 and 6 G2. One patient had distant metastases (Table 2). Two patients showed cellular atypia only, and in neither did repeated aspiration biopsy 3 months later reveal evidence of carcinoma. The palpation of slight induration, judged to be benign, was noted in 16.5% of all the cases (Table 1).

Table 1. Results of screening for carcinoma of the prostate

Age (years)	Randomly selected men (n)	Acceptance rate (%)	Slight induration (%)	Nodular consistency		Aspiration biopsy		Confirmed cancer	
				(n)	(%)	(n)	(%)	(n)	(%)
50–54	357	73.5	10.6	4	1.5	4	1.5	0	0
55–59	360	76.3	16.7	6	2.2	6	2.2	2	0.7
60–64	393	81.4	14.4	15	4.7	14	4.4	2	0.6
65–69	384	79.6	23.5	20	6.5	20	6.5	9	2.9
Total	1494	77.8	16.5	45	3.9	44	3.8	13	1.1

In 24 cases, carcinoma of the prostate was suspected by both clinicians, in 10 cases by the GP alone, and in 10 cases the urologist alone. Of the 13 diagnosed carcinomas 8 were suspected by both clinicians, 1 carcinoma by the GP alone and 4 by the urologist alone. The overall positive predictive value was 0.30 (13/44), 0.27 for the GP and 0.35 for the urologist. The sensitivity and specificity of the method cannot be calculated because the true prevalence of the disease is unknown.

The 13 patients found to have carcinoma were offered further investigation and appropriate treatment (Table 2). Ten patients were treated with radical prostatectomy and one received radiotherapy. One patient with advanced disease was given endocrine treatment and another had concurrent disease and was not treated (Table 2).

The direct cost including salaries, material, and rent was estimated to be U.S.$23 900, or U.S.$21 per examination (Table 3). The cost of aspiration biopsy was U.S.$5852, or U.S.$132 per examination (44 cases).

The total direct cost for the screening programme, including fine needle aspiration biopsy and travelling expenses, was U.S.$36 911, or U.S.$2839 per detected cancer (U.S.$3355 per case treated by a potentially curative method).

The indirect cost is represented by the participating men's loss of working and leisure time. For the 500 men (43%) examined during working time the expenses were calculated from the average earnings including social benefits.

For those men examined outside working hours and for retired men, the indirect cost was calculated from figures obtained from the British Ministry of Transport [6]. This was calculated as U.S.$10 600, or U.S.$9 per man examined, including the indirect cost of the fine needle biopsies such as loss of working time and travel expenses. The total cost for the screening programme was U.S.$47 511, or U.S.$4319 for each cancer that could be treated by a potentially curative method. Nearly all the men gave a brief interview concerning distress immediately after the examination. There was

Table 2. Results of the pretreatment investigations (13 cases), operative lymph node staging (11 cases), histologic classification of the specimen (10 cases) and treatment (13 cases). Tumour classification according to WHO (International Union against Cancer [3])

Patient	Pretreatment classification				Postoperative classification			
Age (years)	T palp	N	M	Grade of malignancy	N	pT	Gleason score	Treatment
66	2	0	0	2	0	3	VII	RP/PRT
56	2	0	0	2	0	–	–	RT
68	2	0	0	2	0	2	V	RP
67	2	0	0	2	0	3	VIII	RP/PRT
66	2	0	0	1	0	3	V	RP/PRT
62	2	0	0	1	0	2	VI	RP
62	2	0	0	2	0	3	VII	RP/PRT
66	2	0	0	1	–	–	–	No treatment
68	1	0	0	1	0	2	VI	RP
68	3	0	1	2	–	–	–	Endocrine
59	2	0	0	1	1	3	VII	RP/PRT
67	2	0	0	1	0	3	VII	RP/PRT
67	2	0	0	1	0	3	VII	RP/PRT

T palp, primary tumour by digital rectal examination; pT, histologic classification of the specimen; RP, radical prostatectomy; RT, radiation therapy; PRT, postoperative radiation therapy.

Table 3. Total cost for the screening programme (U.S. $, 1987 prices) by duplicate digital rectal examination (DDRE)

Cost items	Total (U.S. $)	Per DDRE (U.S. $)
Direct cost		
DDRE	23 900	21
Fine needle aspiration	5 852	5
biopsy	7 159	6
Travel cost		
Indirect cost		
Loss of working and leisure time	10 600	9
Total screening cost	47 511	41

a clear difference in perception of the examination between the older (>65 years) and younger men (Table 4).

The men with suspected malignancy answered two further questionnaires, one concerning the week before fine needle biopsy and the second immediately after receiving the result of the fine needle biopsy. Of the 44 men who

Table 4. Judgment by the men selected regarding distress in connection with screening by DDRE

Age (years)	Major distress (%)	Minor distress (%)	No distress (%)
50–54	3.9	14.7	81.4
55–59	4.7	18.0	77.3
60–64	3.8	13.3	82.0
65–69	0.7	8.4	91.0
Total	3.2	13.4	83.4

were recalled for fine needle biopsy 25 were anxious during the week of waiting, and 18 were more nervous than normal. While waiting for the cytologist's report 30 felt anxiety and most were more nervous than normal.

Discussion

The early diagnosis of prostatic carcinoma is a prerequisite for cure. Unfortunately, symptoms appear late and self-examination is not possible. Screening programmes have been suggested, but have never before been studied in controlled trials. Contrary to other studies our population of 1494 men was randomly selected from a total population of 9026, so that the remaining 8432 who were not examined served as a control group. The follow-up time, however, is still too short to enable any conclusions to be drawn.

The present study shows that nearly 80% of all men invited agreed to participate. The digital rectal palpation was readily tolerated, with less than 10% of the men expressing negative attitudes to the screening. The cost of screening per treated patient was less than that with other screening procedures for malignancy – for example, mammography for breast cancer [2]. This investigation shows that a programme for early detection of prostatic carcinoma can be organised as an integral part of the routine health service. It is possible that with further training the GP's positive predictive value will increase to the level attained by the urologist.

The expected incidence of prostatic cancer, derived from the national cancer register, would be 0.15% a year, compared with the 1% we detected. This figure is seven times the clinical incidence is Sweden for the same age groups [7]. We treated 11 of 13 patients with potentially curative methods. In view of the final staging of the tumours it is most likely that we found and treated clinically significant carcinomas. Staging lymphadenectomy was done in all patients who received potentially curative treatment (ten had radical prostatectomy and one had radiotherapy). One had minimal lymph node disease. Capsular invasion was found in seven cases after radical prostatectomy. These men received postoperative radiotherapy, although its value and the clinical significance of such invasion are not known.

If larger populations are to be examined, major demands will be made on the relevant sector of the health system, and staffing and resources of the urology, oncology and cytology departments will be affected.

Screening for carcinoma of the prostate by a urologist or a GP using digital rectal examination is a cost-effective method of early diagnosis. There is, however, no evidence from randomised studies that early detection of prostatic carcinoma and the application of potentially curative treatment prolongs survival; the only prospective study with a placebo control group could reach no definite conclusions [5].

References

1. Cooner WH, Mosley BR, Rutherford CL et al. (1990) Prostate cancer detection in a clinical urological practice by ultrasonography, digital rectal examination and prostate specific antigen. J Urol 143:1146–1154
2. Håkansson S, Jonsson E (1984) Screening för cancer – underlaget dåligt för kalkyl om kostnader och nytta. Läkartidningen 81:4734–4742
3. International Union against Cancer (1987) Urological tumours. In: Hermanek P, Sorbin LH, (eds) TNM classification of malignant tumours, 4th, fullyrev. edn. Springer, Berlin Heidelberg New York, pp 121–126
4. Lee F, Littrup PJ, Torp-Pedersen ST et al. (1988) Prostate cancer: comparison of Transrectal US and digital rectal examination for screening. Radiology 168: 389–394
5. Madsen PO, Graversen PH, Gasser TC et al. (1988) Treatment of localised prostatic cancer. Radical prostatectomy versus placebo. A 15 year follow-up. Scand J Urol Nephrol Suppl 110:95–100
6. McGuire A, Henderson J, Mooney G (1988) Cost benefit approach in theory. In: The economics of health care. Routledge and Kegan Paul, London, pp 74–101
7. National Board of Health and Welfare (1990) The Cancer Registry. Cancer incidence in Sweden 1986. Liber, Stockholm, p 31
8. National Bureau of Statistics (1988) Causes of Death 1986. National Bureau of Statistics, Stockholm, p 57

Clinical Staging in Carcinoma of the Prostate: Current Aspects

B.J. Schmitz-Dräger, T. Ebert, and R. Ackermann

Urologische Abteilung, Heinrich-Heine-Universität, Moorenstrasse 5, W-4000 Düsseldorf, FRG

Introduction

Radical prostatectomy (RP) and megavolt radiotherapy are assumed to be effective in the treatment of localized carcinoma of the prostate (CaP). Several studies have demonstrated that where there is extracapsular and seminal vesicle infiltration, or lymph node metastases, unfavorable results are associated with local treatments [4,15,19,30,65]. Reports by some authors, however, in patients with locally advanced disease or lymph node metastases suggest a beneficial effect with RP even in more advanced tumor stages [46,62,68,69]. Tumor debulking by inductive antiandrogen treatment with subsequent RP represents another approach to curing patients with metastatic disease. The advent of new therapeutic strategies and the introduction of modern imaging modalities has led to the question of whether current staging procedures still satisfy clinical requirements. The aim of this manuscript is to analyze the accuracy of modern clinical staging modalities and to evaluate their possible impact on the treatment of patients with CaP.

Classification of T Category

Digital Rectal Examination

Despite the introduction of tumor marker determinations and modern imaging modalities, digital rectal examination (DRE) remains the first step in the diagnosis of CaP. Salo and coworkers reported on the correlation of clinical tumor stage, obtained by DRE, with the pathohistological stage after RP [57] (Table 1). They observed understaging by DRE in 47% of 32 patients. Recent results by Ebert and coworkers, who investigated 62 patients prior to RP, yielded a correct local tumor stage in only 21% of the patients, while understaging was observed in 77% [18]. In contrast to

Recent Results in Cancer Research, Vol. 126

Table 1. Digital rectal examination: staging error in clinically localized CaP

Study	No. of patients	Understaging[a] (%)	Extracapsular tumor (%)
Catalona and Stein [11]	82	20 (24)	24
Salo et al. [57]	32	15 (47)	53
Cordes et al. [14]	30	8 (27)	43
Palken et al. [49]	23	6 (26)	26
Ebert et al. [18]	62	32 (52)	60

[a] Extracapsular or seminal vesicle invasion not detected.

other trials using the Whitmore-Jewett staging system, this study was based on the UICC classification [31,36]. Better correlations between DRE and the pathohistological stage of the primary tumor have been observed by other investigators [11,14,49].

The use of different classification systems may be one important reason for these differences. The application of the Whitmore-Jewett system to the data of Ebert and coworkers resulted in a correct tumor stage in 46% of patients, and understaging in 52% (Table 1). Furthermore, Table 1 suggests that understaging was less frequent in those investigations with fewer locally advanced tumors.

Transrectal Ultrasound

Since real-time transrectal ultrasonography (TRUS) was introduced in the diagnosis of CaP by Watanabe and coworkers, numerous investigations have been performed in order to determine the value of this technique in the localization and evaluation of extent of the malignant lesions [66]. Initial reports on the diagnostic accuracy were as high as 80%. These reports have been followed by others which found this technique to be of only limited value [1,13,16,17,24,25,52,55,57].

It is clear that only by comparing the TRUS findings with the respective pathohistological examination can reliable information be provided. The key investigation was reported by Dähnert and coworkers. They performed an ex vivo study on 64 prostates immediately after RP [16]. Using a 5-MHz linear array transducer, they found that 38 glands (59%) had echopenic or hypoechoic lesions which were subsequently identified to be tumor lesions. Tumor areas were isoechoic in the remaining 26 specimens. In contrast, Salo and associates reported 30% hyperechoic tumor lesions in 20 RP specimens examined ex vivo by TRUS [58]. In this investigation 40% of the lesions were found to be hypoechoic. The conclusion from these studies is that a 70% sensitivity represents the upper limit of TRUS in the detection of CaP.

Information regarding capsular penetration or seminal vesicle infiltration, however, may be obtained in an even higher percentage of patients.

Initial results of TRUS examination before RP were reported by Resnick and coworkers. They investigated 23 patients with localized CaP [55]. The extent of the lesion and seminal vesical infiltration was recognized by TRUS in all patients. Pontes and coworkers evaluated 31 patients with clinically localized CaP by TRUS before RP [52]. Capsular and seminal involvement were predicted with 89% and 100% sensitivity, respectively. Other investigators reported less favorable results. Salo and coworkers were able to uncover tumor invasion beyond the capsule or seminal vesicle infiltration in 14 out of 21 patients (67%) [57]. Andriole and associates recognized extracapsular tumor extension with TRUS in four out of seven cases (57%) [1]. Cordes and coworkers reported a sensitivity of 54% in the prediction of extracapsular and seminal vesicle infiltration in 13 patients [14]. Similar results were obtained by Ebert and coworkers [18]. They detected extracapsular tumor growth in 16 out of 30 patients (53%) with TRUS. Summarizing these findings, capsular penetration was predicted in most trials with approximately a 60% sensitivity. It should be mentioned, however, that false positive findings of capsular penetration were also observed in some investigations [18,52,57].

While most authors detect tumor invasion beyond the capsule with TRUS with a 60%–85% sensitivity, there are remarkably conflicting results concerning sensitivity in the detection seminal vesicle involvement. While Resnick and coworkers report a 100% sensitivity, other investigations yield a sensitivity of less than 30% [18,52,55,57].

Magnetic Resonance Imaging

Numerous studies have been performed that analyze the clinical impact of magnetic resonance imaging (MRI) on the diagnosis and staging of CaP [3,7–9,24,26,33–35,37,38,41,47,51,53]. Nevertheless, controversy still surrounds the question as to whether tumor foci can be visualized by MRI. According to some authors, tumor tissue yielded a hyperintensive signal in T2-weighted images [7,35,38]. Other authors found a heterogenous pattern [8]. Tumor nodules of low signal intensity were detected by a third group of investigators [3,9,37]. Mukamel and associates even state that they "were unable with MRI to define intraprostatic lesions (cancerous or benign), to separate the capsule from the prostatic parenchyma, to show capsular and seminal vesicle invasion" [47]. This opinion was also shared by Ling and associates and Poon and associates [41,53].

With this background it seems amazing that only a few investigations are based on a histomorphological comparision with surgical specimens. Hricak and associates reported correct local staging of CaP with MRI in 37 out of 45 patients (83%) when transverse T1- and T2-weighted images supplemented by additional coronal or sagittal images were studied [34]. They concluded that local staging by MRI was significantly better than clinical staging or staging by computed tomography (CT) scan. Unfortunately no data on the prediction of capsular penetration or seminal vesicle infiltration were given in this report. Mukamel and associates investigated 20 patients with CaP prior to RP [47]. In only 2 out of 10 patients with CaP stage C was extracapsular infiltration indicated by MRI. Since classification by MRI was correct in another 10 patients with the tumor confined to the prostate, a 60% accuracy can be calculated. In a prospective study recently conducted at our institution, capsular penetration was predicted by MRI in 13 out of 24 cases (54%) [18]. With false positive results in 2 out of 13 patients (15%), the diagnostic accuracy observed (65%) was similar to that found by Mukamel and coworkers [47].

Computed Tomography

CT scan is frequently used for the evaluation of pelvic lymph nodes as well as for local tumor staging. The inability of this technique to visualize intraprostatic structures does not permit the localization of the malignant lesion within the prostate. Nevertheless, CT scan may disclose extracapsular tumor spread.

The efficacy of CT scan in local tumor staging was addressed by Mukamel and associates. They investigated 20 patients with CaP before PR [47]. In only 1 out of 10 cases (10%) was extracapsular and/or seminal vesicle invasion predicted by CT scan. Cordes and coworkers detected extraprostatic tumor by CT scan in 5 out of 13 cases (38%) [14]. Golimbu and coworkers reported positive CT scans in 2 out of 11 patients (18%), with tumor extension in either the seminal vesicles or the periprostatic tissue [28]. In our own series, CT scan was positive in 3 out of 25 patients (12%) with extracapsular tumor growth [18].

Tumor Markers

The determination of prostatic acid phosphatase (PAP) and prostate-specific antigen (PSA) has been found to be valuable, especially in the follow-up of patients with CaP. Recent investigations have revived the discussion about a possible contribution of marker determination in the staging of CaP [20,49,64].

Stamey and Kabalin investigated 102 men with clinically localized CaP before RP [65]. A close correlation between PSA and PAP serum levels and tumor stage was observed. All patients with CaP stage B had PSA serum levels exceeding 2.5 ng/ml, while in 162 male controls PSA levels under 2.5 ng/ml were found. It should be mentioned that this control group was not age matched, and none of these normal subjects suffered from benign prostatic hyperplasia (BPH). It must be anticipated that patients with BPH will exhibit higher PSA serum values. Therefore, PSA levels do not reliably discriminate between patients with BPH and early CaP [12]. The introduction of higher cut-off values than those proposed by Stamey and coworkers may be required for routine clinical use. Therefore, the value of marker determination in the diagnosis of early CaP as well as in the discrimination between localized and advanced CaP appears to be limited.

Classification of N Category

Bipedal Lymphangiography

Detailed studies by several investigators on the value of bipedal lymphangiography in the staging of pelvic lymph nodes demonstrated a low sensitivity for the modality. Bipedal lymphangiography yielded positive results in approximately 50% of patients with metastatic lymph nodes [32,39,42,63]. The key reason for these results is clearly that presacral and hypogastric lymph nodes are not consistently visualized by this examination. Löning and coworkers reported positive lymphangiograms in 17 out of 28 patients (61%) with no histological finding of lymph node metastases [42]. They concluded that lymphangiography alone can not reliably assess the regional lymph node status in clinically localized CaP.

Computed Tomography

Today CT scanning is widely accepted as the primary examination for the clinical staging of pelvic lymph nodes. In contrast to lymphangiography, which analyses the internal architecture of lymph nodes, only enlarged lymph nodes can be visualized by CT scan. The rate of false positive results due to benign lymph node enlargement may be even higher than the approximately 10% reported by several investigators, as the presence of enlarged lymph nodes, mistaken for metastases, may lead to the exclusion of patients from lymphadenectomy and RP [2,60,67]. The sensitivity of CT scanning in the detection of lymph node metastases has been reported to range from 12.5% or 14% [2,60] to 66% in the series of Weinerman and coworkers [61]. Patient selection may have contributed to these discrepant results since patients with significantly enlarged pelvic lymph nodes by CT scan may have been excluded from subsequent lymph node dissection.

Pelvic Lymphadenectomy

Pelvic lymphadenectomy is the most accurate method of staging pelvic lymph nodes. Despite a false negative rate for frozen sections of between 8% and 19% in patients with macroscopically normal appearing pelvic lymph nodes, pelvic lymphadenectomy as a separate staging procedure is only rarely performed [10,21,45,56,59]. The morbidity associated with staging lymphadenectomy has lead to the proposal of a limited pelvic lymphadenectomy restricted to the hypogastric lymph nodes and the lymph nodes medial to the external iliac vessels. The efficacy of this modification is controversial. Some reports suggest that a limited lymph node dissection may fail to identify a considerable proportion of patients that have lymph node metastases [22,27,44]. In contrast, investigations by Bruce and coworkers indicate that a limited lymphadenectomy would have detected 91% of patients with lymph node metastases [6].

Tumor Markers

The question of whether the determination of PSA and PAP enables the clinician to discriminate between patients with localized CaP and patients with lymph node metastases can be analysed using the data recently accumulated by Stamey and Kabalin [64]. Significant differences between stage B and stage D1 disease were found for PSA serum levels but not for PAP serum levels. Looking at the range of PSA values reported by Stamey and Kabalin, CaP stage D1 was never correlated with PSA values below 10 ng/ml. These findings were confirmed by an investigation by Palken and associates [49]. In both studies, however, PSA levels surpassing 10 ng/ml were observed in a considerable number of patients with CaP stage A and B, thereby limiting the value of this marker in tumor staging.

Classification of M Category

Bone Scan

Currently, the bone scan with technetium-99 m methylene diphosphonate is the most sensitive technique for detecting osseous metastases. Investigations by Schaffer and Pendergrass [61] and Reikeras and coworkers [54] have shown that patients with metastatic CaP may have abnormal bone scans with normal bone radiographs [54,61]. Although the sensitivity of this investigation is hard to assess, a false negative rate of less than 2% was reported. Since bone scanning isotopes are preferably incorporated in all areas of increased bone turnover, a low specificity for this investigation must be anticipated. Therefore, in questionable cases additional radiographs must be obtained in order to have definitive information concerning an area of increased isotope uptake.

Serum Parameters

Alkaline phosphatase (AP) elevations can occur in patients with osseous metastases. AP is not a specific tumor marker since the heat stabile isoenzyme is produced in the liver and is increased in patients with hepatobiliary disorders. Perturbations of bone metabolism may also result in an elevated AP serum level. Because of its inferior sensitivity in the detection of bone metastases as compared to bone scans, determination of AP is only of limited value in the staging of CaP despite its predictive value in the follow-up of patients with metastatic CaP [61].

Data concerning PSA serum values in patients with metastasized CaP can be obtained from the report by Stamey and Kabalin [64]. The analysis of PSA values of 35 patients with metastatic CaP showed PSA levels of more than 25 ng/ml in all cases. This PSA range, however, broadly overlaps with the ranges found for patients with stage B, C, and D1 disease.

Conclusions

There is a consensus that RP and megavolt radiotherapy are treatment options that should be reserved for patients without distant metastases. Therefore, the primary goal of tumor staging in CaP is to exclude the presence of distant metastases. For fundamental reasons, the sensitivity of most examinations aimed toward the demonstration of distant metastases must remain undetermined. Nevertheless, it is assumed that, in the asymptomatic patient, chest X-ray, bone scan, and imaging of the liver by ultrasound or CT scan, provides sufficient information about the presence or absence of distant metastases. The determination of serum parameters, i.e., PSA, AP, PAP, and hepatic enzymes, is certainly valuable for the follow-up of patients with CaP, but only will improve the accuracy of tumor staging under certain circumstances [50,54,61].

The aim of staging the primary tumor is to identify those patients who may benefit from local treatment modalities. It seems amazing that up until now this population has not been characterized. Several investigators have demonstrated better survival rates after RP for patients with CaP confined to the prostate as compared to patients with CaP stage C or with pelvic lymph node metastases [4,19,30]. Prospective randomized trials, however, that prove the efficacy of RP in CaP stage C or D1 have not yet been reported. Therefore, at this point in time, the treatment of CaP depends upon subjective decisions rather than upon scientifically proven concepts. Consequently, staging requirements differ considerably. If an aggressive treatment modality is utilized, the only question is whether or not the tumor can be surgically removed. Further staging examinations seem to be unnecessary.

If RP or radiotherapy is to be applied exclusively to those patients with tumors confined to the prostate, more detailed and precise information about the tumor staging is required. It seems necessary to ask whether currently available measures satisfy clinical needs. Despite the results of recent investigations which demonstrate the close correlation between tumor burden and PSA serum levels, until today marker determination has only limited value in tumor staging for CaP [49,64]. Most comparative evaluating trials on the efficacy of DRE versus different imaging modalities demonstrate a higher sensitivity for TRUS or MRI [14,18,34,57]. The latter techniques seem to be superior to CT scanning in the prediction of extracapsular or seminal vesicle involvement [14,18,57]. This improved sensitivity, however, is paralleled by a decrease in specificity. This could lead to the exclusion of some patients with a potentially curable tumor from definitive treatment. Therefore, the contribution of imaging modalities to local tumor staging and subsequent treatment decisions must be considered critically. Further improvement of imaging modalities is necessary.

Similar considerations apply to pelvic lymphadenectomy as an adjunct to RP or prior to radiotherapy. Since the therapeutic efficacy of pelvic lymphadenectomy has never been proven, tumor staging remains the rationale for this procedure. This modality was correlated with significant morbidity until the advent of limited pelvic lymphadenectomy. The necessity of this procedure has been questioned by several investigators [5,23,40,48]. Nicholson and Richie concluded from their data that a rigidly defined clinical B1 lesion may represent an indication for RP without pelvic lymphadenectomy [48]. In contrast, other investigators reported positive lymph nodes in up to 20% of patients with a clinical stage B1 CaP [5,29,40,43]. Freiha and associates identified two groups of patients. One group had a less than 8% probability of having lymph node metastases, while the other group had a greater than 92% probability of having lymph node metastases [23]. The identification of these groups was based on serum acid phosphatase levels, and the stage and grade of the primary tumor. They concluded that pelvic lymphadenectomy for staging could be omitted in these two groups of patients. The low accuracy of clinical staging and biopsy-based tumor grading, however, argue against the reliability of this classification at least for the group of "low risk" patients. Therefore, if RP is to be restricted exclusively to patients with CaP confined to the prostate, pelvic lymphadenectomy with frozen sections is mandatory. It should be kept in mind, however, that the examination of frozen sections will provide false-negative results in 10%–20% of cases.

The poor predictive value of CT, MRI, ultrasound, and bipedal lymphangiography in the detection of pelvic lymph node metastases, as well as the low morbidity associated with a limited pelvic lymphadenectomy, argue against a major role for imaging techniques in lymph node staging. If, however, criteria could be defined which identify a group of patients which almost certainly have lymph node metastases those patients could be excluded from pelvic lymph node dissection.

Outlook

The need for more precise localization and determination of the extent of CaP is most clearly demonstrated by the numerous attempts during the last few decades to improve the diagnostic accuracy of DRE. Current investigations are mainly directed toward the improvement of the available imaging techniques. The better resolution afforded by 7.5 HMz TRUS transducers or the high resolution surface coils available for MRI, may improve the sensitivity and specificity of these techniques. Studies evaluating the efficacy of positron emission tomography (PET) in the diagnosis of CaP are underway. Interesting preliminary results, however, should always be followed by careful prospective clinical trials in order to confirm the value of new modalities in the diagnosis and staging of CaP.

References

1. Andriole GL, Kavoussi LR, Torrence RJ, Lepor H, Catalona WJ (1988) Transrectal ultrasonography in the diagnosis and staging of carcinoma of the prostate. J Urol 140:758–760
2. Benson KH, Watson RA, Spring DB, Agee RE (1981) The value of computerized tomography in evaluation of pelvic lymph nodes. J Urol 126:63–64
3. Bezzi M, Kressel HY, Allen KS, Schiebler ML, Altman HG, Wein AJ, Pollack HW (1988) Prostatic carcinoma: staging with MR imaging at 1.5 T. Radiology 169:339–346
4. Boxer RJ, Kaufman JJ, Goodwin WE (1977) Radical prostatectomy for carcinoma of the prostate 1951–1976. A review of 329 patients. J Urol 117:208–211
5. Brendler CB, Cleeve LK, Anderson EE, Paulson DF (1980) Staging pelvic lymphadenectomy for carcinoma of the prostate: risk versus benefit. J Urol 124:849–850
6. Bruce AW, O'Cleireachain F, Morales A, Awad SA (1977) Carcinoma of the prostate: a critical look at staging. J Urol 117:319–322
7. Bryan PJ, Butler HE, Nelson AD, LiPuma JP, Kopiwoda SY, Resnick MI, Cohen AM, Haaga JR (1986) Magnetic resonance imaging of the prostate. AJR 146:543–548
8. Buonocore E, Hesemann C, Pavlicek W, Montie JE (1984) Clinical and in vitro magnetic resonance imaging of prostatic carcinoma. AJR 143:1267–1272
9. Carrol CL, Sommer FG, McNeal JE, Stamey TA (1987) The abnormal prostate: MR imaging at 1.5 T with histopathologic correlation. Radiology 163:521–525
10. Catalona WJ, Stein AJ (1982) Accuracy of frozen section detection of lymph node metastases in prostatic carcinoma. J Urol 127:460–461

11. Catalona WJ, Stein AJ (1982) Staging errors in clinically localized prostatic cancer. J Urol 127:452–456
12. Caty A, Gosselin P, Cazin JL, Dehaut JM, Adenis L (1988) Significance of PSA and PAP in patients with or without prostatic cancer. Am J Clin Oncol Suppl 2:63–64
13. Clements R, Griffiths GJ, Peeling WB, Roberts EE, Evans KT (1988) How accurate is the index Finger? A comparison of digital and ultrasound examination of the prostatic nodule. Clin Radiol 39:87–89
14. Cordes M, Tunn UW, Neidl K, Haasner E (1987) Prostatakarzinom. Stadieneinteilung durch transrektale Prostatasonographie und Computertomographie mit histopathologischer Korrelation. Fortschr Röntgenstr 146: 412–414
15. Correa RJ Jr, Gibbons RP, Cummings KB, Mason JT (1977) Total prostatectomy for stage B carcinoma of the prostate. J Urol 117:328–329
16. Dähnert WF, Hamper UM, Eggleston JC, Walsh PC, Sanders RC (1986) Prostatic evaluation by transrectal sonography with histopathologic correlation: the echopenic appearance of early carcinoma. Radiology 158:97–102
17. Denis L, Appel L, Broos J, Declerq G (1980) Evaluation of prostatic cancer by transrectal ultrasonotomography and CT scan. Acta Urol Belg 48:71–77
18. Ebert T, Schmitz-Dräger BJ, Bürrig K-F, Miller S, Pauli N, Kahn T, Ackermann R (1991) Accuracy of imaging modalities in staging the local extent of prostate cancer. Urol Clin North Am 18:453–457
19. Elder JS, Jewett HJ, Walsh PC (1982) Radical perineal prostatectomy for clinical stage B2 carcinoma of the prostate. J Urol 127:704–709
20. Ercole CJ, Lange PH, Mathisen M, Chiou RK, Reddy PK, Vessella RL (1987) Prostatic specific antigen and prostatic acid phosphatase in the monitoring and staging of patients with prostatic cancer. J Urol 138:1181–1184
21. Fowler JE, Togerson L, McLeod DG, Stutzman RE (1981) Radical prostatectomy with pelvic lymphadenectomy: Observations on the accuracy of staging with lymph node frozen sections. J Urol 126:618–619
22. Fowler JE, Whitmore WF (1981) The incidence and extent of pelvic lymph node metastases in apparently localized prostatic cancer. Cancer 47:2941–2945
23. Freiha FS, Pistenma DA, Bagshaw MA (1979) Pelvic lymphadenectomy for staging prostatic carcinoma: is it always necessary? J Urol 122:176–177
24. Friedman AC, Seidmon EJ, Radecki PD, Lev-Toaff A, Caroline DF (1988) Relative merits of MRI, transrectal endosonography and CT in diagnosis and staging of carcinoma of prostate. Urology 31:530–537
25. Fujino A, Scardino PT (1985) Transrectal ultrasonography for prostatic cancer: its value in staging and monitoring the response to radiotherapy and chemotherapy. J Urol 133:806–810
26. Gevenois PA, van Regemorter G, Ghysels M, Delepaut A, van Gansbeke D, Struyven J (1988) IRM de la prostate. Imagerie calculée. J Radiol 89:275–279
27. Golimbu M, Morales P, Al-Askari S, Brown J (1979) Extended pelvic lymphadenectomy for prostatic cancer. J Urol 121:617–620
28. Golimbu M, Morales P, Al-Askari S, Shulman Y (1981) CAT scanning in staging of prostatic cancer. Urology 18:305–308
29. Grossman IC, Carpiniello V, Greenberg SH, Malloy TR, Wein AJ (1980) Staging pelvic lymphadenectomy for carcinoma of the prostate: review of 91 cases. J Urol 124:632–634
30. Grups JW (1988) Langzeitergebnisse der radikalen Prostatektomie. Urologe [A] 27:348–351
31. Hermanek P (1986) Neue TNM/pTNM-Klassifikation und Stadieneinteilung urologischer Tumoren ab 1987. Urologe [B] 26:193–197

32. Hoekstra WJ, Schröder FH (1981) The role of lymphangiography in the staging of prostatic cancer. Prostate 2:433–436
33. Houston ST, Jones LW, Waluch V (1988) Nuclear magnetic resonance imaging in detecting and staging prostatic cancer. Urology 31:171–175
34. Hricak H, Dooms GC, Jeffrey RB, Avallone A, Jacobs D, Benton WK, Narayan P, Tanagho EA (1987) Prostatic carcinoma: staging by clinical assessment, CT, and MR imaging. Radiology 162:331–336
35. Hricak H, Williams RD, Spring DB, Moon KL Jr, Hedgcock MW, Watson RA, Crooks LE (1983) Anatomy and pathology of the male pelvis by magnetic resonance imaging. AJR 141:1101–1110
36. Jewett HJ (1949) Radical perineal prostatectomy for cancer of the prostate: an analysis of 190 cases. J Urol 61:277–280
37. Kahn T, Bürrig K-F, Schmitz-Dräger B, Fürst G, Stork W, Mödder U (1989) Magnetic resonance tomography of prostatic cancer. Histopathologic correlation. Radiologe 29:144–149
38. Küper K, Hess CF, Griebel J, Peter K (1986) Die Darstellung des Prostatakarzinoms in der Kernspintomographie bei 1,5 Tesla. Fortschr Röntgenstr 144:428–434
39. Liebner EJ, Stefani S, Uro-Oncology Group (1980) An evaluation of lymphography with nodal biopsy in localized carcinoma of the prostate. Cancer 45:728–734
40. Lieskovsky G, Skinner DG, Weisenburger T (1980) Pelvic lymphadenectomy in the management of carcinoma of the prostate. J Urol 124:635–638
41. Ling D, Lee JKT, Heiken JP, Balfe DM, Glazer HS, McClennan BL (1986) Prostatic carcinoma and benign prostatic hyperplasia: inability of MR imaging to distinguish between the two diseases. Radiology 158:103–107
42. Loening SA, Schmidt JD, Brown RC, Hawtrey CE, Fallon B, Culp DA (1977) A comparison between lymphangiography and pelvic node dissection in the staging of prostatic cancer. J Urol 117:752–756
43. McCullough DL (1980) Surgical staging of carcinoma of the prostate. Cancer 45:1902–1905
44. McLaughlin AP, Saltzstein SL, McCullough DL, Gittes RF (1976) Prostatic carcinoma: incidence and location of unsuspected lymphatic metastases. J Urol 115:89–94
45. Middleton AW (1981) Pelvic lymphadenectomy with modified radical retropubic prostatectomy as a single operation: technique used and results in 50 consecutive cases. J Urol 125:353–356
46. Morales P, Golimbu M (1980) The therapeutic role of pelvic lymphadenectomy in prostatic cancer. Urol Clin North Am 7:623–628
47. Mukamel E, Hannah J, Barbaric Z, deKernion JB (1986) The value of computerized tomography scan and magnetic resonance imaging in staging prostatic carcinoma: comparison with the clinical and histological staging. J Urol 136:1231–1233
48. Nicholson TC, Richie JP (1977) Pelvic lymphadenectomy for stage B_1 adenocarcinoma of the prostate: justified or not? J Urol 117:199–201
49. Palken M, Cobb OE, Warren BH, Hoak DC (1990) Prostate cancer: correlation of digital rectal examination, transrectal ultrasound and prostate specific antigen levels with tumor volumes in radical prostatectomy specimens. J Urol 143: 115–1162
50. Paulson DF, Uro-Oncology Research Group (1979) The impact of current staging procedures in assessing disease extent of prostatic adenocarcinoma. J Urol 121:300–302

51. Phillips ME, Kressel HY, Spritzer CE, Arger PH, Wein AJ, Marinelli D, Axel L, Gefter WB, Pollack HM (1987) Prostatic disorders: MR imaging at 1.5T. Radiology 164:386–392
52. Pontes JE, Eisenkraft S, Watanabe H, Ohe H, Saitoh M, Murphy GP (1985) Preoperative evaluation of localized prostatic carcinoma by transrectal ultrasonography. J Urol 134:289–291
53. Poon PY, McCallum RW, Henkelman MM, Bronskill MJ, Sutcliffe SB, Jewett MA, Rider WD, Bruce AW (1985) Magnetic resonance imaging of the prostate. Radiology 154:143–149
54. Reikeras O, Due J, Sundsfjord JA (1988) Bone scintigraphy, radiographic survey and prostatic acid phosphatase in patients with prostatic carcinoma. A comparison of sensitivity. Int Urol Nephrol 20:51–54
55. Resnick MI, Willard JW, Boyce WH (1980) Transrectal ultrasonography in the evaluation of patients with prostatic carcinoma. J Urol 124:482–484
56. Sadlowski RW, Donohue DJ, Richman AV, Sharpe JR, Finney RP (1983) Accuracy of frozen section diagnosis in pelvic lymph node staging biopsies for adenocarcinoma of the prostate. J Urol 129:324–326
57. Salo JO, Kivisaari L, Rannikko S, Lehtonen T (1987) Computerized tomography and transrectal ultrasound in the assessment of local extension of prostatic cancer before radical retropubic prostatectomy. J Urol 137:435–438
58. Salo JO, Ranniko S, Makinen J, Lehtonen T (1987) Echogenic structure of prostatic cancer imaged on radical prostatectomy specimens. Prostate 10:1–9
59. Saltzstein SL, McLaughlin AP (1977) Clinicopathologic features of unsuspected regional lymph node metastases in prostatic adenocarcinoma. Cancer 40: 1212–1221
60. Sawczuk IS, deVere White R, Gold RP, Olsson CA (1983) Sensitivity of computed tomography in evaluation of pelvic lymph node metastases from carcinoma of bladder and prostate. Urology 21:81–84
61. Schaffer DL, Pendergrass HP (1976) Comparison of enzyme, clinical radiographic, and radionuclide methods of detecting bone metastasis from carcinoma of the prostate. Radiology 121:431–434
62. Schmidt JD, McLaughlin AP III, Saltzstein SL, Garcia-Reyes R (1982) Risk factors for the development of distant metastases in patients undergoing pelvic lymphadenectomy for prostatic cancer. Am J Surg 144:131–134
63. Spellman MC, Castellino RA, Ray GR, Pistenma DA, Bagshaw MA (1977) An evaluation of lymphography in localized carcinoma of the prostate. Radiology 125:637–644
64. Stamey TA, Kabalin JN (1989) Prostate specific antigen in the diagnosis and treatment of adenocarcinoma of the prostate. I. Untreated patients. J Urol 141:1070–1075
65. Veenema RJ, Gursel EO, Lattimer JK (1977) Radical retropubic prostatectomy for cancer: a 20-year experience. J Urol 117:330–331
66. Watanabe H, Igari D, Tanahashi Y, Harada K, Saitoh M (1975) Transrectal ultrasonotomography of the prostate. J Urol 114:734–739
67. Weinerman PM, Arger PH, Coleman BG, Pollack HM, Banner MP, Wein AJ (1985) Pelvic adenopathy from bladder and prostate carcinoma: detection by rapid sequence computed tomography. AJR 140:95–99
68. Zincke H, Utz DC (1983) Radical surgery for stage D1 prostate cancer. Semin Urol 1:253–260
69. Zincke H, Utz DC, Taylor WF (1986) Bilateral pelvic lymphadenectomy and radical prostatectomy for clinical stage C prostatic cancer: role of adjuvant treatment for residual cancer and in disease progression. J Urol 135:1199–1205

Radical Prostatectomy for Carcinoma of the Prostate: Long-Term Results

H.G.W. Frohmüller and M.P. Wirth

Urologische Klinik und Poliklinik (Direktor: Prof. Dr. H.G.W. Frohmüller), Universität Würzburg, Josef-Schneider-Strasse 2, W-8700 Würzburg, FRG

Introduction

Since 1969 radical prostatectomy has been the preferred treatment in patients with localized prostate cancer at our institution. It is regarded as the most efficacious treatment modality since complete removal of the cancer offers the possibility to cure patients with localized disease. Many studies published in the last two decades have demonstrated that radical prostatectomy leads to long-term disease-free survival rates.

Patients and Methods

Between January 1969 and July 1989, 303 patients with prostatic cancer were treated by radical prostatectomy at the Department of Urology, University of Würzburg Medical School. The mean age of the patients at the time of operation was 61.5 years (39–77 years). In 270 cases, radical prostatectomy was performed utilizing the retropubic approach [1] and in 33 cases the perineal approach was used [2,3]. Staging pelvic lymphadenectomy was carried out in all but the first seven patients. Histological analysis of the lymph nodes was routinely performed intraoperatively by frozen section technique.

Of these 303 patients, 105 are at risk for more than 10 years. One patient was lost during follow-up; the longitudinal data of 104 patients were therefore evaluable. Of these, 79 patients had pathological stage B lesions (pT1–2pN0M0), 17 had stage C (pT3pN0M0), and 8 had stage D1 (pT1–3pN1M0) lesions according to the staging system of Whitmore [4] or the 1979 TNM system of the UICC [5] respectively. In order to permit a comparison of our results with those reported in the literature, the TNM classification of 1979 was utilized in this study. This means that only tumors penetrating through the capsule of the prostate were classified as stage pT3. Those tumors that are only infiltrating the apex or the prostatic capsule, are classified as stage pT2.

All survival data presented in this study are observed data. No statistical analysis was applied to determine actuarial survival rates.

Results

Long-Term Survival Data

Ten years after radical prostatectomy 69 of the 104 patients were alive. Thus, the 10-year survival rate of all 104 patients, including those with stage D1 (pT1–3pN1–2M0) disease, was 66.3%. A total of 79 patients with stage B (pT1–2pN0M0) prostate cancer had a 10-year survival rate of 73.4%, which is higher than the expected actuarial survival rate of an age-matched male population (69.9%). In 17 of the 104 patients, in whom a stage C (pT3pN0M0) prostate cancer had been diagnosed, the 10-year survival rate decreased to 52.9%. Two out of eight patients (25%), who were treated by radical prostatectomy on the assumption of a clinical stage B or C lesion but, in fact, had microscopic nodal disease on pathological examination (stage D1, pT1–3pN1–2M0), were still alive after 10 years. The overall survival and the stage-related survival rates are presented in Table 1.

Of the 104 patients who underwent radical prostatectomy, 61 (58.6%) lived for 10 or more years without any evidence of their disease. Ten year disease-free survival rates were 67.1% (53 out of 79) for patients with stage B lesions, and 47.1% (8 out of 17) for patients with stage C lesions. None of the patients with stage D1 tumors were free of disease 10 years after operation (Table 1).

Discussion

From January 1969 to July 1989, 303 patients with prostatic cancer were treated by radical prostatectomy at our institution. The long-term results of

Table 1. Ten year survival rates after radical prostatectomy (n = 104)

No. of patients	Stage	Observed overall survival		Disease-free survival		Tumor-related survival	
		(n)	(%)	(n)	(%)	(n)	(%)
79	A or B	58	73.4	53	67.1	69	87.3
17	C	9	52.9	8	47.1	11	64.7
8	D1	2	25.0	0	0	6	75.0
Total		69	66.3	61	58.6	86	82.7

104 patients who could be followed for at least 10 years were reported. It could be demonstrated that patients with stage B prostate cancer have a survival expectancy following radical prostatectomy which is slightly better than that of an age-matched male population (73.4% and 69.9%, respectively). This unexpected result may be due to either or both of these reasons: Firstly, patients who are treated by radical prostatectomy usually have no severe concomitant disease and have for this reason a better life expectancy than age-matched males in general; secondly, patients who undergo radical surgical procedures for a malignant disease are routinely followed up more thoroughly than others.

Table 2 compares the survival data of our study with those of others reported in the literature. The results of most of these studies are quite similar to those presented here. However, Schroeder and Belt [8] and Elder et al. [14] did not perform a staging lymphadenectomy. This renders an accurate staging of the tumor and a comparison with our results impossible because the percentage of patients with stage D1 lesions included in their studies cannot be determined. This might account for the lower survival rates reported by Schroeder and Belt [8] and Elder et al. [14] compared with other series.

Since the introduction of radiotherapy as a treatment modality for prostatic cancer by Bagshaw et al. in 1965 [16], numerous studies have been published presenting long-term survival data after external beam or interstitial radiation therapy of carcinoma of the prostate. On the basis of these publications radiation therapy has become the main competitor of radical prostatectomy for potentially curative treatment of prostate cancer.

Table 2. Longitudinal studies following radical prostatectomy

Source	Year	Tumor stage	10-Year survival rate (%)
Gilbertson [6]	1971	B	69
Culp and Meyer [7]	1973	B	72
Schroeder and Belt [8]	1975	B	61[a]
Boxer et al. [9]	1977	B	67[a]
DeVere White et al. [10]	1977	B	64
Correa et al. [11]	1977	B	79
Walsh and Jewett [12]	1980	B	75
Zincke et al. [13]	1981	B	71
Elder et al. [14]	1982	B	65[a]
Gibbons et al. [15]	1984	B	74
Present study	1990	B	73
Schroeder and Belt [8]	1975	C	36[a]
Boxer et al. [9]	1977	C	29[a]
Elder et al. [14]	1982	C	40[a]
Present study	1990	C	53

[a] No lymphadenectomy performed.

In most of the studies dealing with radiation therapy statistical methods are used (e.g., Kaplan-Meier technique [17]) to determine expected survival rates [18–27]. The actuarial 5-year and 10-year survival rates of these studies are given in Table 3. Most of these studies present 5-year survival rates following radiation therapy of prostate cancer which are almost equal to those obtained after radical prostatectomy. However, the 10-year survival rates, if evaluated, demonstrate a decrease in the survival expectancy of patients treated by radiotherapy compared with that following radical prostatectomy.

Paulson and the Uro-Oncology Research Group [29] reported a prospective randomised study that compared radical prostatectomy and radiation therapy. With first evidence of treatment failure as the end point for determination of efficacy, radical prostatectomy was shown to be more effective than megavoltage irradiation in establishing disease control. From 1969 to 1982, at the University of Würzburg Medical School, 39 patients with stage A to C (T1–3pN0M0) prostatic cancer, who either refused radical prostatectomy or were unable to undergo this operation for medical reasons, were treated by external beam radiation [31,32]. The observed survival rate of these patients was found to be 48.4% at 7 years, compared to 78.8% for surgically treated patients at the same follow-up time (Fig. 1).

Table 3. Longitudinal studies following radiotherapy for prostate cancer

Source	Year	Tumor stage	Survival rates	
			5-Year (%)	10-Year (%)
Bagshaw and Ray [18]	1983	A or B	79[a]	57[a]
Sause et al. [19]	1986	A or B	75[a]	–
De Laney et al. [20]	1986	A or B	86[a]	–
Bagshaw [21]	1986	A or B	81[a]	60[a]
Paulson (ed) [28]	1987	A or B	65	46
Harisiadis et al. [22]	1978	A–C	62[a]	40[a]
Whitmore [23]	1980	A–C	71[a]	–
Garett et al. [29]	1984	A–C	72	–
Kurup et al. [24]	1984	A–C	78[a]	–
Whitmore [25]	1984	A–C	65[a]	–
Harisiadis et al. [22]	1978	C	58[a]	33[a]
Bagshaw and Ray [18]	1983	C	56[a]	36[a]
Sause et al. [19]	1986	C	65[a]	–
Bagshaw [21]	1986	C	61[a]	36[a]
Paulson (ed) [28]	1987	C	42	26
Pilepich et al. [26]	1987	C	64[a]	–
Zagars et al. [27]	1987	C	72[a,b]	47[a,b]

[a] Expected (actuarial) survival.
[b] Adjuvant hormonal treatment.

The main reason why radiotherapy does not achieve the same 10-year and 15-year survival rates as radical prostatectomy may be due to the heterogeneity of prostate cancer [33]. In one and the same tumor one can find radiosensitive cell clones and radioresistant clones. To date, it is not possible to select purely radiosensitive tumors. Neither tumor grade nor tumor stage nor DNA ploidy show a clear correlation with radiosensitivity or radioresistance [34]. By performing a radical prostatectomy all malignant cells within the prostate are excised, at least theoretically. Radiotherapy, however, is not really capable of destroying all cancer cells in the prostate, especially not in patients with a high tumor volume [33]. Viable tumor cells were found by Kabalin et al. [35] in 25 of 27 biopsy specimens obtained 5.2 years after radiation therapy for prostate cancer. The clinical significance of viable tumor cells after radiotherapy is controversial. However, it can be assumed that these persistent cancer cells might account for the higher progression rates observed in patients who underwent radiotherapy for treatment of prostatic cancer than in those who were treated by radical prostatectomy.

Based on the data presented in the literature and in this study, it can be concluded that radical prostatectomy currently affords the best results in curing localized carcinoma of the prostate. Further studies are necessary to determine whether an adjuvant hormonal or chemotherapeutic treatment after radical prostatectomy in patients with stage C and D1 tumors may reduce progression rates and lead to a survival rate which is comparable to that of an age-matched male population.

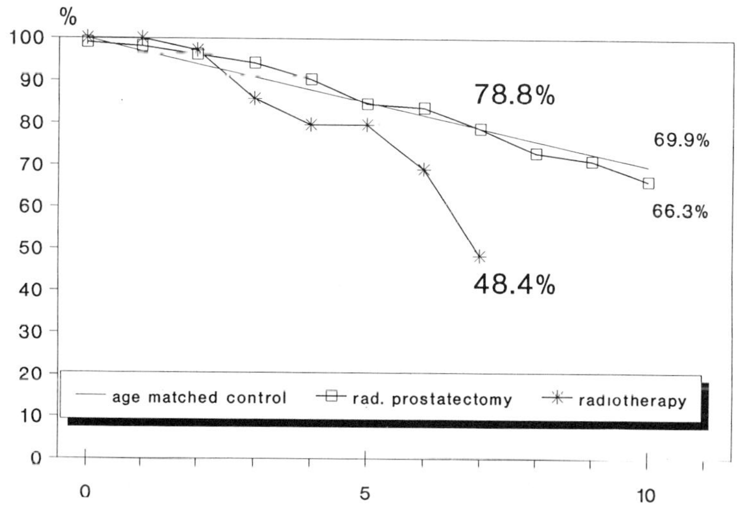

Fig. 1. Comparison of observed survival rates following radical prostatectomy and radiotherapy for carcinoma of the prostate. (Department of Urology, University of Würzburg Medical School)

References

1. Millin T (1945) Retropubic prostatectomy; new extravesical technique: report of 20 cases. Lancet ii:693
2. Young HH (1905) The early diagnosis and radical cure of carcinoma of the prostate. Being a study of 40 cases and presentation of a radical operation which was carried out in four cases. Bull Johns Hopkins Hosp 16:315
3. Belt E, Ebert CE, Surber AC Jr (1939) A new anatomical approach in perineal prostatectomy. J Urol 41:482
4. Whitmore WF Jr (1956) Symposium on hormones and cancer therapy. Hormone therapy in prostatic cancer. Am. J. Med. 21:697
5. Spiessl B, Scheibe O, Wagner G (eds) (1979) TNM Klassifikation maligner Tumoren, 3rd rev. edn. Springer, Berlin Heidelberg New York, p 134
6. Gilbertsen VA (1971) Cancer of the prostate gland. JAMA 215:81
7. Culp OS, Meyer JJ (1973) Radical prostatectomy in the treatment of prostatic cancer. Cancer 32:1113
8. Schroeder FH, Belt E (1975) Carcinoma of the prostate: study of 213 patients with stage C tumors treated by total perineal prostatectomy. J Urol 114:257
9. Boxer RJ, Kaufman JJ, Goodwin WE (1977) Radical prostatectomy for carcinoma of the prostate: 1951–1976, a review of 329 patients. J Urol 117:208
10. DeVere White RD, Paulson DF, Glenn JR (1977) The clinical spectrum of prostate cancer. J Urol 117:323
11. Correa RJ Jr, Gibbons RP, Cummings KB, Mason JT (1977) Total prostatectomy for stage B carcinoma of the prostate. J Urol 117:328
12. Walsh PC, Jewett HJ (1980) Radical surgery for prostate cancer. Cancer 45:1906
13. Zincke H, Fleming TR, Furlow WL, Myers RP, Utz DC (1981) Radical retropubic prostatectomy and pelvic lymphadenectomy for high stage cancer of the prostate. Cancer 47:1901
14. Elder JS, Jewett HJ, Walsh PC (1982) Radical perineal prostatectomy for clinical stage B2 carcinoma of the prostate. J Urol 127:704
15. Gibbons RP, Correa RJ Jr, Brannen GE, Mason JT (1984) Total prostatectomy for localized prostatic cancer. J Urol 131:73
16. Bagshaw MA, Kaplan HS, Sagerman RH (1965) Linear accelerator supervoltage radiotherapy. VII. Carcinoma of the prostate. Radiology 85:121
17. Kaplan EL, Meier P (1958) Non-parametric estimation from incomplete observation. J Am Stat Assoc 53:457
18. Bagshaw MA, Ray GR (1983) External beam radiation therapy of prostate carcinoma. In: Skinner DG (ed) Urological cancer. Grune and Stratton, New York
19. Sause WT, Richards RS, Plenk HP (1986) Prostatic carcinoma: 5-year follow-up of patients with surgically staged disease undergoing extended field radiation. J Urol 135:517
20. De Laney TF, Shipley WU, O'Leary MP, Biggs PJ, Prout GR Jr (1986) Preoperative irradiation, lymphadenectomy and 125-iodine implantation for patients with localized carcinoma of the prostate. Int J Radiat Oncol Biol Phys 12:1779
21. Bagshaw MA (1986) Current conflicts in the management of prostatic cancer. Int J Radiat Oncol Biol Phys 12:1721
22. Harisiadis L, Veenema RJ, Senyszyn JJ, Puchner PJ, Tretter P, Romas NA, Chang CH, Lattimer JK, Tannenbaum M (1978) Carcinoma of the prostate: Treatment with external radiotherapy. Cancer 41:2131
23. Whitmore WF Jr (1980) Interstitial radiation therapy for carcinoma of the prostate. Prostate 1:157

24. Kurup P, Kramer TS, Lee MS, Phillips R (1984) External beam irradiation of prostate cancer. Experience in 163 patients. Cancer 53:37
25. Whitmore WF (1984) Interstitial I-125 implantation in the management of localized prostatic cancer. In: Kurth KH, Debruyne FMJ, Schroeder FH, Splinter TAW, Wagener TDJ (eds) Progress and controversies in oncological urology. Liss, New York, p. 513
26. Pilepich MV, Zinninger MSN, Johnson RJ, Sause WT, Perez CA, Hanks GE, Krall JM, Martz KL, Russ HH (1987) Prognostic significance of nodal involvement in locally advanced (stage C) carcinoma of prostate – RTOG experience. Urology 30:535
27. Zagars GK, von Eschenbach AC, Ohnson DE, Oswald MJ (1987) Stage C adenocarcinoma of the prostate. An analysis of 551 patients treated with external beam radiation. Cancer 60:1489
28. Paulson DF (1987) The management of prostatic malignancy. In: Dekernion JB, Paulson DF (eds) Genitourinary cancer management. Lea and Febiger, Philadelphia, p. 107
29. Paulson DF and the Uro-Oncology Research Group (1982) Radical surgery vs. radiotherapy for adenocarcinoma of the prostate. J Urol 128:502
30. Garett PG, Herman JG, Rawlings GA, Hawkins NV, Gospodarowicz MK, Keen CW, Rider WD (1984) Radical external beam radiation therapy for prostate carcinoma. J Can Assoc Radiol 35:139
31. Egert-Rink, R (1990) Strahlentherapie des Prostata-Carcinoms. Dissertation, University of Würzburg
32. Grups JW, Wirth M, Frohmüller H (1986) Lokal begrenztes Prostatakarzinom. Vergleich radikale Prostatektomie versus Strahlentherapie. Fortschr Med 104:651
33. Müller H-A, Ackermann R, Frohmüller HGW (1980) The value of perineal punch biopsy in estimating the histological grade of carcinoma of the prostate. Prostate 1:303
34. Alken CE, Dhom G, Kopper B, Rehker H, Detz R, Kopp S, Ziegler M (1977) Verlaufskontrolle nach Hochvolttherapie des Prostatacarcinoms. Urologe [A] 16:272
35. Kabalin JN, Hodge KK, McNeal JE, Freiha FS, Stamey TA (1989) Identification of residual cancer in the prostate following radiation therapy: Role of transrectal ultrasound guided biopsy and prostate specific antigen. J Urol 142:326

Adjuvant Therapy After Radical Prostatectomy

H.G.W. Frohmüller and M.P. Wirth

Urologische Klinik und Poliklinik (Direktor: Prof. Dr. H.G.W. Frohmüller), Universität Würzburg, Josef-Schneider-Strasse 2, W-8700 Würzburg, FRG

Introduction

In order to assess the significance of adjuvant forms of therapy after radical prostatectomy for carcinoma of the prostate it is necessary to define those subgroups of patients who would possibly benefit from such adjuvant treatment. It seems obvious that in patients where radical prostatectomy leads to survival rates comparable to actuarial survival rates, adjuvant forms of treatment are unnecessary. This is the case for patients with stage A (pT1pN0M0) and B (pT2pN0M0) prostate cancer as shown by the data of our own institution. In patients with stage A and B disease 73.4% are alive after radical prostatectomy according to an observed 10-year follow-up, whereas the actuarial 10-year survival rate of age-matched males is 69.9%. From these data it can be concluded, therefore, that in patients with stage A and B prostate cancer an adjuvant therapy after radical prostatectomy is not needed.

Stage C (pT3pN0M0) Prostate Cancer

The observed 10-year survival rate after radical prostatectomy in patients with stage C prostate cancer treated at our institution is lower than the actuarial survival rate. Out of 17 patients with stage C (pT3pN0M0) carcinoma of the prostate, 9 (53%) were alive 10 years after radical prostatectomy. The actuarial survival rate of an age-matched male population, however, is 69.9%. The survival data in our study is in accordance with the literature (Table 1). The fact that patients with stage C prostate cancer have a reduced life expectancy because local control of the tumor by radical prostatectomy alone is not sufficient in all of these patients appears to indicate the need for adjuvant treatment modalities in at least a subgroup of patients with stage C disease. In the search for selection criteria for these subgroups of patients

Table 1. Results of radical prostatectomy in stage C prostate cancer

Authors	No. of patients	Survival rates		Statistics
		10 Years (%)	15 Years (%)	
Elder et al. [1] (1982)	35	40	13	Observed
Anscher und Prosnitz [2] (1987)[a]	113	63	21	Computed
Schellhammer [3] (1988)	13	54	11	Observed
University of Würzburg (1990)[a]	17	53	–	Observed

[a] With pelvic lymph node dissection.

Table 2. Seminal vesicle infiltration and tumor recurrence in stage C prostate cancer

Authors	Year	Total no. of patients	Positive seminal vesicles			Negative seminal vesicles		
			n	Recurrence		*n*	Recurrence	
				(*n*)	(%)		(*n*)	(%)
Paulson et al. [4]	1986	143	30	7	23	113	8	7
Bosch et al. [5]	1987	29	12	5	42	17	3	18
Catalona et al. [6]	1988	9	3	1	33	6	1	17
Total		181	45	13	29	136	12	9

who have a higher risk of tumor progression, various parameters have been studied.

Invasion of the seminal vesicles has been detected as a poor prognostic sign. Paulson et al. [4], Bosch et al. [5], and Catalona et al. [6] were able to demonstrate that in patients with seminal vesicle involvement tumor recurrence was more frequent than in patients with tumor-free seminal vesicles (Table 2).

Positive tumor margins are another poor prognostic sign according to the results reported by Paulson and coworkers [4] and Bosch and coworkers [5] (Table 3).

Tumor grade is also known to be of prognostic value. Patients with high grade tumors have a significantly lower life expectancy than those with well-differentiated tumors (Table 4).

Another important parameter appears to be DNA ploidy. The results at our institution demonstrate that the probability of non-progression after radical prostatectomy in stage C prostate cancer is higher in patients with diploid tumors than in those with aneuploid or tetraploid malignancies

Table 3. Positive margins and tumor recurrence after radical prostatectomy

Authors	Year	Positive Margins			Negative Margins		
		n	Recurrence		*n*	Recurrence	
			(*n*)	(%)		(*n*)	(%)
Paulson et al. [4]	1986	29	5	15	99	10	10
Bosch et al. [5]	1987	12	5	42	17	2	12

Table 4. Survival after radical prostatectomy – significance of tumor grade

Authors	Year	Stage	Tumor grade					
			Low		Intermediate		High	
			(*n*)[a]	(%)[b]	(*n*)[a]	(%)[b]	(*n*)[a]	(%)[b]
Culp [7]	1968	B	21	90	41	61	–	–
Boxer et al. [8]	1977	A–C	75	56	38	53	8	0
Zincke et al. [9]	1981	A–D1	31	96	268	77	41	60
Schellhammer [3]	1988	A–C	15	87	15	73	8	25

[a] Total number of patients.
[b] Percentage of patients alive after 10 years.

Table 5. DNA ploidy, tumor stage, and recurrence after radical prostatectomy

Stage	Diploid		Tetraploid		Aneuploid	
	No. of patients	Recurrence (*n*)	No. of patients	Recurrence (*n*)	No. of patients	Recurrence (*n*)
C	31	3	5	4	6	3

(Table 5). All patients in this study were observed for a minimum of 6 years and a maximum of 10 years. Results after radical prostatectomy for prostate cancer in patients with a diploid tumor show a computed 10-year survival rate which is comparable to that of the age-matched male population. Patients with aneuploid or tetraploid tumors, however, have a statistically significant diminished survival rate.

From these data it can be concluded that possible indications for adjuvant therapy after radical prostatectomy in stage C disease could be DNA aneuploidy, positive tumor margins, seminal vesicle invasion, and high tumor grade.

Table 6. Results of radical prostatectomy in stage C prostate cancer with adjuvant therapy

Authors	Total no. of patients	Adjuvant therapy		Survival rates		Statistics
		Form	*n*	10 years (%)	15 years (%)	
Jewett et al. [10] (1968)	17	DES	4	–	6	Observed
Schröder and Belt [11] (1975)	213	DES[b]		36	20	Computed
Boxer et al. [8] (1977)	94	DES[b]		29	–	Computed
Zincke et al. [9] (1981)[a]	50	DES/ radiotherapy[c]		70	–	Computed
Flocks [12] (1973)	69	AU 198	69	67	28	Observed
Anscher and Prosnitz [2] (1987)[a]	46	DES/ radiotherapy	46	90	90	Computed

DES, diethylstilbestrol.
[a] With pelvic lymph node dissection.
[b] Not all patients received DES.
[c] Not all patients received DES or radiotherapy.

Another important parameter which has be taken into account is the prostate-specific antigen (PSA). If after radical prostatectomy PSA is still present in the serum of the patient, residual tumor has to be suspected and additional forms of treatment seem necessary.

Ten-year survival rates between 40% and 63% after radical prostatectomy in stage C disease without adjuvant therapy have been reported (Table 1). The 15-year survival rates range between 11% and 21%. However, Elder et al. [1] and Schellhammer et al. [3] did not perform a pelvic lymphadenectomy, and therefore D1 cases are possibly included in their studies. Reliable data on adjuvant forms of therapy are not available to date, since in almost all studies concerning this subject not all patients received adjuvant therapy (Table 6). Furthermore, in some of these studies the survival rates were observed and in others they were computed. Only Zincke et al. [9] performed a staging lymphadenectomy in this adjuvant treatment setting. In their study Zincke et al. [9] included patients, however, who received radiotherapy in addition to hormonal treatment. In all these studies using adjuvant hormonal therapy, the 10-year survival rates are no better than those observed after radical prostatectomy alone, with the exception of the data presented by Zincke et al. [9].

In patients who received additional local radiotherapy after radical prostatectomy, Flocks [12] reported a 10-year survival rate of 67% and a 15-year survival rate of 28% by administering additional Au-198 treatment. Anscher and Prosnitz [2] published computed 10- and 15-year survival rates of 90% for 46 patients after additional external radiotherapy. These data are unmatched in the literature. Reviewing the data of Anscher and Prosnitz [2] more precisely, it can be seen that in those 46 patients who received additional local radiotherapy, local tumor recurrences (4%) were significantly lower than those for the 113 patients reported in the same article who underwent a radical prostatectomy without additional treatment. The latter group exhibited 25% local tumor recurrences, which according to reports in the literature is a high percentage following radical prostatectomy in stage C prostate cancer. When the frequency of distant metastases is considered, a completely different picture emerges. Patients receiving radical prostatectomy as the sole treatment modality were found to have distant metastases in 15% of the cases. Following radical prostatectomy and local radiotherapy, however, distant metastases were reported in 28% of the patients. Since distant metastases are the most important prognostic parameter in these patients, the excellent survival data for adjuvant local radiotherapy in stage C prostate cancer after radical prostatectomy as reported by Anscher and Prosnitz [2] are difficult to explain.

The fact that local radiotherapy after radical prostatectomy has only a minimal influence on the prognosis is demonstrated by our own data for 17 patients with stage C disease who had an observed follow-up of more than 10 years after radical prostatectomy. In 8 of these 17 patients a tumor progression was observed. All eight patients showed distant metastases, and in two of these eight cases local tumor recurrence was also present. For this reason, local radiotherapy would have been of no value because the prognosis for these two patients depends on the distant metastases that were demonstrated and not on local tumor recurrences.

From the data presented here it can be concluded that additional hormonal therapy after radical prostatectomy could have a positive influence on patients with stage C prostate cancer. Local radiotherapy, however, seems to be of little therapeutic value.

Stage D1 (pT1-3pN1M0) Prostate Cancer

After radical prostatectomy was performed for local tumor control in patients with locoregional lymph node metastases, the observed survival rate in our study was definitely lower than that for the age-matched male population. Of eight patients with stage D1 disease only two were alive 10 years after radical prostatectomy. Immediately following radical prostatectomy all eight patients received adjuvant therapy by means of a bilateral orchiectomy. For the two patients with a metastasized tumor who were alive 10 years after surgery, this form of treatment seems to have been advantageous. The data from the Mayo Clinic, which is the institution with the greatest experience in treating patients with stage D1 disease by radical prostatectomy, demonstrate that early orchiectomy is the single most important factor that influences the prognosis. Zincke [13] reported in 1989 that patients with D1 disease who were treated by orchiectomy after radical prostatectomy showed a computed 10-year survival rate of almost 80%. In patients not treated by orchiectomy a computed 10-year survival rate of only approximately 50% could be demonstrated. From these data it can be concluded that an adjuvant form of therapy is necessary for patients with stage D1 disease treated by radical prostatectomy. Orchiectomy alone, however, seems to be of limited value only.

Chemotherapy in Prostate Cancer

Up to the present time chemotherapy has been shown to reveal unsatisfactory results in patients with advanced hormone-refractory prostate cancer. As adjuvant treatment chemotherapy has been performed only sporadically in the past. According to our present knowledge, the objective responses, partial remissions and, very rarely, complete remissions following monochemotherapy in pretreated patients with metastasized prostate cancer can be expected in a maximum of 30% of the cases [14].

Prospects

In order to achieve better results in locally advanced prostate cancer in the future, prospective randomized trials should be done to prove the value of adjuvant forms of therapy after radical prostatectomy. The retrospective studies available to date do not do this. In carefully controlled prospective randomized trials it should be investigated whether better tumor control can be achieved by additional hormonal therapy or chemotherapy or a combination of both after radical removal of the prostate. The fact that chemotherapy has not achieved favorable results in patients with far-advanced, pretreated tumors with diminished bone marrow function should be taken into consideration. Such forms of therapy should not, however, be excluded for locally advanced prostate cancer in randomized trials.

References

1. Elder JS, Jewett HJ, Walsh PC (1982) Radical perineal prostatectomy for clinical stage B2 carcinoma of the prostate. J Urol 127:704
2. Anscher MS, Prosnitz LR (1987) Postoperative radiotherapy for patients with carcinoma of the prostate undergoing radical prostatectomy with positive surgical margins, seminal vesicle involvement and/or penetration through the capsule. J Urol 138:1407
3. Schellhammer PF (1988) Radical prostatectomy. Patterns of local failure and survival in 67 patients. Urology 31:191

4. Paulson DF, Stone AR, Walther PJ, Tucker JA, Cox EB (1986) Radical prostatectomy: anatomical predictors of success or failure. J Urol 136:1041
5. Bosch RJ, Kurth KH, Schroeder FH (1987) Surgical treatment of locally advanced (T3) prostatic carcinoma: early results. J Urol 138:816
6. Catalona WJ, Miller DR, Karoussi LR (1988) Intermediate term survival results in clinically understaged prostate cancer patients following radical prostatectomy. J Urol 140:540
7. Culp OS (1968) Radical perineal prostatectomy: its past, present and possible future. J Urol 98:618
8. Boxer RJ, Kaufman JJ, Goodwin WE (1977) Radical prostatectomy for carcinoma of the prostate: 1951–1976, a review of 329 patients. J Urol 117:208
9. Zincke H, Fleming TR, Furlow WL, Myers RP, Utz DC (1981) Radical retropubic prostatectomy and pelvic lymphadenectomy for high stage cancer of the prostate. Cancer 47:1901
10. Jewett HJ, Bridge RW, Gray JF, Shelley WM (1968) The palpable nodule of prostatic cancer: results 15 years after radical excision. J A M A 203:403
11. Schröder FH, Belt E (1975) Carcinoma of the prostate: a study of 213 patients with stage C tumors treated by total perineal prostatectomy. J Urol 114:257
12. Flocks RH (1973) The treatment of stage C prostatic cancer with special reference to combined surgical and radiation therapy. J Urol 109:461
13. Zincke H (1989) Extended experience with surgical treatment of stage D1 adenocarcinoma of prostate. Urology 33 [Suppl]:27
14. Wirth MP (1990) Wertigkeit der Chemotherapie in der Behandlung des fortgeschrittenen Prostatakarzinoms. In: Frohmüller HGW, Wirth MP (eds) Behandlung des fortgeschrittenen Prostatakarzinoms. Springer, Berlin Heidelberg New York

Endocrine Therapy of Advanced Prostatic Cancer

R. Berges, H. Schulze, and T. Senge

Abteilung für Urologie, Marien-Hospital, Rhein-Ruhr-Universität,
Widumer Strasse 8, W-4690 Herne, FRG

Introduction

Even today, the majority of patients suffering from prostatic cancer have developed advanced stages of disease at the time of diagnosis. In fact, 40% already suffer from metastatic disease and have a 50% risk of dying within the first 2 years after diagnosis. Although there is already 50 years experience of hormonal treatment for advanced prostatic cancer the prognosis is still poor and a curative therapy is not available.

Prostatic tissue is a target for androgen action. Physiologic regulation and biological feedback are performed by the hypothalamic-hypophysial-testicular axis. Dihydrotestosterone (DHT), synthesized from testosterone by the 5-alpha-reductase enzyme, is the most important androgen in the prostatic cell. It has the highest affinity of all androgens to its specific intracellular receptor [1]. The role of androgens deriving from the adrenal gland is the subject of controversial discussion since these weak androgens can be converted to DHT via a sulfatase enzyme within the prostatic cell [2].

Since 1941, when Huggins and Hodges discovered the crucial effect of steroidal hormones on prostatic cancer, different options for hormonal intervention have been discussed and tried [3]. Androgen withdrawal is the common goal of these therapies. Androgen suppression leads to the stabilization or remission of metastatic disease in about 80% of patients for a limited period of time [4]. This high response does not, however, disguise the fact that half of the patients with metastatic disease at the time of diagnosis will die within the first 2 years. Moreover, 10% will die within the first 6 months [5].

Androgen deprivation today can be achieved by:

1. Orchiectomy
2. Altering the biologic feedback regulation, i.e., suppressing luteinizing hormone (LH) secretion

3. Direct suppression of androgen biosynthesis
4. Competitive inhibition of androgen binding to their receptors within the prostatic cell

Since the treatment of metastatic disease in prostatic cancer is always palliative, discussions about side effects, efficacy, and the optimal time for onset of therapy are ongoing today [6].

Orchiectomy

Bilateral orchiectomy is regarded as the "golden standard" to which other forms of therapy are compared. It leads to a reduction of circulating testosterone from about 500 ng/ml to 50 ng/ml within the first 24 h. Long-term studies have shown that there is no compensatory stimulation of adrenal androgen production. This 90% reduction of circulating testosterone causes an effective blockade of androgen-dependent metabolic pathways in prostate cancer cells.

The major advantage of orchiectomy is that it causes a constant and lasting androgen suppression. There are no cardiovascular side effects and feminization does not occur [4]. Moreover the costs of this highly effective therapy are very low and patient acceptance has always been excellent. However, as observed for most kinds of antiandrogen treatment, so-called hot flushes often accompany orchiectomy. These hot flushes more likely occur with orchiectomy or treatment with luteinizing hormone-releasing hormone (LHRH) analogs and nonsteroidal antiandrogens than during therapy with estrogens and steroidal antiandrogens like cyproterone acetate [7–11].

The pathophysiologic mechanism of hot flushes is not clear, but there is evidence that intrahypothalamic opioids are responsible for this phenomenon [12,13]. Negative feedback regulation of sex hormones on hypothalamic LHRH secretion is permitted by substances such as opioids or gamma-aminobutyric acid. These substances inhibit action of adrenergic neurons leading to a depletion of catecholamine secretion. Catecholamines are essential for LHRH production in responsible neurons within the hypothalamic nuclei (Fig. 1) [14–16]. Catecholamines are also known to be involved in body temperature regulation [17]. Low sex hormone levels result in low levels of inhibitory opioids leading to a rise in hypothalamic catecholamines. As neurons responsible for thermoregulation are located in close anatomic relation to neurons producing LHRH within the hypothalamus, high catecholamine levels might also stimulate these neurons, thus causing hot flushes (Fig. 2).

This theory would also explain the fact that steroidal antiandrogens like cyproterone acetate ease hot flushes [18,19]. Its progestin-like action at hypothalamic receptors maintains opioid secretion and therefore limits catecholamine release.

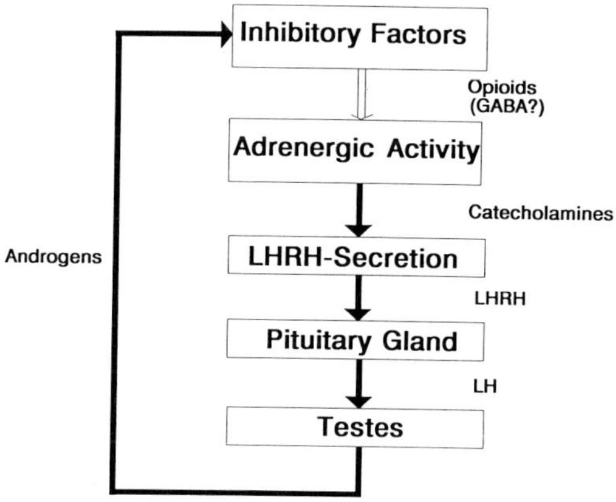

Fig. 1. Endocrine feedback of androgen regulation. ⇒, inhibition; →, stimulation. (Radlmaier [65])

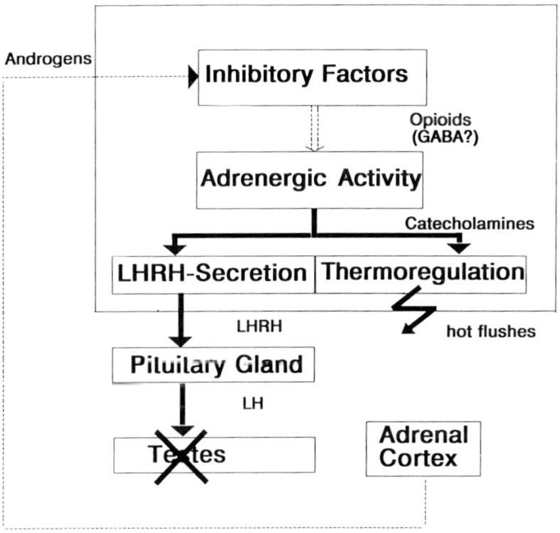

Fig. 2. Possible relation between endocrine regulation and hot flushes after orchiectomy. ⇒, inhibition; →, stimulation. (Radlmaier [65])

Suppression of LH Secretion

Estrogen

Estrogen influences androgen metabolism via a variety of different actions. Besides suppression of pituitary LH secretion, it increases the amount of sex hormone-binding globulin and reduces gonadal testosterone biosynthesis. Furthermore, estrogen stimulates the pituitary prolactin secretion and also inhibits synthesis of DNA in cancer cells of the prostate when given in very high dosages [20,21]. Studies like VACURG I and II have shown that estrogens suppress serum testosterone concentration to castration levels [22].

Diethylstilbestrol (DES) is the best studied and most commonly used estrogen in the treatment of advanced prostatic cancer. Ethinyl estradiol, chlorotrianisene, Premarin, and polyestradiol phospate – used as intramuscular depot – have also been shown to be effective.

Estramustine phosphate is a chemical linkage of estradiol-17-phosphate and N-Lost. This substance combines the benefits of estrogens in the therapy of advanced prostatic cancer with a cytotoxic agent. Randomized clinical trials, however, have so far failed to show that treatment with estramustine phosphate is any more beneficial than treatment with bilateral orchiectomy or DES [23,24].

In general, estrogen therapy can be accompanied by severe cardiovascular complications, gynecomastia, malfunction of erythropoiesis and adrenal cortex, disorders of lipid and protein metabolism, or peripheral edema. A number of clinical trails have been established to find the best dosage regimen needed to achieve the most effective androgen depression with the fewest possible side effects [22,25,26]. Other studies have clearly demonstrated the potentially morbid outcome of estrogen therapy showing that up to 25% of patients treated with estrogens had to be hospitalized due to serious cardiovascular side effects; no such events were observed in patients treated by orchiectomy [27]. Thus, administration of estrogens seems an unfavorable therapy.

LHRH Analogs

During recent years a new treatment concept for advanced prostatic cancer has been developed, which introduces supraactive LHRH analogs [28,29]. While intermittent action of LH-RH on the pituitary is crucial for its stimulating effect on LH and follicle-stimulating hormone (FSH) secretion, a constantly high level of this peptide leads to the exhaustion of LH receptors with consequent gonadotropin deprivation. This in turn is followed by suppression of serum androgen concentrations to castration levels [30–32].

When first introduced, the only possible route of administration was in the form of a nasal spray (buserelin, nafarelin) or as a daily subcutaneous injection (leuprolide, nafarelin), whereas today intramuscular or subcutaneous depot forms are available (buserelin, Decapeptyl, leuprolide, goserelin acetate), making monthly and 3-monthly administrations possible. These drugs are well tolerated and obviate the need for surgical intervention. Many men prefer this treatment to orchiectomy when given the choice [33].

Randomized clinical trails comparing DES and leuprolide have shown no difference between these two agents regarding improvement of performance status, improvement of bone pain, median time to progression, and median survival time [34,35]. Thus, LHRH analogs do not appear superior to standard hormonal therapy. Considering the high cost of this therapy the benefits compared to orchiectomy are questionable.

There is an initial enhancement of pituitary secretion with a consequent peak of serum testosterone after the onset of treatment; the use of LHRH analogs leads to a remarkable rise in acid phosphatase enzyme activity as a specific marker of prostatic function. This phenomenon is called "flare up" and is characterized by an increase in biochemical parameters such as prostatic acid phosphatase and prostate-specific antigen and a worsening of clinical parameters such as obstructive dysuria and bone pain. Even paraplegia may occur, which makes pretreatment with antiandrogens advisable [36–38]. To obtain a sufficient prostatic acid phosphatase (PAP) suppression following treatment initiation antiandrogen pretreatment for 7 days is required [39,40].

Progestins

Due to its antigonadotrope action, cyproterone acetate is a potent LH suppressor, leading to testosterone deprivation. Its main action, however, is based on its direct antiandrogene effect which is discussed later [41–43]. There are no other substances available in this class today of clinical importance.

Inhibition of Androgen Synthesis

Testosterone is synthesized from cholesterol and acetate in several biochemical steps involving five different enzymes. Inhibition of any of these enzymes could therefore be significant in the treatment of prostatic cancer. Drugs like aminoglutethimide, ketoconazole, or spironolactone have been used in this connection. Because of their side effects (spironolactone – hypokalemia; aminoglutethimide – adrenal and thyroid insufficiency; ketoconazole – hepatotoxicity and sepsis), these drugs are not standard therapies for prostatic cancer [44–46].

Inhibition of Androgen Binding

Androgens must bind to specific intracellular receptors in order to release their stimulating effect on prostatic tissue. This effect can be suppressed by competitive inhibition by antiandrogenic drugs. Two different kinds of

Table 1. Different action of steroidal and nonsteroidal (pure) antiandrogens

Steroidal antiandrogens (cyproterone acetate)	Nonsteroidal antiandrogens (e.g., flutamide)
Blockade of androgen receptors and additional gestagenic effect	Selective blockade of androgen receptors
↓	↓
Complete androgen depression combined with gonadotropin depression	Selective testosterone depression leading to activation of gonadotropin secretion
↓	↓
Testosterone depression to castration levels	Rise in serum testosterone level
Impotence	No impotence
Moderate gynecomastia	Gynecomastia
No hot flushes	Hot flushes

drugs are available in this class: steroidal and nonsteroidal antiandrogens (Table 1).

Steroidal Antiandrogens

The most important steroidal antiandrogen today is cyproterone acetate. Due to its progestin-like activity it causes LH suppression with a decrease in serum testosterone, which leads adversely to impotence in most patients. Its major effect, however, relates to direct antiandrogen activity in the prostatic cell [43,47,48]. It has been shown that cyproterone acetate thereby has less cardiovascular and thromboembolic side effects than estrogens [49].

Although patients treated with cyproterone acetate often become impotent, there are fewer complaints of hot flushes during this therapy [38]. In fact, cyproterone acetate may even be used for treatment of hot flushes, as mentioned above.

Nonsteroidal Antiandrogens

Nonsteroidal antiandrogens have no endocrine side effects and are therefore called pure antiandrogens [50,51]. They are competitive inhibitors to androgen receptors in all androgen-dependent tissues, acting on the hypothalamus as well as on the pituitary gland [52–54]. During treatment these substances lead to enhanced secretion of LHRH, and LH, respectively, followed by a rise in serum testosterone level. However, studies have shown this testosterone increase to be limited [55]. On the other hand, erectile function is maintained in these patients.

Well-known nonsteroidal drugs include flutamide, nilutamide, and casodex [56]. Flutamide is metabolized to its active form, hydroxyflutamide, during first passage through the liver [57]. It might be hepatotoxic, and cause gynecomastia and occasional diarrhea.

Nonsteroidal antiandrogens, however, have until now not been used as monotherapy in randomized clinical trials because of major concern that the rise in serum testosterone level caused by these drugs may neutralize their antiandrogenic effect [48,51,56,58].

So far, we have considered androgen deprivation in the endocrine treatment of prostatic cancer with reference to gonadal androgens. What, however, is the influence of adrenal androgens on prostatic cancer cells? Their role is still a matter of discussion.

As already mentioned, weak adrenal androgens like dehydroepiandrostrone sulfate or androstenedione can be converted to DHT in prostatic cells. Thus, it has been suggested that simultaneous suppression of adrenal and testicular androgens (so-called complete androgen suppression) may be more effective than testicular androgen deprivation alone. For this reason many studies have been designed which, on the one hand, combine either orchiectomy or LHRH analogs with antiandrogens, or, on the other hand, compare orchiectomy with treatment with LHRH analogues and antiandrogens [59–62]. Some of these studies suggest that complete androgen suppression may be more beneficial regarding time to progression than suppression by pharmacologic agents which do not effect adrenal androgens. However, there is still no overwhelming evidence to indicate that complete androgen suppression has more impact on survival rates than orchiectomy [63,64].

Although we have 50 years of experience of hormonal treatment for advanced prostatic cancer the optimal time for onset of treatment is still unknown.

The beneficial effect of antiandrogen treatment in patients with symptomatic metastatic disease is well established. However, whether early onset of endocrine therapy benefits progression rate and survival time in asymptomatic patients is not yet clear. Clinical data indicate that a delay in the onset of therapy until symptoms occur has no negative effect on the outcome of treatment. In practice, however, this is still very uncommon and not often used.

References

1. Wilbert DM, Griffin JE, Wilson JD (1983) Characterization of the cytosol androgen receptor of the human prostate. J Clin Endocrinol Metab 56:113
2. Farnsworth WE (1975) Human prostatic dehydroepiandrosterone sulfate sulfatase. In: Goland M (ed) Normal and abnormal growth of the prostate III. Thomas, Springfield

3. Huggins C, Hodges CV (1941) Studies on prostatic cancer. I. The effect of castration, of estrogen and of androgen injection on serum phosphatase in metastatic carcinoma of the prostate. Cancer Res 1:293
4. Scott WW, Menon W, Walsh PC (1980) Hormonal therapy of prostate cancer. Cancer 45:1929
5. Blackard CE, Byar DP, Jordan WP Jr (1973) Veterans Administration Cooperative Urological Research Group: Orchiectomy for advanced prostatic carcinoma: a reevaluation. Urology 1:553
6. Chisholm GD (1985) Treatment of advanced cancer of the prostate. Semin Surg Oncol 1:38
7. Charig CR, Rundle JS (1988) Flushing: long-term sideeffects of orchidectomy in the treatment of carcinoma of the prostate. J Urol 139:478A
8. Debruyne FMJ, Denis L, Lunglmayer G (1988) Long-term therapy with a depot luteinizing hormone-releasing hormone analogue (Zoladex) in patients with advanced prostatic carcinoma. J Urol 140:775
9. Frödin T, Alund G, Varenhorst E (1985) Measurement of skin blood-flow and water evaporation as a means of objectively assessing hot flushes after orchidectomy in patients with prostatic cancer. Prostate 7:203
10. Iversen P, Hvidt V, Krarup T, Rasmussen F, Rose C (1988) The Danish prostate cancer group: Zoladex plus flutamide vs. orchiectomy in patients with advanced prostatic cancer – a phase III study. Gynecol Endocrinol 2 [suppl 1]: 59
11. Leuprolide study group (1984) Leuprolide versus diethylstilbestrol for metastatic prostate cancer. N Engl J Med 311:1281
12. Vugt DA van, Bakst G, Dyrenfurth I, Ferin M (1983) Naloxone stimulation of luteinizing hormone secretion in the female monkey: influence of endocrine and experimental conditions. Endocrinology 113:1858
13. Wehrenberg WB, Wardlaw SL, Frantz AG, Ferin M (1982) β-endorphin in hypophyseal portal blood: variations throughout the menstrual cycle. Endocrinology 111:879
14. Dyer RG, Grossmann R: Mansfield S, Diez-guerra FJ, Bicknell RJ, Hollingsworth S (1988) Opioid peptides inhibit noradrenergic transmission in the preoptic area to block LH secretion: evidence from neonatally and rogenised rats. Brain Res Bull 20(6):721
15. Kalra SP, Simpkins JW (1981) Evidence for noradrenergic mediation of opioid effects on luteinizing hormone secretion. Endocrinology 109:776
16. Terasawa E, Krook C, Hei DL, Gearing M, Schultz NJ, Davis GA (1988) Norepinephrine is a possible neurotransmitter stimulating pulsatile release of luteinizing hormone-releasing hormone in the rhesus monkey. Endocrinology 123:1808
17. Cox B, Lomax P (1977) Pharmacologic control of temperature regulation. Annu Rev Pharmacol Toxicol 17:341
18. Claes H, Vanderbussche L, Vereecken RL (1987) Treatment of advanced carcinoma of the prostate by LHRH agonists. In: Murphy GP et al. (eds) Prostate cancer, part A: research, endocrine treatment, and histopathology. Liss, New York, p 229
19. Eaton AC, McGuire N (1983) Cyproterone acetate in treatment of post-orchidectomy hot flushes. Lancet (8363):1336
20. Altwein JE (1983) Estrogens in the treatment of prostatic cancer. In: Pavone-Macaluso M, Smith PH (eds) Cancer of the prostate and kidney. Plenum, New York, p 317
21. Sinha AA, Blackard CE, Seal US (1977) A critical analysis of tumor morphology and hormone treatments in the untreated and estrogen-treated responsive and refractory human prostatic carcinoma. Cancer 40:2836

22. Byar DP (1973) The VACURG studies of cancer of the prostate. Cancer 32:1126
23. Benson RC, Gill GM (1986) Estramustine phosphate compared with diethylstilbestrol. Am J Clin Oncol 9:341
24. Smith PH, Robinson M, Richards B (1984) Estramustine phosphate in prostate cancer. EORTC-Experience: preliminary results of protocol 30762. Urology 23 [Suppl]:64
25. VACURG (1967) Carcinoma of the prostate: treatment comparisons. J Urol 98:516
26. Prout GR Jr, Kliman B, Daly JJ, MacLaughlin RA, Griffin PP, Young HH (1976) II. Endocrine changes after diethylstilbestrol therapy: effect on prostatic neoplasm and pituitary gonadal axis. Urology 7:148
27. Henriksson P, Edhag O (1986) Orchidectomy versus oestrogen for prostatic cancer. Cardiovascular effects. Br Med J 293:413
28. Schally AV, Redding TW, Comaru-Schally AM (1983) Inhibition of prostate tumors by agonistic and antagonistic analogs of LH-RH. Prostate 4:545
29. Schally AV, Arimura A, Coy DH (1980) Recent approaches to fertility control based on derivatives of LHRH. In: Murson PL, Glover E, Deizfaluny E (eds) Vitamins and hormones. Academic, San Diego, p 257
30. Santen RJ, Bardin CW (1973) Episodic luteinizing hormone secretion in man. Pulse analysis, clinical interpretation, physiological mechanisms. J Clin Invest 52:2617
31. Borgmann V, Nagel R, Schmidt-Gollwitzer M (1982) Langzeitsuppression der gonadalen Testosteronproduktion durch den LHRH-Agonisten Buserelinacetat (Hoe 766) beim fortgeschrittenen Prostatakarzinom – eine neue Therapieform? Aktuel Urol 13:200
32. Wenderoth UK, Jacobi GH (1985) Langzeitergebnisse mit dem GnRH-Analogon Buserelin (Suprefact) bei der Behandlung des fortgeschrittenen Prostatakarzinoms seit 1981. Aktuel Urol 16:58
33. Cassileth BR, Soloway MS, Vogelzang NH (1989) Patients' choice of treatment in stage D prostate cancer. Urology 33 [Suppl]:57
34. Sharifi R, Lee M, Ojeda L, Ray P, Stobnicki M, Guinan P (1985) Comparison of leuprolide and diethylstilbestrol for stage D2 adenocarcinoma of the prostate. Urology 26:117
35. Leuprolide study group (1984) Leuprolide versus diethylstilbestrol for metastatic prostate cancer. N Engl J Med 311:1281
36. Aulitzky W, Frick J, Joos H, Hauser W (1987) Inhibition of LHRH-analogue induced testosterone surge by blockade of androgen receptors. J Urol 137:256A
37. Labrie F, Dupont A, Belanger A (1987) Flutamide eliminates the risk of disease flare in prostatic cancer patients treated with a LH-RH-agonist. J Urol 138:804
38. Schroeder FH, Lock TM, Chadha DR, Debruyne FM, Karthaus HF, De-Jong FH, Klijn JG, Matroos AW, De-Voogt HJ (1987) Metastatic cancer of the prostate managed with buserelin versus buserelin plus cyproterone acetate. J Urol 137:912
39. Schulze H, Senge T (1990) Influence of different types of antiandrogens on luteinizing hormone-releasing hormone analogue induced testosterone surge in patients with metastatic carcinoma of the prostate. J Urol 144:934
40. Kuhn J-M, Billebaud T, Navratil H, Moulonguet A, Fiet J, Grise P, Louis J-F, Costa P, Husson J-M, Dahan R, Bertagna C, Edelstein R (1989) Prevention of the transient adverse effects of a gonadotropin-releasing hormone analogue (buserelin) in metastatic prostatic carcinoma by administration of an anti-androgen (nilutamide). N Engl J Med 321:413
41. Mowszowicz I, Bieber DE, Chung KW (1974) Synandrogenic and anti-androgenic effect of progestins: comparison with non-progestational antiandrogen. Endocrinology 95:1589

42. Poyet P, Labrie F (1985) Comparison of the antiandrogenic/androgenic activities of flutamide, cyproterone acetate and megestrol acetate. Mol Cell Endocrinol 42:282
43. Neumann F, Hümpel M, Senge T, Schenck G, Tunn UW (1982) Cyproterone acetate – biochemical and biological basis for treatment of prostatic cancer. In: Jacobi GH, Hohenfellner R (eds) Prostate cancer, vol 3. Williams and Wilkins, Baltimore, p 269
44. Trachtenberg J (1984) Ketoconazole therapy in advanced prostatic cancer. J Urol 132:61
45. Williams G, Kerle DJ, Ware H (1986) Objective response to ketoconazole therapy in patients with relapsed progressive prostatic cancer. Br J Urol 58:45
46. Worgul TJ, Santen RJ, Samojlik E (1983) Clinical and biochemical effect of aminoglutethimide in the treatment of advanced prostatic carcinoma. J Urol 129:51
47. Neumann F, Habenicht UF, Schacher A (1984) Antiandrogens and target cell response – different in vivo effects of cyproterone acetate, flutamide and cyproterone. In: McKerns KW, Aakvaag A, Hansson V (eds) Regulation of target cell responsiveness, vol 2. Plenum, New York, p 489
48. Knuth UA, Hano R, Nieschlag E (1984) Effect of flutamide or cyproterone acetate on pituitary and testicular hormones in normal men. J Clin Endocrinol Metab 59:983
49. de Voogt HJ, Smith PH, Pavone-Macaluso M, de Pauw M, Suciu S, members of the European Organization for Research of Treatment of Cancer Urological Group (1986) Cardiovascular side effects of diethylstilbestrol, cyproterone acetate, medroxyprogesterone acetate and estramustine phosphate used for the treatment of advanced prostatic cancer: results from european organization for research on treatment of cancer trials 30761 and 30762. J Urol 135:303
50. Egle N, Altwein JE (1979) Postpuberal castration and prostatic carcinoma. Urology 6:471
51. Raynaud JP, Bonne C, Moguilewski M (1984) The pure antiandrogen RU 23908 (Anandron), a candidate of choice for the combined antihormonal treatment of prostatic cancer: a review. Prostate 5:299
52. Bruchovsky N (1980) Molecular action of androgens and antiandrogens. In: Hammerstein J, Lachnit Fixon U, Neumann F (eds) Androgenization in women. Excerpta Medica, Amsterdam, p 7
53. Walsh PC, Korenman SG (1971) Mechanism of androgenic action: effect of specific intracellular inhibitors. J Urol 105:850
54. Neumann F (1981) Antiandrogens: pharmacology. In: Antihormone. Bedeutung in der Urologie. Klin Exp Urol 3:89
55. Lund F, Rasmussen F (1987) Flutamide versus stilbestrol in the management of advanced prostatic cancer. Br J Urol 61:140
56. Asscheman H, Gooren LJG, de Voogt HJ (1987) Clinical and endocrinological results of short term Anandron treatment. Eur J Cancer Clin Oncol 23:1231
57. Peets EA, Henson MF, Neri R (1974) On the mechanism of the antiandrogenic action of flutamide in the patient. Endocrinology 94:532
58. Furr BJA (1988) ICI 176334: a novel non-steroidal, peripherally-selective antiandrogen. In: Smith PM, Pavone-Macaluso M (eds) Management of advanced cancer of prostate and bladder. Liss, New York, p 13
59. Crawford ED, Blumenstein BA, Goodman PJ, Davis MA, Eisenberger MA, McLeod DG, Spaulding JT, Benson R, Dorr FA (1990) Leuprolide with and without flutamide in advanced prostate cancer. Cancer 66:1039
60. Keuppens F, Denis L, Smith P, Carvalho AP, Newling D, Bond A, Sylvester R, De Pauw M, Vermeylen K, Ongena P, the EORTC Group (1990) Zoladex and flutamide versus bilateral orchiectomy. Cancer 66:1045

61. Iversen P, Christensen MG, Friis E, Hornbøl P, Hvidt V, Iversen HG, Klarskov P, Krarup T, Lund F, Mogensen P, Pedersen T, Rasmussen F, Rose C, Skaarup P, Wolf H (1990) A phase III trial of Zoladex and flutamide versus orchiectomy in the treatment of patients with advanced carcinoma of the prostate. Cancer 66:1058
62. Iversen P, Suciu S, Sylvester R, Christensen I, Denis L (1990) Zoladex and flutamide versus orchiectomy in the treatment of advanced prostatic cancer. Cancer 66:1067
63. Schulze H, Isaacs J, Senge T (1987) Inability of complete androgen blockade to increase survival of patients with advanced prostatic cancer as compared to standard hormonal therapy. J Urol 137:909
64. Schulze H, Isaacs JT, Coffey S (1987) A critical review of the concept of total androgen ablation in the treatment of prostate cancer. In: Murphy GP et al. (eds) Prostate cancer, part B: imaging techniques, radiotherapy, chemotherapy and management issues. Liss, New York, p 1
65. Radlmaier A (1990) Theoretisches Modell über die Entstehung von Hitzewallungen. In: Nagel R (ed) Aktuelle Standortbestimmung der konservativen Therapie des Prostatakarzinoms. De Gruyter, Berlin, p 163

Chemotherapy of Prostatic Carcinoma

H. Rübben, K. Rembrink, and T. Johnston

Urologische Abteilung, Universitätsklinikum Essen, Hufelandstrasse 55, W-4300 Essen 1, FRG

General Aspects

In 1941 Huggins and Hodges [22] demonstrated the effect of androgen deprivation on prostatic carcinoma in experimental studies and by clinical observation. Since then, androgen deprivation has become the standard treatment of metastatic prostate cancer. About 25% of patients do not respond to this treatment initially and another 50% show a progression of the disease in the follow-up [47]. That means that more than 75% of patients need an additional treatment.

Systemic chemotherapy seems to be a reasonable treatment at this stage of disease, but despite a large number of clinical studies investigating various drugs, combinations, dosages, and treatment schedules, systemic chemotherapy is not widely accepted. Two factors contribute to this state of affairs:

1. Patients with hormone-resistant prostate cancer have been treated for several years already; the tumor is far advanced and the functional capacity of the organs of the patients have been reduced by age and tumor. The reduction in organ function especially applies to renal clearance, vital capacity of the lung, bone marrow function, liver function, and activity of the endocrine system (Fig. 1). Therefore cytostatic drugs with a high organ toxicity cannot be given or only in reduced dosages.
2. The growth of the tumor is determined by its doubling time, proliferation rate, and cell loss. Tubiana and Malaise [69] showed different doubling times of various tumors with measurable metastases and evaluated the proliferation rate and cell loss by cell kinetic methods. The embryonic carcinoma of the testis show a very high proliferation rate of 44% and high cell loss of 42%, resulting in a tumor doubling time of 27 days; this accounts for the clinical observation that the embryonic carcinoma of the

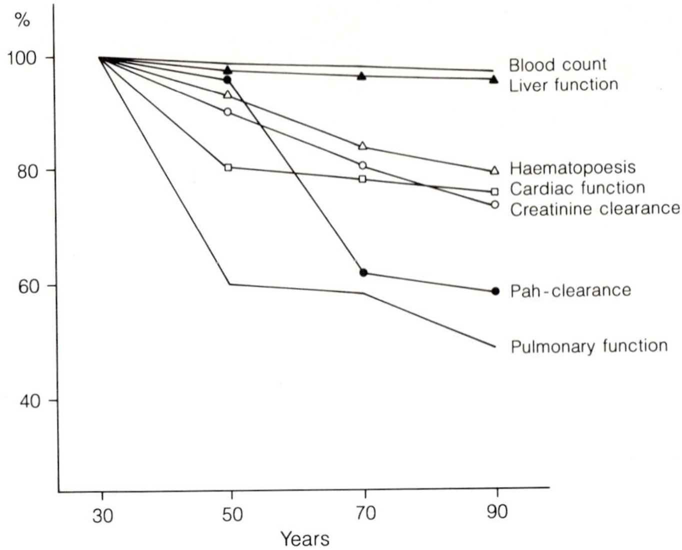

Fig. 1. Reduction of Organ Function Depending on Age. *Pah*, para-aminohippuric acid

testis can be treated by systemic chemotherapy very successfully. In contrast, the adenocarcinoma shows a proliferation rate of only 3% and a cell loss of 2% only, resulting in a tumor doubling time of 83 days. Chemotherapy mainly attacks proliferating cells and can thus hardly influence the adenocarcinoma of the prostate (Table 1).

Nonproliferating cells (G0 phase) can be transformed into the growth fraction by a so-called recruitment by means of operative, chemotherapeutic, or radiation treatment. Only cells in the growth fraction can be destroyed by chemotherapy. After each treatment, recruitment has to be waited for, until the growth fraction is restored for cells of the G0 phase. Only at this stage is a further application or treatment with cytostatic drug reasonable. This general treatment schedule is complicated by the fact that various drugs are only active in selected phases of the cell cycle (e.g., G1, G2 phase, or mitosis).

Even if it could be evaluated that an active cytostatic drug destroyed 30% of all proliferating cells of the adenocarcinoma, the proliferation rate of the tumor could only be reduced from 3% to 2%. A partial remission after treatment could not be observed, as the cell loss rate is merely 2%. The only expected change would be an increased tumor doubling time of 100 days.

The aim of the therapy is to prolong the life of the patient. This can only be achieved by a complete remission. So far this has not been possible in prostatic carcinoma. A partial remission – the reduction of the tumor by more than 50% – does not increase survival time.

Table 1. Doubling time, proliferation rate, and cell loss rate in embryonic carcinoma and adenocarcinoma. (Tubiana and Malaise [69])

	Embryonic carcinoma	Adeno-carcinoma
Doubling time (days)	27	83
Proliferation rate (%)	44	3
Cell loss rate (%)	42	2

Table 2. Results of prospective phase II studies treating bidimensionally measurable metastases in patients with hormone-resistant prostatic carcinoma. (Logothetis et al. [39]; Jones et al. [26–28]; Ghosn et al. [20]; Graf-Dobberstein et al. [21])

Substance	No. of patients	Partial response (%)
ADM + 5-FU	9 Ml s PUL	88
+MMC	12 Ml m	33
VDS	27	19
MMC	31	29
EPI	30	10
IFO	15	0
DDP/MMC	20 (N2–3)	41

ADM, adriamycin; 5-FU, 5-fluorouracil; MMC, mitomycin; VDS, vindesine; EPI, epirubicin; IFO, ifosfamide; DDP, cisplatinum; s PUL, solitary pulmonal metastases; m, multiple metastases, N2–3, lymph node metastases with a diameter of more than 2 cm.

Clinical Aspects

Regarding the above-mentioned cell kinetic aspects, partial remission of a prostatic carcinoma cannot be expected after systemic chemotherapy. This, however, was not the case in numerous clinical phase II studies which tested the efficacy of various agents treating bidimensionally measurable soft tissue metastases (Table 2 and Fig. 2).

If the efficacy of chemotherapy is accepted from this clinical point of view, it should be compared with other treatment schedules in the light of treatment results and side effects.

Antiandrogens, antiprolactins, aminoglutethimide and other substances have not been proven effective. An objective response was obtained in less than 20% of the patients. Even short-term therapy with high dose fosfestrol only leads to a remission of pain in 22%–33% of patients.

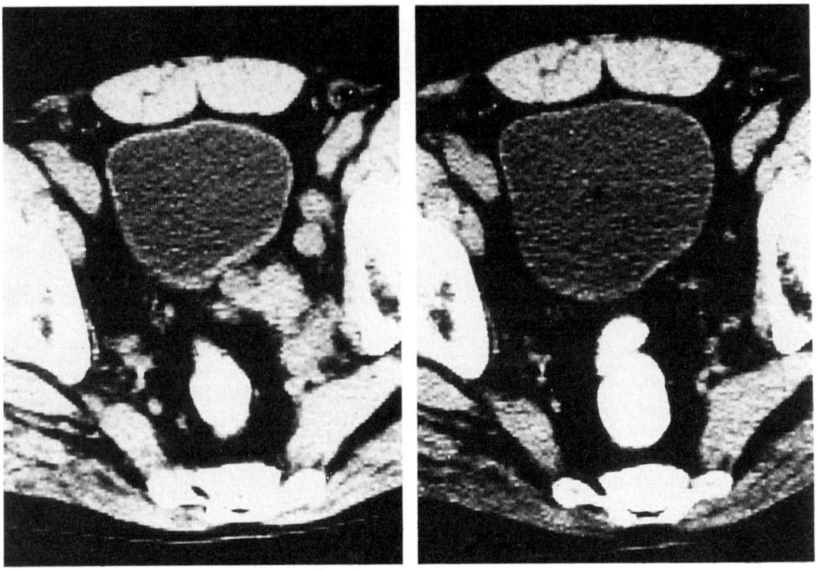

Fig. 2. Partial remission of lymph node metastases in a 61-year-old patient with a hormone-resistant prostate cancer after treatment by cisplatinum and mitomycin C

Table 3. Treatment of the hormone-resistant prostatic carcinoma; protocol 100 (NPCP)

Therapy	No. of patients	Tumor progression rate (%)
Standard (endocrine)	36	81
5-Fluorouracil	33	64
Cyclophosphamide	41	54

A prospective randomized study of the National Prostate Cancer Project (NPCP) demonstrated that in cases of tumor progression under endocrine treatment a readjustment of this endocrine treatment results in a progression rate of about 80% compared with a progression rate of about 60% after chemotherapy (Table 3).

This study indicates that it is reasonable to perform chemotherapy instead of readjusting endocrine treatment.

In large numbers of phase II and phase III studies the effectiveness of cytostatic drugs has been tested for treatment of hormone-resistant prostate cancer. Drugs resulting in a tumor progression rate of less than 75% are listed in Table 4. A partial remission is expected in 20%–30% of the cases.

The overall response rates ranged from 11% after etoposide to 50% after 5-fluorouracil administration; according to the NPCP criteria, stable disease

Table 4. Treatment of hormone-resistant prostate cancer; effectiveness of monochemotherapy regarding tumor progression rate and partial response. (Beckley et al. [3]; De Wys et al. [15]; Schmidt et al. [57,58]; Yagoda [71]; Murphy et al. [48]; Rübben and Altwein [56]; O'Bryan et al. [50,51]; Torti et al. [66]; Yagoda et al. [72]; Merrin [43]; Rossof et al. [54]; Qazi and Khandekar [53]; Moore et al. [45,46]; Mittleman et al. [44]; Kuss et al. [33]; Ansfield et al. [2])

Substance	Tumor progression rate		Partial response	
	No. of patients	(%)	No. of patients	(%)
5-Fluorouracil	95	58–64	14	28
Cyclophosphamide	136	62–66	–	–
Adriamycin	93	73	49	22
Cisplatinum	171	47–64	117	21
Estramustine	206	42–66	59	20
Dacarbazine	88	61–73	–	–
Methotrexate	58	59	–	–
Mitomycin	20	69	–	–
4 Epirubicin	33	49	–	–

Table 5. Response rates after monochemotherapy of hormone-resistant prostate carcinoma. (Eisenberger [18])

Substance	No. of patients	Response rate	
		%	(Range)
Cisplatinum	117	27	(0–44)
Doxorubicin	139	31	(14–84)
Estramustine	398	30	(20–74)
Etoposide	28	11	(4–40)
5-Fluorouracil	24	50	(14–100)
Epirubicin	253	41	(0–51)

for 12 weeks was sufficient to be regarded as a response. The tremendous differences between the data of various authors for the same drugs are remarkable ([18]; Table 5).

The variation of data may be explained by different doses and intervals between application as well as by different response criteria. The ideal interval between applications and the dose-response relation are not known for most of the cytostatic agents.

Some authors try to increase the efficacy of treatment by combining effective substances. A selection of clinical studies is shown in Table 6.

These results encouraged the belief that better response rates could be achieved with combination chemotherapy than with monochemotherapy. However, in prospective randomized studies this hope was not confirmed. Not one of the above-listed drugs proved the superiority of combination

Table 6. Results of combination chemotherapy of hormone-resistant prostatic carcinoma regarding tumor progression and partial response. (Merrin et al. [42]; Ihde et al. [23]; Izbicki et al. [25]; Lloyd et al. [37]; Soloway et al. [62]; Smalley et al. [61]; Al Sarraf [1]; Straus et al. [65]; Berry and MacDonald [6]; Logothetis et al. [38]; Kennealey et al. [31]; Paulson et al. [52]; Rübben [55]; Eagan et al. [16])

Substance	Tumor progression rate		Patial response	
	No. of patients	(%)	No. of patients	(%)
CTX + 5-FU	13	35	13	8
CTX + ADM	22	50	93	13
CTX + ADM + 5-FU	29	31	–	–
CTX + ADM + DDP	16	37	–	–
CTX + ADM + MTX	12	25	–	–
CTX + DDP + Pred	22	50	22	0
ADM + 5-FU + MMC	32	38	–	–
ADM + 5-FU + MMC	24	42	–	–
EST + 5-FU	25	12	25	0
MEL + MTX + 5-FU + VC + Pred	84	61	84	4

CTX, cyclophosphamide; 5-FU, 5-fluorouracil; ADM, adriamycin; DDP, cisplatinum; MTX, methotrexate; Pred, prednisone; MMC, mitomycin; EST, estramustine; MEL, melphalan; VC, vincristine.

chemotherapy over monochemotherapy. The following combinations have been compared with monotherapy (Table 7).

The substance of choice is presently not known. Monochemotherapy should be performed in preference to combination chemotherapy; however, when choosing a substance the side effects and concomitant diseases of the patient should be taken into consideration (Table 8).

A prospective phase II study in our clinic to prove the efficacy of mitomycin 10 mg/m^2 and prednisolone 250 mg at day 1 and cisplatinum 70 mg/m^2 at day 2 given in three cycles with a 28-day interval demonstrates the difficulty in interpreting the results of chemotherapy in hormone-resistant prostate cancer. This phase II study included 20 patients with bone metastases and 20 patients with additional bidimensionally measurable lymph node metastases.

Of the patients with bone metastases, 19% showed a reduction in the number of metastases documented by bone scan. Of the patients with lymph node metastases, 41% showed a partial response, i.e., a reduction of the tumor-infiltrated areas by more than 50% as documented by computed tomography (Table 9).

A total of 70% of the patients died from the tumor within 8 months. The subjective remission rate is remarkable: Improvement of pain score according to WHO criteria could be documented in 69% of patients (Table 10).

Table 7. Results of randomized trials comparing mono- and polychemotherapy in patients with hormone-resistant prostatic carcinoma. (Soloway et al. [63,62]; Smalley et al. [61]; Chlebowski et al. [11]; Muss et al. [49]; Torti et al. [67]; Kasimis et al. [29])

Substance	No. of patients	Partial response (%)	Overall response (%)
EST vs	38	3	18
VC vs	42	3	12
EST + VC	41	0	17
EST vs	50	0	14
DDP vs	51	0	18
EST + DDP	48	0	29
5-FU vs	49	4	14
CTX + ADM + 5-FU	52	4	12
CTX vs	15	0	53
CTX + ADM + 5-FU	12	0	50
CTX vs	17	0	53
CTX + MTX + 5-FU	15	7	53
ADM vs	20	5	45
ADM + DDP	17	12	65
CTX vs	16	0	56
5-FU + ADM + MMC	15	7	40

EST, estramusrine; VC, vinctistine; DDP, cisplatinum; 5-FU, 5-fluorouracil; CTX, cyclophosphamide; ADM, adriamycin; MTX, methotrexate; MMC, mitomycin.

Table 8. Side effects of a selection of cytostatic agents

Substances	Toxicity										
	Alopecia	Anaphylactic reaction	Cardial	Dermatologic	Gastrointestinal	Hepatic	Myelodepression	Neurologic	Pulmonary	Renal	Ototoxic
Doxorubicin	+	(+)	+	(+)	+		+				
4 Epirubicin	+	(+)	(+)	(+)	+		+				
Cisplatinum							(+)	+	(+)	+	+
Cyclophosphamide	+		(+)	(+)	+		+			+	
Dacarbazine	(+)	(+)			+	(+)	(+)	(+)			
Fluorouracil	(+)			(+)	+		+	(+)			
Methotrexate	(+)			(+)	+		+	(+)	(+)	(+)	
Mitomycin				(+)	+		+		(+)	(+)	
Estramustine			(+)		(+)						

Parentheses indicate less pronounced side effect.

Table 9. Partial remission, duration of response, and progression of patients with hormone-resistant prostate cancer treated with mitomycin, prednisolone, and cisplatinum

	Partial remission (%)	Duration of response (months)	Tumor progression rate (%)
M1 oss	19	7	63
N3	41	7	47
Total	30	7	55

Table 10. Subjective response (pain score) according to WHO criteria: difference ≥1 grade

69%:	Improvement
7%:	No change
24%:	Deterioration

In our study, even the side effects were acceptable; serious side effects were only observed in bone marrow depression, but this could be treated by blood transfusion (see Table 11).

This evaluation does not reflect the fact that these patients are not treated by systemic chemotherapy alone, but – depending on concomitant observations – also receive blood transfusions, excellent adjustment of pain medication, psychological treatment, and physical therapy. Viewed in this light, the disappointing results of the European Organization for Research and Treatment of Cancer (EORTC) are more easily understood; comparing the pre- and posttreatment period, pain score, fatigue, and nausea are not improved by systemic chemotherapy with mitomycin or estramustine [19]. The results clearly show that the interpretation of data concerning systemic chemotherapy should be a very careful one, especially regarding the subjective response of the patient.

Selection Criteria

Chemotherapy of the hormone-resistant prostate cancer is evaluated in only a very small number of patients. Subjective and objective remission is seldom, duration of response is limited, and treatment is not without side effects. For this reason, several authors have tried to identify those patients who will gain from the therapy and thereby avoid unnecessary treatment to other patients (Scott et al. [59] and Edsmyr et al. [17] demonstrated that

Table 11. Side effects according to WHO grade[a]

	WHO grade		
	1 (%)	2 (%)	3 (%)
Gastrointestinal tract	50	40	–
Kidneys	10	3	–
Liver	2	2	–
Erythrocytes	10	7	13
Leukocytes	12	9	4
Extravasation	–	–	6

[a] Cessation of therapy: 11%.

systemic chemotherapy is not superior to endocrine treatment in patients with well-differentiated prostatic carcinoma). Clinical data show that polydifferentiated tumors respond as frequently to endocrine treatment as do well-differentiated tumors; unfortunately, one cannot conclude from this data that polydifferentiated tumors respond favorably to chemotherapy. (This group of patients showed a poor prognosis anyway.) Most metastatic prostate cancers are poorly differentiated; therefore differentiation is not a helpful parameter in selecting a therapy. In the same way parameters such as endogenous receptors, immunohistochemical staining, cytophotometry, and cytological regression rate have not helped in the selection of patients who will profit from chemotherapy [8,9,35,60,68].

More important than the above-mentioned parameters are the general condition and pain score of the patient. Patients in a more or less good general condition and without pain have a favorable prognosis [48]. The Berry parameter [7] defines those patients with parameters for lactale dehydrogenase (LDH) >200 u/l, SGOT >50 u/l, and serumalbumin <3.5 g/dl as having a favorable prognosis. Prostatic phosphatase <5 u/l and prolactine <10 ng/dl defines patients with a very short survival time – in those patients systemic chemotherapy may not be indicated. But the most helpful criterion for evaluation of efficacy is the therapy itself. If the therapy does not result in a subjective or objective remission within 6 weeks it should be stopped [23,54].

Alternative Therapeutic Modalities

Besides chemotherapy, more recent methods have been be analysed for efficacy in hormone-resistant cancer. Local radiotherapy of isolated symptomatic bone metastasis shows good subjective response rates with a duration of about 6 months. Interstitial radiotherapy is of limited response duration (Table 12).

Table 12. Radiotherapy for hormone-resistant prostate cancer – subjective response rate (pain) and its duration. (Benson et al. [4]; Keen [30]; e.c. Rathert, personal communication, 1988)

Therapy		No. of patients	Subjective response rate (%)	Duration (months)
Local radiotherapy	30 gy	62	77	6
Half body irradiation	8.9 gy	58	72	4
89 Strontium		14	79	3

Table 13. New treatment modalities for patients with hormone-resistant prostate cancer: objective response und subjective response. (Witjes et al. [70]; Mahler et al. [41]; Labrie et al. [34]; Craig and Crawford [12]; Chang et al. [10]; Denis et al. [13]; De Smedt et al. [14]; Linehan et al. [36])

Therapy	No. of patients	Objective response (%)	Subjective response (%)
Ketoconazole	28	ND	50
Fosfestrole	24	0	33
Aminoglutethimide + hydrocortisone	119	3	ND
Trimetrexate	31	17	ND
Amonafide	17	12	ND
Imidazole derivatives	17	53	12
Suramin	24	46	ND
Diphosphonates	17	ND	94
Hypophysectomy	100	ND	71

ND, no data.

Other modalities include hormone manipulation and the evaluation of newer chemotherapeutic substances. Imidazole derivatives showed objective partial response in 53% of 17 patients, suramin in 46% of 24 patients. Objective response rates achieved by other methods are disappointing. When reading the subjective response all given data should be studied very carefully (Table 13).

Open Questions

The very limited results in systemic chemotherapy of hormone-resistant prostate cancer open a bundle of questions.

Tx	R	mitomycin C: 10 mg/m^2 for 4 weeks
N3	a	
M1	n	epirubicin: 25 mg/m^2 for 1 week
hormone	d	
resistant,	o	
symptomatic	m	estramustine: 560 mg/patient per day

Fig. 3. Scheme of the prospective randomized phase II study in hormone-resistant prostate cancer

Table 14. Hormone-resistant prostate cancer. (Rübben [55])

Substance	No. of patients	Tumor progression		
		6 Months (%)	12 Months (%)	24 Months (%)
FAM	24	42	66	100
EST[a]	26	32	76	88
EST[b]	22	54	84	100

FAM, 5-fluorouracil + adriamycin + mitomycin; EST, estramustine.
[a] Intermittent application.
[b] Permanent application.

Which substance is most effective with special regard to the quality of life of the patient? Concerning this question the urooncological research group is presently performing a randomized phase II study comparing mitomycin C, epirubicin, and estramustine (Fig. 3).

Is the permanent application of cytostatic drugs superior to cytostatic treatment only at the time of symptomatic disease?

A randomized phase II study with 72 patients offered the conclusion that characteristic differences are not to be expected if estramustine is applied only at the time of painful disease compared to permanent application (Table 14).

This question will be answered in a prospective randomized phase II study (Fig. 4).

Indication for Chemotherapy at an Earlier Stage of Disease

The data shown in this review has considered systemic chemotherapy in patients with hormone-resistant prostate cancer. It should, however, be considered whether initiation of chemotherapy at an earlier stage of disease could improve treatment results. In an NPCP study, patients with stable disease under hormone therapy were treated with estramustine phosphate

Tx	R	– epirubicin continously
N3	a	25 mg/m^2 per week
M1	n	
hormone	d	– epirubicin
resistant,	o	25 mg/m^2 per week ×8
symptomatic	m	repeated at progress

End of therapy if progress resistant to epirubicin

Fig. 4. Scheme of a prospective randomized phase II study for treatment of hormone-resistant prostate cancer

Table 15. Treatment of stable prostatic carcinoma under hormonal therapy; efficacy of hormone therapy in comparison to chemohormonal combination therapy (NPCP, protocol 600; [48])

Substance	No. of patients	Tumor progression rate (%)
DES	52	33
DES + CTX	49	20
DES + EST	60	27

DES, diethylstilbestrol; CTX, cyclophosphamide; EST, estramustine.

or a cyclophosphamide. The results showed no benefits from additional chemotherapy (Table 15; [48]).

In 1984 Isaacs [24] had already demonstrated in animal experiments that chemohormonal combination therapy is beneficial in untreated prostatic carcinoma. He explained his experimental results by the observation that the progression of a prostatic carcinoma is not due to adaptation of the tumor to a given therapy but to clonal selection. That means that already at the beginning of tumor growth hormone-sensitive and hormone-resistant tumor clones are present and that resistant clones are not sufficiently affected by hormonal manipulation (Fig. 5).

Clinical studies of Logothetis et al. [40] may have confirmed this experimental data: The combination of endogenous deprivation with adriamycin, mitomycin, and cisplatinum showed a survival time of 4.4 years free of progression in 32% of 28 patients. The authors considered these results better than those obtained by historical control groups. But phase II studies can not document the superiority of a treatment over standard or control therapy and, in view of this, the data of Logothetis should not be cited. No published phase III study has documented any advantage for chemohormonal combination therapy (Table 16).

The study of Benson [5], however, must be considered in more detail. The progression rate after application of estramustine was superior by 15%

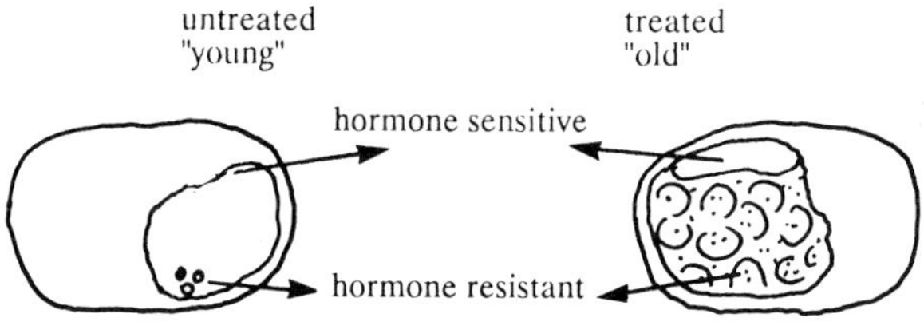

Fig. 5. Progression of prostatic carcinoma under endocrine therapy might be due to clonal selection and not to adaptation of the tumor to therapeutic modalities. (Animal experiments of Issacs [24])

Table 16. Treatment of untreated metastatic prostatic carcinoma: efficacy of hormonal therapy compared to chemohormonal combination therapy regarding tumor progression

Substance	No. of patients	Tumor progression rate (%)	Author
FLU vs	101	69	de Kernion et al. [12a]
EST	102	62	
ZOL vs	131		Kühn and Weißbach [32]
EST 560	122		
DES vs	83	19	
CTX + DES vs	77	12	NPCP 500
CTX + EST	86	17	
DES/orch. vs	88	20	
EST vs	97	22	NPCP 1300
CTX, 5-FU, DES	90	19	
DES vs		85	Benson et al. [5]
EST 900		70	

FLU, flutamide; EST, estramustine; ZOL, zoladex; DES, diethylstil-bestrol; CTX, cyclophosphamide; orch., orchiectomy; 5-FU, 5-fluorouracil.

over treatment by diethylstilbestrol 36 months after initiation of therapy. This observation is cited frequently as evidence of the efficacy of chemohormonal combination therapy. However, interpretation of the data can also lead to a controversial conclusion. The survival rates of both treatment groups were no different after 36 months, indicating that 70% of the patients treated initially by estramustine phosphate were overtreated. If a progression under DES therapy is observed, administration of estramustine can be initiated at this time of disease and is equieffective to the primary administration of chemotherapy (Table 17).

Table 17. Treatment of untreated metastatic prostatic carcinoma by diethylstilbestrol or estramustine phosphate – progression and survival rates after 36 months. (Benson et al. [5])

Substance	No. of patients	Progression rate (%)	Survival rate (%)
Diethylstilbestrol	79	70	39
Estramustine phosphate	78	85	40

Conclusion

The standard therapy of the untreated metastatic prostate cancer is androgen suppression. To date, there are no conclusive data that systemic chemotherapy is very helpful when administered at this stage of disease. Thus systemic chemotherapy is indicated only at the stage of hormone-resistant and progressive prostatic carcinoma. Chemotherapeutic agents which show a subjective or objective remission include 5-fluorouracil, cyclophosphamide, cisplatinum, 4-epirubicin, mitomycin, or estramustine phosphate. Combination chemotherapy is not superior to the application of one drug only. Selection criteria to define patients who will gain from chemotherapy are lacking. If the tumor does not respond to treatment within the first 6 weeks of treatment, the therapy should be stopped. There is an urgent need for evaluation of new therapeutic substances which not only influence proliferating cell population but also increase cell loss in untreated or progressive prostatic carcinoma.

References

1. Al-Sarraf M (1980) Combination of cytoxan adriamycin and cis-platinum (CAP) in patients with advanced prostatic cancer. Proc Am Assoc Cancer Res Am Soc Clin Oncol 21:198
2. Ansfield FJ, Schroeder J, Curreri AR (1962) Five years clinical experience with 5-fluorouracil. JAMA 181:295
3. Beckley S, Wajsman Z, Slack N, Mittelman A, Murphy G (1980) The chemotherapy of prostatic carcinoma. Scand J Urol Nephrol Suppl 55:151
4. Benson RC Jr, Gill GM, Cummings KB (1983) A randomized double blind crossover trial of diethylstilbestrol (DES) and estramustine phosphate (Emcyt) for stage D prostatic carcinoma Semin Oncol 10(3) [Suppl 3]:43
5. Benson RC Jr, Gill GM (1986) Estramustine phosphate compared with diethylstilbestrol. Am J Clin Oncol 9(4):341
6. Berry J, MacDonald RN (1982) Cisplatin, cyclophosphamide and prednisolone therapy for stage D prostatic cancer. Cancer Treat Rep 66:1403
7. Berry WR, Laszlo J, Cox E, Walker N, Paulson D (1979) Prognostic factors in metastatic and hormonally unresponsive carcinoma of the prostate. Cancer 44:763

8. Bichler K-H, Flüchter SH (1983) Kriterien in der Progression des Prostatakarzinoms. In: Faul P, Altwein J (eds) Aktuelle Diagnostik und Therapie des Prostatakarzinoms. Erasmus, Santa Barbara, p 213
9. Buttyan R, Olsson CA (1984) Androgen receptor assays in advanced prostatic cancer. Urol Clin North Am 11:311
10. Chang AY, Bennet JM, Pandya KJ, Asbury R, McCune C (1989) A study of aminoglutethimide and hydrocortisone in patients with advanced and refractory prostate carcinoma. Am J Clin Oncol 12(4):358
11. Chlebowski RT, Hestorff R, Sardoff L, Weiner J, Bateman JR (1978) Cyclophosphamide (NSC 26271) versus the combination of adriamycin (NSC 123127), 5-fluorouracil (NSC 19893), and clyclophosphamide in the treatment of metastatic prostatic cancer. A randomized trial. Cancer 42:2546
12. Craig J, Crawford E (1989) Phase II trial of amonafide in advanced prostate cancer: a Southwest Oncology Group Study. Proc Annu Meet Am Soc Clin Oncol 8:A573 (abstr)
12a. deKernion JN, Murphy GP, Priore R (1988) Comparison of flutamide and emcytin hormone-refractory metastatic prostatic cancer. Urology 31:312–317
13. Denis L, Mahler C, Ongena P, Janssen P, De Coster R, Bruynseels J (1990) A phase I/II study with R75251, a novel imidazole derivative in M1 prostate cancer patients in relapse. 9th Congress of the European Association of Urology (EAU), 13–16 June 1990, Amsterdam, p 187 (abstr no 98)
14. De Smedt E, Denis L, Mahler C, Van Oosterom A, Bruyn E (1990) Suramin: a phase I/II study in patients with prostatic cancer in relapse. 9th Congress of the European Association of Urology (EAU), 13–16 June 1990, Amsterdam, p 379 (abstr no 196)
15. De Wys WD, Bauer M, Colsky J, Cooper RA, Creeck R, Carbone PP (1977) Comparative trial of adriamycin and 5-fluorouracil in advanced prostatic cancer – progress report. Cancer Treat Rep 61:325
16. Eagan RT, Hahn RG, Myers RP (1976) Adriamycin (NSC 127127) versus 5-fluorouracil (NSC 19893) and cyclophosphamide (NSC 26271) in the treatment of metastatic prostate cancer. Cancer Treat Rep 60:115
17. Edsmyr F, Adersson L, Könyves I (1982) Estramustine-phosphate: experimental studies and clinical experience. In: Jacobi GH, Hohenfellner R (eds) Prostate cancer. Williams and Wilkins, Baltimore
18. Elsenberger MA (1988) Chemotherapy for prostate carcinoma. NCI Monogr 7:151
19. Fossa SD, Aaronson NK, Newling D and the members of the EORTC Genitourinary Group (1990) Advanced hormone resistant prostatic cancer: preliminary observations on subjective morbidity and palliation. 9th Congress of the European Association of Urology (EAU), 13–16 June 1990, Amsterdam, p 4 (abstr no 2)
20. Ghosn M, Droz JP, Mahjoubi M, Theodore C, Azab M, Ostronoff M (1989) Phase II trial of ifosfamide (IFO) in hormone-refractory metastatic cancer of the prostate (HRMCP). Proc Annu Meet Am Soc Clin Oncol 8:A549 (abstr)
21. Graf-Dobberstein C, Rübben H, Deutz FJ, Otto U, Block T, Arregui R (1989) Prospektive multizentrische Phase II Studie zur Wirksamkeit von Cisplatin und Mitomycin in der Therapie des hormonrefraktären Prostatakarzinoms. 41th Congress of the German Society for Urology, 4–7 Oct 1989, Freiburg. Book of abstracts, p 184
22. Huggins C, Hodges CV (1941) Studies on prostatic cancer: I. The effects of castration, of estrogen, and of androgen injection on serum phosphatases in metastatic carcinoma of the prostate. Cancer Res 1:293

23. Ihde CD, Brunn PA, Cohen MH, Dunnick NR, Eddy JL, Minna JD (1980) Effective treatment of hormonally-unresponsive metastatic carcinoma of the prostate with adriamycin and cyclophosphamide. Cancer 45:1300
24. Isaacs JT (1984) The timing of androgen ablation therapy and/or chemotherapy in the treatment of prostatic cancer. Prostate 5:1
25. Izbicki RM, Amer MH, Al-Sarraf M (1979) Combination of adriamycin and cyclophosphamide in the treatment of metastatic prostatic carcinoma. Cancer Treat Rep 63:999
26. Jones WG, Fossa SD, Denis L, Coninx P, Glashan RW, Akdas A, De Pauw M (1983) An EORTC phase II study of vindesine in advanced prostate cancer. Eur J Cancer Clin Oncol 19:583
27. Jones WG, Bono AV, Verbaeys A, De Pauw M, Sylvester R, and members of the EORTC (1986) Can the primary tumour be used as the sole parameter for response in phase II chemotherapy studies in metastatic prostate cancer? An EORTC genito-urinary group report. World J Urol 4:176
28. Jones WG, Fossa SD, Bono AV, Croles JJ, Stoter G, De Pauw M, Sylvester R (1986) Mitomycin C in the treatment of metastatic prostate cancer: report of an EORTC phase II study. World J Urol 4:182
29. Kasimis B, Miller JB, Kaneshiro CA et al. (1985) Cyclophosphamide versus 5-fluorouracil, doxorubicin and mitomycin C (FAM) in the treatment of hormone resistant metastatic carcinoma of the prostate: a preliminary report of a randomized trial. J Clin Oncol 3:385
30. Keen CW (1980) Half body radiotherapy in the management of metastatic carcinoma of the prostate. J Urol 123:713
31. Kennealey GT, Marsh JC, Walsh DA et al. (1978) Treatment of advanced carcinoma of the prostate with estramustine and 5-fluorouracil (FU). Proc Am Assoc Cancer Res Am Soc Clin Oncol 19:394
32. Kühn M, Weißbach L (1990) Primary therapy of metastasized prostate carcinoma with depot GnRH analogue (Zoladex) versus estramustine phosphate (EMP; Estrazyt). 9th Congress of the European Association of Urology (EAU), 13–16 June 1990, Amsterdam, p 254 (abstr no 132)
33. Kuss R, Khoury S, Richard F et al. (1980) Estramustine phosphate in the treatment of advanced prostatic cancer. Br J Urol 52:29
34. Labrie F, Dupont A, Bélanger A, Cusan L, Brochu M, Turina E, Pinault S, Lacourciere Y, Emond J (1989) Anti-hormone treatment for prostate cancer relapsing after treatment with flutamide and castration – addition of aminoglutethimide and low dose hydrocortisone to combination therapy. Br J Urol 63:634
35. Leistenschneider W, Nagel R (1980) Zytologisches Regressionsgrading und seine prognostische Bedeutung beim konservativ behandelten Prostatakarzinom. Acta Urol 11:236
36. Linehan WW, LaRocca R, Stein CY, McClellan W, Cooper M, Weiss G, Choyke P, Cassidy J, Ulrich M, Myers CE (1990) Use of suramin in treatment of patients with advanced prostate carcinoma. 58th Annual Meeting of the American Urology Association (AUA) 13–17 May 1990, New Orleans, p 131 (abstr no 220)
37. Lloyd RE, Jones SE, Salmon SE et al. (1976) Combination chemotherapy with adriamycin (NSC-123127) and cyclophosphamide (NSC-26271) for solid tumors: a phase II trial. Cancer Treat Rep 60:77
38. Logethetis CJ, Von Eschenbach AC, Samuels ML, Trindade A, Johnson DE (1982) Doxorubicin, Mitomycin and 5 FU (DMF) in the treatment of hormone resistant stage D prostate cancer: a preliminary report. Cancer Treat Rep 66: 57

39. Logothetis CJ, Samuels ML, Von Eschenbach AC, Trindade A, Ogden S, Grant C, Johnson DE (1983) Doxorubicin, Mitomycin-C, and 5-Fluorouracil (DMF) in the treatment of metastatic hormone refractory adenocarcinoma of the prostate, with a note on the staging of metastatic prostate cancer. J Clin Oncol 1:368
40. Logothetis CJ, Hossan E, Dexeus FH, Sella A, Amato RJ, Kilbourn RG (1990) Primary combined chemotherapy and hormonal therapy for the treatment of patients with advanced metastatic adenocarcinoma of the prostate. 58th Annual Meeting of the American Urology Association (AUA) 13–17 May 1990, New Orleans, p 481 (abstr no 309)
41. Mahler C, Keuppens F, Nowé P, Van Erps, Denis L (1988) Prospective study of diethylstilbestrol-diphosphate in the treatment of prostatic cancer in relapse. Urol Int 43(1):47
42. Merrin C, Etra W, Wajsman Z, Baumgartner G, Murphy G (1976) Chemotherapy of advanced carcinoma of the prostate with 5-fluorouracil, cyclophosphamide and adriamycin. J Urol 115:86
43. Merrin CE (1979) Treatment of genitourinary tumors with cisdichlorodiammineplatinum (II): experience in 250 patients. Cancer Treat Rep 63:1579
44. Mittleman A, Shukla SK, Murphy GP (1976) Extended therapy of stage D carcinoma of the prostate with oral estramustine phosphate. J Urol 115:409
45. Moore GE, Bross IDJ, Ausman R et al. (1968) Effects of 5-fluorouracil (NSC-19893) in patients with cancer. Eastern Clinical Drug Evaluation Programm. Cancer Chemother Rep 52:641
46. Moore MR, Troner MB, Desimone P et al. (1986) Phase II evaluation of weekly cisplatin in metastatic hormone resistant prostate cancer: a Southeastern Cancer Study Group trial. Cancer Treat Rep 70:541
47. Murphy G (1978) Management of advanced cancer of the prostate. In: Skinner DF, De Kernion JB (eds) Genitourinary cancer. Saunders, Philadelphia
48. Murphy GP, Slack NH, and participants in the National Prostatic Cancer Project (1984) Current status of the National Prostatic Cancer Project treatment protocols. In: Denis L, Murphy GP, Prout GR, Schröder F (eds) Controlled clinical trial in urologic oncology. Raven, New York, p 13
49. Muss HB, Howard V, Richards F, White DR, Jackson DV, Cooper MR, Stuart JJ, Resnick MI, Brodkin R, Spurr CL (1981) Cyclophosphamide versus cyclophosphamide, methotrexate, and 5-fluorouracil in advanced prostatic cancer a randomized trial. Cancer 47:1949
50. O'Bryan RM, Luce JK, Talley RW, et al. (1973) Phase II evaluation of adriamycin in human neoplasia. Cancer 32:1
51. O'Bryan RM, Baker LH, Gottlieb JF et al. (1977) Dose-response evaluation of adriamycin in human neoplasia. Cancer 39:1940
52. Paulson DF, Berry WR, Cox EB et al. (1979) Treatment of metastatic endocrine-unresponsive carcinoma of the prostate gland with multiagent chemotherapy: indicators of response to therapy. J Natl Cancer Inst 63:615
53. Qazi R, Khandekar J (1983) Phase II study of cisplatin for metastatic prostatic carcinoma. An Eastern Cooperative Oncology Group study. Am J Clin Oncol 6:203
54. Rossof AM, Talley RW, Stephens RL, Thigpen T, Samson MK, Groppe C, Eyre HJ, Fisher R (1979) Phase II evaluation of cis-dichlorodiammineplatinum (II) in advanced malignancies of the genitourinary and gynecologic organs, Southwest Oncology Groups study. Cancer Treat Rep 63:1557
55. Rübben H (1986) Chemotherapie beim Prostatakarzinom. Onkologisches Forum für Chemotherapie – Schwerpunkt Prostatakarzinome. 2:11
56. Rübben H, Altwein JE (1987) Das fortgeschrittene Prostatakarzinom – Ein therapeutisches Dilemma? Urologe [A] 26:7

57. Schmidt JD, Gibbons RP, Johnson DE, Prout GR, Scott WW, Murphy GP (1976) Chemotherapy of advanced prostatic cancer. Urology 6:602
58. Schmidt JD, Scott WW, Gibbons RP, Johnson DE, Prout GR, Loening SA, Soloway MS, Chu TM, Gaeta GF, Slack NH, Saroff J, Murphy GF (1979) Comparison of procarbazine, imidazole-carboxamide and cyclophosphamide in relapsing patients with advanced carcinoma of prostate. J Urol 121:185
59. Scott WW, Gibbons RP, Johnson DE, Prout GR, Schmidt JD, Saroff J, Murphy GP (1976) The continued evaluation of the effects of chemotherapy in patients with advanced carcinoma of the prostate. J Urol 116:211
60. Seppelt U, Sprenger E (1980) Zellkern-DNS-Analysen durch Einzelzell-Fluoreszenz-Zytophotometrie an Prostatakarzinomen vor und während der Therapie. Verh Dtsch Ges Urol 32:68
61. Smalley RV, Bartolucci AA, Hemstreet G, Hester M (1981) A phase II evaluation of a 3-drug combination of cyclophosphamide, doxorubicin and 5-fluorouracil and or 5-fluorouracil in patients with advanced bladder carcinoma or stage D prostatic carcinoma. J Urol 125:191
62. Soloway MS, Shippel RM, Ikard M (1979) Cyclophosphamide, doxorubicin hydrochloride, and 5-fluorouracil in advanced carcinoma of the prostate. J Urol 122:637
63. Soloway MS, De Kernion JB, Gibbons RP, Johnson DE, Leoning SA, Pontes JE, Prout GR, Schmidt JD, Scott WW, Chu TM, Gaeta JF, Slack NH, Murphy GP (1981) Comparision of estramustine phosphate and vincristine alone or in combination for patients with advanced hormone refractory, previously irradiated carcinoma of the prostate. J Urol 125:664
64. Soloway MS, Beckley S, Brady MF et al. (1983) A comparison of estramustine phosphate, versus cis-platinum alone versus estramustine phosphate plus cis-platinum in patients with advanced hormone refractory prostate cancer who had extensive irradiation to the pelvis or lumbosacral area. J Urol 129:56
65. Straus MJ, Parmelee J, Olsson C, De Vere White R (1978) Cytoxan, adriamycin and methotrexate (CAM) therapy of stage D prostate cancer. Proc Am Assoc Cancer Res Am Soc Clin Oncol 19:314
66. Torti F, Aston D, Lum BL et al. (1983) Weekly doxorubicin in endocrine refractory carcinoma of the prostate. J Clin Oncol 1:477
67. Torti FM, Shortliffe L, Carter SK et al. (1985) A randomized study of doxorubicin versus doxorubicin plus cisplatin in endocrine-unresponsive metastatic prostatic carcinoma. Cancer 56:2580
68. Trachtenberg J, Walsh PC (1982) Correlation of prostatic nuclear androgen receptor content with duration of response and survival following hormonal therapy in advanced prostatic cancer. J Urol 127:466
69. Tubiana M, Malaise E (1976) Comparison of cell proliferation kinetics in human and experimental tumors: response to irradiation. Cancer Treat Rep 60(12):1887
70. Witjes FJ, Debruyne FM, Fernandez-del-Moral P, Geboers AD (1989) Ketoconazde high dose in management of hormonally pretreated patients with progressive metastatic prostate cancer. Dutch South-Eastern Urological Cooperative Group. Urology 33(5):411–415
71. Yagoda A (1979) Phase II trials with cis-diammine-dichloroplatinum (II) in the treatment of urothelial tumors. Cancer Treat Rep 63:1565
72. Yagoda A, Watson RC, Natale RB et al. (1979) A critical analysis of response criteria in patients with prostatic cancer treated with cis-diamminedichloride platinum II. Cancer 44:1553

Organ-Sparing Excision of Renal Cell Carcinoma

M. Marberger and R. Simak

Urologische Universitätsklinik, Alser Strasse 4, 1090 Wien, Austria

Introduction

Since Robson et al.'s pioneering work in 1969 [20], perifascial tumor nephrectomy with en bloc removal of kidney, tumor, fatty capsule, and adrenal is the accepted treatment of nonmetastasizing renal cell carcinoma (RCC); controversy only exists regarding the need for and extent of retroperitoneal lymph node dissection. Its results are well documented but with a lack of effective adjunctive treatment have not improved substantially in the last 20 years (Fig. 1; [10]).

The principle of radical nephrectomy with primary interruption of renal circulation prior to any manipulation of the tumor-bearing kidney has its limitations where the contralateral kidney is absent, predamaged, or also contains a tumor so that a radical approach results in renal insufficiency. At the other end of the spectrum, in the case of the minute peripheral tumor detected by chance in an asymptomatic patient, the need for nephrectomy may appear questionable in spite of a normal contralateral kidney. These tumors are detected far more frequently where ultrasonography and computerized tomography are freely available. Their natural history is unclear, and they are frequently difficult to differentiate from benign lesions. Organ-preserving tumor excision can therefore be employed for two different indications:

- An imperative indication of RCC in a solitary kidney or bilateral RCC
- An elective indication of small peripheral tumors of questionable malignancy in the presence of a normal contralateral kidney

RCC in the Solitary Kidney

This clinical entity is rare and accounts for less than 2% of patients in most large clinical series [9], but the figure is rising with the better resolution of

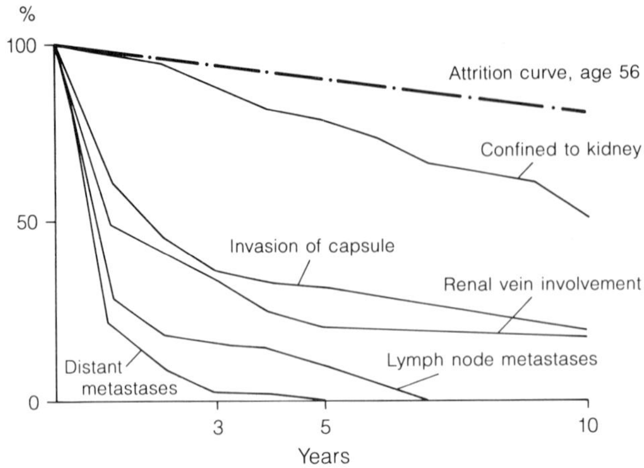

Fig. 1. Actuarial survival of 306 patients with RCC treated with radical nephrectomy. (From Marberger and Hohenfellner [10])

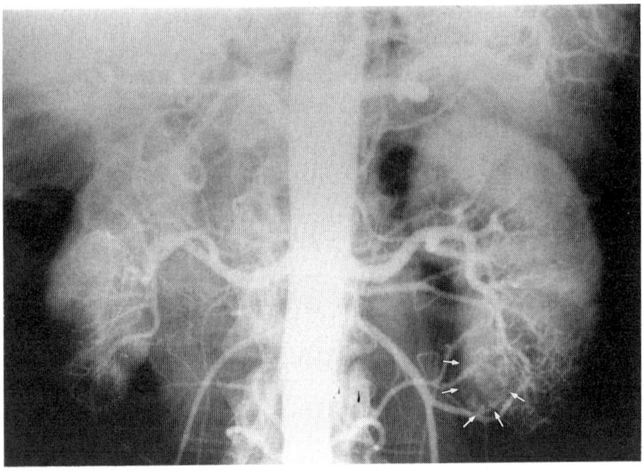

Fig. 2. Angiogram of patient evaluated for hypertension with contracted right kidney; the small RCC on the left lower pole (*arrows*) was diagnosed by chance

modern imaging techniques and their liberal use in all types of renal disease. Unilateral tumors in the presence of a diseased contralateral kidney must also be considered in this category, albeit the degree of renal damage of the remaining kidney may render the indication for conservative surgery elective (Fig. 2). If split radioisotope function studies signalize a contribution of less than 30% of the total renal function from this kidney, significant renal insufficiency is to be expected with contralateral tumor nephrectomy. Even with a lesser degree of renal damage of the nontumor-bearing kidney, conservative surgery should be aimed at with peripheral RCC, as most renal pathology has a natural tendency to deteriorate.

Although patients may occasionally survive without treatment for years [23], the disease usually rapidly leads to death of the patient [28]. Occasional long-term survivors have been reported after external beam radiation [8], chemotherapy [7], or hormonal treatment [9], but there can be no doubt that only surgical excision gives any chance of reliable cure.

As radical nephrectomy, followed by chronic hemodialysis renders the patient anephric, allotransplantation would appear the simplest solution without compromising the principles of surgical radicality. In our experience, patients subjected to acute renal insufficiency and hemodialysis suffer significant morbidity, and fatal complications seem to occur more frequently than in patients undergoing gradual renal insufficiency prior to admittance into a dialysis program [11]. Early renal allotransplantation would alleviate this problem, but the data provided by Penn [18] clearly suggests a waiting period of at least year: of 16 patients subjected to transplantation within a year after radical nephrectomy, 9 died of the tumor within an observation period of 6–59 months, whereas none of 11 patients transplanted after 12 months succumbed to RCC. Moreover, for patients subjected to chronic hemodialysis and/or renal allotransplantation where there is no malignant disease, the 5-year survival rate barely exceeds 58% [3,4].

Wherever enough renal parenchyma may be preserved to sustain an adequate renal function, organ-preserving tumor excision is clearly the treatment of choice, and radical nephrectomy is regarded as a last resort.

Provided there is no tumor involvement of the main branch of the renal vein or vena cava, partial nephrectomy in situ is feasible. The surgical technique is crucial for success. In the presence of any nonresectable metastases, the intervention is useless; the preoperative workup must therefore include axial computerized tomography of the abdomen and chest and a bone scan. With any larger tumor we routinely obtain a renal angiogram (Fig. 3A,B), and this is even supplemented by retrograde renal phlebography if there is any suggestion of renal vein involvement. Precise delineation of renal vascular anatomy is extremely helpful during partial nephrectomy, and in the context of renal tumors, provides a valid indication for obtaining renal angiography.

The patient is well hydrated prior to the operation to ideally have a urine output of >100 ml/h. With the transperitoneal approach we have experienced a case of severe intraperitoneal urine extravasation, which proved difficult to manage. Since the principles of perifascial tumor nephrectomy [20] cannot be maintained, we now routinely use an extended flank incision. Gerota's fascia is opened and the kidney exposed, carefully preserving the fatty capsule in the region immediately adjacent to the tumor (Fig. 3C). The structures of the renal pedicle are carefully dissected, in particular the renal artery with its main branches, the renal vein, and the ureter and snared with vessel loops. If a major branch can be identified that feeds predominantly

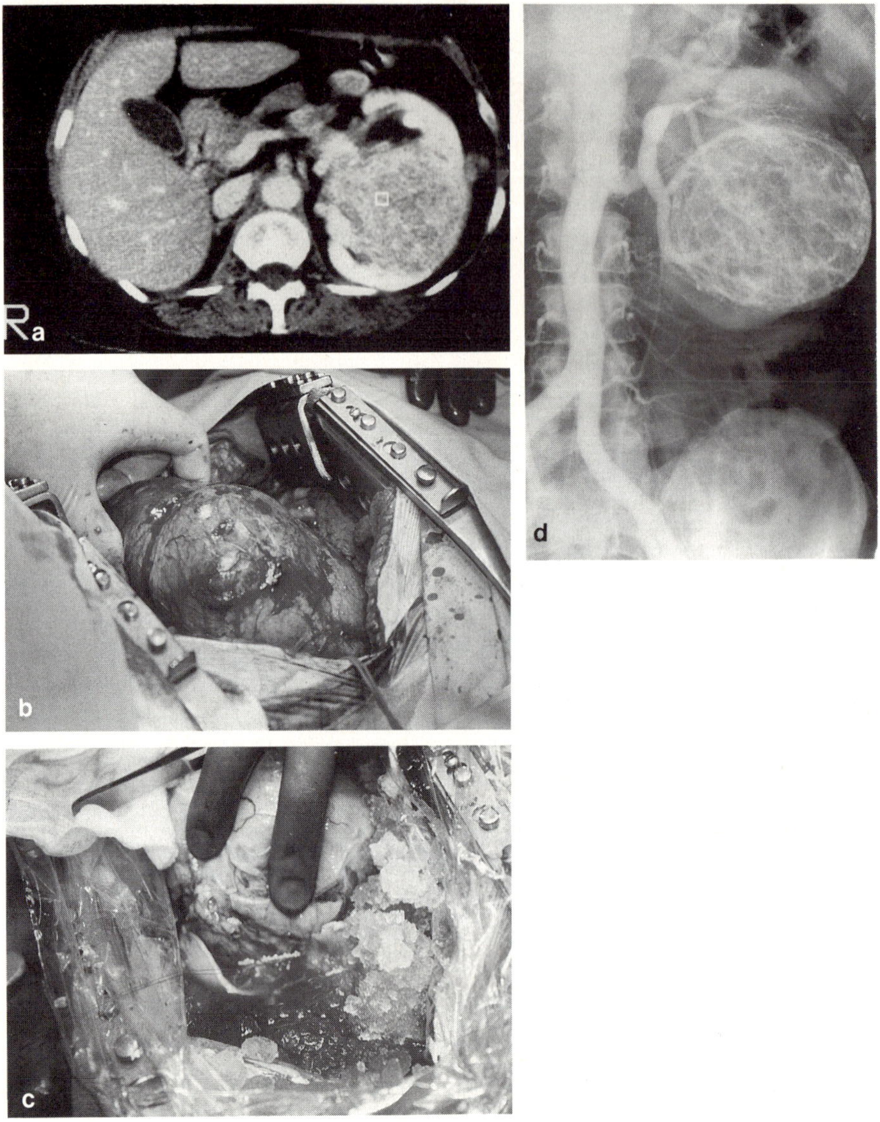

Fig. 3a–h. Surgical technique for tumor excision from solitary kidney. **a** Axial computerized tomography. **b** Renal angiogram. **c** Exposure of kidney. **d** Regional hypothermia and renal ischemia, developing cleavage plane outside pseudocapsule. **e** Partial resection of involved collecting system. **f** Parenchymal defect after removal of the tumor; watertight closure of the collecting system. **g** Closure of parenchymal defect with running suture along renal capsule. **h** Kidney after re-establishing renal circulation. **e–h** see p. 93

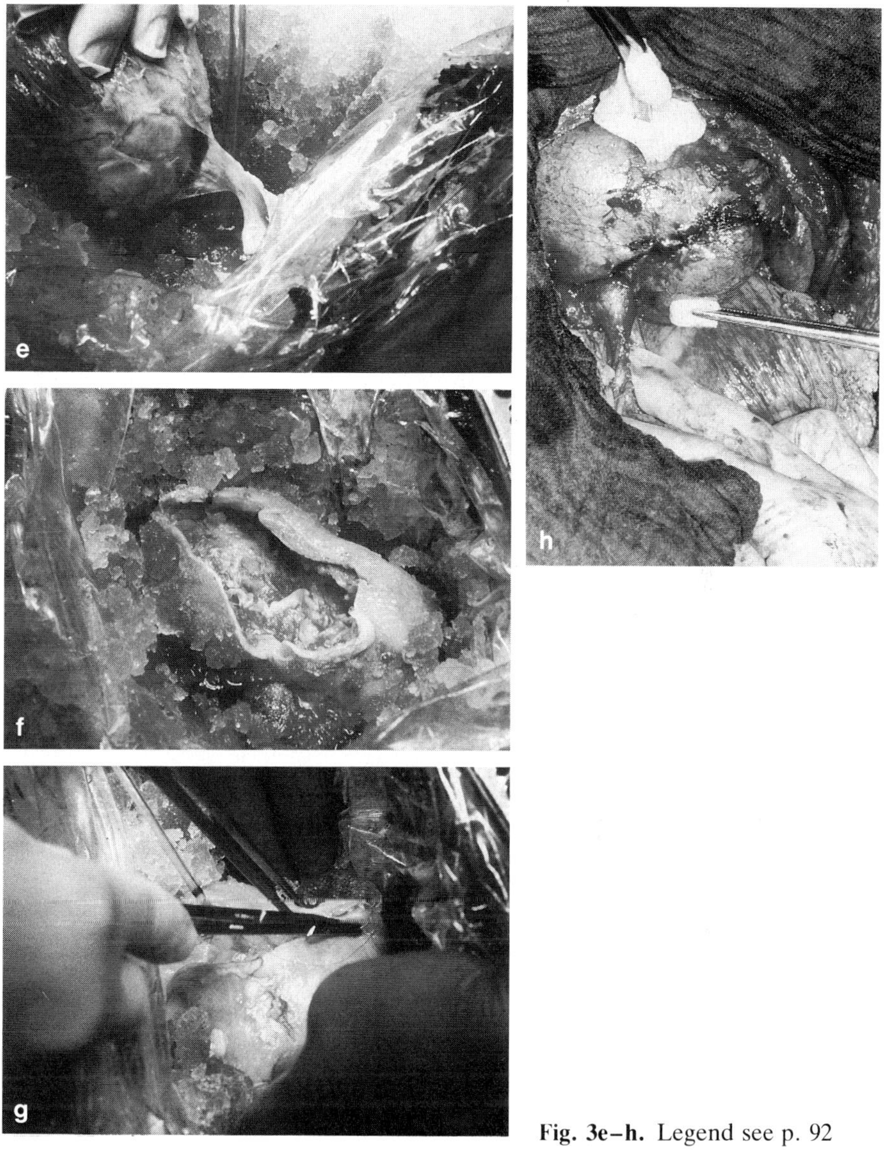

Fig. 3e–h. Legend see p. 92

the tumor, it is clamped with a soft bulldog clamp; if only the tumor and no normal renal parenchyma blanches, this vessel is ligated.

As a bloodless field is mandatory for precise dissection, and any time stress on the surgeon because of threatening ischemic damage of the kidney strictly to be avoided, we routinely perform the procedure in renal ischemia and regional hypothermia. The technique has been described in detail elsewhere [13,14]. In short, the kidney is suspended with umbilical tapes and the renal artery and any accessory arteries are clamped with soft bulldog clamps; unless a major branch of the renal vein is opened, venous backflow is so low that the renal vein need not be occluded. The sealed end of a standard polyvinyl intestinal bag is opened and the bag placed around the kidney, so that the purse-string band of the bag can be closed around the renal pedicle to form an open-ended dam around the kidney. Sterile ice which was obtained by freezing double-packed sterile bags of saline in a standard refrigerator, is crushed with an orthopedic hammer on a side table after removing the sterile wrapping. The space between kidney and intestinal bag is filled with ice slush. This is continued by an assistant throughout the ischemic period. Rapid intravenous administration of 12.5 g mannitol, immediately prior to clamping the renal artery and again as renal circulation is re-established further promotes renal protection.

The tumor is excised from the kidney in a plane outside its pseudocapsule, so that the tumor is still covered by a thin coat of healthy parenchyma. In general, this plane can be developed by blunt dissection once the fibrous capsule is incised sharply about 5–10 mm away from the visible border of the tumor at the surface of the kidney (Fig. 3D). Using loop magnification, larger vessels can usually be identified, transsected and suture ligated with 5-0-PDS sutures. The collecting system does not limit the line of resection; if the tumor invades the collecting system, it is excised, but care is taken to facilitate its reconstruction or closure by selecting the line of excision appropriately (Fig. 3E). If a peripheral calix loses its connection to the pelvis in this manner, the related parenchyma is likewise excised rather than risking urinary extravasation. Where a major segment of the collecting system has to be removed because of tumor invasion (e.g., a central tumor), the collecting system has to be reconstructed rather than resorting to more resection of renal tissue.

Once the tumor is removed, it is sent in toto to the pathologist for frozen sections to control the completeness of tumor excision. We found it more reliable to have the pathologist sample the specimen for a complete coat of healthy tissue along the resection plane rather than taking random biopsies from the kidney after the tumor is removed. In our experience, tumor invading the tissue can usually also be seen in the bloodless field with loop magnification so that the resection can be extended. After sampling for complete removal, the entire specimen is dipped in India ink and step sectioned for final histology.

While the frozen sections are evaluated hemostasis is finalized by precise suture ligation of all identifiable transsected vessels (Fig. 3F). The collecting system is closed watertight with 5-0 polyglycolic acid sutures. Where the collecting system has to be reconstructed, e.g., the remnants of the upper pole calices reanastomosed to the pelvis, this is also attempted in a watertight fashion; if necessary, flaps can be fashioned from the renal pelvis to close defects. If the collecting system was opened, an internal self-retaining

Table 1. Complications of in situ excision of in RCC versus work bench excision in the EIRSS study. (From Marberger et al. [12])

Complication	In situ[a]	Ex situ[b]
	n	*n*
Hemorrhage	2	2
Renal artery/vein thrombosis	–	2
Femoral vein thrombosis	–	1
Temporary renal failure, dialysis	2	5
Temporary urinary extravasation	3	3
Gastrointestinal hemorrhage	1	–
Ureteric stenosis	1	1
Secondary nephrectomy	1	2
Patients with complications	8 (17%)	9 (43%)
Operative mortality	1 (2%)	0

EIRSS, European Intrarenal Surgical Society.
[a] Total of 46 patients.
[b] Total of 21 patients.

ureteral stent is routinely inserted. With large defects, a thin nephrostomy tube is also inserted. In most cases, the large defect created by excision of the tumor can be closed by simply folding the ischemic kidney over the defect and approximating the incisional edges with a running 4-0-PDS suture (Fig. 3G). Where the defect is too large, a free patch of peritoneum is glued into the raw surface of the defect with fibrin adhesive (Tissucol, Immuno, Austria), and sewn in place with running sutures to the remaining capsule, with anchoring sutures in the renal parenchyma to prevent it from being lifted off the defect. This technique provides reliable hemostasis at a minimum loss of renal tissue. It is most important to close the defect before renal circulation is re-established as it is almost impossible to achieve good hemostasis and reconstruction with a blood-filled parenchyma without resorting to deep parenchymal sutures with a potentially large loss of renal function (Fig. 3H).

In our experience, this surgical technique permits in situ excision of virtually all renal tumors that have not invaded the renal pedicle. Theoretically, complete excision of the kidney, extracorporeal excision of the tumor, and reconstruction of the kidney (work bench surgery), and subsequent autotransplantation reduces the risk of tumor cell spillage. Experience, however, has show that this approach results in a significantly higher morbidity (Table 1). It is clearly only indicated with direct tumor involvement of the renal vein where the difficult reconstructive procedures can usually only be performed ex situ under the ideal conditions of bench surgery. Tumors of this type without metastases are rare [12]. The three patients we treated in

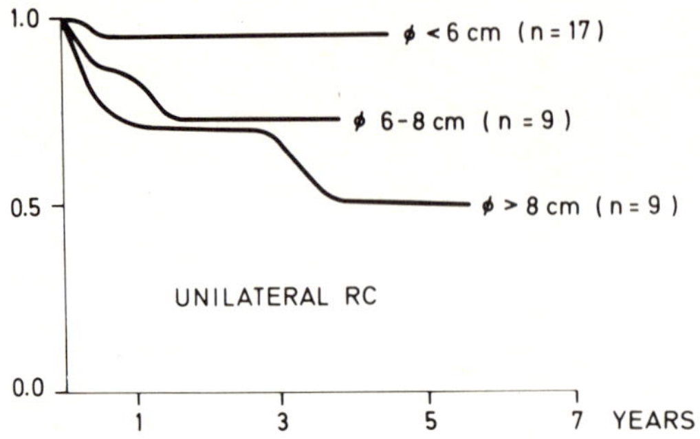

Fig. 4. Suvival after excision of RCC from solitary kidneys in relation to tumor diameter. (From Marberger [15])

this manner all succumbed to the tumor within 2 years in spite of what appeared to be radical tumor excision. We do not routinely perfom a radical lymphadenectomy in conjunction with excisional surgery of renal tumors as the merits of this approach are even questionable with radical tumor nephrectomy [19]. If there is any indication for lymph node involvement, e.g., computerized tomography, the nodes are sampled by frozen section prior to partial nephrectomy. If they prove to be positive, it is useless to excise the tumor.

The results obtained by this approach are excellent [5,12,17,24,25]. RCC in the solitary kidney of patients whose contralateral kidney was removed because of renal cancer probably only represents an asynchronous form of bilateral RCC. Where the contralateral kidney was removed for a benign condition or was congenitally absent, the survival rates reported after excisional surgery appear to be identical to that of radical nephrectomy (Fig. 4). The number of locally recurrent tumors in all reports lies below 15% and may at least in some cases represent unknown small satellite tumors that were missed at partial nephrectomy, especially when genetically defined RCC is treated (Table 2).

A word of caution must be issued: these reports usually comprise a highly selected group of patients treated by specialists in the field; the data was analyzed retrospectively and the surgical techniques and patient selection criteria vary considerably. The usual 5-year follow-up may also be too short to determine the progression rate of some tumors with a low biological aggressiveness. Moreover, it is consistently difficult to detect a local recurrent tumor at an early stage, and a precise follow-up – including regular ultrasonography at short intervals, computerized tomography and occasionally even angiograms – is needed; in most reports this was not performed during follow-up. This may also explain why so far simple enucleation of the tumor within the pseudocapsule [16] seems to yield the same survival results in spite of violating surgical rules. Nevertheless, there can be no doubt that the survival achieved with excisional surgery compares

Table 2. Local recurrent tumor after organ-sparing excision of RCC

Author	Year	No. of patients	Recurrence (%)
Jacobs et al. [6]	1980	59	10
Marberger et al. [12]	1981	67	9
Topley et al. [25]	1984	23	13
Smith et al. [24]	1984	32	9
Novick et al. [17]	1989	100	9
Furrer et al. [5]	1990	39	5

favorably with the results of radical nephrectomy in the presence of a normal contralateral kidney; it is the technique of choice at least for tumors confined to the kidney.

Bilateral RCC

The need for conserving healthy renal parenchyma is as mandatory with bilateral tumors as with tumors in solitary kidneys. The same surgical principles are therefore employed. As a rule, the two sides should not be treated simultaneously as, e.g., transient renal insufficiency in the immediate postoperative period is usually balanced by the contralateral kidney even if it is tumor bearing. The kidney with the tumor that appears to be simpler to remove is routinely approached first; if the procedure is successful, the decision to remove the contralateral kidney in case of a more extensive tumor can then be met with greater ease. Nevertheless, wherever possible, bilateral partial nephrectomy remains the treatment of choice [29].

Considerable debate has been devoted to the question whether the tumor in one kidney is a metastasis of the tumor in the contralateral kidney or whether each represents an independent autonomous process [26]. Wickham [28] considered the former to be true for asynchronous bilateral RCC, and considered this group to have a more unfavorable prognosis. Schiff et al. [22] concluded the contrary and noticed a more aggressive behavior of synchronous bilateral RCC. The issue remains unsettled, but without doubt the data on the prognosis must be considered in view of the small numbers of patients analyzed in both series. The multicenter study of the European Intrarenal Surgical Society [12] observed a significantly larger patient group and found no statistically significant difference between synchronous and asynchronous disease (Fig. 5).

The bilateral renal masses observed routinely in patients with van-Hippel-Lindau Syndrome undoubtedly belong to this category as >50% prove to be RCC at histological investigation [2]. All renal masses in patients with this

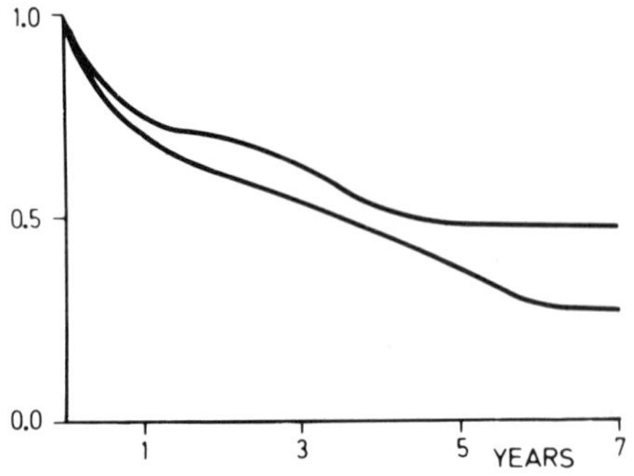

Fig. 5. Survival after excision of RCC in bilateral synchronous (*upper curve*: $n = 18$) and asynchronous (*lower curve*: $n = 18$) tumors, EIRSS series. (From Marberger et al. [12])

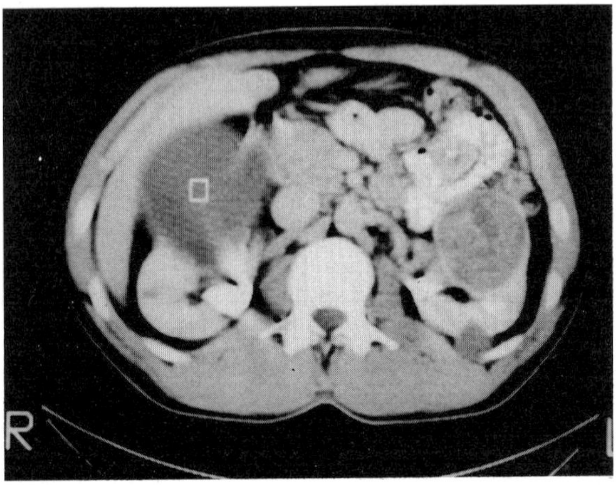

Fig. 6. Hippel-Lindau disease: multiple bilateral kidney tumors, of which the two largest proved to be RCC at excision

autosomal-dominant disease must therefore be excised, preserving as much renal parenchyma as possible (Fig. 6). In contrast, the multiple angiomyolipomas frequently detected in patients with tuberous sclerosis can clearly be defined as such by the demonstration of fat within the lesions by computerized tomography; they need not be excised.

Organ-Preserving Tumor Excision with a Normal Contralateral Kidney

As far back as 1950 Vermooten [27] suggested organ-preserving tumor excision even in the presence of a normal contralateral kidney. The approach was abandoned with the standardization of radical perifascial tumor nephrectomy. In recent years, it has regained interest due to the good results of excisional surgery in solitary kidneys and bilateral tumors. Moreover, with renal ultrasonography and axial computerized tomography the number of small peripheral tumors diagnosed has risen dramatically within the last decade, and most are detected at a stage where they are still asymptomatic. In 1980, 39% of patients treated for RCC at the Rudolfstiftung, Vienna, were staged as T1–T2, and 11% were asymptomatic at the time; in 1989 the respective figures had increased to 57% T1–T2 tumors and 23% asymptomatic patients. This trend will continue with further refinement of modern imaging techniques.

This results in a diagnostic dilemma: small peripheral solid tumors are extremely common in the renal cortex and usually benign. In an extensive autopsy study, Apitz [1] found an overall "14.9% solid nodules" in kidneys which were mainly benign mesenchymal tumors or adenomas. Although the majority of his nodules were minute, some of them would probably be picked up today as by chance finding at ultrasonography. Angiomyolipomas can clearly be defined by their fat content with computerized tomography, but for all other renal parenchymal tumors imaging techniques cannot provide a final diagnosis. As the tumors grow slowly, follow-up investigations at short intervals are usually inconclusive (Fig. 7). Fine needle aspiration biopsy has proven highly unreliable in our experience. It is extremely difficult to ascertain the precise position of the tip of the very flexible needle within the small tumor at the time of aspiration, even under computerized tomography guidance. Core needle biopsies obtained with special positioning belts under computerized tomography guidance are somewhat more reliable, yet it is usually still impossible to identify an adenoma versus a low grade RCC. Even when the total tumor is excised and available for frozen sections, an immediate histological diagnosis is frequently unreliable: in 14 small peripheral tumors investigated in this manner, the final tumor diagnosis was changed after viewing the paraffin sections in 4 of 14 patients, among them one false positive and one false negative for RCC. In essence, small peripheral cortical tumors are extremely common, and we are picking up more and more as our imaging technique improves, but we have no reliable way of clearly defining their dignity other than with excision and definitive histology at present. Should the kidney be removed for every small peripheral tumor detected?

The key to the problem obviously lies in the natural history of small RCC which presumably are the origin of larger RCC at some point. Of the 36 kidneys with a RCC in a solitary kidney in the EIRSS series [12] only one patient with a tumor <6 cm had lymph node metastases, and none had major vein involvement. The specimens of 62 patients were available for reevaluation by the reference pathologist, Dr. R. Pugh (Institute of Urology, London). Although only the appearance of spindle cells was found to have a statistically significant impact on prognosis, these were never observed in

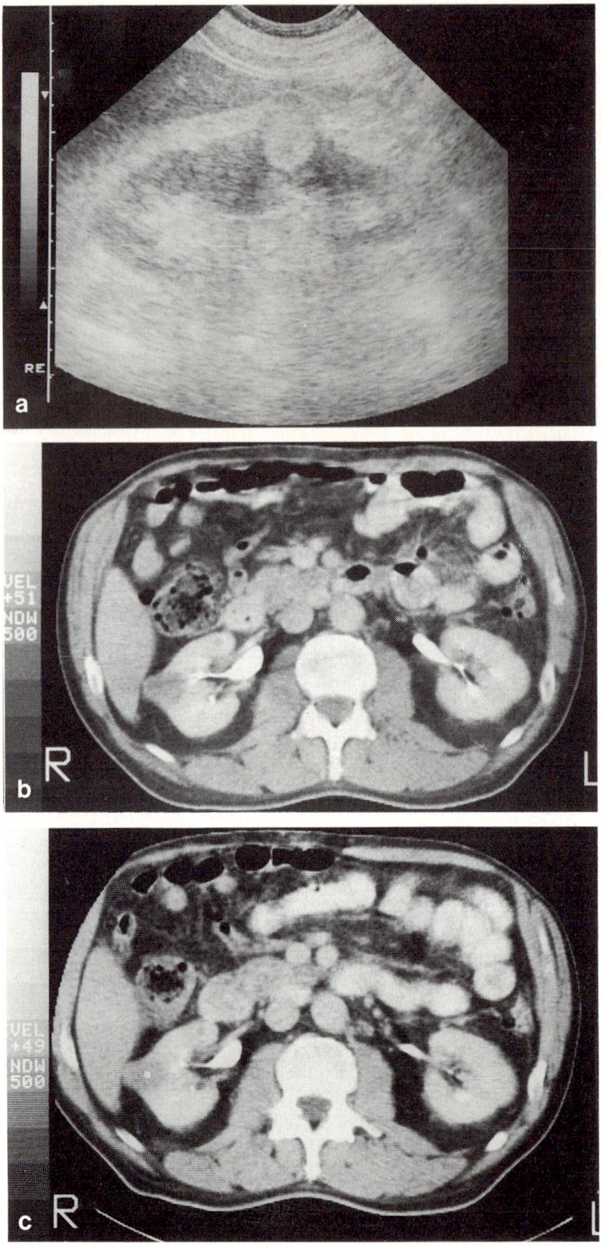

Fig. 7a–c. Peripheral solid renal tumor, 2.5 cm in diameter, diagnosed by chance in asymptomatic patient. **a** Sonogram October 1989. **b** Axial CT October 1989. **c** Axial CT April 1990, showing no change. Tumor proved to be RCC

tumor smaller than 6 cm in diameter. In general, the elements usually considered to signalize a better prognosis dominated, mainly oncocytic elements and clear cells [15]. In this context, the only patient that succumbed in the group of patients with tumors <6 cm in diameter excised from a solitary kidney died from a complication of renal artery thrombosis after autotransplantation, with no evidence of residual tumor. RCC detected as a small solitary peripheral tumor is obviously usually low-grade disease with a rather low tendency to disseminate rapidly.

The key to safe surgical excision of these tumors lies in establishing a plane of excision free of tumor. Almost all small RCC seem to be confined by a pseudocapsule which is formed by the compression of renal parenchyma as the tumor grows. In the most detailed study of this pseudocapsule available today, Rosenthal et al. [21] conclusively showed that the tumors tend to infiltrate the layers of the pseudocapsule at a very early stage, and already in tumors of a diameter of 4 cm the capsule may be perforated by tumor cells along perforating vessels. If the tumor is "shelled out" of the pseudocapsule as is easily possible along this "natural" cleavage plane, there is a considerable risk of leaving behind tumor cells. Despite this tendency to infiltrate this pseudocapsule, the small tumors show little tendency to spread diffusely outside the capsule; if excision is performed outside the capsule with a margin of healthy parenchyma several millimeters thick around the tumor, complete excision can be achieved.

Applying this concept to the clinical situation requires careful patient selection. Although larger tumors may frequently also be excised safely, we limit this approach to isolated peripheral tumors of a maximum diameter of 4 cm. Central tumors are considerably more difficult to excise safely with a good margin of healthy tissue without embarrassing the hilar structures and are therefore not suited for this approach. Potential multiplicity of tumors remains a theoretical problem. We routinely obtain high-quality axial computerized tomography of both kidneys with 4 mm cuts. If there is any evidence for multiplicity of tumors, the conservative approach is abandoned, although this has proven extremely rare in our experience. Of course, multiple tumors other than the tumor first detected can be missed even with this technique, but this is also true for the contralateral kidney which remains after unilateral nephrectomy.

The surgical technique is identical to that outlined for tumors in solitary kidneys. Although small tumors can be removed without clamping the renal artery and without regional hypothermia, we feel optical control of the plane of dissection is an important safety feature. In at least two apparently well-confined tumors, we noticed tongue-like tumor extensions and small tumor satellites of several millimeters in diameter, and the kidney was removed. Infrared coagulation [5] has been recommended to destroy any potential tumor remnants in the pseudocapsule after enucleation, but the depth of destruction with this technique is undefinded. A good margin of healthy tissue as established by step sectioning of the entire tumor

Table 3. Final histology of patients will "solid" peripheral renal mass <4 cm, approached by conservative surgery. (Rudolfstiftung, Vienna, 1980–1988)

	No. of patients
Segmental arterial emboli	1(1)[a]
Calcified cyst	4(1)[a]
Multilocular cystic dysplasia	1
Angiomyolipoma	6(1)[a]
Adenoma	15
Genuine hypernephroma	1
Lymphangioma	1
Oncocytoma	7(1)[a]
Renal cell carcinoma	17(2)[a]
Total	53(6)

[a] The numbers in parenthesis indicate the number of kidneys ultimately removed.

appears the only reliable parameter. We removed one kidney after definitive histology suggested that the tumor reached the plane of resection; the kidney proved to be tumor free when examined after nephrectomy. The decision to excise a tumor rather than removing the entire kidney puts heavy responsibility on the urologist. With nephrectomy these small tumors probably have an almost 100% chance of cure, which must also be achieved with local excision without taking any risks. Surgical removal is the patient's only reliable chance of cure.

Strictly adhering to these principles, the senior author (Marberger) subjected 52 patients to this approach during a 10-year period. All had what was considered to be a solid lesion of questionable dignity in a peripheral position of the kidney, with a maximum diameter of 4 cm. Careful preoperative evaluation, which frequently included fine needle aspiration biopsies, usually failed to give a clear diagnosis and did not demonstrate any signs of metastasis or multifocal disease. In two patients the kidney was removed at surgery because the pseudocapsule of the tumor seemed grossly perforated by the tumor and in a further two patients it was removed because the tumor seemed multifocal at resection. The two former kidneys were shown to have low-grade RCC perforating the pseudocapsule; of the latter two, one showed multiple small oncocytomas and the other presented multiple arterial infarctions (incidentally, fine needle aspiration biopsy of this kidney had suggested a low-grade RCC).

Table 3 gives the final histology. Most striking was the high percentage of benign lesions, which brought the number of benign tumors removed surgically at this department during the same period to 13% of all renal tumors treated. With the exception of the cystic lesions, which were all cysts with calcified walls and a bloody content, the surgeon could usually not define the nature of the lesion at the time of surgery. Applying the standard rules of treating solid tumors, all patients would have been subjected to radical tumor nephrectomy. All RCC excised were only low-grade pT1–2 tumors removed with a clean margin of healthy tissue. One patient 72 years old at the time of surgery succumbed to an unrelated cause 8 years later, without any evidence of recurrent disease. One patient was lost to follow-up. All other patients were followed up between 1 and 8.7 years (mean 3.8 years). No evidence for recurrent tumor or metastasis was noted in any patient. In concludion, organ-sparing tumor excision appears to be a valid alternative in a highly selected group of patients with small solitary peripheral tumors of the kidney. Many of these tumors will ultimately prove to be benign and this number probably increases as more smaller tumors are detected and removed. Provided the lesion is excised outside the pseudocapsule, with a healthy margin of normal tissue in a bloodless field with careful pathologic evaluation, this is as safe as tumor nephrectomy.

References

1. Apitz K (1943) Die Geschwülste und Gewebsmißbildungen der Nierenrinde III. Die Adenome. Virchows Arch [A] 311:321–359
2. Bennington JL, Beckwith JB (1975) Tumors of the kidney, renal pelvis and ureter. In: Firminger HJI (ed) Atlas of tumors pathology, 2nd ser fasc 12. Armed Forces Institute of Pathology, Washington DC, pp 93–199
3. Dreikorn K, Röhl L (1981) Nierentransplantationen. In: Hohenfellner R, Zingg E (eds) Urologie in Klinik und Praxis. European Dialysis and Transplantations Association Report, 1990. Thieme, Stuttgart
4. European Dialysis and Transplantation Association (1990) Report on the 27th Congress of the European Dialysis and Transplantation Association, Vienna
5. Furrer A, Riedmiller H, Schärfe T, Thüroff JW, Hohenfellner R (1990) Organerhaltende Chirurgie bei Nierentumoren. Urol Poster 2:7–8
6. Jacobs SC, Berg SI, Lawson RK (1980) Synchronous bilateral renal cell carcinoma: total surgical excision. Cancer 46:2341–2345
7. Johnson DE, Von Eschenbach A, Sternberg J (1978) Bilateral renal cell carcinoma. J Urol 119:23–24
8. Malamet M (1965) Bilateral renal cell carcinoma: a 14 year survival. J Urol 94:348–355
9. Malek RS, Utz DC, Culp OS (1976) Hypernephroma in the solitary kidney: experience with 20 cases and review of the literature. J Urol 166:553–556
10. Marberger M, Hohenfellner R (1979) Das Nierenkarzinom. Nieren-Hochdruckkr 6:253
11. Marberger M (1980) Organerhaltende Chirurgie beim Nierenkarzinom. Akt Urol 11:325–334
12. Marberger M, Pugh RCB, Auvert J, Bertermann H, Constantini A, Gammelgaard PA, Petterson S, Wickham JEA (1981) Conservative surgery of renal carcinoma: The EIRSS experience. Br J Urol 53:528–531
13. Marberger M (1988) Organerhaltende Nierentumorexcision. Akt Urol 19(2): 58–66
14. Marberger M (1983) Regional hypothermia In: Marberger M, Dreikorn K (eds) Renal preservation. Williams and Wilkins, Baltimore

15. Marberger M (1986) Conservative surgery for renal adenocarcinoma. In: de Kernion JB, Pavone-Macaluso, M (eds) Tumors of the kidney. Williams and Wilkins, Baltimore, pp 157–172
16. Novick AC, Zincke H, Neves RS, Topley HM (1986) Surgical enucleation for renal cell carcinoma. J Urol 135:235–238
17. Novick AC, Streem S, Montie JE, Pontes JE, Siegel S, Montague DK, Goormastic M (1989) Conservative surgery for renal cell carcinoma: a single-center experience with 100 cases. J Urol 141:835–839
18. Penn I (1977) Transplantation in patients with primary renal malignancies. Transplantation 24:424
19. Pizzocaro G, Piva L, Salvioni R (1983) Lymph node dissection in radical nephrectomy for renal cell carcinoma: is it necessary? Eur Urol 9:10–12
20. Robson CJ, Churchill BM, Anderson W (1969) The results of nephrectomy for renal cell carcinoma. J Urol 101:297–301
21. Rosenthal CL, Kraft R, Zingg ES (1984) Organ preserving surgery in renal cell carcinoma: tumor enucleation versus partial kidney resection. Eur Urol 10: 222–228
22. Schiff M Jr, Bagley DH, Lytton B (1979) Treatment of solitary and bilateral renal carcinoma. J Urol 121:581–583
23. Schorn A, Marberger M (1984) Long-time survival of untreated bilateral renal cell carcinoma with supradiaphragmatic vena caval thrombus. J Urol 131: 108–109
24. Smith RB, Dekernion JB, Ehrlich RM, Skinner DG, Kaufmann JJ (1984) Bilateral renal cell carcinoma and renal cell carcinoma in the solitary kidney. J Urol 132:450–454
25. Topley M, Novick AC, Montie JE (1984) Long-term results following partial nephrectomy for localized renal adenocarcinoma. J Urol 131:1050–1952
26. Vermillion CD, Skinner DG, Pfister RC (1972) Bilateral renal cell carcinoma. J Urol 108:219–222
27. Vermooten V (1950) Indications for conservative surgery in certain renal tumors: a study based on the growth pattern of the clear cell carcinoma. J Urol 64:200–208
28. Wickham JEA (1975) Conservative renal surgery for adenocarcinoma. The place of bench surgery. Br J Urol 47:25–36
29. Zincke H, Swanson SC (1982) Bilateral renal cell carcinoma: influence of synchronous and asynchronous occurrence on patient survival. J Urol 128: 913–915

Renal Cell Carcinoma with Vena Caval Extension: Operative Management

S.D. Boyd, D.G. Skinner, and G. Lieskovsky

Department of Urology, School of Medicine, University of Southern California, Los Angeles, CA 90033, USA

Introduction

Renal cell carcinoma, unless it is localized, remains a difficult tumor to cure. There were nearly 23 000 new cases of renal cell cancer diagnosed in the United States alone in 1989, and there were almost 10 000 deaths directly related to this disease. For the majority of patients, radical nephrectomy with early ligation of the renal artery and vein and regional lymph node dissection remains the primary therapy for potentially curable disease. The natural history of this tumor is quite unpredictable except for patients presenting with metastatic disease for whom 5-year cures are rarely achieved. The best cure rate is seen in patients with cancer confined to the kidney and 5-year survivals may approach 70%. Once the tumor invades the perinephric fat the 5-year survival drops to 50%. Regional lymph node involvement which occurs in approximately 20% of patients at the time of diagnosis further decreases survival, and spread to contiguous viscera or distant metastases results in only an occasional cure without effective adjuvant therapy.

Direct extension of renal cell carcinoma into the vena cava has been found by Skinner et al. [3] and Kearny et al. [1] in 4%–10% of patients undergoing nephrectomy to treat cancer. Skinner et al. [2] recognized that venous extension by tumor thrombus was a potentially curable lesion provided that complete removal could be achieved. This paper will present our current approach to these patients with renal cell carcinoma tumor thrombus extending into the vena cava. Our preoperative evaluation, surgical management, results, and conclusions will be discussed.

Materials and Methods

From 1972 to 1988, a select group of 57 patients with renal cell carcinoma and vena caval tumor thrombus were evaluated. Of these patients, 56 were considered operative candidates and underwent surgical exploration. These patients ranged from 31 to 76 years with an average age of 59 years. There were 36 men and 21 women in the overall group. The tumors were on the right side in 37 patients and on the left side in 20 patients. Hematuria was present in 37% of patients. Symptoms or signs of vena caval obstruction were present in 41% of patients. A varicocele was noted in 34% of men. All patients since 1982 had preoperative computerized tomography (CT) of the abdomen and chest. All patients underwent an inferior venacavagram or magnetic resonance imaging (MRI) to evaluate the vena cava. Superior venacavography or echocardiography was performed if there was inferior vena caval occlusion or if the uppermost extent of the vena caval thrombus was unclear on MRI. Arteriograms were not routinely performed as part of the preoperative evaluation.

Patients were divided into three groups according to the level of the vena caval tumor thrombus: Level I with tumor thrombus extension below the insertion of the hepatic veins; level II with tumor thrombus extension within the intrahepatic vena cava but not into the atrium; and level III with intra-atrial tumor thrombus extension. This anatomical grouping of patients was based on the cephalad extent of the thrombus and helped determine the operative approach. The presence of metastatic disease was also determined by CT of the chest and bone scan. Forty-three patients had no evidence of metastatic disease before the operation and 13 patients had metastatic disease prior to operation as determined by CT scan.

Surgical Technique

A right thoracoabdominal surgical approach was primarily used. A left thoracoabdominal incision was employed when the primary tumor was on the left side and there was only minimal inferior vena caval involvement. In general, a high right thoracoabdominal approach is necessary for removal of a large intracaval thrombus extending above the hepatic veins. This high incision permits intrathoracic control of the vena cava from the right side. The incision is usually begun over the seventh or eighth rib, beginning at the mid-axillary line and extending across the costochondral junction to the epigastrium and down the midline to below the umbilicus. For exceptionally large tumors originating on the left side, the incision may be extended across the midline to the left subcostal region using a T-shaped inferior midline extension for better exposure.

After the peritoneum is entered, the abdomen is explored. The peritoneal attachments to ascending colon and small-bowel mesentery are divided so that the bowel, duodenum, and pancreas can be retracted away on the superior mesenteric artery pedicle. This facilitates early ligation of the renal artery on the involved side. The vena cava is next exposed and all of the lumbar veins are ligated. Rummel tourniquets can then be loosely applied on the distal vena cava and on the contralateral renal vein (for tumors on the right side) or contralateral renal artery (for tumors on the left side).

A vertical anterior incision in the pericardium, parallel to the phrenic nerve above the diaphragm, is used to gain control of the intrathoracic inferior vena cava in patients with level II or III thrombi. A Rummel tourniquet is applied gently around the vena cava at this point but is not tightened. The vena cava is palpated at this level and an atrial purse-string suture is placed if the tumor thrombus extends into the atrium.

The kidney and Gerota's fascia are mobilized completely with an en bloc dissection of the retroperitoneal nodes. The only remaining points of attachment are the renal vein and the intracaval tumor thrombus. If the tumor thrombus extends only to the subhepatic vena cava (level I), a Rummel tourniquet is placed around the vena cava at this level instead of at the intrathoracic level. If the tumor thrombus extends into the intrahepatic vena cava (levels II and III), a Fogerty vascular clamp is placed on the superior mesenteric artery and two vascular clamps are placed on the porta hepatis. The Rummel tourniquet is tightened on the distal vena cava. The Rummel tourniquet on the contralateral renal vein or artery is secured and, if there is no intra-atrial tumor thrombus, the tourniquet on the intrathoracic vena cava is tightened.

If there is an intra-atrial tumor, an atriotomy is performed by an assisting cardiothoracic surgeon. An index finger is inserted through the purse-stringed atriotomy to explore gently the extent of the tumor within the atrium and, if feasible, to push the thrombus out of the atrium and back into the vena cava before the intrathoracic vena caval tourniquet is tightened. If the intra-atrial tumor is extensive or adherent to the wall of the atrium, cardiopulmonary bypass is required.

A vertical incision is made in the vena cava circumscribing the insertion of the ipsilateral renal vein and extending cephalad almost to the level of the lowest hepatic vein insertion. With gentle dissection with the index finger or with a Kittner dissector, the entire intracaval thrombus can be extracted and removed en bloc with the involved kidney and resected lymph nodes. Vena caval resection is rarely required as the tumor thrombus seldom invades the vena cava; if it does, there is no evidence that resection of the vena cava improves survival.

With the operating table in the Trendelenburg position, the vena cava is irrigated and closed with two layers of running 4-0 arterial silk sutures, making certain that all air is evacuated by flushing the vena cava with a brief release of the Rummel tourniquet on the contralateral renal vein or artery. The proximal vena caval Rummel tourniquet is then released, followed by the contralateral renal vein tourniquet, the vascular clamps on the porta hepatis and superior mesenteric artery, and lastly the distal inferior vena caval tourniquet. The pericardotomy is left open. The thoracoabdominal incision is closed in the usual manner. Any solitary lung or liver nodules are removed.

Patients routinely receive a 20 g infusion of mannitol before the contralateral renal vein or renal artery is occluded. If the vena cava has been occluded completely preoperatively or if clot is present distally, a partially occluding vena caval clip is applied just distal to the insertion of the renal vein to help prevent postoperative embolism.

Results

Of the 56 patients who were explored, 53 underwent radical nephrectomy with en bloc vena caval tumor thrombectomy. Three patients with known metastatic disease and debilitating symptoms secondary to vena caval obstruction were found to be inoperable. Eight patients were found to have intra-atrial involvement, three of which required cardiopulmonary bypass to facilitate tumor extraction from the atrium. Seventeen additional patients were found to have metastatic disease during the operation, but complete resection of all tumor was possible in 40 patients. The metastatic disease was found in the lung (47%), contiguous viscera (29%), liver (12%), or mediastinum (12%).

The average operative time was 3¾ h. The average liver and renal warm ischemia time was 14 min and ranged from 8 to 20 min. Vena caval ligation or clipping was performed in 17 patients (32%). There were three intraoperative deaths: two due to exsanguination and one due to massive tumor pulmonary emboli. The average blood loss was 5000 cm^2 with an average replacement of 10 units. The blood loss was naturally highest for the level III tumors. These numbers are currently somewhat lower due to the judicious use of the cell saver with multiple filters. There were four other perioperative deaths due to sepsis and multiple organ system failure, leaving an operative mortality rate of 13% (7 of 53 patients). The remaining patients had a median postoperative hospital stay of 12 days (5–58 days). Postoperative complications occurred in 41% of patients and included hepatic dysfunction (11%), renal failure requiring temporary dialysis (8%), sepsis (8%), pancreatitis (2%), arrhythmia (2%), prolonged ileus (2%), adrenal insufficiency (2%), and pleural effusion (2%).

Histological examination revealed 28 patients with clear cell-type tumors, 21 with clear cells and granular cells, and 4 with spindle cell-type renal cell carcinoma. Pathological staging of the 53 patients showed 9 with no evidence of other spread outside of the kidney, 9 with invasion into the fat alone, 8 with positive regional lymph nodes, and 27 patients with metastatic disease.

Table 1. Five-year survival rates

	No. of patients	Actuarial[a] (%)	Actual (%)	No. of patients followed 5 years
All patients	53	29	28	25
Level I	21	35	42	12
Level II	24	18	20	10
Level III	8	0	0	3
Complete resection	40	34	33	21
−Fat, −nodes	9	40	38	8
+Fat	9	47	50	6
+Nodes	8	60	33	3
Stage III	26	47	41	17
Stage IV	27	0	0	8

[a] By methods of Kaplan and Meier.

Table 2. Effect of metastases on survival

	Survival rate (%)		
	1-Year	3-Year	5-Year
No evidence of presperative metastases (43)	60	39	33
Evidence of presperative metastases (13)[a]	10	0	0

[a] Median survival 4 months; longest 13 months.

The 46 patients who survived the perioperative period have been followed for 12–122 months, with a median follow-up of 64 months. Tables 1 and 2 show the actuarial Kaplan-Meier 5-year survival rates for all 53 patients and the actual 5-year survival rates for the 25 patients in this series operated on before 1984. Several factors show an impact on survival. These include the level of the vena caval tumor extension, the extent of local tissue involvement, whether a complete excision of all tumor was possible, and the presence of metastatic disease. Of note, there were no 5-year survivals for patients with level III thrombi or those with stage IV metastatic disease. In the absence of metastatic disease to other organs, the survival with perinephric fat or nodal involvement is comparable to the survival seen in those with only intrarenal neoplasm with vena caval extension. The overall 5 year actuarial and actual survival rates for patients without evidence of distal metastatic disease was 47% and 41% respectively.

Of the 40 patients with complete excision of all grossly recognized tumor, the 5-year actuarial and actual survival rates were 34% and 33%. For those patients without metastatic disease recognized before the operation, the 5-year actuarial and actual survival rates were 32% and 33%. For those 13 patients with metastatic disease diagnosed before operation, 3 were unresectable and only 5 of the other 10 patients had a complete resection of all grossly apparent disease. The median survival of these 10 patients was 4 months and the longest survivor lived 13 months.

More than one half of our 53 patients (27) had metastatic disease apparent at the time of surgery. None of these patients lived 5 years, but the actuarial and actual 1-year survival rates were 37% and 26%. Of these 27 patients with metastatic disease, 13 underwent complete resection of all apparent tumor and had a mean survival of 9 months; the longest survivor lived 21 months.

Discussion

The key to successful removal of the intracaval tumor thrombus lies in good preoperative delineation of the level of the thrombus and the degree of vena caval obstruction. While this can still be well defined by venacavography, we now use MRI because it is noninvasive. The major advantages of MRI are that thrombosis versus patency in the vena cava can be determined without the use of an intravenous contrast agent, and that MRI can acquire images of the vena caval tumor thrombus in saggital, coronal, and axial planes. Because MRI images are less exacting in the beating heart, we use echocardiography, as well as superior venacavography with cardiac angiography if atrial involvement is expected.

The use of CT scanning of both the abdomen and chest is extremely important in these patients. Even with modern techniques, the magnitude and risks of this type of surgery remain high. The presence of known metastatic disease before operation prevents successful curative surgery. None of our tumor thrombus patients with distant metastatic disease survived 5 years. This does not mean, however, that surgery in this group cannot offer palliation. Those patients with debilitating symptoms due to the vena caval thrombus can be provided with an improved quality of life for several months by excising the tumor. Because the median survival of those patients undergoing surgery with known metastatic disease before operation is less than 1 year, however, this surgery should be considered only if the symptoms of vena caval obstruction are severely debilitating or if death is imminent without surgical intervention. Most patients with evident preoperative metastatic disease should not undergo an attempt at surgical resection. The use of this surgery in patients with level III tumors also must be proposed mainly for palliation. Again in this group we have not had any 5-year survivors. These patients routinely have extremely severe symptoms and surgery can offer tremendous short-term relief. The greatest success of this surgery is going to be in patients where all disease can be excised. Long-term follow-up now shows a 5-year survival rate of 34% for those patients where complete tumor resection is possible and a survival of 47% for those patients without metastatic disease to other organs. With these numbers in mind, an aggressive and optimistic approach is advocated for those patients with renal cell carcinoma with vena caval thrombus without preoperative evidence of metastatic disease. Surgery is not indicated in patients with known metastatic disease unless it is a very young patient who really wants to buy another year of life with surgery. Hopefully in the future, more effective adjuvant chemotherapy can further change our approach to this disease.

References

1. Kearney GP, Waters WB, Klein LA et al. (1981) Results of inferior vena caval resection for renal cell carcinoma. J Urol 125:769
2. Skinner DG, Pfister RF, Colvin R (1972) Extension of renal cell carcinoma into the vena cava: the rationale for aggressive surgical management. J Urol 107:711
3. Skinner DG, Vermillion CD, Colvin RB (1972) The surgical management of renal cell carcinoma. J Urol 107:705

The Role of Cytokines in the Therapy of Renal Cell Carcinoma

T. Ebert, B.J. Schmitz-Dräger, and R. Ackermann

Urologische Universitätsklinik, Moorenstrasse 5, W-4000 Düsseldorf, FRG

Renal cell carcinoma (RCC) is the most common malignant tumor of the kidney and the third most common urologic cancer [12]. Since the onset of symptoms frequently occurs late and clinical signs may vary considerably, correct diagnosis is often established only when the disease is advanced. In fact, at initial presentation approximately 50% of patients already have metastatic disease [2,7,14]. The median duration of survival in patients with overt distant disease at presentation is 10 months, with a 2-year survival rate of 10%. There is no effective systemic therapy, as renal cancer has proven refractory to both hormonal treatment [13] and chemotherapy [19]. There are, however, recent preliminary reports of responses to immunotherapy.

Although immunologic approaches are generally considered the most recent strategies in the therapy of malignant diseases, the first immunotherapeutic efforts were undertaken more than 100 years ago. Busch [3] in 1866 reported the inoculation of erysipelas in a patient with malignant sarcoma. He observed subsequent shrinking of the tumor. It was, however, not until recently that modern immunobiologic research tools finally enabled investigators to examine the exact cellular and humoral mechanisms involved in host-tumor interactions.

Spontaneous regression of metastases is still regarded as the major rationale for immunotherapeutic efforts in RCC. Although this event is quite rare, it is found more frequently only in malignant melanoma [5]. Immunologic phenomena are believed to mediate these occurrences of regression [18]. Consequently, attempts have been made in patients with metastatic RCC to induce tumor regression by various immunologic methods (Fig. 1). The most extensively studied drugs are the interferons (IFNs) and interleukin-2 (IL-2).

IFNs are glycoproteins produced by leukocytes (IFN-alpha), fibroblasts (IFN-beta), or lymphocytes (IFN-gamma). They have antiviral and antineoplastic activity which seems to be mediated by direct tumor cytotoxicity and/or indirect cytotoxicity through alteration of immune responses [11].

Recent Results in Cancer Research, Vol. 126

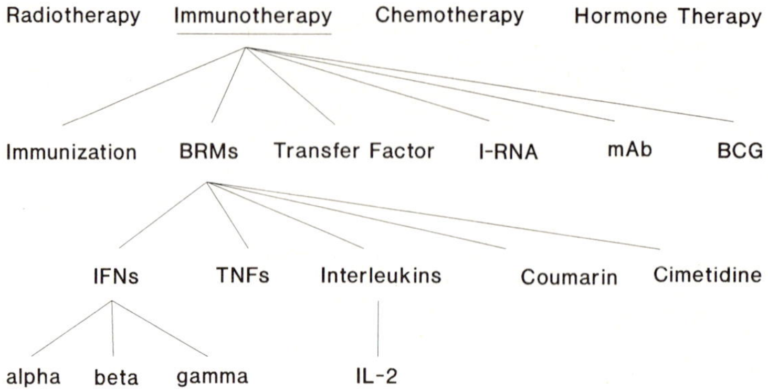

Fig. 1. Immunotherapeutic investigations in RCC. *mAb*, monoclonal antibodies; *BCG*, bacille Calmette-Guérin; *IFN*, interferon; *TNF*, tumor nacrosis factor; *IL-2*, interleukin-2; *BRM*, biological response modifier

Table 1. IFN-alpha in the therapy of RCC. (de Mulder [4])

Author	No. of patients	Complete and partial response rate (%)
Einzig (1984)	62	11
Quesada (1985)	73	26
Umeda (1986)	153	15
Schnall (1986)	22	5
Krown (1987)	37	14
Buziad (1987)	22	23
Otto (1988)	42	17
Figlin (1988)	19	26
Total	430	17

IFN-alpha was the first IFN to be produced by recombinant techniques. For this reason, the vast majority of clinical trials conducted so far have utilized this type of IFN. Tumor remission rates (complete response, CR, plus partial response, PR) range between 5% and 26% (Table 1).

It has been shown that an intermediate daily dose of $3–10 \times 10^6$ IU IFN-alpha given for at least 3 months yields the highest response rates (Table 2). Severe side effects are associated with this form of therapy (Table 3). Subcutaneous or intramuscular administration seems to abrogate significantly the most severe adverse effects of IFN-alpha. A subgroup of patients have been identified who respond best to IFN-alpha (Table 4).

IFN-beta has been evaluated in only a few phase I/II trials in patients with metastatic RCC. Although higher doses could be used due to decreased morbidity, treatment with IFN-beta does not appear to be superior to IFN-alpha treatment [8].

Table 2. Dose–response rate of IFN-alpha. (Horoszewicz and Murphy [10])

Daily dose ($\times 10^6$ U)	No. of patients	Complete and partial response rate (%)
<3	189	12
3–10	317	19
>10	196	16
>10 dose escalation	410	14

Table 3. Main side effects of IFNs

Fatigue	Headaches
Fever	Exanthema
Chills	Pruritus
Arthralgia	Dizziness
Anorexia	Leukopenia
Nausea	

Table 4. Responder characteristics

- Limited tumor burden
- Lung metastasis
- Prior nephrectomy
- Disease free interval >1 year
- High performance status
- No prior chemo-/radiotherapy

IFN-gamma has the most immunomodulatory properties of all the interferons. It has been used in only a few clinical trials, and is apparently the most toxic IFN. Response rates were generally lower than those achieved with IFN-alpha or IFN-beta. In our department, 20 patients with metastatic RCC received IFN-gamma. One (5%) PR was observed during a follow-up period of 3–30 months.

IL-2 is a small protein which is important in the proliferation and maturation of a variety of effector cells including B and T lymphocytes, natural killer cells, and lymphokine-activated killer (LAK) cells. It has been demonstrated in vitro and in vivo that IL-2 can induce antitumoral activity in peripheral blood lymphocytes [15]. It is, however, not yet known which cells within the immune system comprise the LAK population [8].

In 1985, Rosenberg and co-workers [16] introduced IL-2 and LAK cells into the immunotherapy of malignancies. Since then a number of trials have been performed worldwide in patients with metastatic RCC using IL-2 alone or in combination with autologous "activated lymphocytes." The excitement

Table 5. IL-2 and LAK cells in the therapy of RCC. (Fisher [6])

	No. of patients	Complete and partial response rate (%)
IL-2 (bolus)	101	19.0
IL-2 (bolus) + LAK	205	26.0
IL-2 (c.i.)	91	15.0
IL-2 (c.i.) + LAK	203	17.0
Total	600	20.2

c.i., continuous infusion.

Table 6. Reasons for toxicity of IL-2

- Related to IL-2
- Related to IFN-gamma
- Related to TNF production
- Related to IL-1 production
- Related to IL-6 production

Table 7. Toxicity of IL-2

High fever, rigors
Vascular leak syndrome
Septicemia
Pruritus
Thrombocythemia
Anemia
Liver dysfunction
Neuropsychiatric dysfunction
Death (4%)
IL-2 + LAK is more toxic than IL-2 alone
Toxicity is dose dependent

over early response rates has subsequently been tempered by clinical reality. Fisher [6] summarized the data on 600 patients treated with IL-2 and LAK combinations (Table 5). In the vast majority of cases the duration of response is limited to a few months.

Severe side effects which can be explained by IL-2-induced lymphokine production (Table 6) are observed (Table 7). Hypotension, weight gain, and impairment of renal and hepatic function occur in nearly every patient. These side effects usually resolve promptly after IL-2 administration is stopped. In almost 20% of patients lung edema develops and may require intubation [17]. More recent reports [1] indicate that the severity of side effects can be decreased by s.c. administration of IL-2.

Various combinations of different cytokines with other anti-tumor drugs have been examined in recent years. Response rates range between 10% and 20% [9]. No single drug combination was found to be clearly superior to any other.

Immunotherapeutic trials in RCC using IFNs and interleukins have induced regression of metastases (CR + PR) in 10%–20% of patients. These response rates were obtained in controlled studies with selected patients. To date, it is not known if response is associated with prolonged survival. No prospective, randomized trial has been performed which demonstrates any survival advantage for those patients treated with adjuvant immunotherapy compared to those receiving no adjuvant treatment. Although we know that certain patients respond to immunotherapy, we still can not identify responders prior to treatment. In order to avoid the administration of potentially harmful drugs to nonresponders, new prognostic parameters (histologic, serologic, immunologic, or molecular-biologic) must be found that identify those patients who might benefit from immunotherapy.

References

1. Atzpodien J, körfer A, Franks CR, Poliwoda H, Kirchner H (1990) Home therapy with recombinant interleukin-2 and interferon-alpha2b in advanced human malignancies. Lancet i:1509
2. Bassil B, Dosoretz DE, Prout GR Jr (1985) Validation of the tumor, nodes and metastasis classification of renal cell carcinoma. J Urol 134:450–454
3. Busch W (1866) VI. Verhandlungen ärztlicher Gesellschaften Berl Klin Wochenschr 23:245–246
4. de Mulder P (1990) Are the achievements with interferon in advanced renal cell carcinoma worthwhile? 9th Congress of the European Association of Urology, June 13–16, Amsterdam
5. Everson TC, Cole WH (1966) Spontaneous regression of cancer. Saunders, Philadelphia, p 4
6. Fisher RI (1990) Interleukin-2, is more less or less more? The problem of dose in immunotherapy. 9th Congress of the European Association of Urology, June 13–16, Amsterdam
7. Giuliani L, Martorana G, Giberti C, Pesatore D, Magnani G (1983) Results of radical nephrectomy with extensive lymphadenectomy for renal cell carcinoma. J Urol 130:664–668
8. Graham SD Jr (1989) Immunotherapy of renal cell carcinoma. Semin Urol 7:215–227
9. Heicappell R, Ackermann R (1990) Rationale for immunotherapy of renal cell carcinoma. Urol Res 18:357–372
10. Horoszewicz JS, Murphy GP (1989) An assessment of the current use of human interferons in therapy of urological cancers. J Urol 142:1173–1180
11. Oldham RK, Smalley RV (1983) Immunotherapy: the old and the new. J Biol Resp Modif 2:1–37
12. Paganini-Hill A, Ross RK, Henderson BE (1983) Epidemiology of kidney cancer. In: Skinner DG (ed) Urological cancer. Grune and Stratton, New York, pp 383–407

13. Pizzocaro G, Piva L, DiFronzo G, Giongo A, Cozzoli A, Dormia E, Minervini S, Zanollo A, Fontanella U, Longo G, Maggioni A (1987) Adjuvant medroxyprogesterone acetate to radical nephrectomy in renal cancer: 5-year results of a prospective randomized study. J Urol 138:1379–1381
14. Robson CJ, Churchill BM, Anderson W (1969) The results of radical nephrectomy for renal cell carcinoma. J Urol 101:297–301
15. Rosenberg SA, Grimm EA, McGrogan M (1984) Biologic activity of recombinant human interleukin-2 produced in Escherichia coli. Science 223:1412–1414
16. Rosenberg SA, Lotze MT, Muul LM, Leitman S, Chang AE, Ettinghausen SE, Matory YL, Skibber JM, Shiloni E, Vetto JT, Seipp CA, Simpson C, Reichert CM (1985) Observations on the systemic administration of autologous lymphokine-activated killer cells and recombinant Interleukin-2 to patients with metastatic cancer. N Engl J Med 313:1485–1492
17. Rosenberg SA, Lotze MT, Muul LM, Chang AE, Avis FP, Leitman S, Linehan WM, Robertson CN, Lee RE, Rubin JT, Seipp CA, Simpson CG, White DE (1987) A progress report on the treatment of 157 patients with advanced cancer using lymphokine-activated killer cells and Interleukin-2 or high-dose Interleukin-2 alone. N Engl J Med 316:889–897
18. Snow BW, Schellhammer PF (1982) Spontaneous regression of metastatic renal cell carcinoma. Urology 20:177–181
19. Yagoda A (1989) Chemotherapy of renal cell carcinoma: 1983–1989. Semin Urol 7:199–206

Superficial Bladder Cancer

Intravesical Instillation Therapy: Alternative Treatments

F.M.J. Debruyne, A.P.M. van der Meijden, and J.A. Witjes

Department of Urology, University of Nijmegen, The Netherlands

Introduction

The high recurrence rate of superficial bladder cancer after transurethral resection (TUR) has stimulated urologists and urological investigators to explore pathways to prevent and/or reduce recurrent tumors. For more than three decades intravesical chemotherapy has been applied, and more recently intravesical immunotherapy has also been proposed. Since 20% of recurrent tumors eventually will progress and/or cause metastases, it is obvious that these forms of intravesical treatment also aim at prevention of progression and/or metastases [1,2].

Different agents and regimens for intravesical therapy have been used but many questions remain unanswered with respect to this treatment. These questions relate to selection of patients, the optimal drug to be used, and the best regimen to be applied. One also wonders if there are alternatives to these forms of treatment, i.e., not yet (sufficiently) explored new forms of treatment. All have the same ultimate goal – to prevent recurrences and progression in all patients with superficial bladder cancer. Of course, we also ask ourselves how the development itself (of superficial bladder cancer) can be prevented, i.e., what is the real etiology of this malignant disease and what are the factors that induce mechanisms associated with recurrence and progression. This article will mainly deal with the first question, namely: what is the current optimal intravesical therapy and what are the future perspectives with respect to this form of treatment.

Current Achievements with Intravesical Chemotherapy

In the first investigations with intravesical chemotherapy relatively toxic drugs such as thiotepa were used [3]. These drugs, although effective, were resorbed through the damaged epithelium (TUR) and caused considerable

Recent Results in Cancer Research, Vol. 126

Table 1. Phase III studies in superficial bladder cancer (EORTC GU group)

Protocol	Therapy	Study closed	No. of patients
30751	VM265/THIO/TUR	October 1978	370
30781	Pyridoxine	November 1981	291
30782	THIO/ADM/CDDP	September 1983	356
30790	EPO/ADM/TUR	September 1983	443
30831	MMC	October 1985	531
30832	ADM	October 1985	448
30845	BCG-RIVM/MMC	October 1986	337
30863	EPI	September 1989	511
30864	MMC (marker)		115
30869	EPI (marker)		39
30861	CIS BCG		111
Total			3552

THIO, thiotepa; TUR, transurethral resection; ADM, adriamycin; EPO, epodyl; MMC, Mitomycin C; BCG, bacillus calmelte-guérin; CIS, carcinoma in sita; EPI, epirubicin; CDDP, cisplatin; RIVM, Rijks-instituut voor Volksgezondheid en Milieuhygiene.

toxicity, leading to, e.g., liver disturbances, some of which had a lethal outcome.

Later on epodyl was widely used. Epodyl did not demonstrate systemic toxicity but the preventive effect was regarded as limited [4]. In the 1970s drugs such as Adriamycin and later mitomycin proved to be effective with only minimal toxicity and side effects. This is due to the very limited resorption through the bladder wall even when the instillations were started soon after TUR of the tumor. Many other drugs have been explored but do not show substantial efficacy. Some of them, such as cisplatin had serious side effects. To date, Adriamycin, mitomycin C and epirubicin have emerged as *the* intravesical agents, and are now widely accepted.

The question as to whether one drug is superior to another or one treatment regimen better than another has been investigated in many prospective clinical trials both in the United States and in Europe. The European Organisation for Research and Treatment of Cancer (EORTC) Genito Urinary (GU) Group has made a considerable contribution in these areas. In different trials conducted since 1975 (Table 1) more than 3500 patients have been randomized. Statistical analyses of these prospective randomized clinical studies have shown that none of the drugs mentioned above is superior to another and that none of the regimens has advantages over another. This means that Adriamycin, mitomycin C, and epirubicin have similar results with respect to prevention and/or delay of recurrent disease and that delayed treatment (2 weeks after TUR) has similar efficacy but less toxicity than early treatment. Likewise, maintenance therapy has no advantages over the traditionally applied 6-weekly instillations. Dosage of the drug plays a minimal role. From all these trials it was clear that none of the chemotherapeutic agents could completely prevent recurrences or affect progression rates. They achieve only a more or less substantial delay in time to first recurrence, decrease the recurrence rate, and they probably can delay progression. The response to the intravesical chemotherapy is, therefore, largely dependent on the intrinsic biological characteristics of the tumor and the surrounding "normal" bladder rather than on the drug used. Therefore, it is clear that in the future intravesical therapy has to be directed and adapted with these characteristics in mind, which will, in fact, constitute the refinement and individualization of therapy. These studies have also shown that some tumors are so "benign" with regard to their natural biological behavior that no additional treatment after TUR is needed. The same principles can be applied to intravesical immunotherapy.

Intravesical Immunotherapy

Morales [5] was the first to use immunostimulatory agents for intravesical treatment and prevention of recurrence of superficial bladder cancer. He used 6-weekly instillations of 10^9 units of bacille Calmette-Guérin (BCG) dissolved in 50 cc saline. He showed that BCG is effective in both the treatment of (residual) superficial bladder cancer and in the prevention of recurrences. Since these initial observations numerous studies using BCG have been performed in order to improve the results obtained so far by intravesical chemotherapy, and to prevent later development of progressive local tumors and/or metastases. These studies were conducted worldwide and again the EORTC GU Group intensively participated in this clinical research.

Different strains of BCG and different regimens have been tried. Results obtained so far show that BCG is indeed effective in both the treatment and prevention of superficial bladder cancer and is even the treatment of choice for carcinoma in situ – an intra-epithelial high-grade superficial tumor with great capacity to develop invasive metastatic disease. However, the price for these results is a substantial toxicity, both locally (drug-induced cystitis) and systemically (symptoms of BCG infection) in spite of the use of strongly attenuated BCG bacilli [6].

It has been postulated that the more severe the side effects (virulence) of BCG instillations, the better its efficacy. Prospective randomized trials have not (yet) clearly substantiated this statement and it has not been clearly demonstrated that one strain of BCG is superior to another. It has, however, been shown that 6-weekly instillations are suboptimal and that certainly in patients with biologically aggressive tumors a second course should be considered.

It is not yet understood how BCG exerts its efficacy. Both direct mucosal effect (inflammation) and immunological mechanisms seem to be involved.

The more severe the local side effects (chemical cystitis), the more effective intravesical BCG treatment appears to be. It has also been shown in vitro and in animal and human studies that specific immunological reactions occur [7,8].

The Role of Prognostic Factors in Directing Intravesical Therapy

Intravesical instillations should only be given if the advantages (delay of recurrence and/or progression) outweigh the risks (complications). As stated above, some tumors are so benign in their natural biological behavior, that additional treatment might not be mandatory. For example a solitary or small pTa grade I bladder tumor has less than 5% chance of recurrence, and therefore should not be treated after TUR [9]. On the other hand for some tumors intravesical instillations might be insufficient.

Several factors influencing prognosis have been studied:

1. Although the importance of *tumor grade and stage* depends on the pathology report [10], high-grade tumors mostly have a higher recurrence rate [11–13] and a higher likelihood of progression [14–16]. Tumor stage also has prognostic significance [17,18]. In a proportional hazards model Loening et al. found that tumor stage had a highly significant ($p = 0.006$) relation to recurrence [13].
2. *Tumor multiplicity* is generally accepted as a risk factor for a higher recurrence rate [11,12,19] and progression [16,18]. Other prognostic factors are tumor size [11,14,18], and the number of tumors [12,18]. Dalesio et al. found in a multivariate analysis that the number of tumors was the single most important prognostic factor [12].
3. *Tumor localization* was found to be of importance in two studies, where tumor localization on the trigone [13] or bladder neck and dome [20] were associated with a worse prognosis. When superficial bladder tumors recur within a short period after primary resection, this also has a significant impact on the future recurrence rate [12], and progression [16].
4. *Reaction to intravesical instillation* provides important information [6,10]. Recurrent tumor at 3-months follow-up was an important prognostic factor in a multivariate analysis of Parmar et al. [10].
5. *Random bladder biopsies* provide information on concomitant abnormalities of the urothelium. Abnormal biopsy results (dysplasia and/or carcinoma in situ, CIS) suggest a higher chance of recurrence [15,21] or progression [14]. In the case of abnormal biopsy results tumor stage and grade are found to be higher [21]. The relation between the biopsy results and the chance of recurrence, however, is in most studies not corrected for other prognostic factors.

In this field the EORTC has also done much work. From multivariate analysis of different EORTC studies it appears that the prognosis with regard to recurrence is related to the number of previous recurrences, the grade of the tumor, and the number of tumors. With regard to the risk of progression into an invasive tumor, recurrence rate, grade, and the size of the tumor appear to be relevant.

With combinations of these factors it is possible to identify subgroups of patients with high or low risks for recurrent disease or progression.

Cellular and molecular biological parameters such as oncogenes, growth factors, DNA profile, chromosomal abnormalities and others are of limited or no direct clinical importance at this time.

Future Perspectives

It is clear now that both intravesical chemotherapy and immunotherapy play an important clinical role in the daily management of superficial bladder cancer. However, the optimal approach and optimal regimen has not yet been clearly defined and the *complete* prevention of recurrence of disease has not yet been achieved. It is questionable whether this goal will be reached in the (near) future. Indeed, as long as the mechanism of actions of the intravesical therapy are not fully understood and its relation to the etiology and biological behavior of (superficial) bladder cancer is not fully elucidated, intravesical therapy will remain a prophylactic palliation rather than an effective treatment.

Future investigations should follow two courses. First, a better understanding of carcinogenesis and progression of bladder carcinoma has to be obtained, upon which knowledge a rationale and effective therapy can be based. Second, a better insight into and understanding of clinical and more importantly biological prognostic factors has to be obtained. This will help us to individualize the therapeutic approach once a (superficial) bladder tumor has developed.

Agents responsible for bladder carcinogenesis can be identified in epidemiological studies. Some of the factors that play a role are recognized but their intrinsic mechanisms are poorly understood. It is clear that substances such as aniline derivatives, tryptophan, or nitrosamines play a role in the onset (initiation) and progression of both superficial and invasive bladder cancers but the molecular and biological background remains fully speculative. Evidently, molecular biological research has now provided some indications regarding genetic changes associated with bladder cancer. One of the most important findings in this respect has been published recently by Tsai et al. and Olumi et al. [22,23] who demonstrated that the allelic loss on chromosome 17P is predominantly found in invasive disease. The candidate tumor suppressor gene P53 is frequently mutated and may therefore be associated with progression [24]. However, these findings are rarely observed in superficial disease and the factors involved in recurrence and progression are unknown. To form metastases, bladder cancer cells must acquire properties for extracellular matrix degradation, escape immune surveillance and obtain an enhanced cell motility and growth autonomy: all phenomenons which are detectable in invasive disease but hardly present in superficial bladder cancer [25]. Even more accessible techniques such as (multiparameter) flow cytometry (FCM) are not in a position to differentiate those superficial tumors with the potential for recurrence and invasiveness. Although FCM can clearly differentiate between diploid Ta G1 and aneuploid T3 G3 tumors, there remains a large gray zone of tumors whose future behavior with respect to recurrence and progression cannot be predicted. Differentiation within and elucidation of this gray zone constitutes the challenge for the near future.

Shall we make further progress through applicable clinical research? The possibilities for advancement and the means to advance are certainly there. The big challenge, however, lies not only in the better prediction of prognosis of a given superficial bladder tumor, but also in the determination of an individualized approach to the tumor. As long as the etiology of bladder cancer cannot be influenced, the individualized selection of intravesical treatment is the best alternative for the future; this will almost certainly lead to the introduction of alternative agents and regimens, which have better results in the prevention of both recurrent disease and progression.

References

1. Zincke H, Utz DL, Taylor WF, Myers RP, Leary FJ (1983) Influence of thiotepa and doxorubicin at time of transurethral surgical treatment of bladder cancer in tumor recurrence: a prospective, randomized, controlled trial. J Urol 129:505
2. Haaff EO, Dresner SM, Kelley DR, Ratliff TL, Shapiro A, Catalona WJ (1985) Role of immunotherapy in the prevention of recurrence and invasion of urothelial bladder tumors: a review. World J Urol 3:76
3. Veenema RJ, Dean AL, Uso AC, Roberts M, Longo F (1969) Thiotepa bladder instillations: therapy and prophylaxis for superficial bladder tumors. J Urol 101:711
4. Riddle PR (1973) The management of superficial bladder tumors with intravesical epodyl. Br J Urol 45:84
5. Morales A, Eidinger D, Bruce AW (1976) Intracavitary bacillus Calmette-Guérin in the treatment of superficial bladder tumors. J Urol 116:180
6. Lamm DL, Steg A, Boccon-Gibod L, Morales A, Hanna MG Jr, Pagano F, Alfthan O, Brosman S, Fisher HAF, Jakse G, Chisholm GD, van der Meijden APM, Debruyne FMJ (1989) Complications of bacillus Calmette-Guérin immunotherapy: review of 2602 patients and comparison of chemotherapy complications. In: Debruyne FMJ, Denis L, van der Meijden APM (eds) EORTC Genito-Urinary Group monograph 6: BCG in superficial bladder cancer. Liss, New York, pp 335–355
7. van der Meijden APM, de Jong WH, de Boer EC, Steerenberg PA, Debruyne FMJ, Ruitenberg EJ (1989) Immunological aspects of intravesical administration of bacillus Calmette-Guérin (BCG) in the guinea pig. Urol Res 17:47–55
8. van der Meijden APM, Steerenberg PA, van Hoogstraten IMW, Kerckhaert JA, Schreinemachers LMH, Harthoorn-Lasthuizen EJ, Hagenaars AM, de Jong

WH, Debruyne FMJ, Ruitenberg EJ (1989) Immune reactions in patients with superficial bladder cancer after intradermal and intravesical treatment with bacillus Calmette-Guérin. Cancer Immunol Immunother 28:287–295

9. Brosman SA, Lamm DL, Meijden APM, et al. (1989) A practical guide to the use of intravesical BCG for the management of stage Ta, T1, CIS, transitional cell cancer. In: Debruyne FMJ, Denis L, van der Meijden APM (eds) EORTC Genito-urinary Group monograph 6: BCG in superficial bladder cancer. Liss, New York, pp 311–323
10. Parmar MKB, Freedman LS, Hargreave TB, Tolley DA (1989) Prognostic factors for recurrence and follow-up policies in the treatment of superficial bladder cancer: report from the British Medical Research Council Subgroup on Superficial Bladder Cancer (Urological Cancer Working Party). J Urol 142: 284–288
11. Cutler SJ, Heney NM, Friedell GH (1982) Longitudinal study of patients with bladder cancer: factors associated with disease recurrence and progression. In: Bonney WW, Prout GR (eds) Bladder cancer, AUA monograph, vol 1, chap 4: Williams & Wilkins, Baltimore, pp 35–46
12. Dalesio O, Schulman CC, Sylvester R, et al. (1983) Prognostic factors in superficial bladder tumors. A study of the EORTC group, GU group, J Urol 129:730–733
13. Loening S, Narayana S, Yoder L, et al. (1980) Analysis of bladder tumor recurrence in 178 patients. Urology 16:137–141
14. Heney NM, Ahmed S, Flanagan MJ, et al. (1983) Superficial bladder cancer: progression and recurrence. J Urol 130:1083–1086
15. Smith G, Elton RA, Beynon LL, et al. (1983) Prognostic significance of biopsy results of normal-looking mucosa in cases of superficial bladder cancer. Br J Urol 55:665–669
16. Lutzeyer W, Rübben H, Dahm H (1982) Prognostic parameters in superficial bladder cancer: an analysis of 315 cases. J Urol 127:250–252
17. Heney NM, Nocks BN, Daly JJ, et al. (1982) Ta and T1 bladder cancer: location, recurrence and progression. Br J Urol 54:152–157
18. Herr HW (1991) Progression of stage T1 bladder tumors after intravesical bacillus Calmette-Guérin. J Urol 145:40–44
19. Morgan JDT, Bowsher W, Griffiths DFR, Matthews PN (1991) Rationalisation of follow-up in patients with non-invasive bladder tumors. A preliminary report. Br J Urol 67:158–161
20. Stephenson WT, Holmes FF, Noble MJ, Gerald KB (1990) Analysis of bladder carcinoma by subsite. Cancer 66:1630–1635
21. Flamm J, Dona S (1989) The significance of bladder quadrant biopsies in patients with primary superficial bladder cancer. Eur Urol 16:81–85
22. Tsai YC, Nichols PW, Hitti AL, Williams Z, Skinner DG, Jones PA (1990) Allelic losses of chromosomes 9, 11, and 17 in human bladder cancer. Cancer Res 50:44–47
23. Olumi AF, Tsai YC, Nichols PW, Skinner DG, Cain DR, Bender LI, Jones PA (1990) Allelic loss of chromosomes 17p distinguishes high grade from low grade transitional cell carcinomas of the bladder. Cancer Res 50:7061–7063
24. Sidransky D, von Eschenbach A, Tsai YC, Jones P, Summerhayes I, Marshall F, Paul M, Green P, Hamilton SR, Forst P, Vogelstein B (1991) Identification of p53 gene mutations in bladder cancers and urine samples. Science 252:706–709
25. Brinquier PP, Schalken JA, Debruyne FMJ (1990) New markers for biological aggressiveness of bladder cancer: Proceedings of the 2nd International Symposium on recent advances in urological cancer diagnosis and treatment, 1990

When Should We Stop Intravesical Treatment in Patients with Superficial Bladder Cancer?

B. von Heyden and G. Jakse

Urologische Klinik, RWTH Aachen, Pauwelsstrasse 30, W-5100 Aachen, FRG

Introduction

Transurethral resection (TUR) is the treatment of choice for patients with superficial bladder cancer (i.e., Ta, T1 tumors). There are few patients in whom cystectomy is indicated as the initial treatment. External beam irradiation or brachytherapy is applied only in special centers. External beam irradiation certainly does not lower the recurrence rate or improve survival compared to TUR [22]. Interstitial radium implants used in a very selected group of patients seem to be highly effective, both in terms of survival and recurrence rate [26].

Carcinoma in situ (Tis) is a disease entity which needs separate consideration not only because of different biological behavior but also because of different treatment strategy. Tis may be diagnosed as the first manifestation of a transitional cell carcinoma of the bladder or in association with a visible tumor; moreover, it is found in patients followed after initial TUR of a Ta, T1 tumor. Most of the authors reporting on the treatment of Tis consider intravesical instillations of chemo- or immunotherapeutic agents as an appropriate first-line treatment modality [3,12,15].

Aim of Intravesical Treatment

Regarding the intravesical application of different drugs for patients with superficial bladder cancer we have to differentiate between the short- and long-term goals of the treatment. Intravesical treatment is applied in patients (a) with residual tumor who cannot be resected completely transurethrally and who are not candidates for cystectomy, (b) with Tis, and (c) with Ta, T1 tumors after complete TUR in order to prevent tumor recurrence and tumor progression. Patients with Tis and Ta, T1 tumors are considered in this report.

Recent Results in Cancer Research, Vol. 126

Carcinoma in Situ

In some 80% of patients with Tis, disease progresses if treated by TUR alone. The initial aim of the intravesical treatment is the complete destruction of Tis, thus preventing tumor progression and the need for more aggressive treatment such as cystectomy. This aim can be achieved in a considerable percentage of patients by means of intravesical chemo-or immunotherapy. However, some 50% of patients will experience a recurrent tumor within the bladder, ureter, and prostatic urethra respectively [15].

Ta, T1 Transitional Cell Carcinoma of the Bladder

The rate of tumor recurrence after TUR depends on several factors such as tumor grade, multiplicity, size, invasion, and concurrent urothelial atypia or Tis. Depending on these prognostic factors the recurrence rate and the occurrence of progressive disease varies considerably. Whereas multiplicity and size are the most important parameters for recurrence rate, tumor progression is influenced mainly by tumor grade, stage, and associated Tis [5,7,13,14,18,20,23].

The aim of intravesical treatment after TUR in Ta, T1 tumors, therefore, is twofold: (a) reduction in recurrence rate and prolongation of disease-free interval, and (b) prevention of tumor progression and death due to progressive tumor.

There is a considerable amount of data which indicate that intravesical treatment after TUR can prolong the disease-free interval and reduce the recurrence rate [18,19,24]. However, the percentage of patients who experience a recurrent tumor is influenced only marginally. The data recorded in terms of tumor progression and tumor death are less impressive.

Fitzpatrik et al., Fossa et al., and Soloway demonstrated that for patients with residual disease who had complete tumor remission with intravesical chemotherapy, the chance of tumor progression and cancer death was considerably lower than for those patients who demonstrated only partial or no response [6,8,24]. Green et al. indicated that patients treated by intravesical chemotherapy after complete TUR of a superficial bladder cancer may have some advantage in terms of tumor progression and cancer death compared to those who do not receive such prophylactic treatment [9]. However, Rübben et al. could not detect any such advantage [23]. But one must remember that the risk of tumor progression is insignificantly low in patients with noninvasive, well to moderately differentiated tumors [14]. Moreover, Kurth et al. demonstrated that patients initially assigned to the no-treatment arm received adjuvant treatment in the case of recurrence; that means, in essence in most studies reported, the authors compare immediate versus delayed adjuvant treatment [14]. This makes it imperative that the data be read carefully and only reports with a meaningful number of patients with a risk of progressive tumor should be considered. A review of the literature by Herr and Jakse on this subject indicates that chemotherapy and immunotherapy with bacille Calmette-Guérin (BCG) can reduce tumor progression and cancer death in this high-risk group [11].

Termination of Intravesical Treatment

Several of the reasons for stopping treatment are self-explanatory and will not be dealt with in detail. Noncompliant patients, patients in whom systemic or local toxicity prevents the programmed application of intravesical treatment, and tumor progression in terms of stage are among those reasons which undoubtedly indicate termination of treatment.

However, there are other reasons for stopping treatment which are less evident, including treatment in recurrent tumors of same stage, second- and third-line treatment, or in the case of tumor persistence in patients with Tis.

Retreatment in the Case of Tumor Recurrence

There are a considerable number of patients who seem to respond to intravesical treatment in terms of prolongation of tumor-free interval and decrease of tumor recurrence rate. How should we deal with such a patient who experiences a new tumor after we have stopped the scheduled chemo- or immunoprophylaxis?

The report of Dalesio et al. indicates that a recurrent tumor is a bad prognostic sign with regard to further tumor recurrences [5]. However, it has been shown by Bretton et al. and Kavoussi et al. that retreatment with BCG in the case of tumor recurrence is effective [2,16]. Bretton et al. reported on 374 patients who were treated with TUR and six weekly instillations of BCG. Of these, 28 patients experienced tumor recurrences which the authors felt were refractory to TUR alone [2]. They gave these patients a second course of BCG. Of these 28 patients 13 subsequently needed a cystectomy, but 15 could be controlled with BCG. Only one died due to progressive disease.

Kavoussi et al. treated 55 patients after complete TUR of Ta, T1 tumors with six .weekly instillations of BCG [16]. Of these, 20 (36%) remained tumor free with an average follow-up of 21 months; 29 patients experienced tumor recurrence and received a second course of BCG. Of these 29 patients, 19 (65%) remained tumor free, but 10 experienced another tumor recurrence and 1 of these died due to progressive disease.

The report of Catalona et al. noted an increased risk of local tumor progression and metastatic disease in patients who received more than two courses of BCG [4]. Tis or involvement of the prostatic urethra were considered as bad prognostic signs. However, we must take into account the fact that this report dealt with a mixed group of patients (Tis, 29; residual disease, 20; and prophylaxis, 29); moreover, the study did not differentiate between patients with persistent or recurrent tumors.

In the search for indicators which would predict that retreatment would be effective and suggest a low risk of tumor progression, Bretton et al. showed that the longer the interval between initial tumor and tumor recurrence, the better the result [2]. The crucial interval determined by this group was some 21 months.

Moreover, it was demonstrated by Herr et al. that the following variables were significant with regard to tumor progression in the case of BCG intravesical treatment: (a) before BCG treatment, there was evidence of T1 tumor and the duration of disease was less than 1 year, (b) at 3 months after BCG treatment, there was evidence of T1 tumor and the duration of disease was less than 1 year, with positive cytology, and (c) at 6 months after BCG treatment, there was evidence of T1 tumor, positive cytology, and positive biopsy for Ta, T1 tumor [12]. Prognostic risk groups were best defined after 6 months following BCG treatment. The risk of tumor progression for a patient with a T1 tumor at 6 months at check cystoscopy was 71% within 1 year of treatment and 100% at 3 years. Therefore, in the case of a recurrent or refractory T1 tumor at 6 months, a retreatment should not be considered; immediate cystectomy is the treatment of choice for such a patient.

Extravesical Tumor

Bretton et al. reported on patients with Tis extending into the prostatic urethra [1]. TUR and intravesical BCG was applied in 23 patients; progression occurred in 10 (44%) of the patients. This is a risk which should not be underrated and contrary to the suggestion of Bretton et al. we strongly recommend cystectomy where there is any involvement of the prostatic urethra [1,15].

Second- and Third-Line Therapy

Koontz et al. reported on 85 patients who failed thiotepa therapy [17]. These patients had Ta, T1, Tis, or Tx residual disease. In 26% of these patients the bladder was cleared of residual disease by 8 instillations of 40 mg mitomycin C. None of these patients showed tumor progression during treatment.

Soloway and Ferry treated 30 patients with Tis, Ta, T1 tumor after they had already received mitomycin C [18], thiotepa [24], or both agents [14] with 6-weekly instillations of BCG [25]. Only 65% (Ta), 25%(T1), and 5 out of 9 with Tis became tumor free with this additional treatment.

However, the authors stress the fact that there is a considerable risk of tumor progression during therapy.

Tumor Persistence

Jakse treated 45 patients with Tis initially with doxorubicin [15]. A complete remission was obtained in 70%. Three of the nonresponding patients were treated with mitomycin C and in all three a complete remission was obtained. An immediate second course of BCG for patients with Tis who do not respond can also be safely given, as shown by Pagano et al. and Kavoussi et al. [16,21]. However, the situation is very different for patients who initially have a T1 tumor. Herr et al. quite nicely demonstrated a significantly increased risk of tumor progression for those patients who still have T1 tumors at 3 months cystoscopy check compared to those who do not have exophytic tumor or Tis and show no malignant cells in the bladder washing [10].

Patients with residual Tis at 6 months certainly have an increased risk for tumor progression, but this occurs only within 2–3 years. For this reason, Herr et al. suggested that a second-line intravesical treatment for these patients may be indicated with reasonable safety [11].

Conclusions

Intravesical adjuvant treatment is probably not necessary for patients with well-differentiated, noninvasive bladder cancer, since the risk of tumor progression is so low that TUR alone seems to be the appropriate treatment. On the other hand for patients with T1, G3 tumors there is a high risk of tumor progression and adjuvant treatment is necessary. Undoubtedly, cystectomy offers the best chance of permanent cure – especially in these patients. However, cystectomy has a considerable impact not only in terms of short-term morbidity but also in terms of permanent changes to quality of life.

The crucial problem with intravesical treatment given as an adjuvant and therapeutic intervention (for patients with Tis) is how to accurately anticipate disease progression at a time when cystetcomy would still cure the patient. At the present time, therapy strategies should be based on prognostic factors which can be determined from large groups of patients whose tumor characteristics have been exactly defined and the follow-up well documented [5,10,18].

However, an individualized approach using specific markers should be the goal we are aiming for – being able to recognize the moment when we have to stop a conservative policy and offer a more aggressive therapy. As shown above, we certainly have to stop intravesical treatment in patients who demonstrate (a) failure to respond to second-line therapy, (b) tumor relapse and adverse prognostic factors, (c) persistence of Tis after second course of BCG or 6 months of chemo- and immunotherapy, and (d) Tis of the prostatic urethra. In these cases the risk of tumor progression and death due to cancer increases significantly and the course of the disease in terms of recurrence rate or tumor-free interval cannot be influenced positively. Other factors such as noncompliance, local and systemic toxicity, and tumor progression during therapy are further considerations which go without saying.

References

1. Bretton PR, Herr HW, Badalament RA. The treatment of superficial transitional cell carcinoma of the bladder with mucosal and/or ductal involvement of the prostatic urethra with intravesical BCG. J Urol 141:853–856
2. Bretton PR, Herr HW, Kimmel M, Whitmore WF, Laudone V, Oettgen HF, Fair WR (1990) The response of patients with superficial bladder cancer to a second course of intravesical BCG. J Urol 143:710
3. Brosman SA (1985) The use of BCG in the therapy of bladder carcinoma in situ. J Urol 134:36
4. Catalona WJ, Hudson MA, Gillen DP, Andriole GL, Ratliff TL (1987) Risks and benefits of repeated courses of intravesical bacillus Calmette-Guérin theapy for superficial bladder cancer. J Urol 137:220
5. Dalessio O, Schulman CC, Sylvester R, dePauw M, Robinson M, Denis L, Smith PH, Viggiano G (1983) Prognostic factors in superficial bladder cancer. J Urol 129:730
6. Fitzpatrick JM, Khan O, Oliver RT, Riddle PR (1979) Longterm follow-up in patients with superficial bladder tumours treated with intravesical Epodyl. Br J Urol 51:545
7. Flamm J, Havelec L (1990) Factors affecting survival in primary superficial bladder cancer. Eur Urol 17:113
8. Fossa SD (1985) The need for long term follow-up studies after treatment of superficial bladder cancer, especially after intravesical chemotherapy. In: EORTC Genitiurinary Group monograph 2, part B: superficial bladder tumors. Liss, New York, pp 31–37
9. Green DF, Robinson MRG, Glashan R, Newling D, Dalesio O, Smith PH (1984) Does intravesical chemotherapy prevent invasive bladder cancer? J Urol 131:33
10. Heney N, Ahmed S, Flanagan MJ, Frable J, Corder M, Hafermann M, Hawkins I (1983) Superficial bladder cancer: progression and recurrence. J Urol 130:1083
11. Herr HW, Badalament RA, Amato DA (1989) Superficial bladder cancer treated with BCG: multivariate analysis of factors affecting tumor progression. J Urol 141:22
12. Herr H, Jakse G, Sheinfeld J (1991) The T1 bladder tumor. J Urol (submitted)
13. Herr H, Pinsky CM, Whitmore WF (1986) Long-term effect of intravesical bacillus Calmette Guérin on flat carcinoma in situ. J Urol 135:265
14. Jakse G, Loidl W, Seeber G, Hofstädter F (1984) Stage T1, grade 3 transitional cell carcinoma of the bladder; an unfavorable tumor. J Urol 137:39

15. Jakse G (1987) Die Stellung der Zystektomie in der Behandlung des in situ Karzinoms der Harnblase. Akt Urol 18:255
16. Kavoussi LR, Torrence RJ, Gillen GP, Hudson MA, Haaff EO, Dresner SM, Ratliff TL, Catalona WJ (1988) Results of 6 weekly intravesical bacillus Calmette-Guérin instillations on the treatment of superficial bladder tumors. J Urol 139:935
17. Koontz W Jr, Heney NM, Soloway MS, Platz C, Hazra TA, Trump DL (1985) Mitomycin C for patients who have failed on thio-tepa. Urology 26 [Suppl 4]: 30–31
18. Kurth K-H, Schroeder FH, Debruyne F, Senge T, Pavone-Macaluso M, de Pauw M, ten Kate F, Sylvester R (1989) Long-term follow-up in superficial transitional cell carcinoma of the bladder: prognostic factors for time to first recurrence, recurrence rate and survival. In: Murphy GP, Khoury S (eds) Therapeutic progress in urological cancer. Liss, New York, pp 481–490
19. Kurth K-H, Sylvester R, dePauw M, tenKate F (1989) Intracavitary treatment of transitional cell carcinoma of the bladder: questions and lessons after 27 years of experience. Prog Clin Biol Res 310:40
20. Lutzeyer W. Rübben H, Dahm H (1982) Prognostic parameters in superficial bladder cancer: an analysis of 315 cases. J Urol 127:250
21. Pagano F, Bassi P, Milam C (1989) Low-dose BCG-Pasteur strain in the treatment of superficial bladder cancer: preliminary results. Prog Clin Biol Res 310:253
22. Quilty PM, Duncan W (1986) Treatment of superficial tumor of the bladder by radical radiotherapy. Br J Urol 58:147
23. Rübben H, Lutzeyer W, Fischer N, Deutz F, Lagrange W, Giani G (1988) Natural history and treatment of low and high risk superficial bladder tumors. J Urol 139:283
24. Soloway MS (1988) The management of superficial bladder cancer, with emphasis on intravesical chemotherapy. In: Raghavan D (ed) The management of bladder cancer. Arnold, London
25. Soloway MS, Ferry A (1987) BCG for treatment of superficial transitional cell carcinoma of the bladder in patients who have failed thiotepa and/or mitomycin C J Urol 137:871
26. van der Werft-Messing B, Hop WCJ (1981) Carcinoma of the urinary bladder treated either by radium implant or by transurethral resection only. Int J Radiat Oncol Biol Phys 7:299
27. Wolf H, Hojgard K (1983) Prognostic factors in local surgical treatment of invasive bladder cancer with special reference to the presence of urothelial dysplasia. Cancer 51:1710

Superficial Bladder Cancer: Primary Treatment with TUR with or without Laser-Coagulation

D. Jocham

Abteilung für Urologie, Medizinische Universität zu Lübeck, Ratzeburger Allee 160, W-2400 Lübeck, FRG

Local treatment of bladder carcinoma includes transurethral resection (TUR) alone, combination therapy with electrocoagulation, laser treatment with thermal effect, or adjuvant local therapy with chemotherapeutics or immunomodulators. To what extent superficial bladder carcinoma can be treated most effectively is not yet known and there are no clear parameters with regard to time to progression, survival time, and quality of life. As far as laser application is concerned, available data indicate a reduction in the local recurrence rate under treatment with the neodymium-YAG laser.

The American urologist and laser specialist J.A. Smith [19] commented on a paper by Staehler published in the Journal of Urology in 1985 [21]: "Although no definitive therapeutic advantages for laser treatment of bladder cancer have been established, speculation has centered around the possibility that tumor recurrence due to implantation of viable cells in remote areas of the bladder would be decreased after laser therapy. If a preliminary electrocautery resection of the tumor is performed before laser treatment this theoretical advantage would seem to be eliminated. Although further investigation is necessary lasers do appear to offer some practical advantages over electrocautery resection and excellent results have been obtained in uncontrolled series" [23]. This statement concerning laser therapy remains valid in the light of the results of prospective randomized studies [3,7].

Experience has proven that the neodymium-YAG laser is also suitable for the treatment of benign condylomata, hemangioma, and tumors outside of the bladder. Laser treatment can also be applied on an outpatient basis [22]. A neodymium-YAG laser costs approximately U.S.$80000, an expenditure which has to be justified, and may be justified for the treatment of bladder cancer considering the ease of application and reduced risk. No hard data exist concerning the curative efficiency of laser application and cost reduction as compared to alternative procedures. National differences in insurance systems also have to be taken into account when assessing costs. In view of

Recent Results in Cancer Research, Vol. 126

Table 1. Transurethral electrocautery resection

Indication	
Diagnostic	
Risk of understaging	TA <2%
	T1 5%
	T2/3 39%
Therapeutic	
Curative	TA, T1, T2 (?)
Palliative	T3/T4
Side effects	
Bleeding	
Perforation	
Obturator nerve stimulation	
Spread of tumor cells	

these considerations all one can do is present a compilation of data available on the treatment of superficial bladder cancer by TUR and thermal laser therapy to make a better assessment of laser therapy.

The basic therapy of superficial bladder cancer (Ta and T1 tumors) is the transurethral electroresection of the tumor. The tumor is removed from the bladder wall or reduced in size. Step resection is performed for tumor staging. Superficial bladder carcinoma can be detected with high accuracy if the TUR is carried out correctly. Understaging of the TA tumor occurs in less than 2% of cases, understaging of the T1 tumor in 5%. In T2 and T3a tumors, understaging occurs in 39% of cases [20] (Table 1).

Curative resection of T2 infiltrating tumors is open to debate. For T3 and T4 tumors, TUR is only diagnostic or palliative. Risks associated with TUR include postoperative bleeding, perforation of the bladder wall, and, in resection at the lateral wall, stimulation of the obturator nerve. Nidation of vital tumor cells in other mucosal areas of the bladder may occur following tumor manipulation. Soloway [20] demonstrated this in previously damaged mucosa.

The prognosis of bladder tumors, especially regarding risk of progression, depends on the tumor stage and grade of malignancy. The TaG1 tumor has a considerably lower local risk of progression, even with multiple recurrences, compared to superficial tumors with a high grade of malignancy (G3) or a marked aneuploidy.

Following TUR, a progressive stage-related recurrence is found in less than 4% of patients with primary Ta tumors, in 19% of those with T1G1 tumors, and in 30%–37% of those with T1G3 tumors [5,11,12] (Table 2). As well as by local progression, the prognosis of bladder carcinoma is also determined by metastatic disease. Less than 1% of exclusively superficial bladder tumors independent of the grade of differentiation are accompanied by metastases during an observation period of 5 years. Patients with tumor infiltration of the lamina propria have a maximal 22% risk of develop-

Table 2. Bladder tumor prognosis correlated with tumor stage and grade

Local tumor progression	Rate (%)
TA	<4
T1/GI	19
T1/GII	30–37
Metastases (5-year follow-up)	
TA (GI–GIII)	<1
T1	22

ing metastases. A higher recurrence rate is observed in patients with a common multifocal tumor growth than in patients with a rare solitary tumor. Severe dysplasia or carcinoma in situ with tumor foci that cannot be treated adequately by TUR may account for this. Lundbeck [11], however, was unable to confirm this evaluation.

According to the literature, the recurrence rate following TUR ranges from 29% to 97% for patients with Ta/T1 tumors within 5 years, with at least 50% recurrence within the first 24 months [6]. Ta tumors have fewer recurrences than T1G3 tumors [12,20].

How does the laser fare compared to other treatment modalities? The demands on a therapeutic procedure which either supplements or replaces TUR are not only the reduction of recurrences but also an improvement in survival rates. As a rule, this can only be achieved by reducing local progression and the risk of developing a metastasis. Special consideration should therefore be given to T1 and T2 tumors.

Before discussing clinical applications of the laser in the treatment of the bladder cancer, some fundamental principles are first described.

New possibilities for laser application in urology were first investigated in the early 1970s, and in this field, three major areas of laser application have become especially important:

1. Thermal destruction of tumors
2. Photodynamic procedures for diagnosis and therapy of tumors
3. Intracorporeal laser lithotripsy – particularly applied to the disintegration of ureteral stones

Laser has been used to achieve thermal tissue destruction for many years. In addition to CO_2 and argon lasers, which are no longer significant in urology, the neodymium-YAG laser is now enjoying increasing popularity. Light transmission of the neodymium-YAG laser (Table 3) through water is high at a wavelength of 1064 nm, facilitating endoscopic application, e.g., in the bladder. Even in tissue, absorption of this laser radiation is comparatively low, resulting in a deep-acting heat effect with little damage to the superficial tissue, provided homogeneous irradiation is guaranteed. The depth of necrosis at a maximum 45 w is 3 mm when applied in the air and 6 mm when

Table 3. Application of the neodym-YAG laser in the therapy of bladder tumors

External	15–25 w
Endoscopic	max. 45 w
Pulse duration	2 s
Depth of necrosis	max. 6 mm

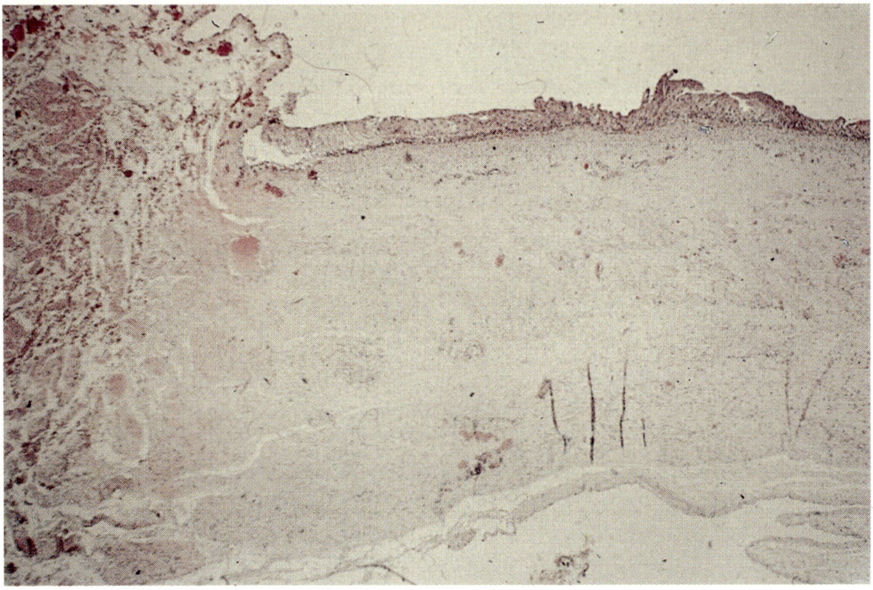

Fig. 1. Thermal denaturation resulting from laser application at 60°–100°C. (By courtesy of E. Keiditsch, M.D., Munich, FRG)

applied under water. The tissue-damaging mechanism is based on thermal denaturation of the tissue.

After reaching temperatures of 60°–100°C, three areas of damage can be detected histologically (Fig. 1): A central necrosis, a hemorrhagic infarction of the surrounding tissue, which is secondarily necrotized and decomposed, and a peripheral edema. In addition to direct thermal denaturation, enzymes are inactivated at 40°–50°C, damaging the cell physiology in the cell borders. The release of photons may additionally result in direct damage to the cell nucleus.

Neodymium-YAG laser monotherapy completely destroys tumors less than 5 mm in diameter (Fig. 2). Histological examination of the tumor tissue is possible even after laser application, although it is more difficult. It is unknown how effectively contact-free laser avoids iatrogenic tumor cell

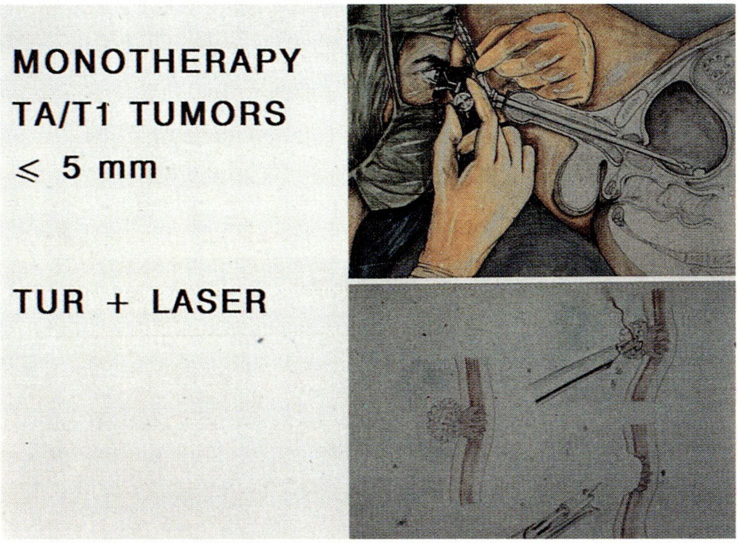

Fig. 2. *Above*: Neodymium-YAG laser monotherapy of bladder tumor. Monotherapy is completely effective in the case of TA and T1 tumors less than 5 mm in diameter. *Below*: Combined TUR and neodymium-YAG laser therapy

implantation; this may well result in a clinically significant decrease in the recurrence rate of bladder tumors.

For larger tumors, laser is applied as a supplement to TUR – especially for the irradiation of the tumor edge and the resection bed. Circular laser treatment or radiation of the resection bed is performed with the object of (a) reducing the recurrence rate, and (b) avoiding lymphogenous spread of tumor cells [7].

Data concerning the avoidance of lymphogenous spread of tumors cells by laser treatment are controversial. According to Keiditsch [9] the spread of ink particles from the bladder wall to regional lymph nodes can be experimentally interrupted following irradiation; Loughlin and colleagues [10], however, did not find any differences between TUR and YAG laser therapy with regard to the spread of an isotope into lymphatics beyond the bladder.

Are there other differences between electrocautery and the neodymium-YAG laser therapy? Histologically different profiles of zones of necrosis following TUR or laser therapy can, in fact, be observed. A more homogeneous, clearly confined zone of necrosis throughout the whole bladder wall is demonstrated only in the case of laser application [7].

Beisland and Kvernebo [2] agree with these findings; they found a sharply defined high-grade interruption in the blood circulation in the bladder wall following the application of the neodymium-YAG laser which was not observed following TUR or coagulation.

Although these results favor the application of the neodymium-YAG laser, an evaluation of the procedure is only useful when based on clinical data, and herewith lie the difficulties. Retrospective and several prospective, randomized studies have been based on recurrence rates within observation periods of a maximum of 2 years and not on the evaluation of the local progression rate, the development of metastases, or survival rates [1,8, 13–15,17,18,21]. There is a total absence of clinically relevant long-term results. What does the literature tell us?

In a retrospective analysis of superficial tumors, Shanberg and colleagues [14] observed no recurrence within 2 years. In infiltrating bladder cancer, the general recurrence rate has not been shown to be statistically significantly reduced. General advantages of laser application have been presented including shorter hospitalization periods, and quicker patient recovery [4].

In the case of infiltrating tumors with unfavorable prognosis, Smith [17] shows pathological biopsy findings following laser therapy in 20%–100% of cases.

Representing the investigations of Beisland, Oslo; Hofstetter, Munich and Luebeck; Meyer, Vienna; and Staehler, Munich, the results of the prospective studies of Beisland and Seland [3] and the results of Hofstetter [7] will be discussed. Beisland and Seland [3] studied 122 patients who were treated either by TUR alone or by mechanic removal of the tumor and subsequent laser irradiation. Tumor recurrence in the treated area was significantly lower in the laser group (4.8%) than in the TUR group (31.6%).

This is even observed when tumor recurrence is related to tumor stage and tumor grade. Local recurrences in the laser group occurred only in the group with infiltrating carcinomas of high malignancy.

As anticipated, the general recurrence rate of the tumors – an essential aspect of disease prognosis – was not reduced.

In 1984, Hofstetter et al. [7] reported interim results of a prospective comparative study of TUR alone, of laser therapy with or without previous TUR, and the application of these procedures in combination with post-operative mitomycin instillation. The small number of patients, the lack of a concluding analysis, and the short observation period of 12 months have been criticized. The results – in the treatment of primary tumors – indicate the superiority of laser therapy compared to TUR, particularly in combination with local cytostasis. The small number of cases does not allow any analysis with regard to tumor recurrence. If results for the different groups are combined, the local recurrence rate was reproducibly lower with laser therapy than with TUR. As mentioned before, it has been shown that the general recurrence rate is not affected – a decisive factor as regards the fate of the patient [7].

When considering the advantages of laser treatment compared to TUR, reduced patient strain and reduced risk of the intervention such as irritation of the obturator nerve and perforation should be mentioned. Postoperative placement of a transurethral catheter following laser therapy is generally not required. Laser application is also conceivable on an outpatient basis, at least for the treatment of small tumors. In the treatment of superficial tumors that are widely spread within the bladder laser application is also a speedier procedure [4].

In conclusion, the American urologist and laser specialist J.A. Smith [19] has qualified his previously mentioned comment on Staehler's 1985 paper [21]:

"Overall, lasers have not revolutionized the treatment of bladder cancer. Nevertheless, there are some potential therapeutic advantages for patients with superficial disease and some well-demonstrated practical advantages. These advantages, coupled with the inevitability of new developments in laser instrumentation and technique, imply that lasers will have a lasting role in the treatment of patients with bladder cancer".

References

1. Beisland HO, Sander S, Fossberg E (1985) Neodymium-YAG laser irradiation of urinary bladder tumors: follow-up study of 100 consecutively treated patients. Urology 25:559–563
2. Beisland HO, Kvernebo K (1986) The microcirculation in Neodymium-YAG laser irradiated and in electrocoagulated urinary bladder tumors evaluated with laser doppler flowmetry. Urol Res 14:149–152
3. Beisland HO, Seland P (1986) A prospective randomized study on neodymium-YAG laser irradiation versus TUR in the treatment of urinary bladder cancer. Scand J Urol Nephrol 20:209–212
4. Fowler CG, Boorman LS (1986) Outpatient treatment of superficial bladder cancer. Lancet i (8471):38 (letter)
5. Heney NM, Ahmed S, Flanagan MJ, Frable W, Corder MP, Hafermann MD, Hawkins IR (1983) Superficial bladder cancer: progression and recurrence. National bladder cancer collaborative group A. J Urol 130:1083–1086
6. Herr HW, Pinsky CM, Whitmore WF Jr, Oettgen HF, Melamed MR (1983) Effect of intravesical bacillus Calmette-Guérin (BCG) on carcinoma in situ of the bladder. Cancer 51:1323–1326
7. Hofstetter AG, Keiditsch E, Schmiedt E, Frank F (1984) Der Neodym-YAG laser in der Urologie. Derzeitiger Stand der klinischen Erfahrungen. Fortschr Med 102:885–890
8. Hofstetter A (1985) Laser – Science fiction oder eine neue Dimension in der Medizin? Urologe [A] 24:310–312
9. Keiditsch E, Stern J, Zimmermann I, Hofstetter A, Frank F, Pensel J, Rothenberger K (1981) Interruption of urinary bladder wall lymph drainage by neodymium-YAG laser irradiation. In: Atsumi K, Nimsakul N (eds) Laser Tokyo '81. Inter Group, Tokyo, pp 10–48
10. Loughlin KR, Kaplan W, Forman R, Laffin S (1986) The relative efficacy of the YAG laser versus electrocautery in preventing radionuclide spread from the bladder wall. J Urol 135:125A
11. Lundbeck F, Jacobsen F, Vaeth M (1989) Impact of the category of early recurrence on tumor progression in bladder tumors treated by transurethral resection. Eur Urol 16:291–294

12. Lutzeyer W, Rübben H, Dahm H (1982) Prognostic parameters in superficial bladder cancer: an analysis of 315 cases. J Urol 127:250–253
13. Malloy TR, Wein AJ, Shanberg A (1984) Superficial transitional cell carcinoma of the bladder treated with the neodymium-YAG laser: a study of the recurrence rate within the first year. J Urol 131:251
14. Shanberg AM, Baghdassarian R, Tansey LA (1987) Use of ND-YAG laser in treatment of bladder cancer. Urology 29:26–30
15. Smith JA JR, Dixon JA (1984) Argon laser phototherapy of superficial transitional cell carcinoma of the bladder. J Urol 131:655–656
16. Smith JA JR, Dixon JA (1984) Neodymium-YAG laser irradiation of bladder hemangioma. Urology 24:134–136
17. Smith JA JR (1986) Treatment of invasive bladder cancer with a neodymium: YAG laser. J Urol 135:55–57
18. Smith JA JR (1985) Laser treatment of bladder cancer. Semin Urol 3:2–9
19. Smith JA JR (1989) Current concepts in laser treatment of bladder cancer. In: Therapeutic progress in urological cancers. Liss, New York, pp 463–469
20. Solowy MS (1989) Diagnosis and management of superficial bladder cancer. Semin Surg Oncol 5:247–254
21. Staehler G, Chaussy C, Jocham D, Schmiedt E (1985) The use of Neodymium-YAG lasers in urology: indications, technique and critical assessment. J Urol 134:1155–1160
22. Warner RS, Bloiso G, Cohen MS (1985) Office laser surgery for urologists. Urology 25:363–364
23. Wolf H, Hojgaard K (1983) Urothelial dysplasia concomitant with bladder tumors as a determinant factor for future new occurrences. Lancet ii:134–136

*Immunocytology as a Possible Marker to Identify Patients Who Require Prophylaxis**

E. Huland, A. Schneider, and H. Huland

Abteilung für Urologie, Universität Steglitz, 1000 Berlin, FRG

Introduction

Many studies have already proved that prophylaxis can influence the natural course of disease in patients who have had a complete transurethral resection (TUR) in the case of a superficial bladder carcinoma. Our studies have focused on the use of mitomycin C bladder instillations after TUR (Tables 1 and 2). In a prospective, randomized unicenter controlled study and in a multicenter study of 425 patients, we have shown that it is possible to (a) reduce the recurrence rate from 60% to 15%, (b) reduce the progression rate from 25% to 4%, and (c) influence the survival time of patients (Table 2). Of particular note in our study is the fact that we included all consecutive patients, and our results show that in our hospital overtreatment – or better, overprophylaxis – is common. Data from our control group showed that at least 40% of patients who were prescribed prophylactic treatment will not develop recurrent tumors, even without prophylaxis. Thus, the most important issue in our own and many other studies is: Can those patients who will not develop a recurrent bladder tumor and who therefore do not need prophylactic therapy be reliably identified?

Most physicians use two information sources to decide whether or not a patient should receive prophylaxis after complete TUR, namely the patient's medical history and the analysis of the removed tumor. Prophylaxis is commonly prescribed for patients who have either already had tumor recurrences in the past or who presently have T1, grade II, and grade III lesions. Prophylaxis is generally not recommended for patients with a first-time occurrence of a small, solitary TA grade I lesion. This common criterion is based on data from numerous studies which have demonstrated that recurrence and progression rates correlate with size, grade, stage, and

* This work is supported by the Deutsche Krebshilfe.

Table 1. Results of study on the use of mitomycin C bladder instillations after TUR

	Control group		Mitomycin C group	
	(*n*)	(%)	(*n*)	(%)
Total no. of patients	31		54	
Recurrent tumor	17	54.8	6	11.1
Progression:				
Stage	8		2	
Grade	4		1	
Recurrences per 100 patient months	3.7		0.552	
Death:				
With metastases	6	19.4	1	1.9
Without tumor	3	9.7	5	9.3
Alive with metastases			1	

Data from the Urology Clinic of the University of Hamburg, FRG.

Table 2. Results of study on bladder carcinomas[a]

	M1	M2	M3	M4
No. of patient	209	98	72	40
Treatment well tolerated	64	68	64	50
Recurrence rate (%)	16	8	17	20
Progression rate (%)	3	1	6	8
Deaths (*n*)	3	1	2	0

[a] Average follow-up time of 30 months.

other parameters of these tumors. Such an approach is, however, in our opinion, not precise enough to predict the course of disease in an individual patient. Studies with a longer follow-up of patients with superficial bladder carcinomas have shown that there is only a gradual difference with respect to recurrence and progression rates among patients who have tumors of different stage and grade (Fig. 1). This suggests that there might be a small group of so-called TA grade 0 tumors that actually pose no real risk of recurrence or progression. Our group therefore concentrated on improving diagnosis in order to accurately identify those patients with a superficial bladder carcinoma who have no risk of recurrent and progressive disease and who therefore need no prophylactic treatment.

We developed several monoclonal antibodies (MAB) directed against tumor-associated antigens in transitional cell carcinomas of the bladder. We named one of them 486P corresponding to the cell line used for immunization. MAB 486P is an IgM antibody which detects glycoprotein in the cell

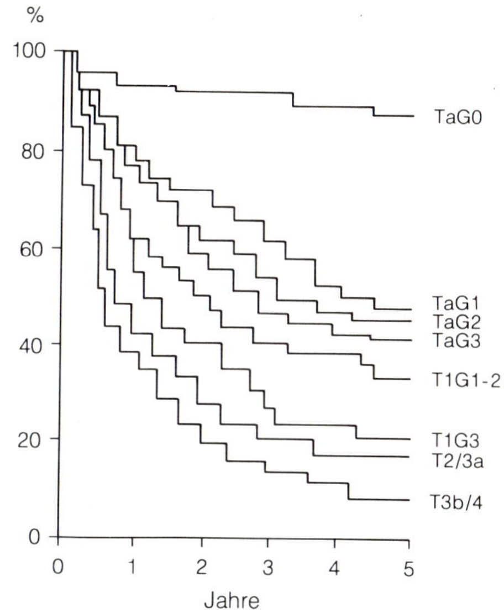

Fig. 1. Recurrence rates over 5 years for patients ($n = 1646$) with superficial bladder carcinomas according to stage and grade of tumor. (From Rübben et al.)

membrane and has a high molecular weight of 200 000 daltons. In immune histological tests it reacts with 90% of superficial bladder carcinomas. MAB 486P was extremely helpful in identifying cells in the urine of patients with bladder carcinoma. Using indirect immune cytology with the help of an alkaline phosphatase reaction, this monoclonal antibody stained all malignant transitional cells in some patients (Fig. 2). In most urine specimens of patients with bladder carcinoma, however, heterogenous cell staining occurs, i.e., some morphologically similar malignant cells react with MAB 486P but some do not (Fig. 3). Due to this heterogenous reaction and to the fact that, in some normal patients, benign cells occasionally react to this MAB, we did a quantitative analysis of urine specimens whereby at least 100 transitional cells were counted in a given urine sample and the percentage of cells that reacted to MAB 486P was calculated (Fig. 4). We thus could see in a control group that half of the patients with benign prostatic hypertrophy (BPH) or urinary tract infection (UTI) do not react with the MAB at all, yet 50% – have positive cells but not more than 30% positive stained cells. We took this as a baseline dividing positive and negative quantitative immunocytology. In our first study we were thus able to show that, of 40 patients with superficial bladder carcinomas, 36 (90%) had a positive quantitative immunocytology according to the MAB 486P. The high rate of 90% sensitivity was much better than the 50% obtained by using standard cytology. The definite advantages of quantitative immunocytology compared to standard cytology were shown in a follow-up

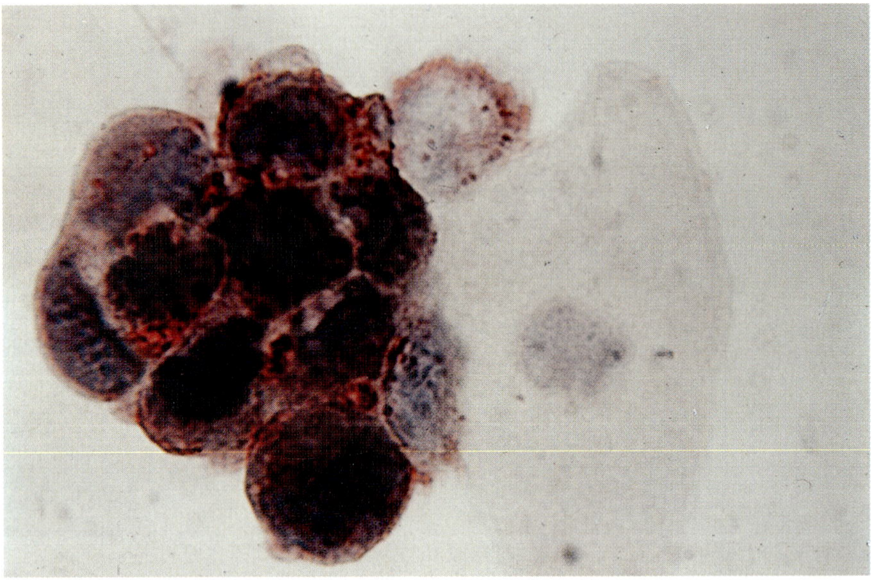

Fig. 2. Staining of all malignant transitional cells with MAB 486P in the urine of patient with bladder carcinoma

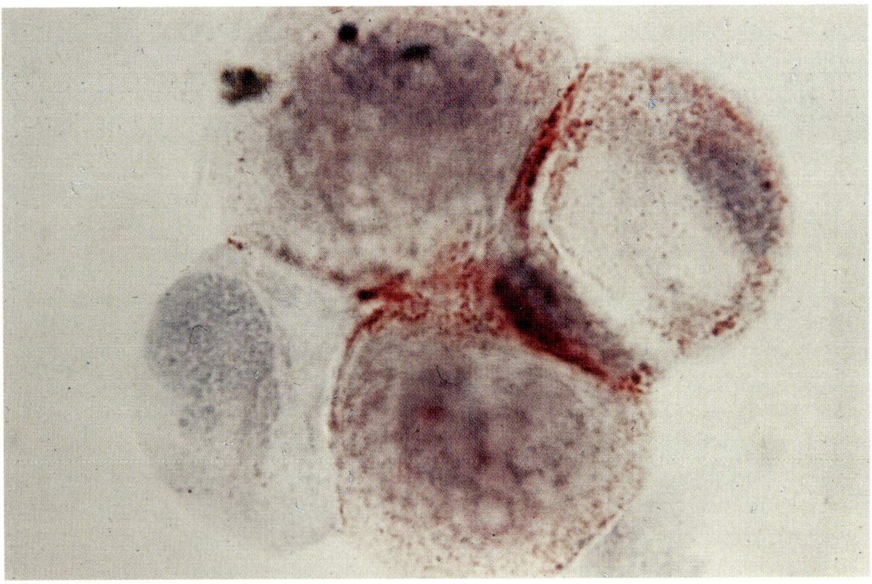

Fig. 3. Heterogenous cell staining with MAB 486P in the urine of patient with bladder carcinoma

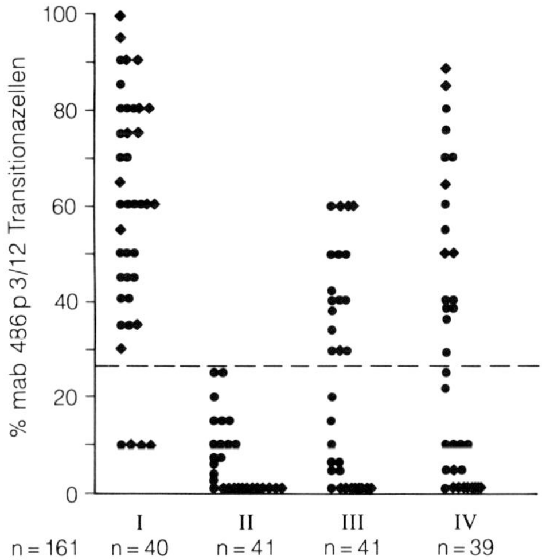

Fig. 4. Analysis of transitional cells in urine specimens in groups I–IV showing percentage of cells that reacted to MAB 486P. *Above dashed line*, positive reaction; *below dashed line*, negative reaction; ◆, positive cytology; ●, negative cytology

Table 3. Sensitivity of standard cytology and quantitative immunocytology in patients with superficial bladder carcinomas

	GI (49)[a] (%)	GII (94)[a] (%)	GIII (98)[a] (%)	Total no. of patients (241) (%)
Positive quantitative immunocytology	91.8	89.4	92.9	91.3
Positive standard cytology	59.2	63.8	84.7	71.4

[a] No. of patients in parentheses.

study on 69 patients with superficial bladder carcinoma. In this prospective study we compare the results of standard cytology, flow cytometry, and quantitative immunocytology, and this study shows that the right positive rate in grade I tumors is as good as in grade II and grade III tumors with MAB 486P (Table 3). The right positive rate with standard cytology was no better than 59% in patients with grade I tumors. In patients with grade I transitional cells of the bladder, we often observe a great number of benign-looking cells which stain intensively with our MAB, indicating that MAB 486P reacts with cells which still are benign according to classic morphological criteria but which have already undergone malignant transformation (Fig. 5).

To prove this hypothesis, we did extensive immune histological mapping studies of six bladders removed because of a T2 transitional cell carcinoma

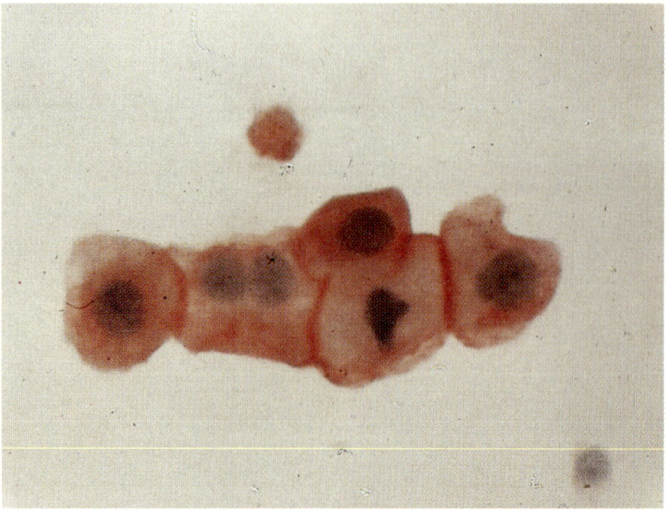

Fig. 5. Positive reaction to MAB 486P in morphologically normal cells

and compared the results with two normal bladders taken from kidney donors. We specifically examined the morphologically normal areas of the bladder distant from the tumor. In these areas we found extensive fields reacting positively to MAB 486P (Fig. 6). We also studied those areas which were morphologically benign but which stained with MAB 486P. Most of these regions already showed loss of blood group antigen and expression of T antigen or and loss of T cryptic antigen (Fig. 6). We therefore have indirect evidence of malignant transformation in those specimens that look morphologically benign but are MAB 486P positive. From this study we conclude that MAB 486P is a good marker for detecting malignant transformation at a very early stage, i.e., even before visible lesions can be seen by morphology or endoscopy. Assuming this is true, quantitative immunocytology should be an excellent marker in the follow-up of patients with superficial bladder carcinomas.

In order to prove this assumption, we started a prospective study 2½ years ago. All patients with superficial bladder carciomas TA and T1, grade I to grade III, were included (Table 4). The tumor was removed by TUR, which was repeated in those with T1 lesions. Four weeks later cystoscopy and cytology were performed to confirm completeness of surgical removal of the tumor. No prophylaxis was given. Urine samples were taken every 4 weeks from all patients. Cystoscopy was done every 3 months.

I will now give a preliminary report on this on-going study. We are supervising 55 patients (44 males, 11 females) whose mean age is 65.8 years. The carcinomas comprised 17 TA lesions and 35 T1 lesions and 3 Cis; 19 grade I lesions, 28 grade II lesions, and 8 grade III lesions. We have

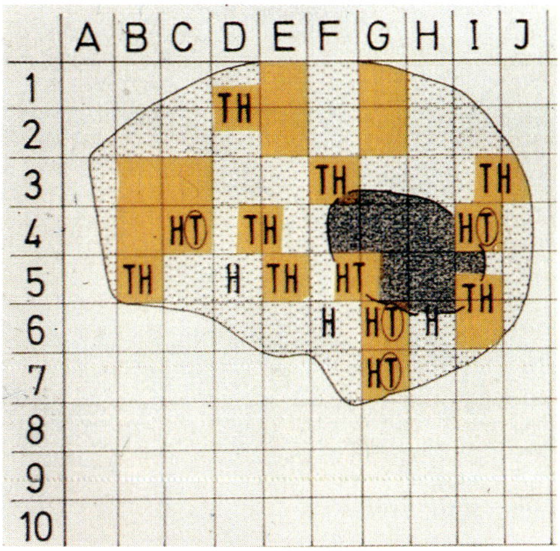

Fig. 6. Mapping study of morphologically benign but MAB 486P positive area surrounding T2, grade III bladder carcinoma. ■, MAB 486P positive; *H*, loss of blood group antigen; *T*, expression of T antigen; Ⓣ, loss of T crypt antigen

Table 4. Prospective study of patients with superficial bladder carcinomas

	Patient characteristics					
No. of patients			55			
Female			11			
Male			44			
Age (years)			64.6 (28–81)			
Mean (months)			21 (6–46)			
	Tumor characteristics					
	Stage			Grade		
	pTA	pT1	pTis	GI	GII	GIII
No. of patients	17	35	3	19	28	8

examined 1200 bladder specimens for cytology and immunocytology, which equates to 13.5 urinary examinations per patient. The follow-up period was 2–38 months with a mean of 18.5 months. The preliminary results show that of 22 patients who never had a positive immunocytology test, 20 have not yet developed a recurrent tumor (Table 5). Immune histological examination of the tumor of the two patients who had recurrent tumor showed that these specific tumor were MAB 486P negative. In summary: about 40% of our patients had negative immunocytology and only 2 developed a recurrent tumor. It appears that quantitative immunocytology (MAB 486P) can reliably identify those 40% of patients who will not develop a recurrent tumor.

33 of the patients had more than one positive immunocytology test in the follow-up (Table 6). 18 of them have already developed a recurrent

Table 5. Quantitative immunocytology in the follow-up of patients with superficial bladder tumor after TUR

	Quantitative immurocytology		Recurrent tumor	Prophylaxis
	No. of patients	%		
Negative immunocytology	22	40	2	0
Positive immunocytology	33	60	18	15

Table 6. Tumor characteristic at beginning of the study

	Stage			Grade		
	pTA	pT1	pTis	GI	GII	GIII
Patients with negative immunocytology in the follow-up (n = 2)	9	13	0	11	11	0
Patients with positive immunocytology in the follow-up (n = 33)	8	22	3	8	17	8

tumor 3–6 months after immunocytology tests became positive. A longer follow-up is needed to see whether the other nine patients will have tumor recurrence. Assuming our results can be confirmed, quantitative immunocytology can be used as a marker to identify those who require prophylaxis.

We also plan to perform another study to test quantitative immunocytology as a marker to determine the required duration of prophylaxis. We will attempt to answer the question: Can quantitative immunocytology (specifically MAB 486P) be a marker to determine the duration and type of prophylaxis required in patients with a superficial transitional cell carcinoma of the bladder? The goal is to use quantitative immunocytology as a marker to enable more individualized prophylaxis in patients with superficial bladder carcinoma rather than as the means to perform merely schematic prophylaxis.

*Image Analysis DNA Cytometry of Bladder Cancer**

M. Stöckle, F. Steinbach, G. Voges, and R. Hohenfellner

Abteilung für Urologie, Universitätsklinik Mainz, Langenbeckstrasse 1, W-6500 Mainz, FRG

Introduction

Invasive bladder carcinoma is characterized by a considerable interindividual heterogeneity. The natural course of the disease and the sensitivity of the tumors to various treatment modalities is thus unpredictable. To overcome this problem and to adopt treatment strategies suited to the individual aggressiveness of the tumors, several characteristics of tumor biology have been defined and tested for their clinical relevance. However, so far most of the results of these studies have been disappointing. Histological grading, for example, which is most commonly used to characterize tumor biology, only differentiated two types of invasive bladder cancer in our investigations, namely GII and GIII. Both types had no significant predictive value for the outcome of the patients after radical surgery for bladder cancer [5]. Histological grading is, in fact, based on the subjective impression of the pathologist only, and not on objective or measurable criteria. Therefore, as demonstrated by Ooms et al. [8], it is not surprising that the intra- and interindividual reproducibility of histological grading is poor.

Searching for more reliable characteristics of tumor biology than those presently available, we started to analyse DNA histograms of bladder carcinomas with the help of an automated image analysis system. Analysis with this system provides more accurate results than flow cytometry [11], which is at present the more commonly used technique for DNA analysis.

Material and Methods

Preparation and Staining for Image Analysis. Paraffin-embedded bladder cancer specimens of 75 radical cystectomies were chosen for analysis, and

* Supported by Grant Sto 177/2-1 of the Deutsche Forschungsgemeinschaft.

5 μm slides and HE staining was used for the identification of tumor areas within the paraffin block. In 65 of the 75 tumors enough tissue was available for image analysis. Tumor areas of at least 3–4 mm in diameter were sufficiently large. A 50-μm section (two sections in relatively small tumors) was dewaxed (Histo-Clear, National Diagnostics, Summerville, New Jersey, USA), rehydrated in alcohol of decreasing concentration, and treated enzymatically with 0.05% pronase in 0.15 *M* phosphate-buffered saline, pH 7.4, at 37°C for 30 min. The cell suspension was then sucked repeatedly through a Pasteur pipette and filtered through a 70-μm nylon mesh. The cells were centrifuged, resuspended in 50% ethanol containing 2% polyethylene glycol and diluted to a concentration of 10 000 per milliliter [2]. Standard smears were prepared by centrifugation of 1 ml of this suspension on glass slides using a cytospin technique. The slides were air-dried and consecutively stained using the Feulgen-SITS technique for quantitative staining of DNA [13].

DNA Image Analysis. Automated image analysis for DNA cytometry was performed using the computer-controlled TV analysis system Leytas (Leiden Television Analysis System) [15]. This essentially consists of a computer-controlled microscope with automated focusing and movement of slides. The microscope image is screened and analyzed using the TV camera.

Image analysis was performed using the Television Analysis System (TAS)-image analysis software program (Fig. 1, Leitz Company, Wetzlar, FRG), which allowed the predominant selection of Feulgen-stained nuclei of epithelial morphology. The DNA content of these nuclei was measured. An extensive artifact rejection, based on mathematical morphological criteria [7] is part of the program. Only single nuclei were analyzed; overlapping nuclei, dirt, artefacts, and most of the stroma and inflammatory cells could be excluded from analysis. Digitized images of the selected cells were stored in a memory buffer allowing visual control of the selected objects (Fig. 2). Selection and measurement are fully automated.

Results

Analysis of the 65 evaluable samples with the image analysis system allowed the distinction of three main types of invasive bladder tumors.

Diploid tumors (Fig. 3a) have a tumor cell peak with the same DNA content (2c) as the nontumorous epithelial cells surrounding the tumor. Some cells show an increased DNA content, reaching a value of 4c, probably corresponding to S- and G2M-phase cells of the mitotic cell cycle. Tumor cells with a higher DNA content than 4c were not found (Fig. 3b). It is almost impossible to distinguish the DNA histograms of these tumors from those of a nontumorous epithelial population. For this reason, all smears of diploid tumors were checked by experienced cytologists to make certain that the histograms were really based on tumor cells.

Fig. 1. Leytas 2 with molecular image analysis computer (MIAC) image analysis system

Aneuploid tumors have a stem-cell line with a DNA content higher than the diploid value. Diploid tumor cells normally were not found. Again, those tumor cells with an increased DNA content, i.e., up to double the value of the stem-cell line, probably represent S- and G2M-phase cells of the aneuploid line. Within the group of aneuploid tumors, the DNA content of the stem-cell line allowed further subtyping:

1. Hypotriploid tumors have a stem-cell line with a DNA content of less than 3c (Fig. 4a).
2. Hypertriploid tumors (Fig. 4b) have a stem-cell line with a DNA content of more than 3c. Unlike hypotriploid tumor cells, more than 20% of all tumor cells found in many of these tumors have a DNA content higher than the stem-cell line, thus suggesting a high proliferation rate.
3. Hypertetraploid tumors, found in only 2 of 65 cases, had a stem-cell line with a DNA content of more than 4c (Fig. 4c).

A third main type of invasive bladder tumor is characterized by a peak in the diploid, tetraploid, and octoploid region (Fig. 5a). Disregarding the diploid cells, we initially interpreted this tumor type as being a special subtype of aneuploidy, calling it tetraploid or hypotetraploid [9]. With further investigation, we learned that this type always had a high content of diploid cells. Comparing ploidy patterns of different areas of these tumors we even found that a near diploid pattern could frequently be demonstrated (Fig. 5b, Table 1). In these cases only a modified cell selection program searching especially for cells with an increased DNA content was able to

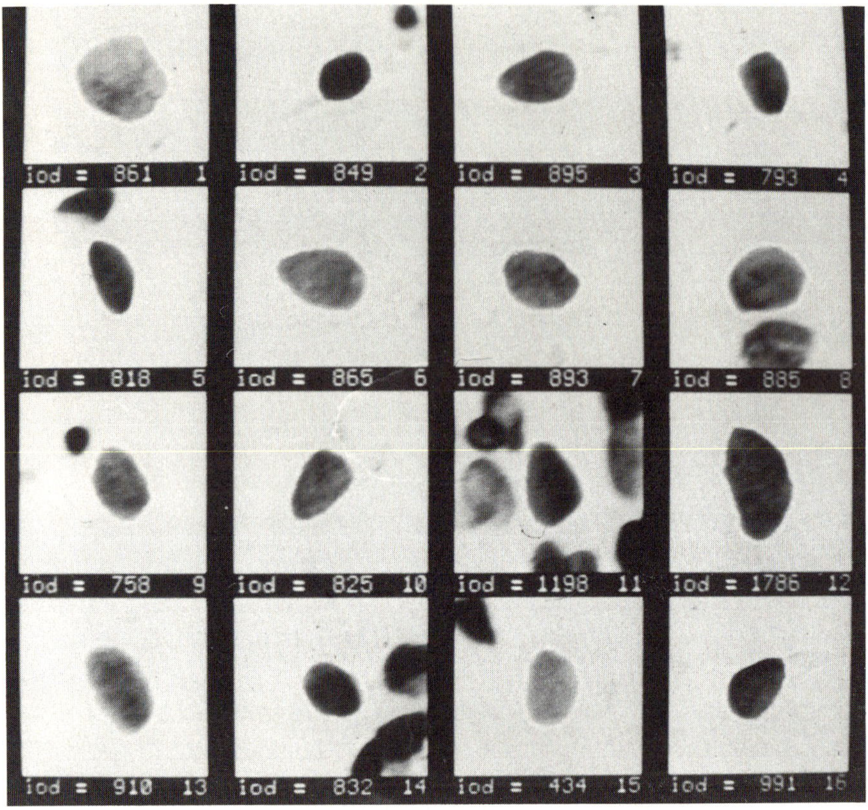

Fig. 2. TV screen with digitized memories of the selected nuclei for visual control and eventual elimination of overlooked artefacts

demonstrate the nondiploid character of the tumor (Fig. 5c). Having thus learned that the percentage of tetraploid and octoploid tumor cells could vary considerably within the same tumor whereas the frequency of diploid tumor cells remained constantly high, we decided to classify this tumor type as polyploid.

The different types of invasive bladder carcinomas as characterized by DNA image analysis demonstrated a significant prognostic impact on the long-term survival of patients after radical cystectomy (Fig. 6).

The prognostic validity of this type of tumor characterization was underlined by a "blind" test: 14 paraffin-embedded specimens of identically treated (radical cystectomy) tumors, all with the same tumor stage (pT2N0) had been sent to us from an outside institution (Department of Urology, University of Padova, Italy) with no information on the clinical outcome of the patients. We found that six of the tumors were diploid. The remaining eight were aneuploid. Of these, two were hypotriploid; the other six showed

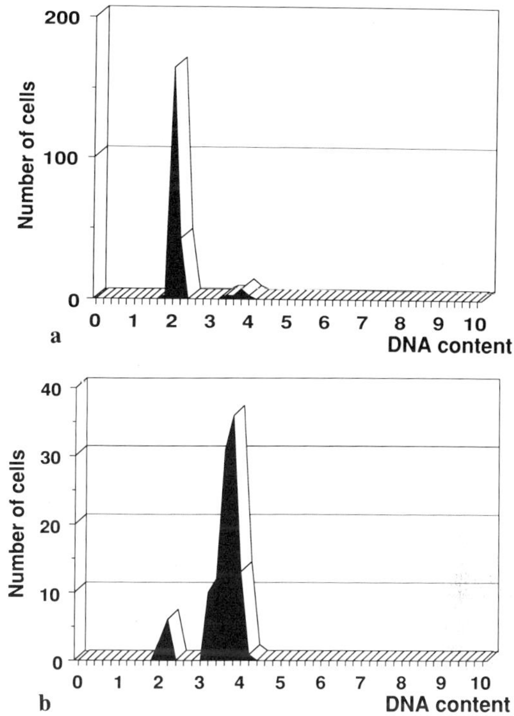

Fig. 3 a,b. DNA content of diploid tumor cells. **a** The stem-cell line at 2c shows the same DNA content as nontumorous epithelial cells which have been analyzed as diploid standard. Proliferating tumor cells reach tetraploid DNA values (4c) during G2M phase of the mitotic cycle. **b** Measurement with altered selection criteria, selectively searching for cells with an increased DNA content. Using the selection program "search for rare events", it can be documented that no cells with a DNA content higher than 4c are present on the slide (compare with Fig. 5c)

higher degrees of aneuploidy. One of these six even had a stem-cell line with a DNA above 6c, a pattern which we had never seen before. Correlation with the clinical outcome revealed that three of the six patients with non-hypotriploid aneuploid tumors had developed tumor progression after cystectomy in spite of the favorable tumor stage. No tumor progression was observed in the group with diploid or hypotriploid tumors.

Tumors with Two Stem-cell Lines. Of the 65 analyzable tumors out of our own cystectomy series [10], only 4 showed two separate stem-cell lines, which in 2 were a combination of a diploid and an aneuploid line (Fig. 7a and b), and in the other 2 comprised two different aneuploid lines. Interestingly, in all four patients the interval between diagnosis and definitive radical surgery of their primary tumor had been rather long. Two of them had had radiotherapy previously with persistent cancer and were finally operated on 1–2 years after radiation. However, these are the only two radiated tumors we have been able to analyze so far. Therefore, one could speculate whether a failed radiotherapy might be a predisposing factor for the occurrence of additional stem-cell lines.

DNA Histogram Patterns During Tumor Progression. Different areas of progressive tumors were analyzed in 14 patients (Table 1). In 11 of these

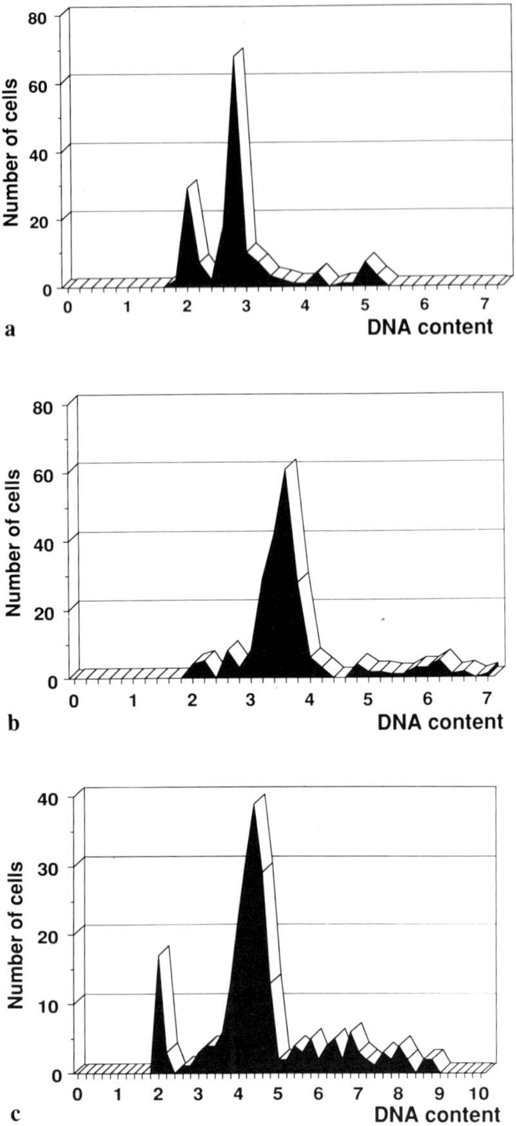

Fig. 4 a–c. Aneuploid tumors characterized by a stem-cell line with a DNA content higher than 2c. Diploid cells are found infrequently and represent non-tumorous epithelial nuclei surrounding the tumor. DNA content of the stem-cell line allows further subtyping of the tumor: **a** hypotriploid tumor; **b** hypertriploid tumor; **c** hypertetraloid tumor

14 cases, the primary tumors could be compared with their lymph node metastases. Heterotopic primary tumors were compared with each other in three cases. In one case, a distant metastasis (adrenal) was compared with the primary tumor and three lymph node metastases. In one case, the DNA histogram of a recurrent tumor in the urethra was compared with the primary bladder cancer treated by cystectomy 1 year earlier.

In all cases with aneuploid tumors, the DNA histogram was found to be a very stable characteristic of each individual tumor. The only exception was

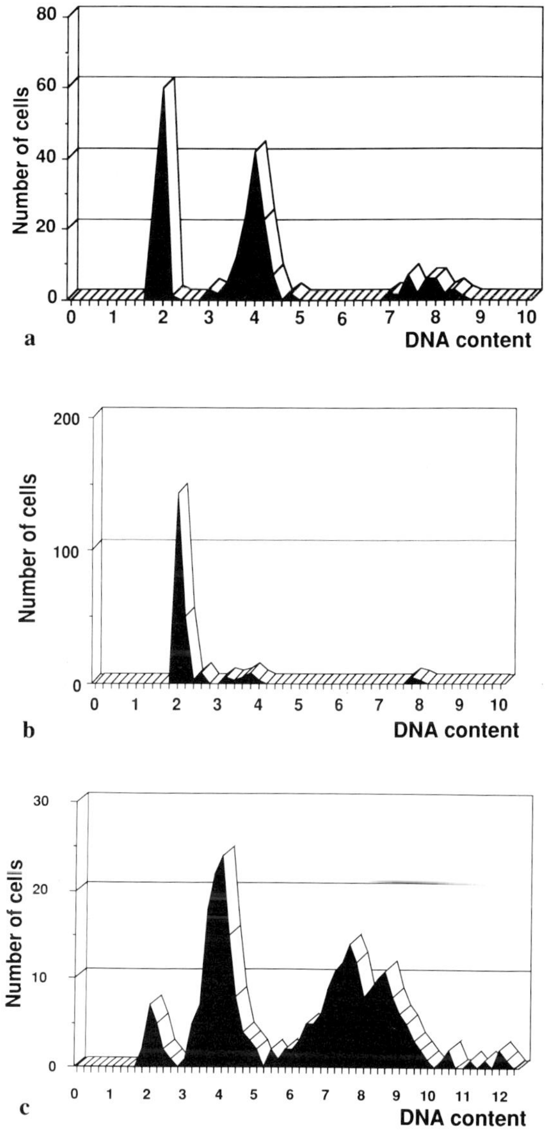

Fig. 5 a–c. Polyploid tumor (**a**) characterized by three peaks in the diploid, tetraploid, and octoploid range, (**b**) with a "nearly diploid" pattern (only three cells in the octoploid range indicate the nondiploid character of the tumor), and (**c**) analyzed according to the same selection criteria as in Fig. 1b (selection search for nuclei with increased DNA content), the nondiploid character of the tumor is now indicated undoubtedly by the presence of an octoploid tumor cell peak

patient no. 1 who had a diploid-hypotriploid combination tumor (Fig. 7a). In this case, the more aggressive aneuploid line seemed to displace the diploid line during tumor progression (Fig. 7b). Polyploid (diploid-tetraploid) tumors on the other hand were characterized by different amounts of tetraploid and octoploid cells in the distinct analyzable parts of the tumor.

DNA Cytometry of Superficial Bladder Cancer. Smears of transurethral biopsies or of urine sediment can also be used for DNA cytometry. Samples

Table 1. Comparative analysis of different parts of 14 tumors. The intraindividual homogeneity of the tumors indicates that invasive bladder cancer is a disease of clonogenic nature and underlines the high reliability of image analysis DNA cytometry (see also Fig. 7a and b)

Patient no.	Tumor area	Ploidy pattern[a]
1	PT	Diploid + 2.9
	LN	Diploid + 2.9
2	PT	Diploid-tetraploid
	LN	Diploid
3	PT	Diploid-tetraploid
	LN	Diploid-tetraploid
4	pT3b[b]	3.1
	pT2[b]	3.1
	LN	3.1
5	PT	2.65
	Urethral recurrence	2.65
6	PT	Hypertetraploid
	LN	Hypertetraploid
7	pT3b[b]	Diploid
	pT2[b]	Diploid
8	PT	Diploid-tetraploid
	3 LN's	Diploid
	Distant metastasis	Diploid-tetraploid
9	PT	3.1
	LN	3.1
10	PT	Diploid-tetraploid
	LN	Diploid
11	PT	2.9
	LN	2.9
12	PT	3.2
	LN	3.2
13	PT	Diploid-tetraploid
	LN	Diploid-tetraploid
14	pT1(−2)[b]	3.4
	pT1[b]	3.4

PT, primary tumor; LN, lymph node metastasis.
[a] Subtype and stem-cell DNA content of aneuploid lines.
[b] pT stages indicate heterotopic bladder tumors of different stages grown at distinct areas of the same bladder.

from 62 consecutive patients with bladder tumors were analyzed in this way; 39 tumors were found to have a non-diploid DNA pattern. Of those patients who seemed to be adequately treated with a transurethral resection of their tumor only (pTaG1 and pTaG2), seven were found to be aneuploid. This group of patients has to be regarded as having a higher risk of tumor recurrence and – more important – of tumor progression and therefore requires careful follow-up examinations at short time intervals.

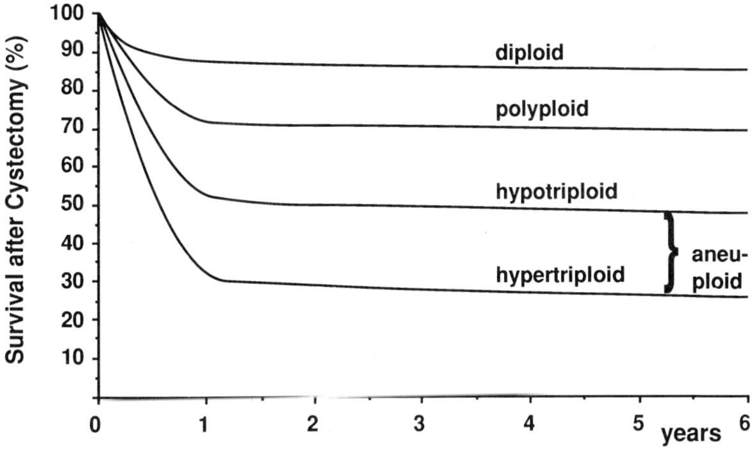

Fig. 6. Survival rates of 65 bladder cancer patients after radical cystectomy with respect to the DNA histogram of their tumors as calculated by multivariate analysis for tumor stage pT3. Even after this stratification for differences in tumor stage, the survival differences are statistically significant ($p = 0.009$, log-rank test). A survival curve for the hypertetraploid tumor type is not shown as it was found only in two patients, both of whom died within a few months following cystectomy, thus indicating its extremely aggressive nature

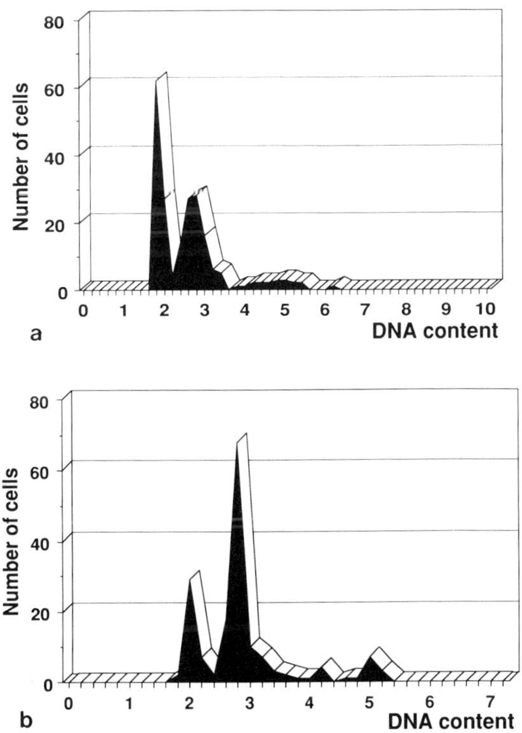

Fig. 7. **a** Diploid-hypotriploid combination tumor. DNA histogram of the primary tumor. **b** DNA histogram of a lymph node metastasis of the primary tumor shown in Fig. 7a. The tumor now is characterized by a majority of hypotriploid cells which seem to overgrow the diploid ones. However, the diploid line shown in Fig. 7a still can be documented

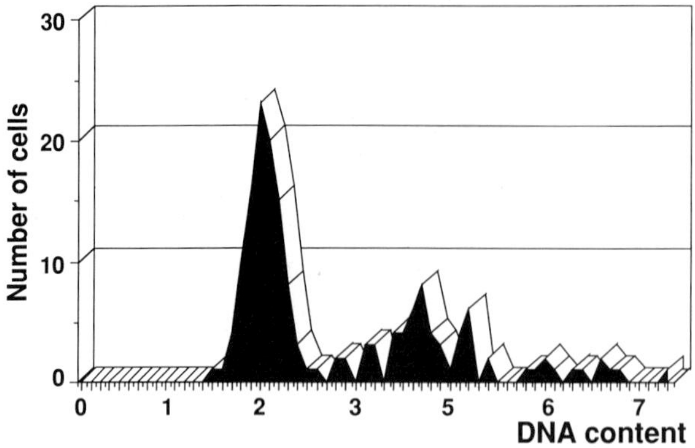

Fig. 8. Urine sediment DNA histogram of a patient who presented a contradictory cytology. The majority of his urinary cells were found to be diploid. However, there is clear-cut evidence of additional aneuploid cells which is well documented by the presence of cells with a DNA content higher than 4c. The cystectomy and urethrectomy specimen of this patient revealed small areas with multifocal carcinoma in situ

A monitoring of aneuploid carcinoma in situ during and after intravesical therapy (e.g., bacille Calmette-Guérin, BCG) is also possible. Patients who continue to have aneuploid cells in their urine despite intravesical treatment could be regarded as candidates for early radical surgery. The urine DNA histogram of such a patient is shown in Fig. 8. This patient initially underwent a transurethal resection of a pT2 carcinoma and postoperatively underwent intravesical instillations with BCG because of concomitant multifocal carcinoma in situ. Random biopsies after BCG treatment revealed moderate dysplasia only; urine cytology during this time was differently interpreted by different investigators. DNA image cytometry of the urine sediment (Fig. 8), however, clearly demonstrated the presence of aneuploid cells. The patient decided to undergo radical cystectomy and urethrectomy. Pathologically, areas of carcinoma in situ were present both in the bladder and in the urethra.

Discussion

Hedley's technique [4] of deparaffinization of "historical" tumor material for cytometric evaluation has provided access to a considerable number of identically treated tumors to study the prognostic relevance of different tumor types as analyzed by the DNA histogram. In our studies, image analysis proved to be superior to conventional flow cytometry for three reasons [11]:

1. The use of smear preparations instead of cell suspensions allowed a morphological recheck of the analyzed cells. The question as to whether

a sample showing a diploid DNA pattern contains tumor cells can therefore be answered easily by cytological control of the smears. The cell suspensions used for flow cytometry allow no morphological control of the analyzed cells.

2. The ability to selectively measure suspicious nuclei during image analysis increases the reliability of this system. Especially in tumors containing high numbers of inflammatory cells such as lymph node metastases, flow cytometry commonly resulted in false diploid histograms, disregarding the low percentages of nondiploid cells. Image analysis, on the other hand, by omitting most of the inflammatory cells from analysis on the basis of their distinct morphology, unequivocally shows the nondiploid character of these tumors by the demonstration of a nondiploid stem-cell line and/or by the presence of cells with a DNA content higher than 4c.
3. Characterization of the polyploid tumor type was not possible with flow cytometry, and to date it has not been described in the literature [1,3,14]. As the nontumorous nuclei cannot be excluded from DNA measurement in flow cytometry, a high diploid peak is always found. Therefore, those polyploid tumors with a large amount of tetraploid and octoploid cells (Fig. 5a) were generally interpreted as aneuploid-tetraploid tumors. Other parts of polyploid tumors showing fewer tetraploid and octoploid elements were judged to be diploid in most instances.

Comparing DNA patterns of various sites of identical tumors, the DNA histogram was found to be a stable tumor characteristic (Table I). This finding is in good agreement with the cytogenetic analysis of Summers et al. [12] and this suggests that invasive bladder cancer is a disease of clonogenic nature.

Ljungberg et al. [6], analyzing DNA patterns of metastasizing renal cell carcinoma, found diploid metastases of an aneuploid primary tumor in some cases, and vice versa. They interpreted their findings as an expression of polyclonal disease, with each clone having the ability to invade and metastasize. However, the existence of a DNA polyploid type of renal cell carcinoma could also be a possible explanation for these controversial findings.

Clinical Perspectives of Image Analysis DNA Cytometry. The various available treatment modalities for invasive bladder cancer, e.g., radical surgery, radiotherapy, or chemotherapy, are completely effective in only a limited percentage of patients, and in clinical practice, no tumor characteristics have been defined which can predict the response of an individual tumor to various treatment modalities. Our present studies are addressing the question whether the DNA histogram of a tumor can be used as a predictor of response to various treatment modalities in order to treat the patients more effectively.

In patients with superficial and transurethrally resectable tumors, DNA cytometry defines a small group of patients characterized by the presence of aneuploid tumor cells. These patients may be carrying a higher risk of tumor recurrence and/or progression and, therefore, careful follow-up studies at short intervals are recommended after transurethral resection of aneuploid superficial tumors.

Summary

In a retrospective analysis, the DNA histograms of 65 paraffin-embedded bladder carcinomas from radical cystectomy specimens (stage pT1–pT4a, pN0, pN1, pN2) were analyzed using an automated image analysis system (Leytas 2).

Automated image analysis was able to characterize invasive bladder carcinoma as being either diploid, polyploid, or aneuploid. Within the group of aneuploid tumors, the DNA content of the stem-cell line allowed further subtyping of the tumors; hypotriploid, hypertriploid, hypertetraploid, and even hyperpentaploid tumors could be distinguished. Comparing different sites of identical tumors, the DNA histogram was found to be a stable and reproducible tumor characteristic. The various tumor types differed significantly in prognosis.

This technique can also be applied to smears of urine sediment or transurethrally resected tumor chips. In the case of superficial tumors, DNA cytometry defines those tumors which are potentially invasive, requiring careful follow-up and/or early aggressive treatment.

References

1. Coon JS, Schwartz D, Summers JL, Miller AW, Weinstein RS (1986) Flow cytometric analysis of deparaffinized nuclei in urinary bladder carcinoma. Comparison with cytogenetic analysis. Cancer 57:1594–1601
2. Driel-Kulker AMJV, Mesker WE, Velzen IV, Tanke HJ, Feichtinger J, Ploem JS (1985) Preparation of monolayer smears from paraffin-embedded tissue for image cytometry. Cytometry 6:268–272
3. Gustafson H, Tribukait B, Esposti PL (1982) The prognostic value of DNA analysis in primary carcinoma in situ of the urinary bladder. Scand J Urol Nephrol 16:141–146
4. Hedley DW, Riedlander ML, Taylor IW, Rugg CA, Musgrove EA (1983) Method for analysis of cellular DNA content of paraffin-embedded pathological material using flow cytometry. J Histochem Cytochem 31:1333–1335
5. Jacobi GH, Klippel KF, Hohenfellner R (1983) 15 Jahre Erfahrung mit der radikalen Zystektomie ohne präoperative Radiotherapie beim Harnblasenkarzinom. Akt Urol 14:63
6. Ljungberg B, Stenling R, Roos G (1986) Prognostic value of deoxyribonucleic acid content in metastatic renal cell carcinoma. J Urol 136:801–804

7. Meyer F (1979) Iterative image transformations for automative screening of cervical smears. J Histochem Cytochem 27:128–135
8. Ooms EC, Anderson WA, Alons CL, Boon ME, Veldhuizen RW (1983) Analysis of the performance of pathologists in the grading of bladder tumors. Hum Pathol 14:140
9. Stöckle M, Tanke HJ, Mesker WE, Ploem JS, Jonas U, Hohenfellner R (1987a) Automated DNA image cytometry: a prognostic tool in infiltrating bladder carcinoma. World J Urol 5:127–132
10. Stöckle M, Alken P, Engelmann U, Jacobi GH, Riedmiller H, Hohenfellner R (1987b) Radical cystectomy – often too late? Eur Urol 13:361
11. Stöckle M, Tanke HJ, Voges G, Riedmüller H, Hohenfellner R (1989) DNA histogram of invasive bladder carcinoma: comparison of flow cytometry and automated image analysis. In: Rübben H, Jocham D, Jacobi G (eds) Investigative urology, vol 3. Springer, Berlin Heidelberg New York
12. Summers JL, Falor WH, Ward R (1981) A 10-year analysis of chromosomes in noninvasive papillary carcinoma of the bladder. J Urol 125:177–178
13. Tanke HJ, van Ingen EMV, Ploem JS (1979) Acriflavine-Feulgen Stilbene staining: a procedure for automated cervical cytology with a television based system (LEYTAS). J Histochem Cytochem 27:84–86
14. Tribukait B, Gustafson H, Esposti P (1979) Ploidy and proliferation in human bladder tumors as measured by flow-cytofluorometric DNA-analysis and its relations to histopathology and cytology. Cancer 43:1742–1751
15. van Jugen EM, Verwoerd N, Ploen JS (1980) LEYTAS-2: a hybrid system for the analysis of cytological preparations, using both hardware and software methods. Microscopica Acta [Suppl] 4:73

Radical Cystectomy: The Potential for Cure and Reconstruction

S.D. Boyd, D.G. Skinner, and G. Lieskovsky

Department of Urology, School of Medicine, University of Southern California, Los Angeles, CA 90033, USA

Introduction

In 1989 there were nearly 50 000 newly diagnosed cases of bladder cancer in the United States and this disease directly accounted for more than 10 000 deaths. Given its prevalence and impact, one would surmise that great strides would have been made in the understanding and selective treatment of this cancer over the last 30 years. In 1959 Dr. Jewett stated: "in no other viscus in the body has the treatment of malignant disease remained so controversial as in the bladder." This comment unfortunately still holds true today.

The sequence and timing of intervention in potentially invasive bladder cancer is dependant on several factors which form the important criteria for successful management. These include the grade of the tumor, the tumor configuration (papillary versus solid), the presence of muscle invasion, and the presence or absence of carcinoma in situ (CIS). The majority of invasive tumors are high grade and invasive at the time of initial presentation. Kaye and Lang in 1982 [4] noted that in 166 consecutive patients with invasive bladder cancer seen over a 10-year period, only 16% had a history of prior bladder tumor. Patients with high-grade bladder cancer require aggressive therapy. Undertreatment regardless of stage should be avoided. The choices of therapy for invasive bladder cancer in 1990 remain as multiple and controversial as they were in 1959. Present options may include definitive external beam radiation, interstitial brachiotherapy with or without external radiation boost, preoperative radiation and radical cystectomy, radical cystectomy with or without postoperative adjuvant chemotherapy, neo-adjuvant chemotherapy and radical cystectomy, neoadjuvant chemoradiation, and neoadjuvant chemotherapy and segmental resection with bladder preservation or salvage.

This paper will deal with the use of radical cystectomy as a definitive treatment for bladder cancer. Histological criteria for radical cystectomy

Recent Results in Cancer Research, Vol. 126

may include any muscle-invading transitional cell carcinoma, grade 3 T1 tumors with CIS, multifocal grade 3 T1 tumors, CIS refractory to intravesical mitomycin and bacille Calmette-Guérin (BCG), CIS involving the prostatic urethra, and multifocal grade 2 tumors felt unsuitable for transurethral resection.

Technique of Radical Cystectomy

There are numerous factors important in reducing the complications of cystectomy. Preoperatively patients should have a short, effective bowel preparation with good hydration. The bowel preparation starts the day before surgery by placing the patient on a clear liquid diet. A laxative is given the morning prior to surgery. Oral neomycin and erythromycin base are then given in multiple doses throughout the day. Intravenous hydration is started the evening prior to surgery and the patient is kept NPO after midnight. No enemas are given unless colon surgery is anticipated.

At surgery the patient is placed in the hyperextended supine position. If a simultaneous urethrectomy in a male is anticipated, the patient is also frog legged so that a second team can perform the urethral dissection, and the entire cystectomy and urethrectomy specimen can be removed en bloc. A midline abdominal incision is used curving just to the left of the umbilicus. The urachal remnant is identified and circumscribed so that it remains en bloc with the bladder. The abdominal contents are mobilized and the small bowel packed high in the abdomen away from the pelvis. The abdomen is thoroughly explored to determine the full extent of the disease. A self-retaining retractor is positioned; we prefer to use a Finochietto retractor.

A thorough lymph node dissection is performed beginning at the bifurcation of the aorta. The limits of the node dissection are the genitofemoral nerves laterally, the circumflex vein and the lymph node of Rosenmüller distally, and the obturator fossa posteriorly. The rationale for a thorough pelvic lymph node dissection includes the fact that 20%–30% of candidates for cystectomy without clinical evidence of metastases will have positive lymph nodes. As will be discussed later, the presence of nodal disease negatively affects survival, but the presence of micrometastases only may carry an improved survival potential. Once a patient has been accurately staged, the presence of deeply invasive tumors and metastatic disease might indicate those who would benefit most from adjuvant therapy. Lymph node dissection does not significantly add to the morbidity or mortality of the radical cystectomy.

Once the lymph node dissection has been completed, the lateral vascular pedicles to the bladder are delineated and divided between ligaclips. This division begins with the superior vesicle artery branch off the hypogastric artery. The posterior pedicle is then delineated after opening the peritoneal fold between the bladder and rectum, staying close to the rectal side in order to clearly develop a plane beneath Denonvilliers' fascia. This will allow the surgeon to very easily define the posterior pedicles which can also be divided between ligaclips. In women the posterior pedicles, including the cardinal ligaments, are clipped and divided for approximately 4–5 cm beyond the cervix. At this point the posterior vaginal wall is opened and the vagina is circumscribed anteriorly. The anterior vaginal wall is dissected from the posterior wall of the bladder down to the region of the urethra. This maneuver further defines the two distal posterior pedicles extending from the bladder to the lateral aspect of the vagina on either side. These pedicles are clipped and divided distally, freeing the bladder from its posterior attachments except for the urethra. In some patients with large, deeply penetrating posterior tumors, it may be advantageous to remove the anterior vaginal wall completely; the lateral vaginal wall is then incised down to the location of the urethral meatus.

After the posterior dissection is completed, attention is turned to the anterior dissection. In men the puboprostatic ligaments are divided. The dorsal venous complex can then be delineated and controlled. The apex of the prostate can be carefully divided from the membranous urethra, and the entire cystoprostatectomy specimen with lymph nodes removed. If a neobladder is to be performed, the anastomotic sutures of 2-0 chromic can be placed in the membranous urethra at this time. These are then draped aside for later anastomosis to the neobladder. It is currently our policy not to routinely perform potency-sparing radical cystectomy. Following the report from our institution by Pritchett in 1988 [7], it is felt that potential nodal disease may be left in the pelvis using the nerve-sparing technique. Because of the differing survival rate for patients who have had a single nodal micrometastasis either excised or left in the pelvis, we do not feel that potency-sparing surgery is the most appropriate therapy.

In women, once the urethropubic ligaments have been divided, a large curved Kocher clamp is placed on the urethra, and the anterior vaginal wall is opened distally and incised circumferentially around the urethral meatus. The vaginal cuff is closed longitudinally with interrupted Dexon or chromic sutures. Preservation of the anterior vaginal wall, when feasible, results in a functional vagina. When the anterior vaginal wall is removed because of the local extent of the primary tumor, closure results in a small, largely obliterated vagina. It is important to suspend the closed vagina from each side of the pectineal (Cooper's) ligament to prevent vaginal prolapse or development of an enterocele.

Results

Between 1971 and 1989, 597 patients with bladder cancer were treated for cure utilizing radical cystectomy. The operative mortality of this group of patients was 1.7% (10/597). Ahlering and Skinner [1] detailed a hypotensive

Table 1. Eligibility criteria for adjuvant chemotherapy trial following radical cystectomy

Pro
Surgically confirmed invasive carcinoma of bladder stage P3, P4 or N+ and M0. No involved lymph nodes above aortic bifurcation
Age >19–74
Signed informed consent
Contra
Prior noncutaneous malignancy within 10 years
Prior chemotherapy or pelvic irradiation
Bilirubin >1.5 mg/ml, SGOT>2 × normal, alkaline phosphatase elevated, WBC <3500, and platelet count <150 000
Serum creatinine >1.0 mg/ml/M^2 within 6 weeks of cystectomy
Karnofsky performance status <50
Medical, social, or psychological factors which would make the patient a poor risk for completion of chemotherapy

technique of anesthesia which lowered the average blood loss from the procedure from 1700 cm^2 to 800 cm^2. The male to female ratio of this group was 4:1. The mean age was 62 years with a range of 30–85 years. Postoperative complications occurred in 11% of patients. The mean hospital stay was 11 days.

Between 1971 and 1978, 100 patients underwent high-dose, short-course preoperative radiation therapy of 1600 rad prior to cystectomy. In 1984, Skinner and Lieskovsky [8] compared these 100 patients with 97 patients who had been treated by surgery alone, and found no significant difference in survival or pelvic recurrence rates. Therefore, we no longer use preoperative radiation routinely for our cystectomy patients, and in fact have not done so since 1980. Results by stage reveal that 160 patients since 1980 had cystectomy for cure with pathological stage P2 or less, with or without CIS. The 5-year survival in this group was greater than 80%. The overall survival for all invasive bladder cancer patients (P3, P4, ±nodes) was 50%. The presence of positive lymph nodes in these patients was 23%; the overall survival for patients with positive lymph nodes was 32%. If less than 10% of the lymph nodes were positive, the 5-year survival was nearly 50%. This data led us to look further at the use of adjuvant chemotherapy as a potential to increase postoperative survival.

Between 1980 and 1989, 241 patients underwent radical cystectomy for invasive disease (P3, P4, ±nodes). Of these, 198 met the eligibility requirements for the adjuvant chemotherapy trial (Table 1); 101 patients were ultimately enrolled and randomized (Fig. 1). Five of these patients were found to have pure squamous cell carcinoma and five were found to have adenocarcinoma of the bladder. That left 91 patients with transitional cell carcinoma of the bladder for study. Chemotherapy primarily employed cisplatinum 100 mg/M^2, doxorubicin 60 mg/M^2, and cyclophosphamide

Protocol
Carcinoma of Bladder
Muscle Invasion, MO
↓
Cystectomy, Pelvic Node Dissection
↓
P_3,P_4,N+
↓
Randomize

Chemotherapy	Observation
4 Cycle PAC	7/80–12/88

Fig. 1. Protocol scheme for invasive bladder cancer adjuvant chemotherapy trial. *PAC*, cispatinum–doxorubicin–cyclophosphamide

600 mg/M^2 (PAC). The initial design was to accrue 75 patients in each arm of the study, chemotherapy or observation. Ultimately 44 patients were randomized to the chemotherapy arm and 47 to observation. Though the trial was designed to have patients undergo 4-monthly cycles of chemotherapy, only 21 of the 44 patients received the full four courses. Two patients received only three courses, 6 received two courses, 4 received one course, and 11 patients in this arm refused chemotherapy.

Interim analyses were performed after 50, 75, and 100 patients had been accrued and followed for a minimum of 6 months. The study was terminated at the third analysis when the superiority of the chemotherapy arm was demonstrated by the stratified Wilcoxon test at $P = 0.0025$ for time to recurrence and $P = 0.0078$ for survival (Figs. 2 and 3). The 3-year tumor-free interval for our patients was 70% (±9%) in the chemotherapy arm and 46% (±11%) in the observation arm. Median survival of all patients was 51 months for the chemotherapy group and 29 months for the observation group. The outcome for males greater than the age of 65 with negative lymph nodes or only one positive lymph node was very good without chemotherapy (Fig. 4). In men less than 65 years of age and women of any age, there was a definite benefit with chemotherapy for patients with negative lymph nodes or only one positive lymph node. There was no benefit to survival with chemotherapy for patients with more than one positive lymph node though there was an increase in time to recurrence or tumor-free interval.

It is now felt that adjuvant chemotherapy definitely has a role in the treatment of bladder cancer. This randomized study represents the first comparative prospective trial designed to assess the potential benefits of adjuvant chemotherapy following cystectomy. Though the numbers in some groups were small, a significant increase in time to progression and in survival for patients with transitional cell carcinoma treated with adjuvant cisplatinum combination chemotherapy was shown. For time to progression, the effect of chemotherapy is most notable in the first 2 years after treat-

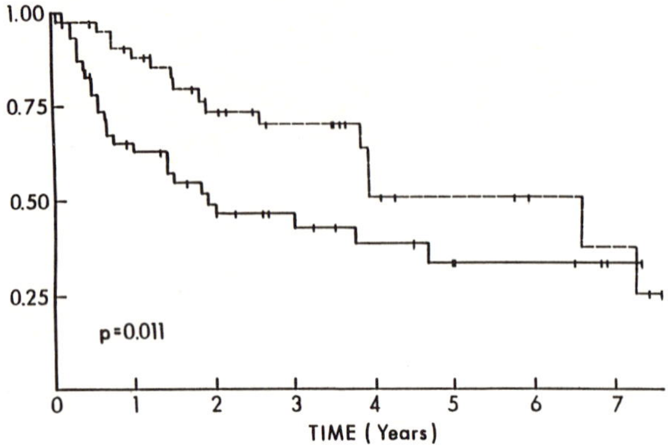

Fig. 2. The effect of adjuvant chemotherapy on progression of bladder cancer following radical cystectomy for 91 patients enrolled in the randomized trial. The *tick marks* on the curves represent patients who were still disease free at the last follow-up. The percentage of patients who are disease free is noted on the *vertical axis*. —, 44 patients receiving chemotherapy; -, 47 patients on observation arm of study

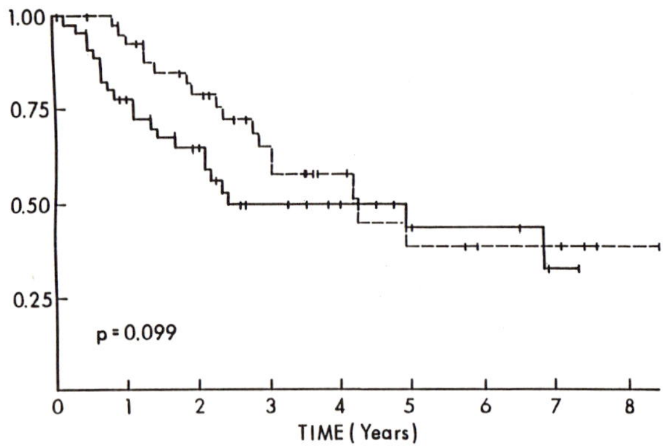

Fig. 3. The effect of adjuvant chemotherapy on survival for 91 patients (same as in Fig. 1) enrolled in the randomized trial. The *tick marks* represent patients who were still alive at follow-up. ---, 44 patients receiving chemotherapy; —, 47 patients on observation arm of study

ment. Subset analyses suggest that the patient who benefits the most from therapy will have no lymph nodes or only 1 lymph node involved. Patients with two or more nodes involved are at risk of early death from disease. In these patients, adjuvant chemotherapy increases the time to progression by an average of 14 months without increasing survival. The goal now is to develop adjuvant chemotherapy protocols that are more acceptable to patients and increasingly effective.

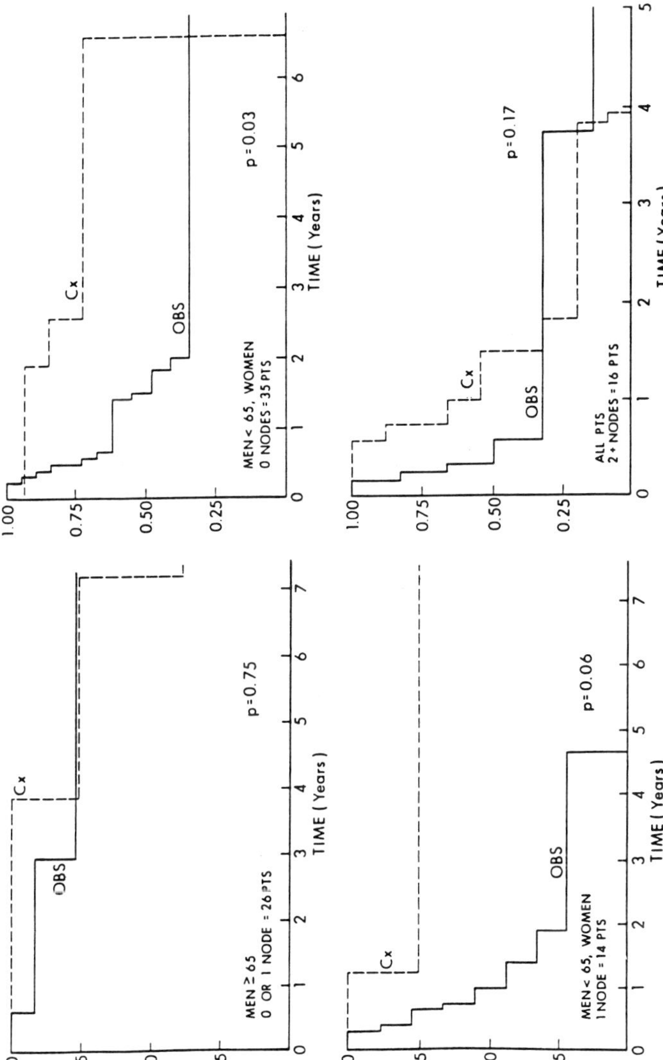

Fig. 4. Comparison of the effect of chemotherapy on time to recurrence with each of four prognostic subgroups. Note that the *horizontal time scale* for patients with two or more positive nodes is expanded. *Cx*, chemotherapy patients; *OBS*, observation patients

Reconstruction

The multiple options for urinary diversion can factor greatly in the patient's choice of treatment. The overall goal of any urinary diversion is to protect the upper renal tracts from significant damage during the life span of the patient. The diversion patient would also like to be able to live as normal an existence as possible. Given the option, most patients would likely choose to continue voiding as naturally as possible following surgery. Until recently, however, the most common form of urinary diversion was the refluxing ileal conduit which requires patients to wear an ostomy appliance. In 1982, Kock [5] published his results on continent cutaneous urinary diversion using detubularized ileum. The key to his diversion was a continent reservoir that could be constructed with extremely low internal pressures. Intussuscepted ileal valves were formed to prevent reflux and provide continence.

Over the last 8 years, we have used the Kock principal to assess various options for urinary diversion. These include cutaneous continent diversion, neobladders, and ileo-anal reservoirs. Cutaneous continent diversion allows a patient to be continent by employing a catheterizable stoma with an underlying continence nipple valve. We have used this form of diversion in over 500 patients with excellent long-term results. The major potential complication of this surgery is incontinence; approximately 10% of patients presently require revisional surgery to ensure continence to the skin. The cutaneous reservoir can have a very large capacity of over a liter of fluid. Most patients find that they only need to catheterize every 6 h during the day, can sleep through the night without catheterizing, and can lead a very acceptable lifestyle.

The Kock neobladder employs the same detubularized reservoir with antireflux valve but needs no continence valve as the reservoir is anastomosed directly to the membranous urethra. This type of surgery is designed for males without presence of tumor in the prostatic urethra. We have used this surgery in over 100 patients and are becoming increasingly convinced that this is the surgery of choice for males who are appropriate candidates. We are also continuing to look at the use of the ileo-anal reservoir as described by Kock and Ghoneim [6] with the hemi-Kock antirefluxing reservoir anastomosed to the rectum. This internal diversion also utilizes an antirefluxing valve in the sigmoid colon to keep the urine contents from refluxing into the colon. We have to date performed this surgery on 13 patients and still consider it at a very experimental stage.

Comparing physical and emotional considerations with regard to continent urinary diversion and ileal conduit diversion, continent diversion surgery takes at least 1 h longer than conduit surgery but early complications should be comparable. The operative mortality should be no different in either group and the overall late complications should also be comparable, though there will be more revisional surgery for the continent cutaneous form of diversion for leakage of urine through the stoma. According to a study by Boyd in 1987 [2], continent diversions have a definite advantage for patients in terms of self-confidence, self-image, personal relations, and sexual desire. There is, however, no question that continent diversion signifies increased surgical complexity and may not be the choice for the surgeon who performs only the occasional cystectomy. Quality of life issues and patient acceptance of the end results, however, will become increasingly important in the future as physicians and patients alike continue to examine the various therapeutic options.

References

1. Ahlering TE, Henderson JB, Skinner DG (1983) Controlled hypotensive anesthesia to reduce blood loss in radical cystectomy for bladder cancer. J Urol 129:953–954
2. Boyd SD, Feinberg SM, Skinner DG, Lieskovsky G (1987) Quality of life survey of urinary diversion patients: comparison of ileal conduits versus continent Kock ileal reservoirs. J Urol 138:1386–1389
3. Jewett HJ (1959) Carcinoma of the bladder: development and evaluation of current concepts of therapy. J Urol 82:92–100
4. Kaye KW, Lange PH (1982) Mode of presentation of invasive bladder cancer: reassessment of the problem. J Urol 128:31–33
5. Kock NG (1982) Urinary diversion via a continent ileal reservoir: clinical results in 12 patients. J Urol 128:469–475
6. Kock NG, Ghoneim MA, Lycke G, Mahran M (1987) Urinary diversion to the augmented and valved rectum: preliminary results with a novel surgical procedure. J Urol 140:1375–1379
7. Pritchett TR, Schiff WM, Klatt E, Lieskovsky G, Skinner DG (1988) The potency-sparing radical cystectomy: does it compromise the completeness of the cancer resection? J Urol 140:1400–1403
8. Skinner DG, Lieskovsky G (1984) Contemporary cystectomy with pelvic node dissection compared to preoperative radiation therapy plus cystectomy in management of invasive bladder cancer. J Urol 131: 1069–1072

Current Concepts in Urinary Diversion: The Ileal Neobladder

R.E. Hautmann

Abteilung für Urologie, Universität Ulm, Prittwitzstrasse 43, W-7900 Ulm, FRG

Introduction

When discussing urinary diversion patient acceptance is the most striking issue. Urinary leakage, the need for a catheter and a stoma, the necessity to wear a bag or appliance, and the constant threat of any type of infection are of low patient acceptance. Perfect continence, a natural voiding behaviour, an internal reservoir, a safe and infection-free upper tract, and an undisturbed body image are of high patient acceptance (Table 1).

The most recent and significant achievements in urinary diversion are the so-called continent diversion and bladder replacement surgery. Before discussing bladder replacement surgery I would like to comment on the current situation and forecast the future of urinary diversion.

Both pouch and bladder replacement are currently experimental surgery and will remain so for the next 10–15 years, the reason for this being that in urinary diversion late complications must be assessed over long time spans. Well-known examples of such complications are the anastomotic carcinoma after Coffey's operation or the incidence of stomal stenosis arising from conduit which increase linearly with time.

As regards the supravesical continent diversion a restricted indication can be predicted due to the 20%–50% need for reoperation resulting from afferent nipple problems, the existence of a stoma, the need for a catheter, and the disturbed body image. Indications for continent supravesical diversion will remain the paraplegic patient and the female requiring diversion (Fig. 1).

Strong arguments in favor of bladder replacement surgery are superior patient acceptance, a completely undisturbed body image and the surgical – technical ease of the procedure, at least for the surgeon with experience in radical prostatectomy.

Of all forms of continent urinary diversion, the neobladder most closely resembles the original organ:

Recent Results in Cancer Research, Vol. 126

Table 1. Patient acceptance in urinary diversion

Low	High
Urinary leakage	Continence
Catheter	Natural voiding
Stoma	Internal reservoir
Bag	Safe upper tract
Reoperation rate	Body image
Infection	
– Pyelonephritis	
– Reservoir	
– Nephrostomy	

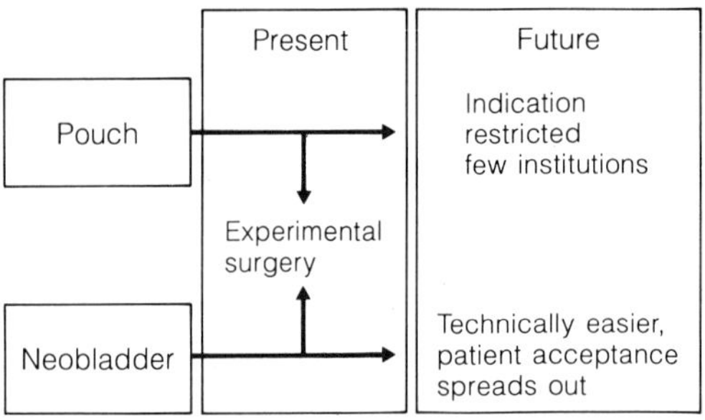

Fig. 1. The future of continent diversion (supravesical) and orthotopic bladder replacement

- The urinary reservoir is placed orthotopically in the small pelvis
- Voiding is achieved by abdominal straining and relaxation of the external urethral sphincter
- Continence is provided by the external urethral sphincter

Camey was the first person to use bladder replacement in a larger number of patients [4]. Unfortunately, his U-shaped tubular reservoir resulted in nocturnal incontinence in the majority of patients [11]. The underlying pathophysiological mechanism is the volume-triggered contractility of the circular bowel musculature, resulting in intermittent high-pressure waves when the reservoir is filled [20]. Whereas reflectory contraction of the external urethral sphincter provides continence during daytime even with intermittent high reservoir pressures, this mechanism fails during sleep, leading to nocturnal incontinence when the fluid within the reservoir exceeds a certain (100–150 ml) volume [4,20].

These specific drawbacks of tubular reservoirs have long since been recognized and overcome, Kock [6,7] being the first to describe the fundamental principles of a detubularized and refashioned bowel reservoir for low-pressure, continent urinary diversion. Meanwhile, a number of modifications have been presented, using different reservoir forms constructed from different parts of the bowel.

We first demonstrated our technique with the ileal neobladder in 1986 [5]. To date, our experience encompasses more than 170 consecutive patients, with 145 patients having follow-up of 3 months and longer.

Preoperative Management

When a patient is selected for radical cystoprostatectomy and ileal neobladder, he should have a tumor negative biopsy of the prostatic urethra. Decompression of dilated upper tracts by a percutaneous nephrostomy for 3–4 weeks may be useful for better tonicization of the distal ureters, but is not mandatory. Before bladder augmentation, diagnostic workup includes an intravenous pyelogram (IVP), voiding cystourethrogram, retrograde urethrogram, urethrocystoscopy, and extensive urodynamic evaluation.

The day before surgery, a central venous catheter is placed, and the patient is hydrated the night before the operation. Although the operation is carried out under general anesthesia, all patients receive an indwelling peridural catheter for better postoperative pain control.

At 12 h before surgery, the bowel is flushed via a nasogastric tube with 10–15 liters of a solution containing 6.14 g/l NaCl, 0.75 g/l KCl, and 2.90 g/l $NaHCO_3$ (Na^+, 139.6 mmol/l; K^+, 10.0 mmol/l; Cl^-, 115.1 mmol/l, HCO_3^-, 34.5 mmol/l). Intravenous broad spectrum antibiotics are started the day before surgery and continued until the second postoperative day.

Operative Technique

Radical cystectomy is carried out in a standard fashion, encompassing removal of the lymph nodes in the region between the external and internal iliac vessels and the obturator nerve. Following descending dissection of the urinary bladder, the apex of the prostate is exposed according to the technique described by Walsh [21] for radical prostatectomy: after ligation and dissection of Santorini's plexus, the urethra is incised immediately distal to the prostatic apex. In the course of the continous separation of the urethra from the prostate, six sutures are anchored in the urethral stump for later anastomosis with the neobladder.

A 60–70 cm segment of ileum is then selected for construction of the neobladder. The aboral site of resection should be approximately 20–30 cm proximal to the ileocecal junction (Fig. 2). It is wise to check at this stage whether the caudal most part of the selected bowel segment can be easily approximated to the pelvic floor. If not, a more orally or aborally situated bowel segment can be chosen, depending on the motility and length of

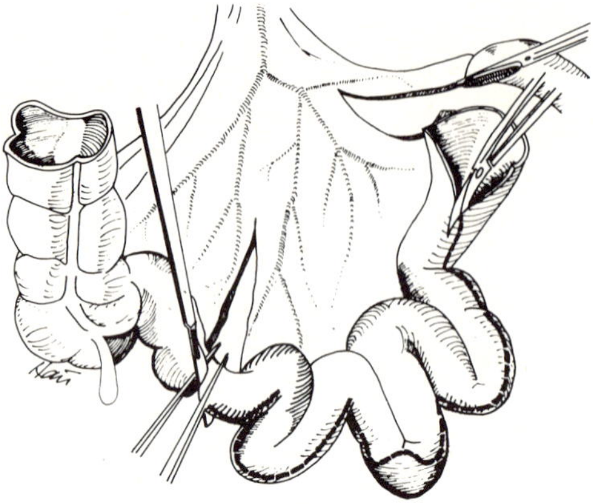

Fig. 2. Selection of ileum segment for ileal neobladder. The distal avascular region of the mesentery between the ileocolic artery and the terminal branches of the superior mesenteric artery is identified and incised, so that a small bowel segment approximately 60–70 cm long is obtained. The distal 20–30 cm of the terminal ileum and Bauhin's valve are saved. The entire bowel segment is incised at its antimesenteric border except for a U-shaped flap which facilitates anastomosis to the urethra

the meso. With this technique, failure to complete the enterourethral anastomosis has never occurred in our series.

Bowel continuity is restored with two rows of interrupted sutures (4-0 Vicryl). The isolated segment is then arranged in a W-shape using Babcock clamps and the bowel is incised longitudinally along the antimesenteric border. At the most dependent part of the segment, previously selected for anastomosis with the urethra, a U-shaped flap is created (Figs. 2 and 3) to gain additional length. The W-shaped bowel plate is then created using running sutures (3-0 Vicryl, Fig. 3).

Urethral anastomosis is prepared by making a pinhole into the U-shaped flap at the caudal most part of the bowel plate. Following transurethral insertion of a 22 Fr three-way catheter, the anastomotic sutures anchored in the urethral stump are connected with the bowel plate as follows: the inner thread is guided through the pinhole with an overholt clamp; the outer thread is stitched through the flap 5–7 mm centrifugal to the pinhole. Thus, optimal mucosa-mucosa adaption is provided. The anastomosis is completed by tying the sutures inside the neobladder (Figs. 4 and 5).

The anterior wall of the reservoir is then partially closed from the pelvic floor to the symphysis using running sutures (3-0 Vicryl).

Implantation of the ureters into the reservoir is in principle performed according to the technique described by Le Duc and Camey [9]. Normally, the right ureter is introduced into the neobladder through the second segment right of the "W". This invariably provides a straight course for the

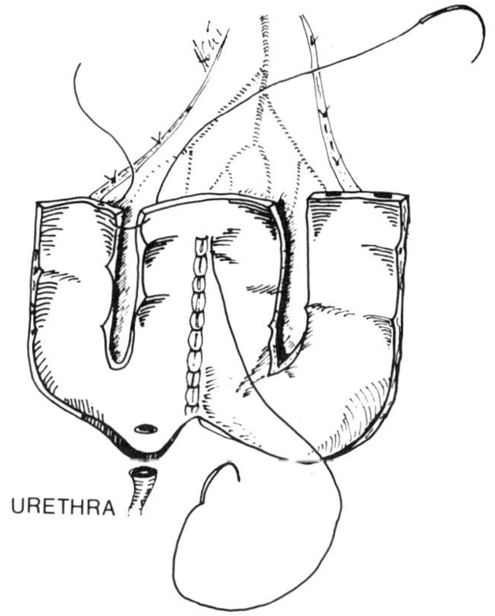

Fig. 3. The small bowel loops have been rearranged in a W-shape. An ileal plate is formed using single-row through-and-through running sutures. Note U-shaped flap for anastomosis to the membranous urethra

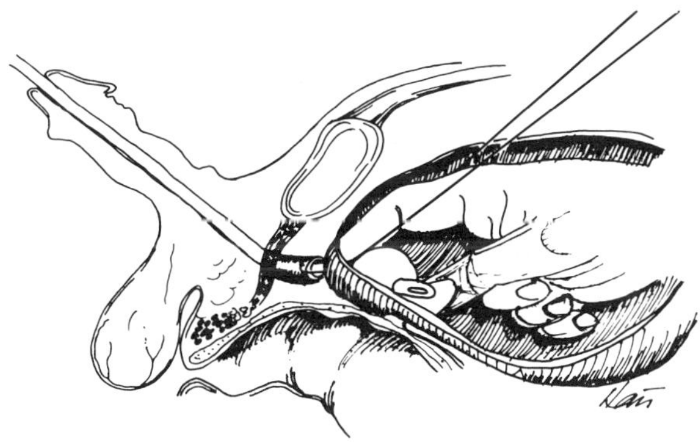

Fig. 4. The ileum neobladder is anastomosed to the urethral remnant with six mattress sutures tied from "inside" the neobladder. The ileum plate is closed to a reservoir with running sutures; the ureteral stents are brought out through the suture line

ureter without interference from the meso of the reservoir. The left ureter can be guided either laterally or medially (after being pulled through under the mesosigmoid colon) to the sigmoid colon, depending on the length of the ureter and the course of the sigma. The implantation site is usually the second segment left of the reservoir. Depending on whether it follows a

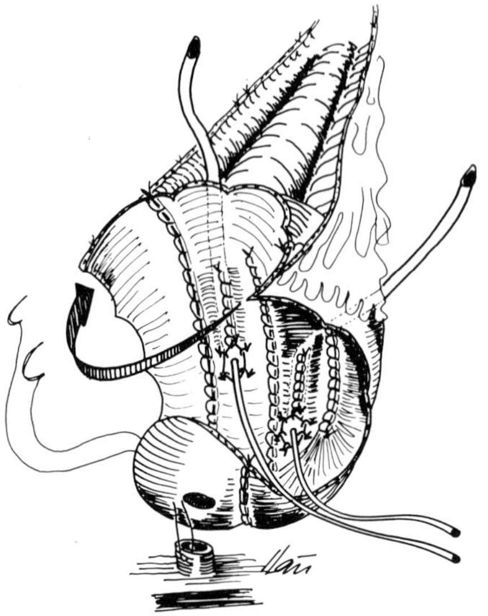

Fig. 5. After implantation of ureters, segments of ileal wall approximately the diameter of a small finger are excised from the U-shaped flap and the ileal neobladder is anastomosed to the urethral remnant with six mattress sutures tied from inside neobladder. The ileal plate is closed to a reservoir with running sutures and ureteral stents are brought out through the suture line

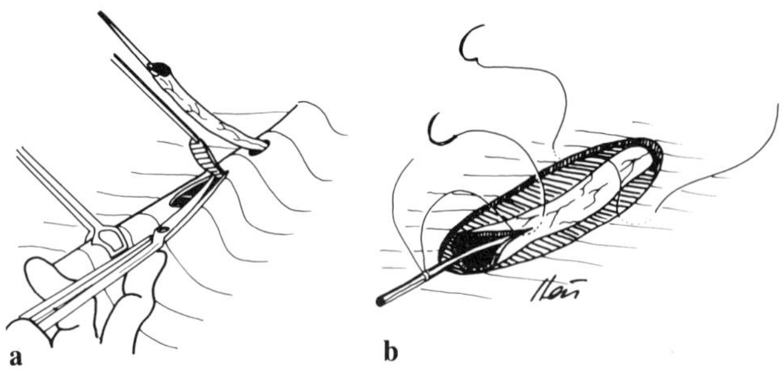

Fig. 6 a,b. Ureteroileal anastomosis. **a** The ureter is pulled through a small incision of the ileum wall at a convenient site. The mucosa is incised to create a mucosal sulcus approximately 3 cm long. **b** The ureter is spatulated and fixed to the ileum wall with two sutures. The ileum mucosa is then closed above the ureter with two to three sutures. A 7 Fr ureteral catheter is placed and secured by a 4-0 plain catgut suture

laterocolic or mediocolic course, the ureter is guided to the wall of the neobladder either behind or in front of the meso. The neo-orifice is spatulated over a length of 5–8 mm and sutured to the ileal mucosa. The mucosal borders of the prepared groove, in which the ureter is embedded, are either closed over the ureter (Fig. 6) or adapted to the ureteral mucosa.

Postoperative Management

As soon as the patient no longer requires pain control via the indwelling peridural catheter, early mobilization is important to prevent thromboembolic and pulmonary complications. All patients are kept on low-dose heparin for 3 weeks. Bowel stimulation is started on the second postoperative day.

Excessive mucus production of the ileal bladder may sometimes cause a problem in the postoperative phase by obstructing the urethral catheter and cystostomy. For this reason, the ileal bladder is rinsed via the cystostomy with 50–100 cm^2 of saline twice a day, starting on the fifth postoperative day. The ureteral stents are routinely removed on the 14th postoperative day. As soon as the urine comes into contact with the ileal bladder mucosa, reabsorption of urine electrolytes may occur. Therefore, the base excess is checked at weekly intervals for the first 4 weeks, and monthly thereafter. Approximately 50% of all patients will need alkalinizing therapy.

The urethral catheter is removed on the 21st postoperative day, after a cystogram has demonstrated complete healing of the ileourethral anastomosis. On the case occasion when leakage of the anastomosis is present, this is treated by prolonged catheter drainage until the leak has closed spontaneously.

After removal of the catheter, the bladder is usually emptied by abdominal straining. The cystostomy is removed as soon as the patient is able to empty the ileal neobladder without significant residual urine.

Patients and Indications

Since April 1986, 161 patients received total bladder replacement or bladder augmentation using the ileum neobladder. In 141 male patients with a mean age of 60.7 years (23–82 years) radical pelvic lymphadenectomy and cystoprostatectomy was performed for bladder carcinoma stages > T1 < pN1M0 and an ileum neobladder for total bladder replacement was created. Eleven patients had had definitive radiotherapy before. Seventeen patients with locally advanced tumors were treated by two courses of neoadjuvant chemotherapy (methotrexate-vincristine-epirubicin-cisplatin, MVEC), 5 patients received adjuvant chemotherapy after cystoprostatectomy and creation of the ileum neobladder, and 11 had palliative chemotherapy following local or systemic tumor progression.

A total of 20 patients (14 females, 6 males) with a mean age of 52.8 years (17–73 years) underwent subtotal bladder resection and bladder augmentation with the ileum neobladder using a slightly modified technique (Table 2); in 4 cases an ileal conduit was changed to a neobladder.

The follow-up period for the 141 patients after cystoprostatectomy was 3–52 months (mean 23.8 months); the follow-up period for the 20 patients

Table 2. Indications for subtotal bladder resection and augmentation using the ileum neobladder

Diagnosis	No. of patients
Tuberculosis	6
Interstitial cystitis	6
Contracted bladder following radiotherapy	3
Ileal conduit to neobladder	4
Urge incontinence after failure of conservative treatment	1

after bladder augmentation averaged 26.3 months (3–46 months). The postoperative follow-up included urinalysis, anlaysis of serum creatinine and electrolytes, and ultrasonography of the kidneys and neobladder (for residual urine). Three months postoperatively 259 renoureteral units (RUU) were evaluated by an IVP. A total of 109 patients have been available for postoperative urodynamic studies and a complete follow-up protocol of their home micturition behavior, single micturition volume, frequency, and continence situation.

Results

Perioperative mortality was 0.5% (due to fulminant pulmonary embolism). In the postoperative phase, 4.5% of patients had to be reoperated on, 3% because of ileus, 1% because of an intraperitoneal abscess, and 0.5% because of a leak at the ileal anastomosis. A prolonged reduction of gastrointestinal function occurred in 13% of patients and was treated conservatively. Of those patients with a leakage at the ileourethral anastomosis 8% had prolonged catheter drainage until the leakage closed spontaneously. In 3% of RUU, transient nephrostomy was necessary because of ureteral obstruction at the ureteroileal anastomosis after removal of the intraoperatively placed ureteral stents. General postoperative complications also included lymphocele (3%), pulmonary embolism (2%), thrombosis (2.5%), and secondary wound healing (2.5%). All these complications were treated conservatively.

Significant late complications after ileum neobladder surgery have been rare. In 8.5% of the patients reoperation was necessary: bowel complications (ileus, intraperitoneal abscess, fistula to colon) occurred in 4%; 4% had ureteral reimplantation because of a stenosis of the ureteroileal anastomosis. One patient had spontaneous rupture of the neobladder due to chronic overextension. Urethral strictures, which were seen in 7.5 of the patients, were treated by urethrotomy or dilatation.

Mild metabolic acidosis occurred in more than 50% of the patients; therefore, all patients routinely received alkalinizing therapy.

Local tumor recurrence and/or distant metastases occurred in 23 patients – the function of the ileum neobladder was affected in only one case of urethral tumor recurrence. A total of 7.5% of patients died from tumor progression. One patient was hospitalized in an outside hospital because of severe pneumonia and metabolic acidosis and died 10 weeks postoperatively at an age of 81 years; one patient died from septicemia of unknown cause; one had myocaridal infarction; and one had apoplexia.

Three months postoperatively 259 RUU were evaluated by IVP – 92.2% of patients in the bladder substitution group and 80.8% of patients in the augmentation group had normal or only minimally dilated upper tracts. The poorer results after augmentation were due to a higher percentage of preexisting obstruction and/or dilatation.

Vesicorenal reflux was found in 6 RUU. In two patients (4 RUU) an ileum conduit was undiverted by anastomosing the intact Wallace anastomosis of the conduit to the ileum neobladder.

Of those patients with a follow-up of more than 3 months postoperatively 78% are totally continent day and night without any spontaneous loss of urine or requirement to wear sanitary pads. Night-time incontinence occurs in 7.7% of patients. Mild stress incontinence with occasional spotting in underwear or moist sanitary pads occurs in 10%; specific therapy is not considered necessary. Only 4.5% of all patients have severe stress incontinence. One of these patients had severe stress incontinence preoperatively.

When the continence situation of different age groups was compared, 89% of patients younger than 70 years were continent day and night, while 50% of patients of age 70 and above were incontinent, indicating that for the older age group achieving control of the external urethral sphincter is much more of a problem than for younger patients.

After the third postoperative month 109 patients underwent repeat urodynamic investigations (Table 3). The maximum bladder capacity was 715 (330–2000) ml. The maximum intravesical pressure at maximum capacity was 26 (8–47) cm H_2O. At half-maximum capacity the average intravesical pressure was 10 (0–27) cm H_2O. The maximum urethral closure pressure was 55 (20–120) cm H_2O. Regarding these parameters, there was no significant difference between patients with radical cystoprostatectomy and patients with bladder augmentation.

Table 3. Urodynamic evaluation of ileal neobladder after third postoperative month

Maximum bladder capacity (ml)	715	(330–2000)
Intravesical pressure at half-maximum capacity (cm H_2O)	10	(0–27)
Intravesical pressure at maximum capacity (cm H_2O)	26	(8–47)
Urethral pressure of external sphincter (cm H_2O)	55	(20–120)
Maximum flow (ml/sec)	24	(5–50)
Meam flow (ml/sec)	16	(2–40)
Residual volume (ml)	18	(0–80)

Continuous urodynamic pressure studies during a period of up to 60 min at maximum and half-maximum capacity revealed "bladder contractions" in 24 of 67 (35%) patients. The average amplitude of these pressure waves was 20 cm H_2O. Only 2 of the 24 patients who exhibited intravesical pressure rises under long-term urodynamic investigation experienced mild night-time incontinence occasionally; the other 22 patients were totally continent day and night. In 2 out of 3 of all patients, there were no active pressure rises measurable within the ileum neobladder during long-term urodynamic investigation.

In the majority of patients, micturition is initiated by abdominal straining, when the patient feels an abdominal sensation resembling the original sensation of a full bladder. Bladder emptying is accompanied by intra-abdominal and intravesical pressure rises up to 110 cm H_2O; reflux does not occur. Evaluation of the micturition protocols revealed a normal frequency of micturition with six times per day (range 3–9) and once per night (rang 0–3). After cystoprostatectomy, the single micturition volume averaged 364 (155–646) ml; after bladder augmentation it averaged 334 (126–775) ml, with marked intraindividual variations. The duration of micturition was between 10 and 366 sec, the maximum uroflow rate between 5 and 50 ml/sec. Residual urine was 18 (0–80) ml on average. Three female patients are on sterile intermittent catheterization, because they are unable to empty their bladders spontaneously following bladder augmentation.

Discussion

The purpose of a conduit – still the form of urinary diversion most widely used – is to provide a nonobstructed urinary flow from the ureteroentero anastomosis to the external stoma. Continent bowel reservoir, however, have the additional task of mimicing the storage function of the urinary bladder. With regard to its viscoelastic capabilities, the bowel wall resembles closely that of the urinary bladder. However, the bowel behaves differently when it comes to fluid filling: in this situation, reflectory contractions of the bowel musculature result in intermittent rises in pressure.

To solve this problem, detubularization and rearrangement of the bowel segment used for a urinary reservoir are mandatory. Kock [6] has demonstrated with his early experiments that the way in which the detubularized bowel parts are arranged does influence the functional outcome of the reservoir: whereas laterolateral closure of a U-shaped bowel plate leads to insufficient dampening of the pressure waves, the craniocaudal closure, as used for the "Kock pouch," eliminates theses pressure peaks almost completely.

A laterolateral closure with our W-shaped reservoir also apparently results in an excellent attenuation of pressure waves. Yet, as with the Kock pouch [13], low-amplitude pressure waves can be observed in a number of patients (35%) as seen from long-term urodynamic investigations [1]. Due to the low amplitude of the pressure waves, function of the reservoir is not necessarily compromised: less than 10% of patients with an ileal neobladder suffer from nocturnal incontinence.

Overall, 78% of patients are completely continent both day and night. In 10%, mild stress incontinence requiring no therapy was found, and 4.5% presented with severe stress incontinence. However, these figures merely reflect a static observation of the patient cohort at a certain point of evaluation. This evaluation covered patients 3–12 months after the operation, and the situation tended to improve during further postoperative treatment. Therefore, it is important to note that none of our neobladder patients has required an artificial sphincter so far.

Comparison with continence results of other neobladder procedures is hampered by contradictory and "semi-quantitative" results in the literature: Kock himself [8] has found 30 of 34 patients with a "urethral" Kock pouch completely continent, whereas the experience of Skinner [16] with the same technique is less favorable: 40% of the patients are only dry at night if they get up for voiding at least twice. Studer [19] states, that "some" of his patients have to wear sanitary pads at night. According to Riedmiller [14], 24 out of 27 patients with a urethral Mainz pouch are completely continent, whereas Broderick [3] has found pressure waves and nocturnal incontinence in 8 of 8 patients with the same reservoir type.

Ureteral implantation with the Le Duc technique has stood the test of time: Reflux (2%) and stenoses requiring operative reintervention (4%) have been rare. The alternative technique – creation of an antireflux nipple – is technically more complicated and may lead to a higher rate of complications, ranging up to 50% in one series in the literature [8].

In principle, all male patients undergoing cystectomy as a curative procedure are candidates for a neobladder. This primary wide range of indications is, however, reduced by oncological, psychological, and age-related restrictions.

The question of oncological contraindications for the neobladder reflects the controversy over simultaneous urethrectomy: Is it justified to leave the urethra in situ when multifocal carcinomas of tumors near the bladder neck have to be dealt with? The literature indicates that the overall risk for urethral recurrencies following cystectomy ranges from 4% to 18% [18], although these results have all been observed in patients with conduits. The effect of a continuing exposure of the urethra to the urinary stream has yet to be defined, which will require yet another 20 years experience with the neobladder technique. To date, our short-term results (average follow-up 23.8 months) indicate a very low rate (1%) of urethral recurrencies. In our opinion, which is also shared by others [10,16], only biopsy-proven tumor involvement of the prostatic urethra is currently an indication for simultaneous urethrectomy, since in this situation the incidence of urethral recurrence seems of dramatically increase to 30%–40% [10,15].

Another restriction on indications for the neobladder is the psychological state of the patient: regular controls are essential because metabolic disturbances, residual urine, or deterioration of the upper tract may occur without any significant symptoms. It is therefore a prerequisite for this operation that the patient has an appropriate understanding of the necessity of being under life-long surveillance.

The only single parameter which has a significant influence on the continence results of the neobladder is the age of the patient: previous evaluations [12] have shown that 90% of patients under 70 years of age have been completely continent, whereas only 50% of patients over 70 have achieved a similar status. It must be kept in mind that a functioning conduit is superior to an incontinent neobladder, and that elderly, sexually less active, patients are not so much adversely affected by a wet stoma as younger ones [2].

Experience with continent urinary diversion in general is still too short lived for a definitive assessment of this technique to be made. Former experiences with other forms of urinary diversion such as ureterosigmoidostomy or the ileum conduit have conclusively demonstrated that, following an initial period of enthusiasm, long-term results – particularly with regard to the upper tract – may be disappointing. The few long-term results available for neobladders [4] are incompletely documented, but suggest a favorable long-term protection of the renal function. Up to 5 years experience with detubularized low-pressure reservoirs confirms this trend so far [14,19,22].

Similarly, valid assessement of the carcinogenesis risk in association with neobladders is naturally hampered by the lack of long-term experience. Due to different potentially relevant factors such as contamination of the reservoir and duration of urine exposure, results obtained with other forms of urinary diversion (ureterosigmoidostomy, ileum conduit) can not be simply transferred to neobladders.

Bladder substitution by continent bowel reservoirs has been a real advancement in modern urology. Short-term experience and patient acceptance of the ileal neobladder has been excellent. Like others [17], to date we consider this technique to be the procedure of choice for male patients requiring radical cystectomy.

References

1. Bachor R, Frohneberg D, Miller K, Egghart G, Hautmann R (1990) Continence after after total bladder replacement: urodynamic analysis of the ileal neobladder. Br J Urol 65:462
2. Boyd SD, Feinberg SM, Skinner DG, Lieskovsky G, Baron D, Richardson J (1987) Quality of life survey of urinary diversion patients: comparison of ileal conduits versus continent Kock ileal reservoirs. J Urol 138:1386
3. Broderick GA, Narayan P, Tanagho E (1989) The dynamics of continence in the detubularized ileocecal reservoir. J Urol 141:304A

4. Camey M (1985) Bladder replacement by ileocystoplasty following radical cystectomy. World J Urol 3:161
5. Hautmann R, Egghart G, Frohneberg D, Miller K (1988) The ileal neobladder. J Urol 139:39
6. Kock NG (1969) Intraabdominal reservoir in patients with permanent ileostomy. Arch Surg 99:223
7. Kock NG, Nilson AE, Nilsson LO, Norlen LJ, Philipson BM (1982) Urinary diversion via a continent ileal reservoir: clinical results in 12 patients. J Urol 128:469
8. Kock NG, Ghoneim MA, Lycke KG, Maahran MR (1989) Replacement of the bladder by the urethral Kock pouch: functional results, urodynamics and radiological features. J Urol 141:1111
9. Le Duc A, Camey M, Teillac P (1987) An original antireflux ureteroileal implantation technique: long-term follow-up. J Urol 137:1156
10. Levinson AK, Johnson DE, Wishnow KI (1990) Indications for urethrectomy in an era of continent urinary diversion. J Urol 144:73
11. Lilien OM, Camey M (1984) 25-year experience with replacement of the human bladder (Camey procedure). J Urol 132:886
12. Miller K, Bachor R, Frohneberg D, Egghart G, Hautmann R (1990) Kontinente Harnbaleitung beim älteren Patienten: Ergebnisse mit der Ileum-Neoblase. Urologe [A] 29:87
13. Norlen L, Trasti H (1978) Functional behaviour of the continent ileum reservoir for urinary diversion. An experimental and clinical study. Scand J Urol Nephrol 49:33
14. Riedmiller H, Köhl U, Thüroff J, Hohenfellner R (1989) Mainz-Pouch – Fünf Jahre klinische Erfahrung. Poster International Symposium, Sept 1989, Nijemegen
15. Schellhammer PF, Bean MA, Whitmore WF (1977) Prostatic involvement by transitional cell carcinoma: pathogenesis, patterns and prognosis. J Urol 118:399
16. Skinner DG, Lieskovsky G, Boyd S (1989) Continent urinary diversion. J Urol 141:1323
17. Skinner DG (1990) Editorial comment. J Urol 143:497
18. Stöckle M, Gökcebay E, Riedmiller H, Hohenfellner R (1990) Urethral tumor recurrences after radical cystoprostatectomy: the case for primary cystoprostatourethrectomy? J Urol 143:41
19. Studer UE, Ackermann D, Casanova GA, Zingg EJ (1989) Three years' experience with an ileal low pressure bladder substitute. Br J Urol 63:43
20. Teigland CM, Röhrborn CG (1987) Harnableitung mittels Ileumschlinge nach Camey – Ergebnisse urodynamischer Studien. Akt Urol 18:11
21. Walsh PC (1986) Radical retropubic prostatectomy. Campbell's Urology 3:2754
22. Wenderoth UK, Bachor R, Egghart G, Frohneberg D, Miller K, Hautmann R (1990) The ileal neobladder: experience and results of more than 100 consecutive cases. J Urol 143:492

*Neoadjuvant Versus Adjuvant Chemotherapy in Invasive Bladder Cancer**

H.I. Scher

Genitourinary Oncology Service, Division of Solid Tumor Oncology, Department of Medicine, Memorial Sloan-Kettering Cancer Center, 1275 York Avenue, New York, NY 19921, USA

Introduction

Invasive bladder tumors represents a range of diseases that vary clinically from minimally invasive with a low metastatic potential, to high grade lesions that metastasize early despite minimal invasion, to large fixed lesions that present with urinary obstruction and pain [1,2]. As the prognosis for each of these groups varies, so should the therapy. Too often, however, a uniform treatment approach is used for all of these clinical scenarios. This approach is becoming outmoded; while the treatment options for invasive bladder tumors are myriad, it is now apparent that chemotherapy is playing an increasingly important role in the management of these malignancies [3]. This is based on the high rate of systemic failure – the major cause of mortality following local treatment modalities – the high response proportion to modern combination regimens, and the possibility that chemotherapy may allow organ preservation. Critical to improving treatment selection for an individual patient is the refinement of our ability to define the biologic behavior of these tumors, both with respect to metastatic risk and sensitivity to treatment. Only then can a specific approach be advised that offers the highest probability of cure with the least morbidity.

When evaluating the role of chemotherapy it is important not to neglect advances in surgical management which have resulted in an improved survival relative to historical series [4–6]. Operative mortality has decreased to less than 2% in most reports. The availability of continent urinary diversions and internal urinary reservoirs has lessened the interest in bladder preservation as a therapeutic goal. Preoperative staging has also been improved, thereby increasing the frequency with which unsuspected nodal spread is discovered pre-laparotomy. Laparoscopic node dissections are also

* Supported by National Cancer Institute (NCI) grant CA-05826.

being investigated in this regard. The net result is a stage migration [7], leading to the inclusion of patients with earlier lesions in higher stage categories, producing an apparent improvement in survival for all stages [8]. A more careful step sectioning of a submitted lymph node has a similar effect, as detection of microscopic tumors upstages a patient with "early" small-volume disease into a more advanced stage. These advances alone have resulted in improved 5-year survival rates for patients treated with surgery as monotherapy [6,9].

The impetus to integrate chemotherapy in the treatment of an invasive bladder tumor initially stemmed from promising results observed in patients with advanced disease. To assess the role of chemotherapy for patients with invasive bladder cancer, however, several questions must be addressed. How effective is "modern" chemotherapy? Should it be routinely administered for all patients with invasive tumors? Is it best used in the pre- or postoperative setting? Can it be used to preserve bladder function without compromising cure? Can survival be improved?

How Effective Is Modern Chemotherapy?

Cisplatin and methotrexate are considered the most active agents in urothelial tract tumors (Table 1). Responses to single agents are generally partial, occur in 15%–30% of cases for a median duration of 3–4 months. Complete remissions are anecdotal. Using combinations based on cisplatin ± methotrexate, however, overall response proportions (complete response,

Table 1. Single agents in the therapy of urothelial tract tumors. (After Scher [3])

Agent	CR+PR		95% Confidence interval (%)
	Cases (100%)	(%)	
Cisplatin			
Single institution	70 (206)	34	28–40
Randomized trials	55 (315)	17	37–55
Carboplatin	21 (186)	15	11–19
Iproplatin	7 (39)	18	6–30
Methotrexate			
"Low" dose	68 (236)	29	23–35
"High" dose	16 (57)	45	37–50
Adriamycin	47 (274)	17	13–22
Vinblastine	6 (38)	16	4–28
Cyclophosphamide	30 (98)	31[a]	22–40
5-Fluorouracil	22 (141)	17[a]	11–25
Mitomycin C	5 (42)	13[a]	2–22
Gallium nitrate	7 (26)	27	11–48

CP, complete response; PR, partial response.
[a] Reflects early trials using varied doses, schedules, and response criteria.

CR + partial response, PR) of up to 70% have been observed, including complete responses in up to 25% of cases [10–14]. Despite encouraging results for phase II trials, only one randomized trial has shown a survival advantage with combination therapy compared to single agent therapy. This is in part the result of the large sample size required to show a benefit based on the efficacy of currently available combinations [15,16]. Seven randomized trials comparing a combination to single agent cisplatin showed no benefit. On careful examination, however, it became clear that the median number of patients entered, 100, was too low to exclude a difference if, in fact, one existed. Stated differently, the beta error was large [16]. Recently, a randomized comparison with 224 evaluable patients treated with either M-VAC (methotrexate, vinblastine, adriamycin, cisplatin), a four-drug regimen developed at Memorial Hospital [10], or single agent cisplatin was reported. The results showed a significant difference in the CR proportion (13% vs 3%), overall response proportion (34% vs 9%, $p = 0.01$), and survival (12.6 vs 8.7 months, $p = 0.02$) in favor of M-VAC [17]. Subsequently, M-VAC was shown to be superior to CISCA (cyclophosphamide, adriamycin, cisplatin), with respect to both overall response proportion (65% vs 46%) and median survival (18.4 vs 9.3 months, $p = 0.0003$) in a second randomized trial [18]. The role of adriamycin in the M-VAC regimen is controversial; however, the median survival for CR patients treated with M-VAC[15] has 38 months in one series [15] compared to 14 months following CMV (cisplatin, methotrexate, vinblastine) [19].

Longer follow-up is needed to determine the proportion of patients enjoying long-term survival. Some estimations however are available from single arm phase II trials. For example, Fig. 1 shows the survival distributions of patients treated with M-VAC at Memorial Sloan-Kettering Cancer Center (MSKCC) with a 74-month median follow-up. The median survival of patients with advanced nodal and metastatic disease was 32.9 vs 12 months, and definitive 6-year survivals 32% vs 17% respectively [19a]. In a trial using CISCA in 92 patients with nodal and metastastic disease, 19% were alive beyond 200 weeks,[11] while 9% of 62 patients treated with CMV were alive at 3.9 years [20]. Comparing M-VAC and cisplatin, with an 18-month median follow-up, 2-year survivals of 20% vs 8% respectively have been observed (P. Loehrer, personal communication).

These data support the recommendation that combination chemotherapy should be considered standard in this disease with a minimum of 3 agents. They also form the basis for investigating chemotherapy in invasive disease, and provide the quantitative basis for estimating the number of patients needed to demonstrate a survival advantage for a strategy that incorporates chemotherapy in one experimental arm of a treatment for invasive bladder cancer.

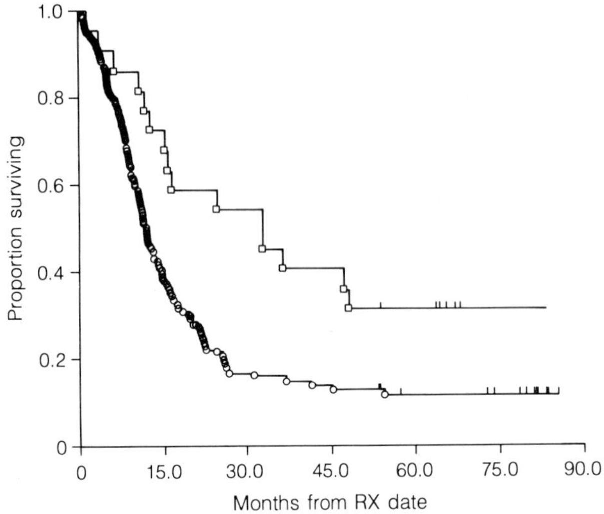

Fig. 1. Survival distributions of patients treated with M-VAC with a 74-month median follow-up. □, 22 patients (7 censored, still alive) with advanced nodal disease; ○, 110 patients (15 censored, still alive) with advanced metastatic disease. *Tick mark* (ℤ) indicates last follow-up. (After Seidman and Scher [19a])

Is Chemotherapy Best Used in the Pre- or Postoperative Setting?

Chemotherapy can be used (a) preoperatively as monotherapy; (b) preoperatively followed by definitive surgery, (c) as part of a planned integrated strategy with radiation therapy, or (d) in the postoperative or adjuvant setting. The latter situation includes both patients treated after a radical cystectomy and pelvic lymph node dissection, and those treated after complete endoscopic removal of an invasive lesion by transurethral resection. Complete endoscopic resection is particularly important in the last example when pursuing an aggressive treatment policy aimed at bladder preservation. Table 2 outlines the advantages and disadvantages of neoadjuvant (preoperative) and adjuvant approaches. Both employ chemotherapy at an early stage, exploit the inverse relationship between tumor burden and curability [21,22], and take into account the increased cure fraction of patients with nodal compared to established metastatic disease (see Fig. 1).

The primary advantage of a neoadjuvant approach is that chemosensitivity is assessed in vivo. This permits treatment of an individual patient as long as they are responding, and can be used to direct the type and need for subsequent therapy to the primary tumor. Response data provide prognostic information, and may make organ preservation possible. The latter may be accomplished by a less extensive operation such as a partial as opposed to a radical cystectomy, or by local radiation therapy with or without chemotherapy [23,24]. Nonresponders can go directly for cystectomy, thus reduc-

Table 2. Comparison of adjuvant and neoadjuvant therapy. (Modified from Scher [3])

	Neoadjuvant	Adjuvant
A. Rationale		
1. Early treatment of micrometastases		
(a) Inverse relationship of tumor burden and cure	+	+
(b) Increased chemosensitivity of small volume	+	+
high growth fraction tumors	+	+
(c) Decreased chance of spontaneous resistance		
B. Factors favoring neoadjuvant therapy		
1. Chemosensitivity determined case by case in vivo		
(a) Response assessment in vivo	+	–
(b) Organ preservation possible	+	–
2. "Downstaging" of the primary tumor can		
(a) Decrease the extent and need for additional therapy	+	–
(b) Convert an "unresectable" to a "resectable" lesion	+	–
(c) Drug delivery not compromised by previous surgery or radiation therapy	+	–
3. Prognostic importance of response in the primary site on prognosis	+	–
4. End point of treatment more precise	+	–
5. Potential for accelerated growth after surgery	+	–
6. Better patient tolerance	+	±
C. Factors favoring adjuvant therapy		
1. Case selection		
(a) Staging error of "T" vs "P"	–	+
(b) Need for therapy based on pathologic as opposed to clinical criteria	–	+
(c) Exposure of patients "cured" by local therapies to cytotoxic agents	–	+
2. Timing of definitive local therapy		
(a) Jeopardizing of curative therapy by prolonged treatment with ineffective agents	–	+
(b) Patient refusal of potentially curative therapy	–	+

ing toxicity. Patient tolerance and the ability to administer adequate doses of chemotherapy may be in the pre- versus postoperative setting. This question, however, is currently under study at M.D. Anderson Hospital (C. Logothetis, personal communication).

Interpreting and comparing reports of neoadjuvant therapy trials can be difficult due to differences in case selection, the techniques for staging and restaging [25], the idiosyncracies of TNM staging system itself [5,26], the response criteria [25], and the chemotherapy doses and schedules. These are outlined in Table 3. Treatment policies may also differ between institutions. At the University of Southern California, radical surgery and creation of an internal urinary reservoir is recommended if muscle invasion is documented

Table 3. Differences in methodology of neoadjuvant trials. (Modified from Scher and Herr [25])

A. *Selection criteria*
   1. Extent of disease
      (a) Invasive but operable
      (b) Invasive but inoperable
      (c) Requiring cystectomy for symptoms: variable risk of metastatic disease
   2. Not "candidates" for cystectomy:
      (a) Medically excluded
      (b) Patient refusal

B. *Extent of staging and restaging*
   1. Clinical
      (a) Cystoscopy and biopsy ±
      (b) Bimanual – pre and post TUR ±
      (c) Degree of resection: complete vs incomplete ±
      (d) Diagnostic scans (CT, MRI, sonography) ±
      (e) Urine cytology ±
      (f) Flow cytometry ±
   2. Pathologic (laparotomy)
      (a) Radical or partial cystectomy
      (b) Transmural bladder biopsy at primary site
      (c) Lymph node dissection vs selective sampling
   3. Staging system
   4. Time to reassessment of response

C. *Criteria of response, evaluation, and reporting*

D. *End points*
   1. Complete response vs "objective" (complete plus partial) response – clinically or pathologically staged
   2. No. of patients who are spared cystectomy:
      (a) Chemotherapy alone
      (b) Chemotherapy followed by radiation therapy
      (c) Chemotherapy followed by partial cystectomy
   3. Time to relapse:
      (a) In the primary site if bladder preserved
      (b) Development of new tumors in the bladder
      (c) Locally in the pelvis
      (d) Distant metastases
   4. Survival

CT, computed tomography; MRI, magnetic resonance imaging.

at any time. Chemotherapy is used in the adjuvant setting for patients with a high risk of metastatic disease based on the pathologic findings at surgery. At MSKCC, a patient with muscle invasive disease would be offered a repeat cystoscopy and biopsy. If no disease were documented the patient would be observed [27]. If muscle invasion is documented on the second cystoscopy, the patient would be considered for neoadjuvant chemotherapy. Staging and restaging procedures can include noninvasive techniques such as computerized tomography, sonography and magnetic resonance imaging as well as invasive procedures such as a cystoscopy and biopsy, or exploratory laparotomy with or without cystectomy. Clearly the latter would be the most stringent and give "inferior" results relative to clinical criteria alone.

The TNM groups patients with widely varied prognoses within the same stage category [5,26]. For example, T4a disease may represent a small lesion near the base of the bladder with no palpable mass, or a large infiltrating tumor with a palpable mass on examination under anesthesia. The presence or absence of a palpable mass has been shown to have prognostic significance [28–30]. The policy to completely resect or not completely resect all visible tumor at the time of the initial cystoscopy has not been standardized. However, when considering bladder preservation, attempts should be made to aggressively resect all visible disease [31]. This strategy may impact adversely on the ability to document response. Response criteria and trial end points can be divided into those that are based on clinical criteria cystoscopy and biopsy, cytology and computerized tomography – or those based on pathologic methods such as exploratory laparotomy with pelvic lymph node dissection, or a partial or radical cystectomy [25]. Chemotherapy doses and schedule vary considerably and few trials report the actual amount of chemotherapy delivered to different individuals within a planned protocol regimen. Dose intensity considerations are usually not included into the reported results [32].

These caveats aside, a review of reported trials does permit some preliminary conclusions. Ample data show that chemotherapy alone can produce complete responses within the bladder, however, reported response proportions using pathologic criteria are uniformly lower than those based on clinical (cystoscopic) criteria. At MSKCC using M-VAC, the proportion of patients who were T0 endoscopically after a median of three cycles of treatment was 48% (13/27; 95% confidence limits, 29%–67%) compared to 23% who were P0 (14/60; 95% confidence limits, 13%–34%) using a pathologic examination of the bladder removed surgically as the end point [33,34]. Complete response proportions are generally higher for combination regimens than for single agents, yet large differences have been observed using the same combination. For example, pathologic complete response proportions of 0 [35] to 60% [36] have been reported with cyclophosphamide, adriamycin, cisplatin (CAP). Response proportions vary inversely with depth of invasion [28,29,33–35]. With M-VAC, 43% of patients with T2 versus only 8% of patients with T4 disease at the start of chemotherapy, achieved a pathologic complete remission [35]. The initial pathology is also important as several groups have shown that carcinoma in situ, nontransitional and mixed histology tumors are less chemosensitive [31,34,35].

The proportion of patients that can be rendered tumor free within the bladder using chemotherapy alone can be estimated using combined data from trials that employed neoadjuvant chemotherapy followed by definitive surgery (Table 4). Ignoring differences in case selection, response criteria and drug delivery the number of trials using the combination, the total number treated, and proportion of complete pathologic responses are illus-

Table 4. "Complete" pathologic response in the bladder using combination chemotherapy. (After Scher [3])

Agent(s)	No. of trials	No. of patients	Complete response at cystectomy	95% Confidence interval (%)
			No. of patients (%)	
DDP/5-FU	1	16	7 (43)	19–68
DDP/MTX	2	75	26 (35)	24–45
CMV	3	71	22 (31)	20–42
CAP	4	83	18 (22)	13–31
M-VAC	7	138	42 (30)	23–38

DDP, cisplatin; 5-FU, 5-fluorouracil; MTX, methotrexate; CAP, cyclophosphamide, Adriamycin, cisplatin; CMV, cisplatin, methotrexate, vinblastine; M-VAC, methotrexate, vinblastine, adriamycin, cisplatin.

trated. Recognizing the heterogeneity within the trials, the observed complete response proportion of 22%–43% shows that for most patients, chemotherapy alone is inadequate treatment of the primary tumor. Whether a partial or radical cystectomy, or whether attempts at bladder preservation using concurrent cisplatin and radiation [24] is optimal at this point is under investigation.

Response to chemotherapy has been shown to have prognostic significance [32,36]. Nevertheless, a cause and effect relationship can not be implied. In a summary series of 125 patients treated with cisplatin based chemotherapy and radical surgery, followed for a median of 25 months (range 21.5–100 months); 91% (53/58) of responders (≤P1 at cystectomy) but only 37% (25/67) of nonresponders (≥P2 at cystectomy) were alive and free of disease. The true role of neoadjuvant therapy can only be assessed when randomized comparisons have been completed [37].

Adjuvant therapy refers to treatment after surgical removal of the bladder where the recommendation to offer chemotherapy is based on pathologic as opposed to clinical grounds. Criteria for treatment vary, but have included the presence of nodal disease or extravesical tumor extension, particularly to adjacent viscera. Using primary surgical therapy first reduces the risk of a resectable tumor becoming unresectable should systemic therapy be ineffective. The number of patients exposed to cytotoxic drugs who may have been cured by surgery is also reduced. Removal of the bladder eliminates the problem of new tumor formation when the bladder is left in situ. This is the treatment policy used by many physicians [38] particularly as the ability to create internal urinary reservoirs is improved.

Most reported trials which evaluated adjuvant chemotherapy have not shown a survival advantage with the addition of chemotherapy [3,8]. This was due to both the small number of patients treated, the limited efficacy of the chemotherapy employed, and difficulties in delivering adequate doses of chemotherapy – a prerequisite for benefit. Two studies using CISCA or CAP were recently reported that did show an apparent survival benefit. Only one was randomized. At the M.D. Anderson Hospital, 339 patients treated by radical surgery were classified on the basis of metastatic risk following surgery. These investigators defined high risk as presence of vascular or lymphatic invasion (37 cases), extravesical tumor spread (50 cases), node positive disease (24 cases), or extension into the adjacent pelvic viscera (22 cases). Of 133 high-risk patients, 71 received and 62 did not receive chemotherapy. The survival distributions of high-risk patients 206 low-risk patients who did not receive chemotherapy were compared. The 5-year survival of the high-risk patients treated with chemotherapy was 70%, similar to that for the low-risk group (76%). In contrast, the survival of the untreated high-risk group was 37% at 5 years. Only patients with extravesical, node positive, or locally extensive disease benefitted from chemotherapy. The patients with vascular or lymphatic invasion were unaffected [39]. These results, while strongly suggestive, can not be considered definitive as the treated versus untreated high-risk populations may not have been comparable [8,40].

At the University of Southern California, 91 patients with P3, P4 ± N+ disease after radical cystectomy and en bloc pelvic node dissection were randomized to receive or not receive four cycles of adjuvant CAP [41]. These 91 patients were derived from a total of 453 patients who underwent radical surgery during the same time period. Of the 453 patients, 241 met the entry criteria of the protocol; 43 of these were medically excluded and 101 agreed to randomization. Of these 101, 10 were excluded with non-transitional cell histologies leaving 91 patients (37% of the eligible group) enrolled. A benefit in favor of adjuvant chemotherapy was demonstrated by comparing the percentage of patients free of disease at 3 years (70% vs 46%, $p = 0.001$) and median survival (4.25 vs 2.41 years); a lower probability of dying from bladder cancer was also observed. The most important prognostic factor was the number of lymph nodes involved and in a subset analysis, the beneficial effect of treatment was not seen in patients with two or more nodes involved. These results must also be viewed cautiously, as the number of patients in each subgroup was small.

Should Chemotherapy Be Administered Routinely to All Patients with Invasive Bladder Cancer?

Outside the protocol setting, the use of chemotherapy for all patients with invasive bladder cancer can not be considered routine or standard therapy. The major problems relate to our inability to accurately predict which patients ultimately require treatment due to a high risk of metastases versus those who are destined to be cured by local strategies alone using currently available techniques. Difficulties arise because the choice of the specific chemotherapy is not based on a predetermined sensitivity to the agents given. Further, the ability to deliver adequate doses of the more aggressive combination programs may be limited in an elderly population with a variety of intercurrent medical problems. Few trials report the total population of patients considered for entry into a protocol study, which makes inferences to a larger population difficult.

When considering treatment in the neoadjuvant setting, it is important to define the therapeutic goal at the outset. In some cases, treatment of micrometastases using the primary tumor as a marker of chemosensitivity is paramount, in others, the preservation of bladder function in a planned multimodality program with radiation therapy [24] is the goal. Questions remain concerning our ability to select treatments for patients based on a clinical staging system that is inaccurate and can result in errors of both under- or overstaging. While understaging occurs more frequently [1], a policy of using chemotherapy to treat micrometastatic disease in a patient with T2 disease is probably inappropriate as the majority of patients will be treated unnecessarily. It may be appropriate to control disease within the primary site as part of an integrated program with radiation therapy, with the aim of bladder preservation (vide infra) [32]. Further, assessing "response" in the bladder can be difficult, particularly as the number of "diagnostic/therapeutic" transurethral resections increases. In some cases, all visible tumor is resected endoscopically, making the true contribution of the chemotherapy. In "eliminating" disease in the bladder even more difficult to assess. While individual patients have been shown to benefit from this approach, a survival benefit has not been proven, and the possibility exists that chemotherapy may adversely affect outcome [40].

These concerns to do not apply to patients with T3b or greater disease, where the metastatic risk is high. The assessment of response in the primary tumor, which can usually be performed after the first 2–3 cycles of chemotherapy, allows an in vivo evaluation of chemosensitivity. Additional treatment of nonresponding patients is probably not justified. At MSKCC, our policy has been to use a minimum of 4 cycles in responding patients prior to definitive surgery, since the probability of achieving complete local control using chemotherapy alone is small. Further in our experience with metastatic disease all long-term survivors received a minimum of four cycles. Concerns have been raised about the risk in delaying definitive treatment if the primary tumor proves to be insensitive to the drugs. In our experience, however, this has not been a significant problem.

Adjuvant therapy is used to delay or prevent recurrent disease in patients with a high risk of metastatic relapse. Problems with an adjuvant approach include the inability to assess response in vivo. Thus, ineffective therapies with the associated toxicity, may be given. This may ultimately compromise the ability to give effective treatment when it is required. The prognostic information gained from response or nonresponse in the primary tumor is lost [3]. Nevertheless, if one considers the observed response proportions and survival with combination chemotherapy, one can infer that chemotherapy would benefit patients with positive lymph nodes after surgery, and that this recommendation should be routine. For example, of 229 patients with N1, N2, and N3 disease at radical cystectomy and pelvic node dissection, the median time to recurrence was 43, 14, and 10 months, and 5-year survival 42%, 11%, and 0% respectively. Advising chemotherapy in this setting would not be inappropriate. Unfortunately, the reported results using adjuvant chemotherapy are not definitive.

The randomized trial outlined was terminated prior to the planned accrual of 75 patients per arm. Second, while a comparison of the medians (50% level) suggested a benefit, a comparison of the 48% level showed a survival benefit for the observation group. The ability to deliver full doses of chemotherapy was a problem, as 11 of 44 (25%) patients randomized to chemotherapy never received it, and only 21 (48%) completed the full four chemotherapy cycles as planned. This raises the question whether other unrecognized prognostic imbalances were present that may have influenced the outcome [8]. Only an evaluation of larger numbers of patients would negate this potential problem. Cautious interpretation is also mandated by the small number of patients in the subgroup analyses. For example, while a benefit was shown for treated patients with one positive lymph node, the total number treated was only 15. Whether this small subgroup is representative of the entire population of patients with urothelial tumors is unknown. Thus, the question whether adjuvant chemotherapy improves the survival of patients with P3, P4, and/or N+ disease requires additional study. Additional confirmatory data are awaited.

Can Chemotherapy Be Used to Preserve Bladder Function Without Compromising Cure?

The increased availability of ileo neobladders and continent reservoirs has led many physicians to suggest that bladder preservation need not be considered a primary treatment end point. However, these can not be constructed in all patients – as low as 10% of males in some series – and are rarely advised for females [42,43]. They are also not free of complications. Of greater import is the fact that they do not change the natural history of the disease, and chemotherapy tolerance is not well characterized. For example, methotrexate reabsorption through a low-pressure reservoir may predispose the patient to significant toxicities at seemingly modest doses [44]. The relative quality of life of a patient who has had a radical cystectomy and ileal conduit, partial cystectomy, ileal neobladder, or who has received radiation therapy has not been adequately studied. For most, retention of a functional "natural" bladder would seem a desirable therapeutic goal.

The use of chemotherapy for bladder preservation can be considered for patients with T2/T3a lesions, particularly if no mass is palpable on examination under anesthesia. The antecedent history of the patient, i.e., presence or absence of in situ disease in the prostatic urethra, functional bladder capacity, and intercurrent illness all affect the decision to proceed with chemotherapy. A patient with a reduced bladder capacity and early invasive (T2) disease is best treated with surgery. For more invasive tumors, chemotherapy can be used, but difficulties arise from our inability to accurately assess which bladders are in fact pathologically free of tumor, a prerequisitie for safe bladder preservation. In our experience, clinical understaging was a significant problem, as 11 of 24 patients thought to have no tumor in the bladder after chemotherapy (endoscopically T0), had residual muscle infiltrating disease at the time of radical surgery. Considering a recent update of our neoadjuvant experience including a 30-month median follow-up, 26 of 95 (28%) patients were free of disease with a functional bladder [45]. In some cases after chemotherapy is administered, only carcinoma in situ will be documented. In selected cases intravesical treatment with bacille Calmette-Guérin (BCG) has been successfully employed. This, however, must be considered controversial.

Shipley et al., showed that local control within the bladder for patients treated with radiation therapy ± chemotherapy was best when the tumor was completely resected endoscopically prior to treatment [24]. Longer follow-up of this approach has shown good 4-year survival distributions in selected cases. However, even in cases where bladder preservation is successful, patients must be maintained on continued surveillance. Longer follow-up of patients who have retained their bladders using neoadjuvant therapy, has shown a continued risk of new tumor formation [24,32]. Whether these patients may have been better served with removal of the bladder at the start is controversial. Studies are ongoing to investigate prognostic factors for survival and for recurrence within the bladder.

Can Survival Be Improved?

Combining the results of reported phase II single-arm trials does not permit definitive conclusions with regard to the impact of chemotherapy on the survival of patients with invasive bladder cancer [46]. While significant progress has been made in the chemotherapy of these tumors, long-term survival is limited to a small proportion. Considering the proportion of complete responses that result from presently available chemotherapy, and bearing in mind that only an increase in the overall complete response rate can impact on the cure fraction, large scale prospective randomized comparisons will be required to show a survival benefit. For example, with overall CR in the range of 15%–20%, of whom only 50% are durable, to show an improvement in the 2-year survival from 50% in a control group to

60% in a chemotherapy-treated group would require 1000 patients. A 15% improvement would require 400 patients [47]. Similarly, to design a study with sufficient power (90%–95%) to show an improvement in the 5-year survival in the range of 6%–13% would require a minimum of 250 deaths. Generally, double this number of patient would have to be entered, treated, and followed for a minimum of 5 years [48].

Restricting entry to patients with a high risk of metastases would decrease the number of patients required to show a benefit for chemotherapy. Despite the importance of these trials, many patients and physicians are reluctant to participate protocol studies [49]. These factors must also be considered when designing protocols to insure that they can answer the question outlined and be completed on a timely basis.

Summary

In a disease where the majority of deaths occur from metastases, improvement in survival requires the integration of systemic therapies. Research efforts must continue to focus on improving case selection criteria, improving complete response proportions, and overcoming drug resistance. Recommending a single treatment plan such as radical surgery, chemotherapy, or radiation therapy for all patients with an invasive bladder cancer is rapidly becoming outdated.

Case selection is being refined by focusing on both clinical and pathologic features of the tumor. The latter include evaluation of NM23 RNA levels [50], or DNA ploidy and T138 surface antigen expression [51], which have been shown to correlate with metastatic potential. The use of hematopoietic growth factors has the potential to improve both the tolerance of chemotherapy [52] and complete response proportions [53], a prerequisite for cure. However, the dose response curves for most of the known active agents are not well defined, and ultimately, new agents and strategies will be required. Drug resistance is a major barrier, but as the mechanisms are unravelled, more selective therapies can be designed. For example, resistance to adriamcyin and vinblastine, two of the agents in the M-VAC regimen are mediated in part by the mdr1 gene [54]. Ongoing studies are attempting to identify prospectively those tumors with high levels of expression which may be more amenable to treatment with drugs that do not act through this mechanism.

The main advantages of the neoadjuvant approach are the ability to perform an in vivo response evaluation and the potential for bladder preservation. In most cases additional therapy for the primary tumor is required as clinical understaging is a significant problem. For some patients, initial surgery with the definition of the prognosis on firm pathologic grounds may represent a better strategy. When this is the case, the recommendation for adjuvant treatment potentially limits therapy to a more restricted population of patients for whom therapy is essential, including, for example, patients with positive lymph nodes at the time of surgery. Ideally, these patients should be entered on clinical trials designed to assess the impact of these strategies survival. Only large scale randomized trials have the potential to minimize the heterogeneity of this patient population.

References

1. Whitmore WF (1983) Management of invasive bladder neoplasms. Semin Urol 1:4–10
2. Droller MJ (1986) Bladder cancer. In: Graham SD (ed) Urologic oncology. New York, Raven, New York, pp 293–321
3. Scher HI (1990) Chemotherapy for invasive bladder cancer: neo-adjuvant vs. adjuvant. Semin Oncol 17:555–565
4. Montie JE, Straffon RA, Stewart BH (1984) Radical cystectomy without radiation therapy for carcinoma of the bladder. J Urol 131:477–482
5. Pagano F, Bassi P, Galetti TP et al. (1991) Results of contemporary radical cystectomy for invasive bladder cancer: a clinicopathological study with an emphasis on the inadequacy of the tumor, nodes and metastases classification. J Urol 145:45–50
6. Skinner DG, Lieskovsky G (1984) Contemporary cystectomy with pelvic node dissection compared to preoperative radiation therapy plus cystectomy in the management of invasive bladder cancer. J Urol 131:1069–1072
7. Feinstein AR, Sossin DM, Wells CR (1985) The Will Rogers phenomenon: stage migration and new diagnostic techniques as a source of misleading statistics for survival in cancer. N Eng J Med 312:1604–1608
8. Tannock I (1990) The current status of adjuvant chemotherapy for bladder cancer. Semin Urol 8:291–297
9. Warner JA, Whitmore WF, Fair WR (1992) Prognosis following cystectomy for node positive bladder cancer. J Urol (submitted)
10. Sternberg C, Yagoda A, Scher H et al. (1985) Preliminary results of methotrexate, vinblastine, adriamycin and cisplatin (M-VAC) in advanced urothelial tumors. J Urol 133:403–407
11. Logothetis CJ, Samuels ML, Ogden S et al. (1985) Cyclophosphamide, doxorubicin and cisplatin chemotherapy for patients with locally advanced urothelial tumors with or without nodal metastases. J Urol 134:460
12. Harker WG, Meyers FJ, Freiha FS et al. (1985) Cisplatin, methotrexate, and vinblastine (CMV): An effective chemotherapy regimen for metastatic transitional cell carcinoma of the urinary tract. a northern California Oncology Group study. J Clin Oncol 3:1463–1470
13. Stoter G, Splinter TAW, Child JA et al. (1987) Combination chemotherapy with cisplatin and methotrexate in advanced transitional cell cancer of the bladder. J Urol 137:663–667
14. Logothetis CJ, Dexeus FH, Chong C et al. (1989) Cisplatin, cyclophosphamide and doxorubicin chemotherapy for unresectable urothelial tumors: the M.D. Anderson experience. J Urol 141:33–37
15. Sternberg CN, Yagoda A, Scher HI et al. (1989) M-VAC for advanced transitional cell carcinoma of the urothelium; efficacy, and patterns of response and relapse. Cancer 64:2448–2458
16. Scher H (1989) Should single agents be standard therapy for urothelial tract tumors? J Clin Oncol 10:694–697

17. Loehrer PJ, Elson P, Kuebler JP et al. (1991) Advanced bladder cancer: A prospective Intergroup trial comparing single agent cisplatin (CDDP) versus M-VAC combination therapy (INT 0078). Proc Am Soc Clin Oncol 10(511): 132
18. Logothetis C, Dexeus F, Sella A et al. (1990) A prospective randomized trial comparing CISCA to MVAC chemotherapy in advanced metastatic urothelial tumors. J Clin Oncol 8:1050–1055
19. Lo Rk, Freiha FS, Torti FM (1989) CMV for metastatic urothelial tumors. In: Johnson DE, Logothetis CJ, von Eschenbach AC (eds) Systemic therapy for genitourinary cancers. Year Book, Chicago, pp 59–63
19a. Seidman AD, Scher HI (1991) The evolving role of chemotherapy for muscle infiltrating bladder cancer. Sem Onc 18:585–595
20. Meyers FJ, Palmer JM, Scudder SA (1989) The multimodality approach to invasive transitional cell carcinoma of the bladder. In: Rous SN (ed) Urology annual 1989. Appleton and Lange, East Norwalk, pp 117–130
21. Goldin A, Venditti JM, Humphreys SR et al. (1956) Influence of the concentration of leukemic inoculum on the effectiveness of treatment. Science 123:840
22. Skipper HE, Schabel FM, Wilcox WS (1964) Experimental evaluation of potential anticancer agents. XII. On the criteria and kinetics associated with curability of experimental leukemia. Cancer Chemother Rep 35:1–9
23. Nativ O, Herr HW, Scher HI et al. (1990) Neoadjuvant chemotherapy and partial cystectomy for invasive bladder tumors. Proc Am Soc Clin Oncol 9: 572
24. Prout GR, Shipley WU, Kaufman DS et al. (1990) Preliminary results in invasive bladder cancer with transurethral resection, neoadjuvant chemotherapy and combined pelvic irradiation plus cisplatin chemotherapy. J Urol 144: 1128–1136
25. Scher H, Herr H (1989) Is chemotherapy a new form of radical treatment in infiltrating tumor localized to the bladder? In: Khoury S, Murphy GP (eds) Therapeutic progress in urological cancers. Liss, New York, pp 563–582
26. Hall RW, Prout GR (1990) Staging of bladder cancer: Is the tumor, node, metastasis sytem adequate? Semin Urol 17:517–523
27. Herr HH (1987) Conservative management of muscle infiltrating bladder cancer: Prospective experience. J Urol 138:1162–1163
28. Gospodarowicz M, Hawkins NV, Rawlings GA et al. (1989) Radical radiotherapy for muscle invasive transitional cell carcinoma of the bladder: failure analysis. J Urol 142:1448–1152
29. Fossa SD, Ous S, Berner A (1991) Clinical significance of the "palpable mass" in patients with muscle-infiltrating bladder cancer undergoing cystectomy after preoperative radiotherapy. Br J Urol 67:54–60
30. Seidman A, Scher HI, Herr H et al. (1991) Survival of patients (Pts) with invasive (T2–4 N0 M0) bladder cancer treated with neoadjuvant methotrexate, vinblastine, adriamycin and cisplatin (MVAC) chemotherapy. 3rd International Congress on Neo-Adjuvant Chemotherapy, p 35
31. Shipley WU, Prout GR, Kaufman D, Perrone TL (1987) Invasive bladder carcinoma: the importance of initial transurethral surgery and other significant prognostic factors for improved survival with full-dose irradiation. Cancer 60 [Suppl 4a]:514–520
32. Hryniuk W, Bush H (1984) The importance of dose intensity in chemotherapy of metastatic breast cancer. J Clin Oncol 2:1281–1288
33. Scher H, Herr H, Yagoda A et al. (1988) Neoadjuvant M-VAC (methotrexate, vinblastine, adriamycin and cisplatin): the effect on primary bladder tumors. J Urol 139:470–474

34. Scher H, Herr H, Sternberg C et al. (1989) Neo-adjuvant chemotherapy for invasive bladder cancer: experience with the M-VAC regimen. Br J Urol 64:250–256
35. McCullough DL, Cooper RM, Yeaman LD et al. (1989) Neoadjuvant treatment of stages T2 to T4 bladder cancer with cis-platinum, cyclophosphamide and doxorubicin. J Urol 141:849–852
36. Logothetis CJ, Samuels ML, Selig DE et al. (1985) Combined intravenous and intra-arterial cyclophosphamide, doxorubicin, and cisplatin (CISCA) in the management of select patients with invasive urothelial tumors. Cancer Treat Rep 69:33
37. Splinter TAW, Scher HI, EORTC-GU group et al. (1990) The prognostic value of the pT-category after combination chemotherapy for patients with invasive bladder cancer who underwent cystectomy. In: Splinter T, Scher HI (eds) Neo-adjuvant chemotherapy of invasive bladder cancer. Liss, New York, pp 219–224
38. Studer UE, Ackermann D, Casanova GA, Zingg EJ (1988) A newer form of bladder substitute based on historical perspectives. Semin Urol 6:57–65
39. Logothetis CJ, Johnson DE, Chong C et al. (1988) Adjuvant cyclophosphamide, doxorubicin, and cisplatin chemotherapy for bladder cancer: an update. J Clin Oncol 6:1590–1596
40. Tannock IF (1990) Endpoints of clinical trials in invasive bladder cancer. Semin Oncol 17:619–627
41. Skinner DG, Daniels JR, Russell CA et al. (1991) The role of adjuvant chemotherapy following cystectomy for invasive bladder cancer: a prospective comparative trial. J Urol 141:459–464
42. Splinter TAW, Jacqmin D, Trevor Roberts J et al. (1989) EORTC-GU Group Study 30851: A phase II study of upfront chemotherapy in patients with invasive bladder cancer. Proc Am Soc Clin Oncol 8:541
43. Kock NG, Nilson AE, Norlen L et al. (1978) Urinary diversion via a continent ileum reservoir: clinical experience. Scand J Urol Neprol 49 [Suppl]:23–31
44. Fossa SD, Heilo A, Bormer O (1990) Unexpectedly high serum methotrexate levels in cystectomized bladder cancer patients with ileal conduit treated with intermediate doses of the drug. J Urol 143:498–501
45. Scher H, Herr H, Sternberg C et al. (1990) M-VAC (methotrexate, vinblastine, adriamycin, and cisplatin) and bladder preservation. In: Splinter T, Scher HI (eds) Neo-adjuvant chemotherapy of invasive bladder cancer. Liss, New York, pp 179–186
46. Geller NL, Scher H, Parmar M et al. (1990) Trial design and statistics: Can we combine available data to evaluate the effects of neoadjuvant chemotherapy for invasive bladder cancer? Semin Oncol 17:628–634
47. Parmar M (1990) Trial design. In: Splinter T, Scher HI (eds) Neo-adjuvant chemotherapy of invasive bladder cancer. Liss, New York, pp 115–118
48. Dalesio O (1990) Trial design and statistics. In: Splinter T, Scher HI (eds) Neo-adjuvant chemotherapy of invasive bladder cancer. Liss, New York, pp 57–64
49. Moore MJ, Tannock IF (1988) How expert physicians would wish to be treated if they developed genito-urinary cancer. Proc Am Soc Clin Oncol 7:118
50. Steeg PS, Bevilacqua G, Kopper L et al. (1988) Evidence for a novel gene associated with low tumor metastatic potential. J Natl Cancer Inst 80:200–203
51. Fradet Y, Tardif M, Bourget L et al. (1990) Clinical cancer progression in urinary bladder tumors evaluated by multiparameter flow cytometry with monoclonal antibodies. Cancer Res 50:432–437
52. Gabrilove JL, Jakubowski A, Scher H et al. (1988) A study of recombinant human granulocyte colony stimulating factor in cancer patients at risk for chemotherapy-induced neutropenia. N Eng J Med 318:1414–1422

53. Logothetis C, Dexeus F, Sella A et al. (1990) Escalated (ESC) MVAC (MTX 30 mg/m^2, adriamycin 60 mg/m^2, vinblastine 4 mg/m^2, cisplatin 100 mg/m^2) with recombinant human granulocytes macrophage stimulating factor [(rhGM-CSF) Schering Corp)] for patients (PTS) with advanced and chemotherapy (CHT) refractory urothelium tumors: a phase I study. J Natl Cancer Inst 82:667–672
54. Ling V (1989) P-glycoprotein and multidrug resistance. Semin Oncol 16:156–165

*Radiochemotherapy for Invasive Carcinoma of the Bladder**

W.U. Shipley, D.S. Kaufman, P.P. Griffin, A.F. Althausen, N.M. Heney, and G.R. Prout Jr.

Departments of Radiation, Oncology, Medicine and Urology, Massachusetts General Hospital Cancer Center, Harvard Medical School, Boston, MA 02114, USA

Introduction

In the past decade there have been substantial improvements in the treatment of patients with muscle-invasive bladder cancer. Patients now experience less short- and long-term morbidity from radical surgery, radical radiotherapy, or a combination of the two techniques [1]. In addition to these technical advances, new investigative tools are now available such as DNA flow cytometry and cell surface markers. In the near future evaluation of somatic point mutations in important tumor suppressor genes in the tumors of individual patients will be possible. However, while these latter factors may well lead to an improved understanding of the histogenesis and heterogeneity of invasive bladder tumors, they have not as yet been shown to be useful in predicting the success or failure of any of the commonly utilized treatment modalities [2].

The local-regional control of a muscle-invading bladder tumor is essential to the successful management of patients with this disease. There is strong evidence, however, that microscopic dissemination of the tumor has already occurred at the time of initial diagnosis in the majority of the patients. Of those patients with muscle-invading tumors that have been treated by radical cystectomy 40%–65% subsequently develop distant metastases [2–4]. Once distant metastases occur, patients usually are not curable, though there have been recent reports of the possible cure of some patients following combination chemotherapy in such situations [5]. Whilst there is as yet no data to suggest that multidrug neoadjuvant systemic chemotherapy will decrease the subsequent development of distant metastases (or increase

---

*This article is also being published simultaneously in Meyer JL (ed) (1991) Radiotherapy, chemotherapy interaction in cancer treatment. Frontiers of Radiation Therapy and Oncology, Vol. 26. Karger, Basel.

Recent Results in Cancer Research, Vol. 126

local long-term cure of the bladder from cancer), randomized trials are now underway to test this hypothesis. Not until data from such randomized trials is available can the efficacy of this approach be known. We strongly support entering all patients on available national or international phase III trials to put this important question to the test.

Data from the 1960s and 1970s has documented that patients then treated with radical radiotherapy had limited potential for survival (compared to those treated by cystectomy) due to other medical conditions. Thus in nonrandomized series the survival benefit for cystectomy (perhaps 10%) very likely is due largely to patient selection. Nevertheless the results with radical radiotherapy *alone* have been disappointing with regard to long-term local-regional control (30%–50%) compared with a 70%–90% freedom from pelvic recurrence for patients treated with radical cystectomy [1]. This failure stimulated several groups to evaluate the combination of concurrent radiochemotherapy in patients who were either unfit for surgery or were unwilling to undergo radical cystectomy. Despite a low incidence of pelvic recurrence, radical cystectomy is a failure as a cancer treatment. Not only does it result in serious compromise in the quality of life of patients due to the inconvenience and discomfort of an artificial stoma, but more importantly the survival of patients with muscle-invading cancer following radical cystectomy is less than 50% because of the high incidence of distant metastases in this group. Three of four randomized phase III trials comparing radical cystectomy combined with preoperative radiotherapy to external beam radiation therapy alone with salvage cystectomy for recurrence report no significant survival benefit from immediate radical cystectomy. Only in the M.D. Anderson trial was preoperative radiotherapy and cystectomy found significantly better than radiotherapy alone, and this was a small series [6]. In the Institute of Urology trial of 194 patients [7], in a Danish trial of 183 patients [8], and in a National Bladder Cancer Collaborative Group A trial of 99 patients (S.J. Cutler, unpublished) there was no statistically significant benefit in survival to those patients treated with planned cystectomy compared with those given external beam radiation therapy alone followed by salvage cystectomy for local recurrence. Review of these data and reports of some success with combination chemotherapy led several groups to study whether combined transurethral surgery, radiotherapy plus chemotherapy might lead to successful bladder preservation in patients presenting with muscle-invading cancer. Results of this approach have led to some encouraging conclusions.

Bladder-Preserving Techniques Other Than or Utilized with Radical Radiotherapy

Transurethral Surgery for Local Cure

Thirteen years ago Barnes and colleagues reported a 27% 5-year survival in 85 patients undergoing transurethral surgery alone for stage T2 or B1 tumors of medium to high grade. However they gave no data on what percentage of these patients had bladders which remained tumor free [9]. Three years ago Herr reported that amongst 217 patients referred with confirmed muscle-invading bladder cancer 9% had already achieved a complete response (CR) based on his cystoscopic reresection [10]. This is especially remarkable because in many of the referred patients the initial surgery was very likely not intended to be a "complete" resection of the tumor. We noted a higher 5-year bladder recurrence-free survival as well as a higher overall survival in the subgroup of patients with T2 or T3 tumors who underwent a "visibly complete" transurethral resection (54%) compared to the group that did not (17%, $P = 0.009$, [4]). Recently Hall at Newcastle-upon-Tyne has reported preliminary results using extensive transurethral surgery followed by adjuvant chemotherapy, resulting in 79% of these patients remaining free of recurrent bladder tumors for 12 months following treatment [11]. These data suggest that thorough transurethral resection can be important both as a treatment modality and as a means of selecting patients for bladder-sparing treatment with cystoscopic resection of tumor followed by radiochemotherapy.

Systemic Multidrug Chemotherapy Alone

Several institutions have recently reported complete CR rates (cystoscopy negative, biopsy negative, and cytology negative) following multidrug systemic chemotherapy in patients with locally advanced bladder carcinoma. With the exception of two early trials that used only single-agent chemotherapy with observed response rates of 7% and 9%, CR rates have ranged from 32% to 52% (Table 1; [12]). Scher and colleagues, however, reported on pathologic staging (by total or partial cystectomy) of clinically complete responding patients to the methotrexate, vinblastine, adriarycin, cisplatin (M-VAC) regimen and documented that 40% had histologically confirmed residual tumor in the cystectomy specimen [13]. Thus 3–4 cycles of M-VAC chemotherapy, even in the completely responding patients, cannot be considered optimal or even adequate therapy. Even in this favorable CR group an additional "boost" to the primary bladder tumor by either surgery or radiotherapy is needed.

Table 1. Primary bladder cancer: complete response rates for chemotherapy alone

Institution	Drugs	No. of patients	Complete response[a] (%)
Memorial Hospital	M-VAC	71	48
National Bladder Cancer group	C ± CTX	30	7
London (Institute of Urology)	M	32	9
Northern California Oncology Group	CMV	12	50
M.D. Anderson	CISCA	28	48
Nice, France	C, 5-FU	19	47
University of Florida	MCV/M-VAC	46	50
University of Chicago	M-VAC	22	32
Mayo Clinic	M-VAC	28	32
European Organisation for Research and Treatment of Cancer	C, M	35	46
Massachuselts General Hospital	MCV	46	52

M, methotrexate; V, virblastine; A, adrianycin; C, cisplatin; CIS, cyclophosphamide; 5-FU, 5-fluorouracil; CTX, cyclophosphamide.
[a] Clinical stage cTO.

Concomitant Systemic Chemotherapy and Radiotherapy

Several groups have used combined radiochemotherapy for patients who are judged unsuitable for cystectomy or in whom a cystectomy was reserved for salvage. In these series the clinical CR rates (biopsy negative and in some series cytology negative) have ranged from 62% to 90% (Table 2; [14–21]). Survival rates of 64% at 4 years for clinical stage T2 patients treated with cisplatin and full-dose radiation therapy and 24% for patients who were clinical stages T3 and T4 were reported by the National Bladder Cancer Group [14]. Jaske and colleagues [15] reported a 46% 4-year survival for a similar series of patients. However in neither of these series was transurethral surgery used systematically in all patients. These survival results are reported as actuarial findings and may, with longer follow-up become less favorable. Based on clinical CR rates alone in nonrandomized trials, radiochemotherapy does seem to be superior to either modality alone. In addition, in all these series no increase (compared to radical radiotherapy alone) was reported in any pelvic complications with the combination of systemic chemotherapy and radiotherapy.

Selection by Response to Initial Treatment of Patients for Successful Bladder Preservation

Because at most only three-quarters of patients will respond to radiotherapy in combination with transurethral surgery and chemotherapy, it is desirable

Table 2. Primary bladder cancer: complete response rates for concomitant drugs and radiotherapy

Institution [Reference]	Drugs	Radiotherapy (Gy)	No. of patients	Complete response[a] (%)
National Bladder Cancer Group [14]	C	60–68	57	79
University of Innsbruck [15]	C ± VM-26	60	22	77
State University of New York, Brooklyn [16]	5-FU ± Mitomycin C	60–65	18	61
University of Erlangen [17]	C	50–56	41	85
Ottawa [18]	C (i.a.)	60	20	90
University of Florida [19]	MCV, C	65	51	66
Massachuselts General Hospital [20]	MCV, C	65	34	74
Radiation Therapy Oncology group [21]	C	40–65	48	65

C, cisplatin; V, vinblastine; M, methotrexate; 5-FU, 5-fluorouracil.
[a] Clinical stage cTO, i.e., biopsy negative with cyrology negative in most series.

to select only these patients for this treatment to full dose who are most likely to respond. While in the future advances in laboratory tumor analysis of surface markers, DNA ploidy, or the persistence of important tumor suppressor genes may identify such subgroups in advance, the best approach at present is to select patients for treatment who are responding completely after the initial combined approach of transurethral surgery, 2–3 cycles of systemic chemotherapy and/or radiochemotherapy to 4000–5000 cGy. Nevertheless there are no data at present to indicate whether such approaches to selective bladder-sparing therapy can be carried out without compromising patient survival (relative to immediate radical cystectomy). Nevertheless such an approach will minimize the need for late salvage cystectomy following full dose (6000–7000 cGy) radiotherapy.

The Selection of Patients for Bladder-Sparing Treatment in the Massachusetts General Hospital Prospective Trial

There is a justifiable concern that a conservative treatment approach to bladder preservation may compromise patients survival in those who fail, or demonstrate local tumor recurrence. In an attempt to prevent recurrences following radiochemotherapy we use a favorable response of the primary tumor to the initial treatment as the criterion to select patients for full-dose bladder-preservation treatment. The goal is to select only those patients that have a very high likelihood of having permanent local control in their bladder. By using early treatment responses as a major selection criterion we have treated approximately two-thirds of our stage T2–T4 protocol patients with bladder preservation and have had encouraging preliminary results. Our treatment schema is to combine transurethral surgery, neo-adjuvant multidrug systemic chemotherapy, and external beam radiotherapy combined with cisplatin. The details of the protocol and early treatment results have been described previously [20]. At the time of entry all patients underwent, in so far as possible, a "visibly complete" transurethral resection of the bladder tumor. Patients then received two cycles of combination systemic MCV (methotrexate 30 mg/M^2, cisplatin 70 mg/M^2, vinblastine 3 mg/M^2) chemotherapy followed by cystoscopic re-evaluation. This was followed by 4000 cGy external beam radiotherapy to the bladder and pelvic lymphatics over 5 weeks, via a four-field box technique. Cisplatin was given on the day prior to starting radiotherapy and again 21 days later. After 4000 cGy radiotherapy and the two courses of cisplatin, cystoscopic re-evaluation was again performed. Radical cystectomy was recommended for patients who had not attained a CR defined as a negative biopsy and a negative urinary cytology. Complete responders, and those patients who were unsuited for cystectomy, were given one additional dose of cisplatin and further radiotherapy (total of 6480 cGy to the bladder tumor volume and a total of 4500 cGy to the whole bladder and to the pelvic lymphatics).

The treatment was moderately well tolerated. There were no cases of treatment-related sepsis or death nor significant renal dysfunction. However, 11 of the 53 patients did not complete the full treatment plan (7 due to chemotherapy-related side effects), and 39% of all patients required some chemotherapy dose reduction. The recently reported status (20 patients) of the 53 patients entered in this protocol from 1986–1988 is shown in Table 3.

By using early treatment response as the major selection criterion we have treated more than two-thirds of our patients with successful bladder preservation and with satisfactory preliminary survival for patients with these advanced tumors (70% at 2 years). The treatment was moderately well tolerated in the 42 patients that completed the planned treatment with no serious postcystectomy complications in this group or amongst the 34 patients who had the full radiochemotherapy. In the latter group there have been no incidences of significant late bowel or bladder irritations or damage. Twelve patients following full radiochemotherapy had either tumor persistence (3 patients) or tumor reoccurrence (9 patients). The three patients with tumor persistence were unfit for cystectomy. In the nine patients who developed tumor reoccurrence two were invasive and seven were superficial transitional cell carcinoma. The two patients with invasive tumors underwent salvage cystectomy with one now free of disease. The seven patients with superficial transitional cell carcinoma were treated with

Table 3. Massachuselts General Hospital protocol 180: invasive bladder cancer

Patient treatment and/or tumor category	No. of patients	Alive (%)	Bladders remaining (%)	Distant metastases (%)	Bladders without any recurrence (%)
All entered, T2–T4	53	72	70	30	43
All treated, T2–T4	42	74	74	26	52
Clinical stage T2	11	91	82	0	64
Clinical stage T3 + T4	31	68	71	35	48
Full chemo- and radiotherapy	34	82	91	21	65
Immediate cystectomy	8	38	0	50	0

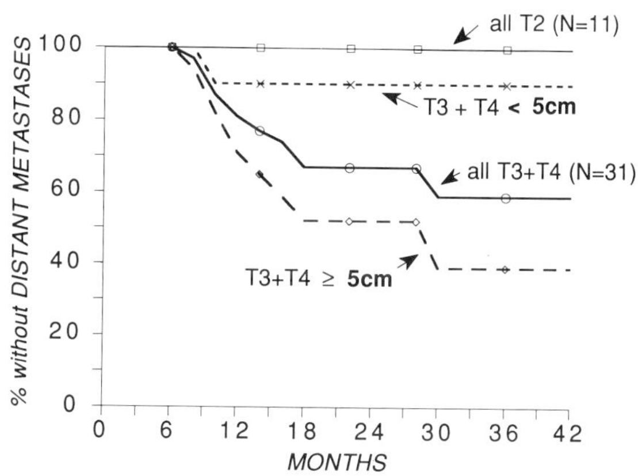

Fig. 1. The probability of remaining free of distant metastases in clinical stage T2 (11 patients) and in all T3–T4 (31 patients) as well in the subgroup who had tumors less than 5 cm in greatest diameter and in the subgroup with tumors 5 cm or more in greatest diameter

topical intravesical therapy and six at the time of this analysis are tumor free within their bladder. The seventh patient refused cystectomy.

Some pretreatment tumor characteristics have predicted significant freedom from systemic disease to date, including tumor stage (T2 vs T3,T4) and initial tumor size within the T3,T4 category (Fig. 1). The response to initial multidrug systemic chemotherapy (combined with transurethral surgery) has been a significant predictor of the survival, distant metastases-free survival and freedom from recurrent invasive cancer (Fig. 2). Amongst the 34 patients treated with full radiochemotherapy the proportion of patients with bladders that have remained tumor free is significantly higher ($P = 0.005$) in those who were in CR after transurethral resection (TUR)-

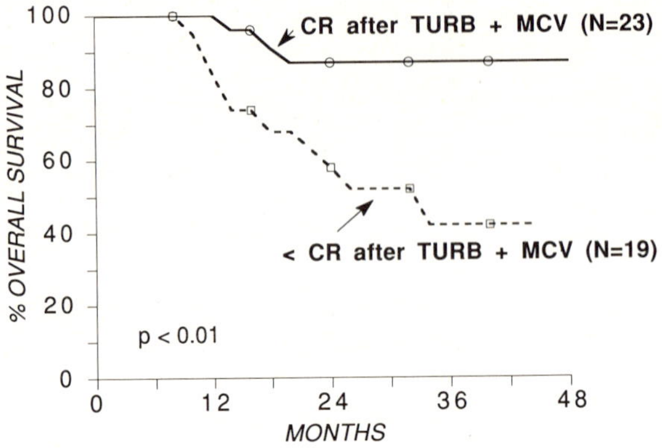

Fig. 2. The overall survival, by actuarial calculation, for those 23 patients with tumors in clinical CR compared to the 19 patients with tumors not in CR following the initial TUR and two cycles of MCV chemotherapy. By the log-rank test this difference has a p value lower than 0.01. *TURB*, transurethral resection of all visible bladder tumor, when possible

MCV therapy (86%) than in those with less than a CR (25%). In addition all of the patients who were in CR following TUR-MCV therapy have bladders remaining free of invasive tumor, while 5 of 12 patients treated with full radiochemotherapy who were not in CR following TUR-MCV had persistent disease or have developed recurrent invasive bladder tumors. We realize, however, that the apparent survival advantage to CR status after the TUR-MCV treatment as shown in Fig. 2 does not ensure a benefit to the whole patient population by this approach. Nevertheless the strong statistical significance of seeing such a response justified our moving ahead to test the possible efficacy of such a neoadjuvant regimen by a randomized trial. Thus, in combination with the Radiation Therapy Oncology Group (RTOG), we have a randomized phase III ongoing trial in which half the patients do, and half the patients do not receive the neoadjuvant MCV chemotherapy. To date this trial has proved acceptable to both patients and physicians and is accruing well.

Conclusions

At this time conclusions about the use of neoadjuvant systemic chemotherapy and/or radiochemotherapy must be preliminary and cautious:

1. Multidrug chemotherapy including both cisplatin and methotrexate produce, in combination with transurethral surgery, and a 30%–50% *clinical* CR rate of the primary tumor. The optimum combination of chemotherapeutic agents has not yet been determined. Systemic combination chemotherapy alone, even in the complete responding patients,

will *not* by itself be adequate to cure the bladder of the primary invasive cancer in the majority of patients. Further therapy by either surgery or radiotherapy seems warranted for the high probability of residual microscopic tumor.
2. Concomitant systemic chemotherapy and external beam radiotherapy produces a high clinical CR rate but the durability of such a remission is unclear. A follow-up of 3–5 years is necessary to be certain that the bladder is sterilized of cancer.
3. Combined cisplatin and external beam radiation therapy does not enhance the toxicity (relative to radiotherapy alone) to either the bladder or the rectum.
4. Debulking transurethral resection of the bladder tumor *seems* to improve the local success rate of patients being considered for bladder preservation with the use of radiochemotherapy.
5. Not all patients respond to radiochemotherapy. Presently the best criterion for patient selection for this treatment is to select those who respond completely to initial transurethral surgery and initial radiochemotherapy. These patients should be then carried to full chemotherapy and radiotherapy doses. However, in the absence of an initial CR immediate cystectomy is recommended. In patients who are complete responders to the full radiochemotherapy treatment permanent and ongoing cystoscopic surveillance is necessary; 20% or more of such patients will develop a new superficial or invasive tumor.
6. Combination chemotherapy plus radiotherapy treatment programs for possible bladder sparing are currently experimental approaches to the treatment of muscle-invading cancer and are best carried out in the setting of an approved clinical trial.

Acknowledgement. The authors are indebted to Shirley Sharff and Jenot W Shipley for their assistance in the preparation of this manuscript.

References

1. Richie JP, Shipley WU, Yagoda A (1989) Cancer of the bladder. In: DeVita VT, Hellman S, Rosenberg SA (eds) Cancer: principles and practice of oncology, 3rd edn. Lippincott, Philadelphia, pp 1008–1022
2. Raghavan D, Shipley WU, Garnick MB, Russell PJ, Richie JP (1990) Biology and management of bladder cancer. N Engl J Med 322:1129–1138
3. Prout GR Jr, Griffin PP, Shipley WU (1979) Bladder cancer as a systemic disease. Cancer 43:2532–2539
4. Shipley WU, Rose MA, Perrone TL, Mannix CM, Heney NM, Prout GR Jr (1985) Full-dose irradiation for patients with invasive bladder carcinoma: clinical and histologic factors prognostic of improved survival. J Urol 134:679–683
5. Sternberg CN, Yagoda A, Scher HI et al. (1988) M-VAC (methotrexate, vinblastine, doxorubicin and cisplatin) for advanced transitional cell carcinoma of the urothelium. J Urol 139:461–469

6. Miller LS, Bladder Cancer (1977) Superiority of preoperative irradiation and cystectomy in clinical stages B_2 and C. Cancer 39:973–980
7. Bloom HJG, Hendry WF, Wallace DM, Skeet RG (1982) Treatment of T3 bladder cancer: Controlled trial of preoperative radiotherapy and radical cystectomy versus radical radiotherapy: second report and review. Br J Urol 54:136–151
8. Barlebo B, Steven K, Sorensen BL et al. (1990) Preoperative radiation (40 Gy) and cystectomy vs. radiotherapy (60 Gy) followed by salvage cystectomy in the treatment of advanced bladder cancer (T2–T4A). A randomized study (DAZECA-8201). J Urol 143:291A
9. Barnes RW, Dick AL, Hadley HL et al. (1977) Survival following transurethral resection of bladder carcinoma. Cancer Res 37:2895–2899
10. Herr HW (1987) Conservative management of muscle-infiltrating bladder cancer: Prospective experience. J Urol 138:1161–1163
11. Hall RR (1986) Transurethral resection and systemic chemotherapy as primary treatment for T3 bladder cancer. In: Yagoda A (ed) Bladder cancer: future directions for treatment. Park Row, New York, pp 117–122
12. Shipley WU, Kaufman DS, Heney NM (1990) Can chemo-radiotherapy plus transurethral tumor resection make cystectomy unnecessary for invasive bladder cancer? Oncology 4:25–32
13. Scher H, Herr H, Sternberg C et al. (1989) Neo-adjuvant chemotherapy for invasive bladder cancer: Experience with the M-VAC regimen. Br J Urol 64:250–256
14. Shipley WU, Prout GR Jr, Einstein AB Jr, Coombs LJ, Wajsman Z, Soloway MS, Englander LS, Barton BA, Hafermann MD (1987) Treatment of invasive bladder cancer by cisplatin and irradiation in patients unsuited for surgery: a high success rate in clinical stage T2 tumors in a National Bladder Cancer Group Trial. JAMA 258:931–935
15. Jaske G, Frommhold H, Nedden DZ (1985) Combined radiation and chemotherapy for locally advanced transitional cell carcinoma of the urinary bladder. Cancer 55:1659–1664
16. Rotman M, Macchia R, Silverstein R et al. (1987) Treatment of advanced bladder carcinoma with irradiation and concomitant 5-Fluorouracil infusion. Cancer 59:710–714
17. Sauer R, Dunst J, Altendorf-Hofman A et al. (1990) Radiotherapy with or without Cisplatin in bladder cancer. Int J Radiat Oncol Biol Phys 19:687–691
18. Eapen L, Stuart D, Danjoux C et al. (1989) Intra-arterial cisplatin and concurrent radiation for locally advanced bladder cancer. J Clin Oncol 7: 230–235
19. Wajsman Z, Klimberg I, Parsons J et al. (1989) Bladder sparing treatment for muscle invasive transitional cell carcinoma: Systemic chemotherapy followed by radiation therapy with adjunctive cisplatin. Proc Am Soc Clin Oncol 8:132
20. Prout GR Jr, Shipley WU, Kaufman DS, Heney NM, Griffin PP, Althausen AF, Bassil B, Nocks BN, Parkhurst EC, Young HH III (1990) Preliminary results in invasive bladder cancer with transurethral resection, neoadjuvant chemotherapy and combined pelvic irradiation plus cisplatin chemotherapy. J Urol 144: 1128–1136
21. Heaney JA, Tester WJ, Porter AT et al. (1990) Combined modality program with possible organ preservation for invasive bladder carcinoma; preliminary results of RTOG protocol 85-12. J Urol 143:290A

Germ Cell Tumors

Treatment of Local Regional Disease

J.P. Donohue

Department of Urology, University Hospital-A112, Indiana University Medical Center, 926 W. Michigan Street, Indianapolis, IN 46202-25250, USA

Nonseminomatous germ cell tumor is an extraordinary tumor in terms of demonstrating the impact of regional lymphadenectomy on survival. Over the years many series have shown that node-positive (pathologic stage II) patients can be cured with local measures in excess of 50% of cases. There are a variety of anatomical considerations that make testicular lymphatic drainage predictable. These relate to features of anatomic descent of the gonad and its associated lymphatics. A number of topographic studies have supported the regional deposition of metastatic disease, and we have a good understanding of this. More importantly, there are biologic considerations that probably lend to the opportunity for cure of testis cancer patients with surgery alone. Many patients will have teratomatous elements which have somewhat lower metastatic potential. What is fascinating, however, is that even undifferentiated germ cell tumors, such as embryonal cancer, are often cured with lymphadenectomy alone. Data recently presented in a multi-institutional cooperative study show that a control group treated with node dissection alone had more than half of the patients as long-term survivors without relapse [1].

Fortunately, the relapsers were rescued at the time of relapse, with only a few exceptions. The overall survival in stage II disease, including all levels of involvement (both stage II-A and II-B), was 98% in the decade from 1979–1989. Naturally, this was achieved with the use of either adjuvant chemotherapy immediately following retroperitoneal lymph node dissection (RPLND), or full-course chemotherapy given only at the time of relapse in the other half of patients (who relapsed after earlier RPLND and observation for stage II disease).

Our experience with clinical staging followed by RPLND surgery is included in Tables 1–4. This spans the recent era since the advent of platinum (1979–1989) as well as experience before the introduction of platinum.

Table 1. Primary RPLND in clinical stage A nonseminomatous germ cell testis cancer – results, 1965–1978

Disease extent (Pathological stage)	Patients		Relapse		Survival (%)	Deaths (*n*)
	n	%	*n*	%		
A	57	66	6	11	100	0
B No adjuvant	9	11	4	44	100	0
B Plus adjuvant[a]	20	23	3	15	95	1
All cases	86		13	15	98.8	1

[a] Non-platinum-based chemotherapy ± abdominal irradiation.
[b] One death from testis cancer.

Table 2. Primary RPLND in clinical stage A nonseminomatous germ cell testis cancer – results, 1979–1989

Disease extent (Pathological stage)	Patients		Relapse		Survival (%)	Deaths (*n*)
	n	%	*n*	%		
A	267	71	29	11	99.2	2[a]
B No adjuvant	65	17	21	32	98.5	1[b]
B Plus adjuvant	46	12	0	0	100	0
All cases	378		50	13	99.2	3

[a] Two deaths, one cancer, one postoperative.
[b] One cancer death.

Table 3. Primary RPLND in clinical stage B – results, 1965–1989

Disease extent (Pathological stage)	Patients		Relapse		Survival (%)	Deaths (*n*)
	n	%	*n*	%		
A	41	24	2	5	100	0
B No adjuvant	54	31	19	35	96.2	2
B Plus adjuvant	79	45	11	14	93.6	5
All cases	174		32	18	95.9	7[a]

[a] Seven deaths: five from cancer, one postoperative, one chemotherapy related.

Table 4. Primary RPLND in clinical stages A and B – overall results, 1965–1989

Disease extent (Pathological stage)	Patients		Relapse		Survival (%)	Deaths (n)
	n	%	n	%		
A	365	57	37	10	99.4	2
B No adjuvant	128	20	44	34	97.6	3
B Plus adjuvant	145	23	14	10	95.8	6
All cases	638		95	15	98.3	11[a]

[a] Eleven deaths: eight from cancer, two postoperative, one chemotherapy related.

A general treatment philosophy when considering retroperitoneal node dissections should be as follows: The surgery should be appropriate (let the punishment fit the crime). Although presently our position with regard to patients with low-stage, low-volume disease is different from that for patients committed to surveillance only, I see the paths converging. Clearly, no one wants to do staging RPLND for its own sake. If the clinical sensitivity of clinical staging were to attain levels of confidence exceeding the 90th percentile, there would be no need to do staging RPLNDs. In the meantime, however, a good interim position is represented by the modified RPLND, particularly when it employs nerve-sparing techniques. Virtually all patients who have a nerve-sparing modified RPLND will ejaculate. Hence, the patient has not been harmed, and he has also been adequately staged. In fact, those who do have microscopic disease have also been given a therapeutic procedure. Currently, the relapse rate locally in patients with nerve-sparing RPLND is entirely satisfactory in our current small and selected series [2].

Regarding high-stage disease, our reviews are generally concordant with those of other groups throughout the world. Disseminated disease requires disseminated treatment first. This implies pre-treatment with systemic chemotherapy which is platinum based. Currently programs employing platinum VP-16, with our without bleomycin, are quite popular and effective. Surgery is reserved for those who have residual clinical disease as noted on the computed tomography (CT) scan. One of our current challenges is to develop predictive criteria for those patients who might have scar tissue and necrosis only in the specimen. There is some analogy with the seminoma date for advanced disease, where it has been found that most patients with a partial remission can be followed because they will have necrosis only in their tissue specimen. However, there are always exceptions to the rule and patients must be followed carefully and relapse must be treated aggressively when and if it occurs.

Another philosophical point that will impact future management of node dissection in testis cancer is that health care providers will want to eliminate qualitative variables in treatment programs as much as possible. Surgery is always a variable in this regard. In order to simplify and codify quantitative aspects of treatment, there will be a move to eliminate initial hospital costs in such things as staging and/or treatment for low-stage disease. Systemic chemotherapy is more easily standardized and quantified. Surgery for low-volume disease, therefore, may be dismantled except in a salvage setting. Unfortunately, this approach will mean that many patients with low-volume disease limited to the retroperitoneum will undergo a highly toxic program that may well have negative long-term consequences on their fertility. Primary germ cell damage is well known in our treatment for advanced disease and only about half of our patients regenerate a satisfactory sperm count after 2–3 years. Hopefully this will not become a treatment excess for low-volume disease patients in the future. It seems a good deal simpler and more direct to treat low-volume disease surgically and eliminate the potential of local relapse in an area (retroperitoneum) difficult to monitor and follow.

My own view is that surgery should be more proactive than reactive in the treatment of low-stage disease. Surgery can cure low-stage disease literally in hours, and when it does not, a cure is still available; and, should the patient relapse, which almost always means pulmonary metastases, it is easily detectable and treatable. Also, this is the only solid tumor that is curable by surgery alone more than 50% of the time when the regional nodes are positive. So, RPLND therapy in the retroperitoneum is now even more feasible and less toxic long term with the development of anatomical nerve-sparing dissections which will preserve ejaculation. At least this gives us an alternative to surveillance which seems to be quite effective and reasonable.

Finally, concerning high-stage disease, there is general agreement that all persistent tumors, after initial chemotherapy, should be removed. We have noted that certain patients with very favorable response criteria can be followed with the expectation of further resolution of their radiographic changes. We hope that we can be more selective in choosing patients for post-chemotherapy surgery. But, in the meantime, if there is any significant concern about the presence of persistent tumor, as a general principle such patients are more safely managed with exploration and RPLND. Currently, using this approach, we have in over 300 postchemotherapy RPLNDs, found 44% of patients with necrosis/fibrosis only, 44% with teratoma, and 12% with cancer. The former two groups are carefully observed postoperatively and the last group with persistent cancer receive two courses of salvage chemotherapy. The overall survival of this group is good (80%–90%), but factors predicting for relapse are bulk of tumor, histology, and site (primary mediastinum) [3,4].

Surgery for massive metastatic disease in testis cancer has become reasonable and appropriate in view of the great advances in chemotherapy for advanced disease [5]. The majority of patients will have their bulky disease reduced in volume and also down-graded histologically into a more mature teratomatous or even necrotic form. Of course, some disease will persist with malignant elements as well [3–11].

From a medical oncologic viewpoint, this information is critical because it will guide subsequent medical management: Those patients with persistent malignant tumor will require further salvage chemotherapy; those with only teratoma or necrosis in the specimen will be carefully monitored.

Often these retroperitoneal tumors are massive. It is of vital importance to resect them completely because local or regional recurrence is quite possible, particularly in the more extensive bulky tumors. Hence, this area represents one of the ultimate surgical challenges. Total tumor extirpation is required with exceptional demands on vascular isolation and preservation. Aggressive ablative purpose must be combined with thoughtful and delicate dissection.

Stage A Nonseminomatous Testis Carcinoma: Rationale and Results of Nerve-Sparing RPLND

The rationale for the use of RPLND in low-stage testis cancer is well grounded. RPLND in this situation is useful in terms of staging because approximately 30% of patients with clinical stage A testis cancer are, in fact, pathologic stage B. In addition, testis cancer is unusual in that removal of metastatic disease from the retroperitoneum is curative in 50%–60% of cases [1]. Hence, the usefulness of RPLND in low-stage non-seminomatous testicular carcinoma is twofold: (a) for staging, and (b) as a therapeutic modality.

The major objection to RPLND in low-stage nonseminomatous carcinoma of the testis has been the fertility consequences of lymphadenectomy. Patients who undergo full bilateral lymphadenectomy universally lose emission and the ability to ejaculate [12]. This major source of morbidity has provided the impetus for the development of surveillance protocols.

With the development of more effective treatment for nonseminomatous testis cancer the issue of morbidity of treatment has become more important. These young men now are likely to be cured of their disease and will demand the same quality of life as their peers. Therefore, the concept of preserving ejaculatory ability without compromising the efficacy of the procedure has become an important issue.

Nodal distribution studies in nonseminomatous testis cancer have afforded an understanding of likely sites of retroperitoneal metastasis in low-volume disease [10]. So-called modified RPLNDs, were developed with these mapping studies in mind. Only the lymphatic tissue possessing a high probability of retroperitoneal metastasis was removed; other lymphatic tissue was left intact. These modified techniques yielded a preservation of emission/ejaculation at approximately the 70% level. This was a definite advancement in the reduction of morbidity of therapy for this disease. However, with increasing knowledge of the anatomy of the sympathetic nervous system in the retroperitoneum, investigators have questioned whether or not ejaculatory ability could be preserved 100% without compromising the efficacy of the operation. We will describe the results of our investigations at Indiana University, on the first 75 patients to undergo prospective nerve-sparing RPLND in low-stage nonseminomatous carcinoma of the testis.

Methods

Patients with a diagnosis of nonseminomatous carcinoma of the testis after radical orchiectomy were subjected to clinical staging by abdominal and chest CT and determination of serum alpha-fetoprotein and beta-human chorionic gonadotropin (beta-HCG). Patients with normal CT scans of the abdomen and chest, and either normal markers or markers falling on the normal decay curve, were considered candidates for nerve-sparing lymphadenectomy.

A midline transperitoneal approach is utilized. After opening the abdomen, the retroperitoneum and nodal tissue along the internal spermatic vessels is palpated. If indeed clinical staging was accurate and no adenopathy is palpated, nerve-sparing RPLND is performed. The posterior peritoneal incisions for nerve-sparing lymphadenectomy are somewhat limited compared to the incision utilized for full bilateral lymphadenectomy.

For a right-sided dissection an incision is made in the posterior peritoneum from approximately the cecum cephalad to the area of Treitz's ligament. Blunt and sharp dissection are then utilized to lift the root of the small bowel off the lymphatic tissue and great vessels, exposing the interaortacaval, precaval, and right pericaval zones. Self-retaining retractors are useful in order to provide adequate exposure. The split-and-roll technique is used to divide the lymphatic tissue overlying the vena cava after careful palpation has disclosed no precaval renal arteries. This lymphatic tissue is then rolled off the vena cava both medially and laterally. Medially in the interaortacaval zone close inspection will reveal sympathetic fibers coursing from the sympathetic chain around lumbar veins into the interaortacaval zone and caudally to the bifurcation of the aorta. These fibers are identified and dissected free of lymphatic tissue prior to the lymphadenectomy. After the prospective dissection of these fibers lymphatic tissue is removed according to the standard templates dictated by mapping studies.

For a left-sided dissection one of two approaches can be utilized. The root of the small bowel may be incised in the area of the duodenum with reflection of the duodenum cephalad, and also to the right of this area with retraction of the mesentery of the left colon both to the left and inferiorly. This exposes the upper interaortacaval zone and the upper-left periaortic zone. The inferior mesenteric artery is left intact. Post-ganglionic sympathetic fibers are identified in the left periaortic zone distally near the area of the inferior mesenteric artery. They are then dissected proximally in a prospective fashion followed by RPLND using the templates for left-sided dissection determined by previous mapping studies. The lower portion of the left periaortic zone is subsequently dissected after the sigmoid colon is flipped medially. This maneuver is also necessary to remove the remaining left internal spermatic vein.

An alternate approach is to mobilize the entire left colon medially as is done for radical nephrectomy, identify the left-sided sympathetic fibers distally in the area of the bifurcation of the aorta, and trace these fibers proximally to the sympathetic chain. After dissection of these fibers prospectively, standard lymphadenectomy is carried out as dictated by the template for left-sided dissection.

Results

Currently there is a 2-year follow-up on the 75 patients subjected to nerve-sparing retroperitoneal lymph node dissection as described [13]. All patients were clinical stage I; 14 patients were pathologic stage II. All 75 patients ejaculated postoperatively. Of the 14 pathologic stage II cases, 4 have relapsed. One patient had a documented retroperitoneal recurrence, was given full-dose chemotherapy, and subsequently underwent post-chemotherapy RPLND. Pathological investigation showed only necrosis. Four of 61 pathologic stage I patients (8%) relapsed, but not in the retroperitoneum.

All patients are currently without evidence of disease. Details concerning the fertility status of these patients will be the subject of a future report.

Figure 1 gives a summary of the improved incidence of postoperative ejaculation related to the changes in operative technique and extent.

Discussion

This report documents the advancement that nerve-sparing RPLND has yielded in the reduction of morbidity in the treatment of nonseminomatous testis cancer. Because of the suspected 60% incidence of abnormal spermatogenesis in patients presenting with this disease prior to therapy, it is unclear whether or not these patients will subsequently have normal fertility [14]. Therefore, we are closely following the fertility status of these

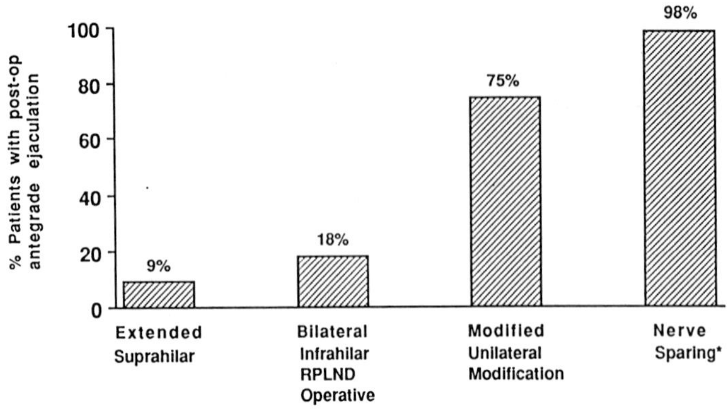

Fig. 1. The incidence of postoperative antegrade ejaculation after primary RPLND in clinical stage A nonseminomatous germ cell testis cancer: the influence of modifications to operative technique. 98% of cases having nerve-sparing PRLND had nerve-sparing modified unilateral RPLND

patients postoperatively. Unlike that of post-chemotherapy RPLND, the perioperative morbidity of staging lymphadenectomy in low-stage testicular cancer is exceedingly low. The average hospital stay is less than 1 week and perioperative complications are exceedingly rare [15]. With the development of a technique which seems to retain the efficacy associated with lymphadenectomy but spares the patient the major source of morbidity, the arguments against lymphadenectomy and in favor surveillance must be reexamined. The reduction of the major source of morbidity with nerve-sparing RPLND, coupled with the data of Hoskins and others which show a relapse rate of more than 30% at 4 years after orchiectomy, argue strongly against surveillance. Also, it is now known that of the 30% of patients who relapse on surveillance protocols, 4%–8% relapse 2 years or more after orchiectomy [16]. A recurrence more than 2 years after lymphadenectomy is, in contrast, very unlikely.

Nerve-sparing lymphadenectomy must now be considered preferred surgical treatment for low-stage nonseminomatous carcinoma of the testis. The increased facility of these nerve-sparing techniques will preserve probably make it possible to ejaculatory ability in selected patients undergoing post-chemotherapy dissection. These nerve-sparing techniques, therefore, make it possible for the modern urologic oncologist to preserve the efficacy of a time-tested operation, while at the same time eliminating its major source of morbidity.

References

1. Williams SD et al. (1987) Immediate adjuvant chemotherapy versus observation with treatment of relapse in pathological stage II testicular cancer. N Engl J Med 317:1433–1438
2. Donohue JP et al. (1988) Preservation of ejaculation following nerve-sparing retroperitoneal lymphadenectomy. J Urol 139:206A (abstr no 176)
3. Loehrer PJ, Williams SD, Clark SA et al. (1986) Teratoma following chemotherapy for nonseminomatous germ cell tumor (NSGCT): a clinicopathologic correlation. J Urol 135:1183
4. Einhorn LH, Williams SD, Mandelbaum I, Donohue JP (1981) Surgical resection in disseminated testicular cancer following chemotherapeutic cytoreduction. Cancer 48:904–908
5. Einhorn LH, Donohue JP (1977) Cis-diammine-dichloroplatinum, Vinblastine and Bleomycin combination chemotherapy in disseminated testicular cancer. Ann Intern Med 87:293–298
6. Donohue JP, Einhorn LH, Williams SD (1980) Cytoreductive surgery for metastatic testis cancer: considerations of timing and extent. J Urol 123:876–880
7. Donohue JP, Rowland RG (1984) The role of surgery in advanced testicular cancer. Cancer 54:2716–2721
8. Loehrer PJ, Sledge GW, Einhorn LH (1985) Heterogeneity among germ cell tumors of the testis. Semin Oncol 12:304–316
9. Ulbright TM, Loehrer PJ, Roth LM et al. (1984) The development of non-germ cell malignancies within germ cell tumors: a clinicopathologic study of 11 cases. Cancer 54:1824–1833
10. Donohue JP, Zachary JM, Maynard BR (1982) Distribution of nodal metastases in nonseminomatous testis cancer. J Urol 128:315–320
11. Donohue JP, Rowland RG, Bihrle R (1988) Transabdominal retroperitoneal lymph node dissection. In: Skinner DG, Leiskovsky G (eds) Diagnosis and management of genitourinary cancer. Saunders, Philadelphia, pp 802–816
12. Donohue JP, Rowland RG (1981) Complications of Retroperitoneal Lymphadenectomy. J Urol 125:338
13. Donohue JP, Foster RS, Rowland RG et al. (1991) Nerve-sparing retroperitoneal lymphadenectomy. J Urol (Submitted)
14. Lange PH, Chang WY, Fraley EE (1987) Fertility issues in the therapy of nonseminomatous testicular tumors. Urol Clin North Am 14:731–745
15. Bihrle R, Donohue JP, Foster RS (1988) Complications of retroperitoneal lymph node dissection. Urol Clin North Am 15:237–242
16. Hoskins P, Dilly S, Easton D (1986) Prognostic factors in stage I nonseminomatous germ cell testicular tumors managed by orchiectomy and surveillance: implications for adjuvant chemotherapy. J Clin Oncol 4:1031–1036

Anatomic Study and Operative Approach to Retroperitoneal Bilateral Nerve-Preserving Lymphadenectomy

K. Colleselli, S. Poisel, W. Schachtner, and G. Bartsch

Abteilung für Urologie, Anichstrasse 35, A-6020 Innsbruck, Austria

Introduction

Seminal emission and ejaculation are primarily under sympathetic control. The efferent fibers originating from T12-L3 travel to the superior and then to the inferior hypogastric plexus (or pelvic plexus). The pelvic plexus also contains parasympathetic fibers originating from the pelvic nerves. According to reports in the literature the percentage of patients presenting with postoperative aspermia ranges between 73% and 88% [9,10,13,14,16].

The topography of the ganglia of the sympathetic trunk and their fibers and their relation to the abdominal aorta and the inferior vena cava were studied by means of anatomic dissection.

Based on the results of this anatomic study a new operative approach and a modified, nerve-preserving surgical technique for small-sized stage B_1 and B_2 retroperitoneal tumors were developed.

Materials and Methods

Different aspects of 12 cadaver specimens sectioned from L2 to and including the sacrum were studied with a magnifying lens:

- Specimens 1–4: Lymphatic tissue was dissected prior to the preparation of the sympathetic nerve fibers.
- Specimens 5–9: Ganglia and fibers of the sympathetic trunk were dissected.
- Specimens 10–12: The abdominal aorta and the inferior vena cava were severed approximately 1 cm above the origin of the inferior mesenteric artery and tipped up in a "split-and-roll" procedure in order to facilitate the preparation of the nerve fibers in the aortocaval and retrocaval zones. Subsequently the operative approach and the resulting modification of the

Recent Results in Cancer Research, Vol. 126

operative technique were studied in six cadavers. In this dissection special attention was paid to the topography of the sympathetic ganglia in relation to the inferior mesenteric artery.

In the anatomic study a total of 18 cadavers of patients between 63 and 78 years of age (12 specimens and 6 complete cadavers) were dissected. The standard fixation method with phenol formalin was used [11].

Anatomy

The sympathetic trunk consists of a series of spindle-shaped ganglia, which are interconnected by short interganglionic rami to form a continous chain of ganglia (Figs. 1 and 2). This trunk extends along the spinal column on either side from the base of the skull to the tip of the coccyx. It is connected with the spinal cord via the rami communicantes, which may behave in a number of different ways. Frequently there are branches which course not only to the corresponding ganglia of the sympathetic trunk, but also run to the adjoining ganglia. In the lumbar region, where the rami communicans are relatively long, their course is particularly variable. They pass beyond the tendinous origins of the psoas major, and a variable number of ganglia can be found at different levels at the anterolateral aspects of the vertebral bodies medial to the psoas major. The right sympathetic trunk is situated dorsal to the inferior vena cava, while the left trunk is located dorsal

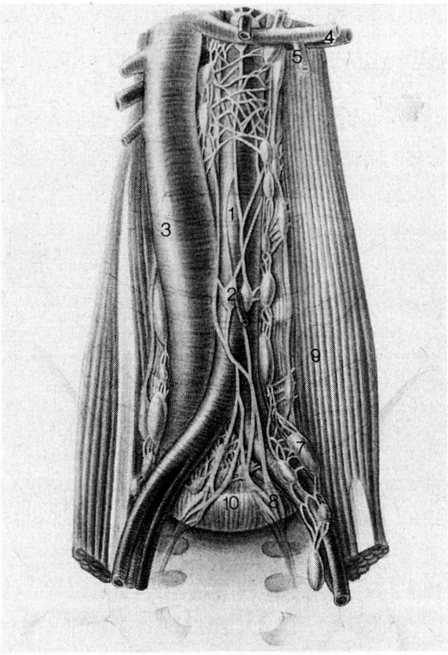

Fig. 1. In specimens 1–4 great difficulties were encountered in the dissection because the lymphatic tissue in the lumbar region had been dissected prior to the preparation of the sympathetic ganglia and fibers. The sympathetic trunk is found as a series of spindle-shaped ganglia which are interconnected by short interganglionic rami to form a continuous chain of ganglia. The right sympathetic trunk is situated dorsal to the inferior vena cava, while the left trunk is located dorsal and lateral to the abdominal aorta. *1*, Abdominal aorta; *2*, inferior mesenteric artery; *3*, inferior vena cava; *4*, renal vessels; *5*, spermatic vessels; *6*, lumbar vessels; *7*, lymph tissue; *8*, sympathetic nerves; *9*, psoas major muscle; *10*, vertebral body

and lateral to the abdominal aorta. The transverse rami connect the two sympathetic trunks. Only rarely are the two trunks arranged symmetrically in the lumbar region. Furthermore, fusions of ganglia are occasionally encountered [2].

The lumbar splanchnic nerves are the continuation of postganglionic fibres of the lumbar ganglia. They travel to the large paraaortic nervous plexuses and eventually reach the superior hypogastric plexus, a continuation of the abdominal aortic plexus which is located on either side of the abdominal aorta between the origins of the superior and inferior mesenteric arteries.

The superior hypogastric plexus is situated ventral to the aortic bifurcation and extends across the common iliac vein in the direction of the promontory (Fig. 2). The right and left hypogastric nerves connect the superior hypogastric plexus to the inferior hypogastric plexus, which is situated on either side of the rectum in the lesser pelvis.

The afferent impulses triggering emission and ejaculation are conveyed via the pudendal nerve, whereas the efferent impulses inducing emission are transported to the target organs via the thoracolumbar sympathetic fibers (T12-L3), the paravertebral ganglia and the superior hypogastric plexus.

The efferent impulses inducing ejaculation are conveyed via the parasympathetic fibers of the pelvic nerves (S2–S3) and cause the bulbocavernous and ischiocavernous muscles to contract, which results in antegrade ejaculation [1,15].

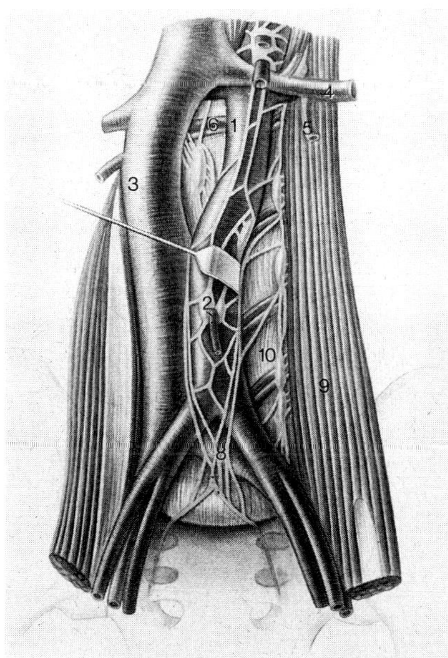

Fig. 2. In specimens 5–9 the lymph vessels and nodes were removed before the sympathetic trunk between the anterior aspect of the psoas major and the left lateral wall of the abdominal aorta, and the right lateral wall of the inferior vena cava was dissected. The L3 ganglion on the left side is seen very clearly because the abdominal aorta is retracted medially. For *abbreviations* see list accompanying Fig. 1

Anatomic Study and Results

In the dissection of specimens 1–4 great difficulties were encountered, mainly because the lymphatic tissue in the lumbar region had been dissected prior to the preparation of the sympathetic ganglia and fibers.

Most lymph vessels are located ventral to the lumbar splanchnic nerves and are hardly distinguishable from these. The difficulties were overcome by embedding the specimens in 72% ethanol and using a magnifying lens with a fourfold magnification. Thus it became possible to assess the topographic relationship between nerves and lymphatics. Subsequently the fibers from the prevertebral plexus to the ganglia of the sympathetic trunk were dissected (Fig. 1).

In specimens 5–9 both the lymph vessels and nodes were removed before the sympathetic trunk between the anterior aspect of the psoas major muscle and the left lateral wall of the abdominal aorta, and the right lateral

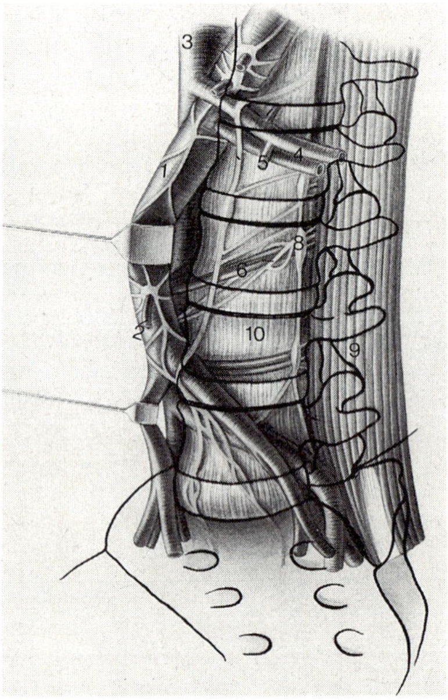

Fig. 3 (*left*). An anatomic preparation similar to Fig. 2, showing in detail the L3 ganglion (located 1 cm cranial to the origin of the inferior mesenteric artery) and the fibers arising from this ganglion. *For abbreviations* see list accompanying Fig. 1

Fig. 4 (*right*). A lateral view of one of the specimens 5–9. The sympathetic trunk is seen between the anterior aspect of the psoas major and the left lateral wall of the abdominal aorta, which is retracted medially. The L3 ganglion and the fibers arising from it can also be seen. For *abbreviations* see list accompanying Fig. 1

wall of the inferior vena cava was dissected. The lumbar vessels were preserved (Figs. 2–4).

Difficulties were encountered when attempting to dissect the fibers interconnecting the right and left sympathetic trunks. On the right side these fibers run dorsal to the inferior vena cava at the level of the third to fourth lumbar vertebrae and then pass into the aortocaval zone. Caudal to the inferior mesenteric artery they unite with the left paraaortic fibers and subsequcntly run to the superior hypogastric plexus. A precise dissection of these fibers along their entire course became possible only after the abdominal aorta and the inferior vena cava had been severed 1 cm above the inferior mesenteric artery. Specimens 10–12 were used for this purpose (Fig. 5).

In addition to the dissection of these fibers connecting the two sympathetic trunks, the topography of the lumbar ganglia was also included in our study. Bearing in mind the results of the study of Whitelaw and Smithwick [15], it was particularly the topography of the third lumbar ganglion which was of interest. The second and third lumbar ganglia are located very closely to each other; they may even fuse in which case there is no interganglionic ramus (Fig. 9).

The lower margin of the third lumbar ganglion is situated approximately a finger's breadth above the origin of the inferior mesenteric artery. If the abdominal aorta is displaced towards the median, the left L3 ganglion is exposed medially to the psoas major muscle on the lateral aspect of the third lumbar vertebra. The right L3 ganglion can be exposed by displacing the

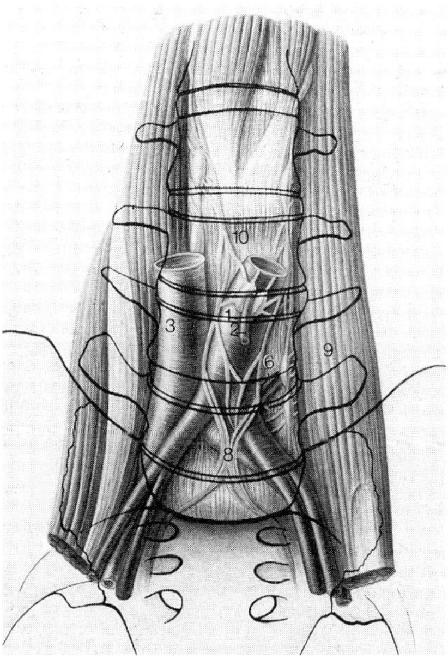

Fig. 5. In specimens 10–12 difficulties were encountered when attempting to dissect the fibers interconnecting the right and left sympathetic trunks. A precise dissection of these fibers along their entire course became possible only after the abdominal aorta and the inferior vena cava had been severed one centimeter above the inferior mesenteric artery. For *abbreviations* see list accompanying Fig. 1

inferior vena cava towards the median and is located slightly ventral to the ventrolateral junction of the third lumbar vertebral body. As mentioned above the two trunks are interconnected by the lumbar splanchnic nerves which come from these ganglia [3].

Modified Operative Technique

On the basis of the anatomic study a hilar and a suprahilar approach (relating to the renal hilum) was developed, leading to the introduction of a modified operative technique.

In the hilar approach the parietal peritoneum is incised from Treitz's ligament to the right iliac vessels. In the suprahilar approach the ascending colon, the right colic flexure, and the transverse colon up to the epiploic foramen are circumcised. Subsequently the duodenum is mobilized and displaced laterally [9].

Bilaterally lymphadenectomy is performed. On the tumor-free side the area of the third lumbar ganglia is preserved. The lymphatics along the internal and external iliac vessels are resected on the side where the tumor is located, followed by total resection of the lymphatics in the aortocaval zone and at the renal hilum.

In right-sided tumors the dissection on the left side extends to approximately a finger's breadth above the inferior mesenteric artery. Caudally the

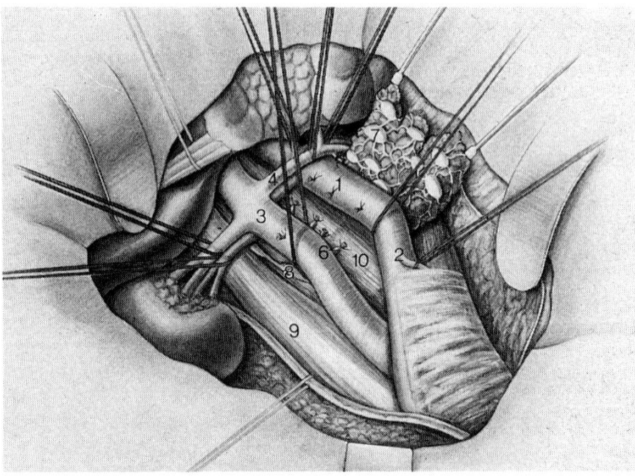

Fig. 6. Overall view of a right-sided tumor in modified bilateral retroperitoneal lymphadenectomy. The dissection on the left side extends to approximately 1 cm above the inferior mesenteric artery. Caudally the nervous tissue anterior of the abdominal aorta, in the paraaortic and aortocaval zones the corresponding lymphatic tissue is preserved. For *abbreviations* see list accompanying Fig. 1

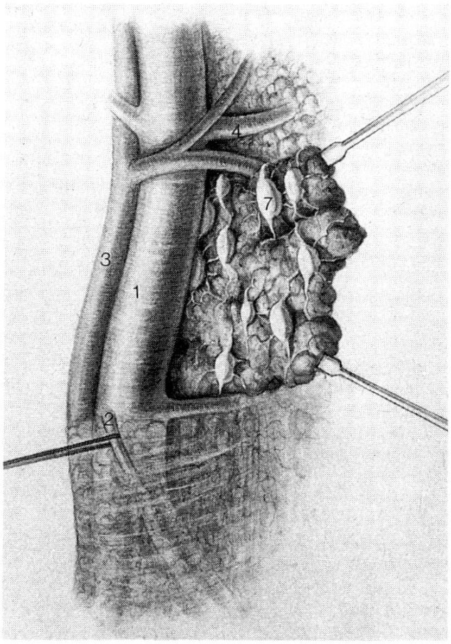

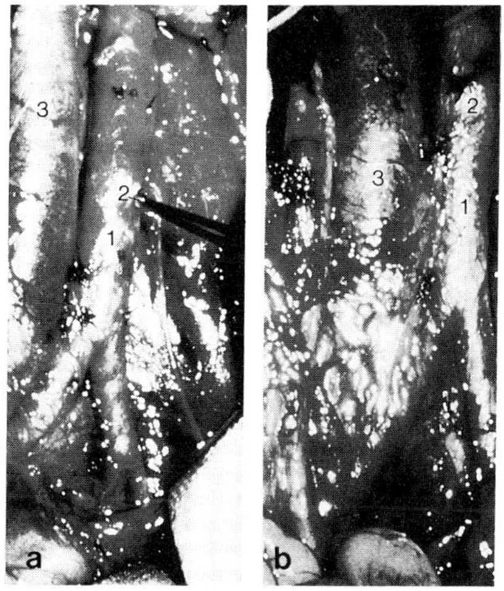

Fig. 7 (*left*). Detailed view of modified bilateral retroperitoneal lymphadenectomy in Fig. 6. The abdominal aorta, the inferior vena cava, the origin of the inferior mesenteric artery and the lymphatic tissue can be clearly seen. For *abbreviations* see list accompanying Fig. 1

Fig. 8 a,b (*right*). False (**a**) and correct (**b**) surgical procedure in modified bilateral retroperitoneal lymphadenectomy in left-sided tumor. In **a** the lymphatic tissue was dissected together with the sympathetic fibers; in **b** precaval, paracaval, and aortocaval zones are preserved

nervous tissue at the anterior wall of the abdominal aorta and in the paraaortic and aortocaval zones is preserved, together with the corresponding lymphatic tissue (Figs. 6 and 7).

In left-sided tumors tissues in the corresponding precaval, paracaval and aortocaval zones are preserved. In these procedures the lumbar vessels should be left intact (Fig. 8).

Discussion

As long ago as 1951 Whitelaw and Smithwick [15] reported that the resection of certain sympathetic ganglia may lead to ejaculatory dysfunction. This condition is also referred to as secondary disorder and has been described by Fowler and de Takets [7].

Whitelaw and Smithwick studied the effects of the various types of lymphadenectomy on ejaculatory function in 161 patients [15]. Regarding surgical procedures in the lumbar region their studies yielded the following observations.

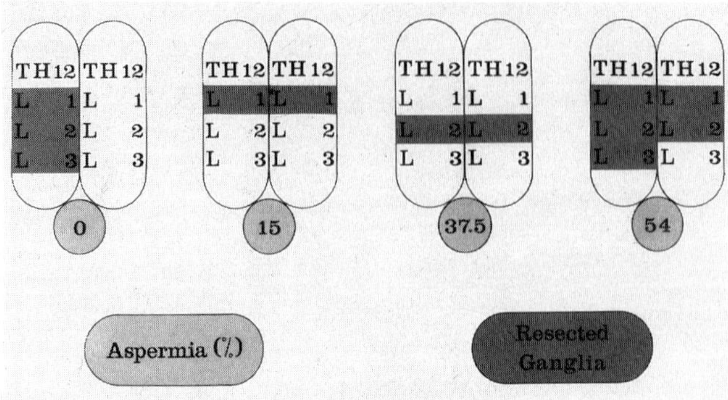

Fig. 9. Ejaculatory dysfunction in relation to resected ganglions according to Whitelaw and Smithwick [15]. In unilateral resection of the L1–L3 ganglia no ejaculatory dysfunction was observed. Bilateral resection of the first lumbar ganglia led to aspermia in 15% of cases, while bilateral resection of L2 resulted in ejaculatory dysfunction in 37.5% of cases. If L3 is preserved unilaterally ejaculation can be maintained in approximately 46% of patients

In unilateral resection of the L1–L3 ganglia no ejaculatory dysfunction was observed. Bilateral resection of the first lumbar ganglia led to aspermia in 15% of cases, while bilateral resection of the second lumbar ganglia resulted in ejaculatory dysfunction in 37.5% of cases. Provided that the third lumbar ganglion is preserved unilaterally, ejaculatory function can be maintained in approximately 46% of patients (Fig. 9).

On the basis of these findings and the results of our anatomic studies we have developed a modified operative technique which aims to preserve emission and ejaculation in patients undergoing bilateral retroperitoneal lymphadenectomy for testicular tumors. This technique also takes the distribution of the tumor volume in stage B_1 and B_2 tumors (according to the mapping of Donohue) into account (Fig. 10) [5,6].

In patients with small-volume retroperitoneal tumor Rowland and Donohue [12] use a split-and-roll technique to separate the lymph node bearing tissue from the sympathetic nerves as they emerge from the posterior side of the vena cava and spread out anteriorly and posteriorly to the aorta.

Our operative approach and modified technique are based on the results of the anatomic study described above. In this study the topography of the third lumbar ganglia and the course of their fibers was of special interest.

As mentioned above, the second and third lumbar ganglia are generally close to each other; the caudal border of the third lumbar ganglion is located approximately a finger's breadth cranial to the origin of the inferior mesenteric artery. On the right side, fibers connecting the right and left sympathetic trunks pass from dorsal to the inferior vena cava into the aortocaval zone. On the left side these fibers first travel in the paraaortic zone, and caudal to the inferior mesenteric artery they join the contralateral fibers. Costa and Furness [4] presumed that a small number of sympathetic

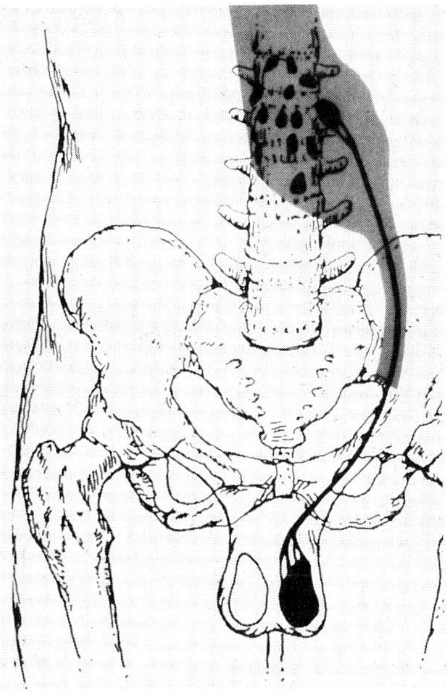

Fig. 10. Distribution of tumor volume in a right-sided tumor (stage B_1 and B_2 tumors according to the mapping of Donohue) is shown in the *shaded area*

fibers reach their target organs via the lumbosacral portion of the sympathetic trunk. These fibers present an additional pathway, which explains the fact that in a small percentage of patients radical bilateral retroperitoneal lymphadenectomy does not entail a loss of seminal emission and ejaculation.

This review of the topography of the sympathetic trunk in the retroperitoneum with regard to lymphadenectomy for testicular tumors has led to the development of a modified operative technique which takes into account both the results of Whilelaw and Smithwick's study and the distribution of the tumor volume in stage B_1 and B_2 tumors according to the mapping of Donohue [5,6,15].

With this modified technique it should be possible to preserve ejaculatory function in approximately 50% of patients with small-sized tumors in the retroperitoneum (B_1 and B_2). As in nerve-preserving lymphadenectomy for stage I tumors, this modification is justified only if the recurrence rate is not higher than that for radical retroperitoneal lymphadenectomy; the results of both techniques must therefore be compared.

References

1. Chiou R-K, Fraley EE, Lange PH (1984) Newer ideas about fertility in patients with testicular cancer. World J Urol 2:26
2. Clara M (1959) Der Sympathicus. In: Das Nervensystem des Menschen (Lehrbuch für Studierende und Ärzte) 3rd edn. Barth, Leipzig VI/E, p 223

3. Colleselli K, Poisel S, Schachtner W, Bartsch G (1990) Nerve-preserving bilateral retroperitoneal lymphadenectomy. J Urol 144:293–298
4. Costa M, Furness S (1973) Observations on the anatomy and amine histochemistry of the nerves and ganglia which supply the pelvic viscera and on the associated chromaffin tissue in the guinea pig. Z Anat Entwicklungsgesch 140:85
5. Donohue JP (1977) Retroperitoneal lymphadenectomy: the anterior approach including bilateral suprarenal-hilar dissection. Urol Clin North Am 4:509
6. Donohue JP, Zachery JM, Maynard SD (1982) Distribution of nodal metastases in nonseminomatous testicular cancer. World J Urol 128:315
7. Fowler EF, de Takats (1949) Side effects and complications of sympath-ectomy for hypertension. Arch Surg 59:1213
8. Javadpour N (1986) Management of stage 2 nonseminomatous testicular cancer. In: Javadpour N (ed) Principles and management of testicular cancer. Thieme, Stuttgart, pp 243–249
9. Lenz P, Meridies R (1972) Fertilitätsstörungen nach retroperitonealer Lymphknotenausräumung wegen teratoider Hodentumoren. Aktuel Urol 4:87
10. Narayan P, Lang P, Fraley EE (1982) Ejaculation and fertility after extended retroperitoneal lymph node dissection for testicular cancer. World J Urol 127:685
11. Platzer W, Putz R, Poisel S (1978) Ein neues Konservierungs-und Aufbewahrungssystem für anatomisches Material. Acta Anat 102:60
12. Rowland GR, Donohue JP (1988) Testicular Cancer: innovations in diagnosis and treatment. Semin Urol 4(3):223
13. Scheiber K, Bartsch G, Biedermann G, Flora G (1985) Störungen der vita sexualis nach Operationen im Retroperitonealraum. Ang Arch 8:71
14. Walsh PC, Kaufmann JJ, Coulsen WF, Goodwin WE (1971) Retroperitoneal lymphadenectomy for testicular tumors. JAMA 217(3):309
15. Whitelaw GP, Smithwick RH (1951) Some secondary effects of sympathectomy with particular reference to disturbance of sexual function. N Engl J Med 245:121
16. Weissbach L, Lange C-E, Rodemund O-E, Zwicker H, Gropp A, Pothmann W (1974) Fertilitätsstörungen bei behandelten Hodentumorpatienten. Urologe [A] 13:80

Management of Early Stages of Testicular Carcinoma: The Current Status

H.-J. Schmoll

Arbeitsgruppe für Intermedizinische Onkologie, Abteilung für Hematologie/ Onkologie, Medizinische Hochschule Hannover, Konstanty-Gutschow-Strasse 8, W-3000 Hannover 61, FRG

Introduction

In the last 15 years a dramatic improvement in cure rates for patients with testicular cancer from less than 50% to 90%, has been achieved by the introduction of chemotherapeutic agents such as bleomycin, ifosfamide, cisplatin, and etoposide. It has been demonstrated that patients with minimal tumor load at the time of diagnosis have an excellent survival chance, exceeding 95%. On the basis of these excellent cure rates several clinical trials were performed to investigate whether chemotherapy could be reduced in terms of number of drugs, number of cycles and dosage without affecting the cure rate. At the same time new forms of supportive treatment with third-generation antiemetics and hematopoietic growth factors have improved the tolerability of chemotherapy and reduced the risk of death owing to sepsis.

A cure rate of more than 95% (Williams 1987a; Stoter 1987) for patients with minimal disease after treatment with cisplatin-based chemotherapy and the possible reduction of chemotherapy-related toxicity offer the opportunity to adopt a new treatment policy in early-stage testicular cancer, where surgery has been the treatment of choice for several decades. These new approaches are, for example, a wait-and-see policy in stage I, primary chemotherapy in stage II, and replacement of radiotherapy by chemotherapy in the early stages of seminoma. However, the risks and benefits of the different treatment modalities have not yet been elaborated in comparative, prospective randomized trials. Therefore, most clinical institutions interested in the treatment of testis cancer still prefer more traditional management with its proven efficacy. It can be expected that future treatment decisions will depend on a number of factors such as individual prognostic factors of the disease, psychological and physical factors of the patient, the standard of the facility at which the treatment will be initiated, the skill of the urologic surgeon and the oncologist, and the new data in the literature.

In the following article some already available data and some suggestions regarding treatment decisions for clinical stage I seminoma and nonseminoma are presented.

Carcinoma in Situ/Testicular Intraepithelial Neoplasia

Testicular seminoma and nonseminoma develop from carcinoma in situ (Cis). The delay between the detection of Cis and the manifest, invasive tumor ranges from 1 to 15 years or longer. The frequency of a Cis in the contralateral testis of patients with unilateral testicular cancer ist 3%–5% by biopsy at the time of orchiectomy for the first cancer (Table 1). This corresponds with the 3%–5% of patients developing secondary testicular cancer in the contralateral testis in historical populations (Dieckmann et al. 1989).

The percentage of secondary testicular cancers in over 1200 patients treated between 1971 and 1991 at our clinic in Hanover is 3.8%. Most likely a Cis could have been detected by biopsy in these patients before manifestation of the secondary tumor. However, none of these patients died of secondary cancer because these tumors were detected early (mostly in stage I) due to routine clinical and sonographic investigation of the contralateral testis and regular registration of tumor markers.

Biopsy of Contralateral Testis

What is the optimal treatment strategy for diagnosis and treatment of Cis today?

Biopsy of the testis does not result in any long-term morbidity. Scar formation may cause problems in follow-up sonographic examinations of the

Table 1. Prevalence of carcinoma in situ (Cis, ITN). (Adapted from Dieckmann et al. 1989)

	Frequency of Cis (%)
Maldescensus testis	2.7–8
Infertility	0.4–1.1
Genetic syndromes	33
Testicular cancer in contralateral testis	4.4–5.7
Testicular cancer + maldescensus of contralateral testis	2.7–23
In testis with testis cancer	63–98

biopsied testis. Since the frequency of testicular intraepithelial neoplasia (TIN) is very low, observed in only 3%–5% of patients, it may be argued that biopsy should only be performed in patients with high risk for Cis, particularly those with pendulum testis, a history of cryptorchism or inguinal hernia, genetic disorders such as adrenogenital syndrome, a family history of testicular cancer, etc. (Table 2). The risk of these patients having Cis and/or developing secondary testis tumor is up to 20 times higher.

Alternatively, close follow-up with routine ultrasonography of the testis by an experienced investigator may increase early detection of the secondary cancer and therefore make biopsy unnecessary; however, such an approach requires three to six investigations a year for 10–15 years, compared to a biopsy of the contralateral testis at the time of initial surgery.

Several treatment options are available for patients with biopsy-proven Cis (Table 2). The overall cure rate has been shown to be the same for the different options. Irradiation results in irreversible damage of the germ cell epithelium causing permanent infertility; therefore, this treatment should only be recommended to patients who have finished their family planning. Sperm banking may be an alternative for those couples who intend further conceptions.

Seminoma

The Development of Chemotherapy

In 1984 Oliver et al. demonstrated that cisplatin, cyclophosphamide, and ifosfamide are each highly effective in the treatment of seminoma. It remains unclear whether bleomycin, the vinca alkaloids, and etoposide have a significant effect. In contrast to nonseminomatous tumors, no dose-response analysis for cisplatin has been performed in seminoma. In nonseminomatous testicular cancer, the cisplatin analogue carboplatin is, if given at a standard dosage, inferior to cisplatin (Bajorin et al. 1991). In seminoma, however, carboplatin monotherapy at a dose of $400 mg/m^2$ every 4 weeks for 4–6 cycles is highly effective and appears to be comparable to single-agent or combination cisplatin chemotherapy, at least for patients with minimal disease, as demonstrated by Horwich et al. (1989) and Schmoll et al. (1991). The cure rate for patients with retroperitoneal lymph node metastases is 100% even in stage IIC with a diameter exceeding 5 cm (Schmoll et al. 1991). Therefore, it can be anticipated that in earlier stages of seminoma (stage IIA/B) carboplatin monotherapy will be as effective as the more toxic regimens currently used. The confirmation of these assumptions by prospective trials will change the therapeutic concept for early-stage seminoma.

Table 2. Options for management of carcinoma in situ of the contralateral testis

	Procedure	Definitive treatment at sign of invasive tumor	Outcome (cure rate, %)	Toxicity
Biopsy → no Cis	–	–	100	Scar
Biopsy → diagnosis of Cis	Observation	Orchiectomy	100	Late infertility; androgen substitution for life
		Organ-preserving tumorectomy (if small tumor)	100	No infertility
		Chemotherapy for stage II	99–100	Potential infertility
No biopsy	Observation	see above	99–100	None; risk of improper follow-up and late diagnosis of secondary tumor
Biopsy All patients High-risk patients only	Irradiation	–	100	Infertility, but no androgen substitution necessary

Seminoma Stage I

Adjuvant Irradiation

Today, adjuvant irradiation of the paraaortic field is the standard treatment for clinical stage I seminoma after orchiectomy, independent of the primary tumor size, yielding cure rates between 94% and 100% in large numbers of patients (Table 3). An optimal radiotherapeutic technique requires a linear accelerator to avoid radiation damage of the surrounding tissue and the contralateral testis. Radiotherapy should be delivered within 3 weeks in fractions of 5 × 2 Gy per week. The risk of secondary cancer induced by radiation with limited radiotherapy is low (Table 4). Infradiaphragmatic irradiation alone, however, carries a danger of inducing a secondary tumor (risk ratio 1.32), indicating that this treatment strategy is not without long-

Table 3. Results of adjuvant irradiation in clinical stage I seminoma

	No. of patients	5-year survival (%)
Vander Werf-Messing (1976)	91	100
Calman et al. (1979)	121	97
Schmitt et al. (1980)	53	98
Thomas et al. (1982)	338	94
Willan and McGowan (1985)	149	98
Thomas (1985)	150	99
Duncan and Munro (1987)	78	95
Zagars and Babaian (1987)	163	97
Fossa et al. (1988)	365	99
Lederman et al. (1989)	79	97
Total	1597	97 (91 – 100)

Table 4. Risk of secondary cancer after irradiation of seminoma (S.D. Fossa, 1990)

	Relative risk (observed/expected)			
	Infradiaphragmatic	Infra- + supradiaphragmatic	Radiotherapy + chemotherapy	Total
Melanoma	1.32	7.7[a]	22.2[a]	3.9[a]
Leukemia	3.53	–	–	2.97
Bronchial carcinoma	1.4	7.7[a]	3.85	2.03[a]
Gastric carcinoma	1.75	8.33[b]	–	1.86
Colon carcinoma	1.10	7.69	–	1.47
Bladder carcinoma	2.0	–	–	1.7
Total	1.32	4.13[a]	2.44	1.58

[a] $p < 0.01$.
[b] $p < 0.05$.

term side effects (Fossa et al. 1990). The acute toxicity of irradiation is relatively low if it is properly delivered using modern techniques with careful planning. The long-term survival rate of 97% in, altogether, 1597 patients (Table 3), however, leaves something to be desired. The reason might be that not all patients were adequately treated with systemic chemotherapy at the time of occurrence of distant metastases.

Adjuvant Chemotherapy

A further option is emerging from the excellent antiseminoma activity of carboplatin. Oliver et al. (1990) treated 26 patients with surveillance and 25 patients with adjuvant cisplatin or carboplatin monotherapy for 2 cycles. No relapse was observed after adjuvant chemotherapy: all patients were free of disease. This trial shows that microscopic disease in clinical stage I seminoma is eradicated by chemotherapeutic regimens with low toxicity.

Although this approach would highly facilitate the follow-up procedure in these patients, the toxicity of systemic cisplatin or carboplatin must be considered if they are to be given to each patient with a stage I seminoma. One aspect is the impairment of sperm epithelium, which might persist in some patients. In those patients with questionable compliance, however, this might be the best treatment option. It is the personal opinion of the author that this strategy should be reserved for such patients.

The current joint MRC/EORTC protocol compares irradiation with adjuvant carboplatin chemotherapy (Fig. 1); final results can be expected in 5 years. A similar trial is planned within the framework of the "Interdisciplinary Working Group for Testicular Cancer" in Germany.

Surveillance

Surveillance is an alternative treatment strategy for patients with stage I seminoma. In a series of 113 patients, Horwich et al. (1990) found a relapse rate of 12% with metastases either in retroperitoneal lymph nodes or in the lung, and very rarely in the bone. All relapsing patients have been rendered disease-free by either radiotherapy or chemotherapy. Four percent of the

Seminoma I, IIA,B
- Irradiation: 25 Gy paraaortic + iliac, ipsilateral
- Chemotherapy: carboplatin ×4; if no response, cisplatin 100 mg/m^2 on day 1 with etoposide 125 mg/m^2 on days 1,2,3

Fig. 1. Current MRC/EORTC protocol for early stage seminoma

patients had a secondary relapse and all could be salvaged by chemotherapy. Alltogether, the rate of patients without progressive or manifest disease was 100%. Only 10% of the patients under surveillance underwent radiotherapy and only 6% received chemotherapy, yielding a total of 16% receiving further treatment, compared with 100% in patients undergoing standard adjuvant irradiation (Horwich et al. 1990). Although long-term follow-up results are not yet available it can be assumed that the cure rate will be similar to the 97% reported for adjuvant irradiation. In a retrospective analysis of patients treated at the Hanover Medical School the relapse rate was 3% in 64 patients treated with adjuvant irradiation and 5% in 31 patients under surveillance, with an eventual no evidence of disease (NED) rate of 100% for both patients groups (Bading, 1992).

Biological Factors for Relapse

The biological factors indicating a high risk of relapse have not yet been identified. These factors could include morphological parameters (e.g., lymphatic or blood vessel invasion), oncogene or suppressor gene expression in the primary tumor (e.g., TIMP 2, p53 or Ki-*ras*, N-*myc*, etc.), or ploidy and proliferative index determined by DNA flow cytometry. These prognostic factors must be investigated to further characterize patients with a high probability of tumor relapse. Knowledge of the prognostic factors could influence treatment decisions, thus improving the results and minimizing the toxicity of treatment.

Seminoma Stage IIA/B

Radiotherapy

The standard treatments for patients with seminoma and retroperitoneal lymph node metastases less than 2 cm (stage IIA) or 2–5 cm in diameter (stage IIB) are irradiation of the paraaortic and paraaortic plus ipsilateral iliac field, respectively. The dosage is 30 Gy given over 3 weeks in stage IIA and 36 Gy given over 4 weeks in stage IIB. 5 × 2 Gy are given per week. The radiation field covers TH 11 to the upper hip joint region of the ipsilateral site (Bamberg et al. 1988). The 5-year survival for stage IIA and IIB ranges from 78% to 100%, since cisplatin-based chemotherapy is available for relapsing patients (Table 6).

Prophylactic mediastinal irradiation is ineffective in improving these results, but adds significantly to the radiation-derived risk for secondary carcinoma, with a risk ratio of 7.7 in patients who had infradiaphragmatic plus supradiaphragmatic radiation (Fossa et al. 1990; Table 4). The mediastinal relapse rate in patients with stage IIA and IIB disease without

Table 5. Mediastinal relapse after abdominal lymph node irradiation without prophylactic mediastinal irradiation in stage IIA and IIB seminoma

	No. of patients	No. of relapses
Doornbos et al. (1975)	6	0
Dosoretz et al. (1981)	8	0
Ball et al. (1982)	13	0
Thomas et al. (1982)	40	0
Herman et al. (1983)	197	8[a]
Huben et al. (1984)	10	0
Willan and McGowan (1985)	20	0
Lester et al. (1986)	6	0
Mason and Kearsley (1988)	11	0
Sagerman et al. (1989)	4	1
Total	315	2.7%

[a] 7/8 salvaged by mediastinal irradiation at relapse.

adjuvant mediastinal radiation is only 2.7% (Table 5); this low relapse rate does not justify prophylactic mediastinal irradiation.

Chemotherapy

Although for most patients still the treatment of choice, the role of irradiation as a primary therapeutic measure in the treatment of stage A/B seminoma must be discussed since the cure rate is well below 100%. The excellent response rates in seminoma stage IIC with single agent carboplatin demonstrated by Horwich et al. (1990) and Schmoll et al. (1991) suggest an excellent response rate in stage IIA/B, too. However, less than 10% of the patients will develop a systemic relapse after irradiation requiring chemotherapy.

Although there are no conclusive data supporting the theory that a distant relapse after primary irradiation leads to a higher level of carboplatin resistance, cure rates of less than 100% in several series with abdominal irradiation as primary treatment mean that careful investigation of this approach in a prospective trial is required. Such a trial is currently being conducted jointly by the EORTC and the Medical Research Council; in the case of an incomplete response or relapse after both modalities, combination chemotherapy with cisplatin plus etoposide is given (Fig. 1). Results from this trial can be expected within 3 or 4 years.

While radiotherapy increases the risk of a secondary carcinoma, an impairment of fertility is the only clinically relevant long-term toxic effect of platinum-polychemotherapy observed so far. It is still unknown whether carboplatin monotherapy will have long-term effects on spermatogenesis.

As long as the results of the ongoing MRC trial are not yet available, radiotherapy will remain the standard treatment in seminoma stage IIA/B and single agent carboplatin must be regarded as an experimental therapy.

Seminoma Stage IIC

The overall survival of patients with a stage IIC seminoma after radiotherapy with or without cisplatin-based chemotherapy ranges from 54% to 82% in most series reported so far (Table 6). Although with modern radiotherapy (wide hockey stick, whole abdominal irradiation including shrinking field technique with 36 Gy) a relapse occurred within the radiation ports only in 3 out of 21 patients (Smalley et al. 1985), this strategy is unable to prevent a distant relapse and is associated with more complications. Therefore, primary radiotherapy of patients with a stage IIC seminoma should be completely abandoned and primary chemotherapy should be the treatment of choice.

Nonseminomatous Germ Cell Tumors

Predictors of Retroperitoneal Lymph Node Metastases

Several clinical and pathological factors have been investigated, mostly in retrospective analyses, to define factors which can predict the involvement of retroperitoneal lymph nodes and/or the development of distant metastases after orchiectomy (Table 7). In a multivariate analysis of the Swedish/Norwegian Testicular Cancer Group (Klepp et al. 1990), vascular invasion

Table 6. 5-year survival of clinical stage II seminoma after primary radiotherapy

	Survival (%)			
	II A (<2 cm)	II A/B	II B (2–5 cm)	II C (>5 cm)
Ball et al. (1982)	78		81	57
Thomas et al. (1982)		87		62
Willan and McGowan (1985)		82		54
Lester et al. (1986)		80		75
Zagars and Babaian (1988)		95		64
Mason and Kearsley (1988)		94		82
Sagerman et al. (1989)		100		100
Lederman et al. (1989)	93		100	75

Table 7. Prognostic factors for metastases in NSGCT derived from surveillance studies

	No. of patients	pT stage	Preorchiectomy tumor markers	Histological subclassification	Vascular invasion
Gelderman et al. (1987)	54	?	no	no	?
Thompson et al. (1988)	36	no	no	no	yes
Dewar et al. (1987)	28	no	?	yes	yes
Sogani and Fair (1988)	102	?	no	yes	yes
Freedman et al. (1987)	259	(yes)	?	yes	yes
Dunphy et al. (1988)	93	?	yes	yes	?
Raghavan et al. (1988)	46	no	no	no	?
Pizzocaro et al. (1987)	85	yes	no	no	yes
Klepp et al. (1990)	279	(yes)	(yes)	yes	yes

?, not evaluated.

was found to be a significant predictor of microscopic involvement of the lymph nodes in clinical stage I nonseminomatous germ cell tumors (NSGCT). Predictors of relapse in pathological stage I tumors were vascular invasion and the absence of teratoma elements. Local tumor stage and the presence of embryonal carcinoma and yolk sac elements were not of significance in predicting either pathological stage II or relapse after pathological stage I. In a combined analysis, significant predictors of either lymph node metastases at retroperitoneal lymphadenectomy (RLA) or of developing distant metastases after a negative lymphadenectomy were vascular invasion and the absence of teratomatous elements, even with normal preorchiectomy α-fetoprotein (AFP) levels. Vascular invasion was also be identified as the most important prognostic factor for the development of metastases in surveillance studies (Table 7). In some trials the histological subclassification was found to be a significant predictor of prognosis. A retrospective analysis of the prospective surveillance trial of the Medical Research Council (Freedman et al. 1987) found vascular invasion, lymphatic invasion, the presence of embryonal carcinoma, and the absence of yolk sac carcinoma to be predictors for the development of metastases. A scoring system based on the histological characteristics of the primary tumor was able to assign patients to low, intermediate, or high risk groups with relapse rates of 0–9%, 10%–15%, and 16–58%, respectively (Table 8).

Table 8. Relapse rate after surveillance strategy in clinical stage I NSGCT according to risk score based on pathoanatomical definitions from MRC trials. (Freedman et al. 1987; Read et al. 1991)

Group (score)	Retrospective trials		Prospective trials[a]	
	Patients (%)	Relapse rate (%)	Patients (%)	Relapse rate (%)
1	3	0	36	16
2	35	9	39	21
3	38	25	23	53
4	24	58		

[a] Groups 3 and 4 combined.

There might be other factors which characterize the biology of a given tumor more precisely. Hyperpentaploidity, a high proliferative index or cytogenetic factors like the number of copies of isochromosome i12p as demonstrated by DNA analysis are among these factors. Further parameters might be of great relevance and must be investigated in future trials, e.g., aberrations of chromosomes 1, 5, 7, 11, 17, 22, the immunohistochemical demonstration of increased proliferation, an overexpression of tumor growth factor (TGF) α, TFG-β or KFGF, the in situ hybridization for oncogenes or suppressor genes as well as the analysis of adhesion molecules, e.g., NCAM, ICAM, collagenase 4, integrin, etc. These factors might better reflect the risk of a given patient than morphological parameters.

Treatment of Stage I NSGCT

Surgical Treatment

Before the advent of chemotherapy, retroperitoneal lymphadenectomy (RLA) was the treatment of choice in patients with clinical stage I disease. However, in 85% of these patients the procedure is purely diagnostic, producing no gain in survival but with the potential risk of loss of ejaculation. Major modifications to this procedure have led to a reduction in fertility disturbance; the modified unilateral RLA has markedly reduced the loss of ejaculation, but this may still occur in about 15% of patients (Fossa et al. 1985; Pizzocaro et al. 1985). Further improvement of this technique has been reported by Donohue (1991), who demonstrated no loss of ejaculation using a subtle nerve-sparing procedure. The relapse rate after radical or modified lymphadenectomy is 10%–26% (Weißbach et al. 1984). Therefore, despite surgery some patients still need chemotherapy. The overall survival of patients who relapsed after a lymphadenectomy has increased from 95% in the pre-cisplatin era to 99% today.

Surveillance Strategy

After a consensus meeting in 1979 headed by the WHO in Lugano, a surveillance strategy for clinical stage I NSCGT was introduced at several centers. The relapse rate after a minimum follow-up of 4 years up to a maximum follow-up of 10 years was 26% in a combined series of 1054 patients (Table 9). The overall survival of 98% is comparable to other treatment strategies. In a recent prospective trial with 373 patients (Read et al. 1991) the survival exceeded 99%. This result probably represents the optimum which can be achieved in a large multicenter setting if chemotherapy is given at the earliest sign of relapse. It is of note that the rate of patients who were lost to follow-up due to patient compliance problems was nearly zero. 26% of the patients required chemotherapy for relapse, while none of them underwent lymphadenectomy. Only one patient died from his tumor, and another died due to bleomycin lung toxicity.

Adjuvant Chemotherapy

In a prospective trial by Gimmi et al. (1990) it was demonstrated that cisplatin, vincristine, bleomycin (PVB) or cisplatin, etoposide, bleomycin (PEB) following orchiectomy completely prevented tumor relapse. In another prospective trial Pont et al. (1990) gave 2 cycles of PEB to 21 patients with vascular invasion, whereas another 26 patients without vascular invasion

Table 9. Relapse rates in the various surveillance trials of clinical stage I nonseminomatous testicular cancer

	No. of patients	Median follow-up (months)	Relapse rate (%)	Survival (%)
Gelderman et al. (1987)	54	29	20	98
Sogani and Fair (1988)	102	40	25	97
Swanson et al. (1987)	82	?	29	99
Rørth et al. (1987)	79	41	30	98
Jewett et al. (1984)	30	16	40	100
Raghavan et al. (1988)	46	40	28	96
Pizzocaro et al. (1987)	85	42	27	99
Freedman et al. (1987) and Read et al. (1991), combined data	658	48	26	98
Total	1054	4–10 years	26	98

underwent orchiectomy only. The relapse rate in the orchiectomy-only group was 4%, indicating that the absence of vascular invasion was a good predictor of having no metastases. In the chemotherapy group (patients with vascular invasion) the relapse rate was 9.5%. The survival was 91%. One patient died from a secondary lung cancer, yielding a disease-related survival of 95%. These preliminary data are now being further investigated in a trial of the Medical Research Council where patients belonging to the Freedman category 4 with a relapse risk above 50% receive 2 adjuvant cycles of chemotherapy with PEB.

What Is the Optimal Therapy?

Since the overall survival approximates 98% in clinical stage I NSGCT independent of the treatment strategy, the relevant factor for a treatment decision is the toxicity. Because of the transient nature of most side effects of all treatments, long-term toxicity is the most relevant parameter influencing the decision for one of the approaches mentioned above. The difference in the toxicity of the various concepts regarding fertility is outlined in Table 10. It should be mentioned that recovery from chemotherapy-induced infertility is very high, while patients on surveillance also have impaired fertility (Hansen et al. 1990) due to the underlying gonadal malfunction. Furthermore, any reduction of toxicity by a nerve-sparing lymphadenectomy may well be lost due to subsequent chemotherapy in 11% of the patients who relapse.

In conclusion, if no nerve-sparing RLA technique is available – which is almost always the case at present – the risk of infertility is higher after a modified lymphadenectomy than with surveillance strategies or adjuvant chemotherapy. Comparing the different therapies with respect to infertility, the surveillance strategy seems to be the least toxic one. For a selected patient who wants to maximally reduce the risk of infertility due to chemotherapy, a nerve-sparing lymphadenectomy can be recommended. On the other hand, for patients with poor personality or social problems that render an adequate follow-up in a surveillance protocol impossible, other treatment concepts should be applied, e.g., adjuvant chemotherapy.

Management of Stage IIA/B

Lymphadenectomy Plus Adjuvant Chemotherapy

In a prospective trial, Williams et al. (1987b) compared RLA only in stage II NSGCT with RLA followed by 2 cycles of adjuvant chemotherapy (PVB). The relapse rate in the surgery-only group was 49%, whereas a relapse rate of only 1% was observed in the adjuvant chemotherapy group.

Table 10. Options for postorchiectomy management in clinical stage I nonseminomatous testicular cancer

Treatment technique	Spontaneous or treatment-associated loss of ejaculation/fertility (long term) (%)	CTx at relapse (%)	Adjuvant CTx (%)	Approximate cumulative risk for infertlity (%)	Overall treatment (%)			
					Surgery	Irradiation	CTx	Both
Adjuvant irradiation	5–10	15	–	5–10	–	100	15	115
RLA, modified	30	15	–	35	100	–	11	111
RLA, nerve sparing	5–10	15	–	6–12	100	–	11	111
Surveillance	5–10	27	–	8–13	<10	–	27	~35
Adjuvant chemotherapy	8–15	2–4	100	15	100	–	100	200

CTx, chemotherapy.

Upon relapse after RLA, the patients received 4 cycles of PVB or PEB. The overall death rates from testicular cancer were not significantly different, with 3% in the RLA-only arm and 1% in the adjuvant chemotherapy arm. No factor, neither T nor N stage, nor vascular invasion nor histology, could be identified as being of prognostic significance.

The conclusion from this trial was for most of the centers that radical or modified RLA followed by adjuvant chemotherapy will yield a 100% cure rate. However, another conclusion is conceivable: if a lymphadenectomy is performed in stage II disease, adjuvant chemotherapy can be omitted since only 50% of these patients will later require chemotherapy due to relapse. Again, this concept cannot be recommended in patients with a questionable compliance.

Primary Chemotherapy Plus Selective Surgery

Another possibility was investigated in some phase II trials with a total of 279 patients who were treated initially with chemotherapy followed by secondary RLA (Table 11). The overall survival was 98%; however, only in 24% of the patients was a secondary RLA procedure of limited extent (only resection of residual tumor) necessary.

What Is the Optimal Therapy

As long as factors predicting the prognosis of a given patient are not available, the treatment decision must be made on the basis of the personal needs of the patients. If the compliance is questionable or if adequate

Table 11. Primary chemotherapy with and without elective lymphadenectomy in clinical stage IIA/B nonseminomatous testicular cancer

	No. of patients	CR/NED (%)	Patients with elective RLA (%)	Tumor death (%)	Survival (%)
Peckham and Hendry (1985)	54	100	22	4	96
Logothetis et al. (1987)	30	100	17	0	100
Socinski et al. (1988)	19	100	32	0	100
Daugaard et al. (1990)	28	100	n.s.	7	93
Pflüger et al. (1990)	26	100	31	0	100
Horwich et al. (1992)	122	99	28	1	98
Total	279	100	25	2	98

CR/NED, complete remission/no evidence of disease.

Table 12. Options in clinical stage II A/B nonseminomatous testicular cancer

		Risk of infertility (%)	Cure rate (%)	Surgery (%)	Chemotherapy (%)
RLA, nerve-sparing	→ CTx at relapse (PEB × 3)	5–15	>95	100	50
RLA, modified	→ CTx at relapse (PEB × 3)	30–70	>95	100	50
RLA	→ adjuvant chemotherapy (PEB × 2)	30–70	>95	100	100
Primary chemotherapy (PEB × 3)	→ resection of residual nodes	5–15	>95	25	100

CTx, chemotherapy.

follow-up is not possible, primary chemotherapy would be most appropriate, because the relapse rate is nearly zero. In a patient who-wants to preserve his fertility as far as possible, a nerve-sparing lymphadenectomy should be performed without adjuvant chemotherapy. It is the author's personal opinion that primary lymphadenectomy followed by adjuvant chemotherapy represents an overtreatment incorporating unnecessary toxicity (Table 12). The criteria for secondary RLA in these patients must be further defined in prospective studies. To look at this question in a prospective way, particularly with respect to toxicity, acute and chronic morbidity, and costs of treatment, frequency and quality of follow-up, relapse rate, and survival, a prospective trial is being performed by the German "Interdisciplinary Working Group for Testicular Cancer."

References

Bading M (1992) Dissertation, Hanover University Medical School

Bajorin DF, Sarosdy MF, Bosl GJ, Weisen S, Heller G (1991) A randomized trial of etoposide + carboplatin (EC) vs. etoposide + cisplatin (EP) in patients (pts) with metastatic germ cell tumors (GC). Proc Am Soc Clin Oncol 10:169 (abstr 535)

Ball D, Barrett V, Peckham MJ (1982) The management of metastatic seminoma testis. Cancer 50:2289–2294

Bamberg M, Sauerwein W, Budach V, Sack H (1988) Gegenwärtiger Stand der Radiotherapie bei Seminomen. In: Höffken K, Schmidt CG (eds) Hodentumoren – Diagnostik und Therapie – Ergebnisse der Internistischen Onkologie, Vol 2. Enke, Stuttgart, pp 55–66

Calman FM, Peckham MJ, Hendry WF (1979) The pattern of spread and treatment of metastases in testicular seminoma. Br J Urol 51:154

Daugaard G, Hansen HH, Rorth M (1990) Treatment of malignant germ cell tumours. Ann Oncol 1:195–202

Dewar JM, Spagnolo DV, Jamrozik KD et al. (1987) Predicting relapse in stage I nonseminomatous germ cell tumours of the testis. Lancet 1:454

Dieckman K-P, Loy V, Huland H (1989) Das Carcinoma in situ des Hodens: Klinische Bedeutung, Diagnostik und Therapie. Urologe [A] 28:271–280

Donohue JP (1991) Results of lymphadenectomy in stage I nonseminomatous germ cell cancer. Proceedings of the international symposium on testicular cancer, San Remo. Acta Med (in press)

Doornbos JF, Hussey DH, Johnson DE (1975) Radiotherapy for pure seminoma of the testis. Radiology 116:401–404

Dosoretz D, Shipley W, Bitzer P, Gilbert S, Parkhust E, Wang W (1981) Megavoltage irradiation for pure testicular seminoma: Results and patterns of failure. Cancer 48:2184–2190

Duncan W, Munro A (1987) The management of testicular seminoma: Edinburg 1970–1981. Br J Cancer 55:443–448

Dunphy CH, Ayala AG, Swanson DA et al. (1988) Clinical stage I non-seminomatous and mixed germ cell tumours of the testis. A clinicopathologic study of 93 patients on a surveillance protocol after orchiectomy alone. Cancer 62:1202–1206

Fossa SD, Ous S, Abyholm T, Norman N, Loeb M (1985) Post-treatment fertility in patients with testicular cancer: II. Influence of cisplatin-based combination chemotherapy and of retroperitoneal surgery on hormone and sperm cell production. Br J Urol 57:210–214

Fossa SD, Aass N, Kaalhus O (1988) Radiotherapy for testicular seminoma stage I: treatment results and long-term post irradiation morbidity in 365 patients. Int J Radiat Oncol Biol Phys 16:383–388

Fossa SD, Kreuser ED, Roth GJ, Raghavan D (1990) Long-term side effects after treatment of testicular cancer. In: Prostate Cancer and testicular cancer. Wiley-Liss, New York, pp 321–330 (EORTC Genitourinary Group monograph 7)

Freedman LS, Parkinson MC, Jones WG et al. (1987) Histopathology in the prediction of relapse of patients with stage I testicular teratoma treated by orchiectomy alone. Lancet 8:294–298

Gelderman WAH, Schraffordt Koops H, Sleijfer DT et al. (1987) Orchiectomy alone in stage I nonseminomatous testicular germ cell tumours. Cancer 59: 578–580

Gimmi G, Sonntag R, Brunner K, Studer U (1990) Adjuvant treatment of high risk (HR) clinical stage I testicular seminoma (TC) with cisplatin (C), bleomycin (B), and vinblastine (V) or etoposide (E). Proc Am Soc Clin Oncol 9:140 (abstr 542)

Hansen SW, Berthelsen JG, von der Maase H (1990) Long-term fertility and Leydig cell function in patients treated for germ cell cancer with cisplatin, vinblastine, and bleomycin versus surveilance. J Clin Oncol 8:1695–1698

Herman JG, Sturgeon J, Thomas FM (1983) Mediastinal prophylactic irradiation in seminoma. Proc Am Soc Clin Oncol 2:133

Horwich A, Dearnaley DP, Duchesne GM et al. (1989) Simple nontoxic treatment of advanced metastatic seminoma with carboplatin. J Clin Oncol 7:1150–1156

Horwich A, Dearnaley DP, Nicholls J, Jay G, Peckham MJ, Hendry WF (1990) Orchiectomy alone for stage I seminoma of the testis. Cancer 65/5:1115–1118

Horwich A, Dearnaley DP, Norman A, Hendry W (1992) Primary chemotherapy for stage II low volume non seminomatous germ cell tumours of the testis. Proc Am Soc Clin Oncol 11:(abstr 601)

Huben R, Williams P, Edson P, Panakan A, Murphy G (1984) Seminoma at Roswell Park 1970 to 1979: an analysis of treatment failures. Cancer 53:1451

Jewett MA, Herman JG, Sturgeon JFG et al. (1984) Expectant therapy for clinical stage A nonseminomatous germ cell testicular cancer? Maybe. World J Urol 2:57–58

Klepp O, Olsson AN, Henrikson H, Aass N, Dahl O, Stenwig AE, Persson BE, Cavallin Stahl E, Fossa SD, Wahlquist L (1990) Prognostic factors in clinical stage I nonseminomatous germ cell tumors of the testis: multivariate analysis of a prospective multicenter study. J Clin Oncol 8:509–518

Lederman GS, Herman TS, Jochelson M, Silver BJ, Chaffey JT, Garnick MB, Richie J, Sheldon ThA, Coleman CN (1989) Radiation therapy of seminoma: 17 years' experience at the Joint Center for Radiation Therapy. Radiother Oncol 14:203–208

Lester SG, Morphis JG, Hornback NB (1986) Testicular seminoma: analysis of treatment results and failures. Int J Radiat Oncol Biol Phys 12:353–358

Logothetis CJ, Swanson DA, Dexeus F, Chong C, Ogden S, Ayala AG, von Eschenbach AC, Johnson DE, Samuels ML (1987) Primary chemotherapy for clinical stage II nonseminomatous germ cell tumors of the testis: a follow-up of 50 patients. J Clin Oncol 5:906–911

Mason BR, Kearsley JH (1988) Radiotherapy for stage 2 testicular seminoma: the prognostic influence of tumor bulk. J Clin Oncol 6:1856–1862

Oliver RTD, Hope-Stone HF, Blandy JP (1984) Possible new approaches to the management of seminoma of the testis. Br J Urol 56:729–733

Oliver RTD, Lore Sh, Ong J (1990) Alternatives to radiotherapy in the management of seminoma. Br J Urol 65:61–67

Peckham MJ, Hendry WF (1985) Clinical stage II nonseminomatous germ cell testicular tumors. Results of management by primary chemotherapy. Br J Urol 57:763–768

Pflüger KH, Ulshöfer B, Rodeck G, Havemann K (1990) First line polychemotherapy in patients with stage II non-seminomatous germ cell tumors (NSGCT). Ann Oncol [Suppl 1]:16 (abstr P9)

Pizzocaro G, Pasi M, Zanoni F, Salvioni R, Milani A, Pilotti S (1985) Relapse pattern of pathologic stage I nonseminomatous germ cell tumors of the testis following orchiectomy and lymphadenectomy. Eur Urol 11:79–82

Pizzocaro G, Zanoni F, Salvioni R, Milani A, Piva L, Pilotti S (1987) Difficulties of a surveillance study omitting retroperitoneal lymphadenectomy in clinical stage I nonseminomatous germ cell tumors of the testis. J Urol 138:1393–1395

Pont J, Höltl W, Kosak D, Machacek E, Kienzer H, Julcher H, Honetz N (1990) Risk-adapted treatment choice in stage I nonseminomatous testicular germ cell cancer by regarding vascular invasion in the primary tumor: a prospective trial. J Clin Oncol 8/1:16–20

Raghavan D, Colls B, Levi J et al. (1988) Surveillance for stage I nonseminomatous germ cell tumors of the testis: the optimal protocol has not yet been defined. Br J Urol 61:522–526

Read G, Stenning SP, Cullen MH, Parkinson MC, Horwich A, Kaye SB, Cook PA (1991) Prospectice study of surveillance for stage I testicular non seminomatous germ cell tumors (NSGCT): a study by the medical research Council. Proc Am Soc Clin Oncol 10:165 (abstr 523)

Rorth M, von der Maase H, Nielssen ES et al. (1987) Orchiectomy alone versus orchiectomy plus radiotherapy in stage I nonseminomatous testicular cancer: a randomized study by the Danish Testicular Carcinoma Study Group. Int J Androl 10:225–262

Sagerman RH, Kotlove DJ, Regine WF, Chnung ChT, King GA, Dalai PS (1989) Stage II seminoma: results of postorchiectomy irradiation. Radiology 172: 565–568

Schmitt G, Sauerwein W, Bamberg M (1980) La radiotherapie des seminomes; resultats du centre universitaire de radiotherapie de Essen. J Eur Radiother 1:37–41

Schmoll H-J, Bokemeyer C, Harstrick A, Dieckmann KP et al. (1991) Single agent carboplatinum (CBDCA) for advanced seminoma: a phase II study. Proc Am Soc Clin Oncol 10:181 (abstr 586)

Smalley SR, Evans RG, Richardson RL, Farrow GM, Earrle JD (1985) Radiotherapy as initial treatment for bulky stage II testicular seminomas. J Clin Oncol 10:1333–1338

Socinski MA, Garnick MB, Stomper PC, Fung CY, Richie JP (1988) Stage II nonseminomatous germ cell tumours of the testis: an analysis of treatment in patients with low volume retroperitoneal disease. J Urol 140:1437–1441

Sogani PC, Fair WR (1988) Surveillance alone in the treatment of clinical stage I nonseminomatous germ cell tumor of the testis. Semin Urol 6/1:53–56

Sogani PC, Whitmore WF, Herr HW, Bosl GJ, Golbey RB, Watson RC, DeCosse JJ (1984) Orchiectomy alone in the treatment of clinical stage I nonseminomatous germ cell tumors of the testis. J Clin Oncol 2/4:267–270

Stoter G (1987) Treatment strategies of testicular cancer in Europe. Int J Androl 10:407–415

Swanson D, Johnson D, von Eschenbach A et al. (1987) Five years experience with orchiectomy (surveillance) for clinical stage I nonseminomatous germ cell testicular tumors. J Urol 137:211A

Thomas G, Rider W, Dembo A, Cummings B, Gospodarowicz M, Hawkins N, Herman J, Keen C (1982) Seminoma of the testis: results of treatment and patterns of failure after radiation therapy. Int J Radiat Oncol Biol Phys 8:165–174

Thomas G (1985) Controversies in the management of testicular seminoma. Cancer 55:2296–2302

Thompson PI, Nixon J, Harvey VJ (1988) Disease relapse in patients with stage I nonseminomatous germ cell tumors of the testis on active surveillance. J Clin Oncol 6:1597–1603

van der Werf-Messing B (1976) Radiotherapeutic treatment of testicular tumors. Int J Radiat Oncol Biol Phys 1:235–248

Weißbach L, Boedefeld EA, Oberdörster W, Vahlensieck W (1984) Therapy in stage I nonseminomatous testicular tumor – a critical review of current strategies. Eur Urol 10:1–9

Willan B, McGowan G (1985) Seminoma of the testis: a 22-year experience with radiation therapy. Int J Radiat Oncol Biol Phys 11:1769–1775

Williams SD, Birch R, Einhorn LH et al. (1987a) Treatment of disseminated germ cell tumors with cisplatin, bleomycin and either vinblastine or etoposide. N Engl J Med 316:1435–1440

Williams SD, Stablein DM, Einhorn LH (1987b) Immediate adjuvant chemotherapy vs. observation with treatment at relapse in pathologic stage II testicular cancer. N Engl J Med 317:1433–1438

Zagars GK, Babaian J (1987) Stage I testicular seminoma: rationale for postorchiectomy radiation therapy. Int J Radiat Oncol Biol Phys 13:155

*The Evolving Role of Chemotherapy of Germ Cell Tumors**

H.I. Scher, D.F. Bajorin, R.J. Motzer, and G.J. Bosl

Genitourinary Oncology Service, Department of Medicine, Division of Solid Tumor Oncology, Memorial Sloan-Kettering Cancer Center, New York, NY 10021, USA

The treatment of germ cell tumors has undergone a marked evolution over the past two decades. With the high rate of cure that is now possible even in advanced stages, research has focused increasingly on tailoring treatment to an individual patient. The aim is to select a treatment approach with the best chance of cure and the least morbidity, which preserves quality of life. For patients with germ cell tumors, the latter involves consideration of both the acute and chronic toxicities of chemotherapy and surgery.

The dramatic increase in complete response (CR) proportions, the prerequisite for cure, came with the development of cisplatin-based combination chemotherapy programs in the 1970s [1–3]. Subsequently, it was shown that maintenance chemotherapy did not improve the long-term outcome [4], that doxorubicin did not increase efficacy when used as part of induction chemotherapy programs [5], and that there was a lower limit of the cisplatin dose–response and survival curve [6].

The focus of clinical trial research shifted as it became apparent that a proportion of patients appeared to be incurable, while others were cured with minimal therapy. Further, as confidence in the long-term chemotherapy results increased, several groups began exploring "observation" policies for stage I disease, reserving chemotherapy only for those who relapsed [7]. As a result, more patients with early stage II and "marker-only" disease, with an inherently better prognosis, were enrolled on "advanced disease" protocols. This resulted in a "stage-migration" [8] and seemingly higher cure rates when, in fact, the outcomes had not changed. At Memorial Sloan-Kettering Cancer Center (MSKCC), a significant decrease in the pretherapy serum tumor marker levels, the number of metastatic sites, and an increase in the probability of achieving a complete remission (vide infra) was observed for patients treated between 1975 and 1982. This stage migration alone resulted in an increase in an apparent increase in the complete response

* This work was supported by Grant CA-05826.

proportions for patients treated on sequential protocols. However, when patients on each protocol were stratified for the probability of complete remission, no difference in CR proportion was observed between protocols [9]. A similar observation was reported by the Medical Research Council (MRC) [10].

The aims of clinical trials evolved further to reflect divergent patient populations. For earlier staged patients with a high probability of cure, the aim became reduction of toxicity without compromising cure, while for more advanced patients, improving cure rates with alternative treatment strategies became the focus. Pivotal studies were completed on the role of adjuvant therapy [11], etoposide was introduced [12,13], and the minimum number of cycles needed for cure defined [14]. The role of bleomycin as induction therapy was also being questioned [15]. In addition, for patients who relapse, curative second-line therapies were developed that included ifosfamide-based combinations and autologous bone marrow transplantation. Surgical management has also changed. More limited retroperitoneal lymph node dissections are being performed which have fewer adverse effects on potency and ejaculatory function [16]. The role of surgery after chemotherapy is likewise evolving [17].

The results for all patients with germ cell tumors treated with chemotherapy can be summarized as follows: 70%–80% of patients treated with cisplatin-based combination chemotherapy and surgical resection of all residual disease achieve a complete remission. Of these, 8%–10% are destined to relapse and require salvage treatment. A consistent 20%–30% do not achieve a complete remission from initial therapy and invariably die of disease. This section will focus on the changing role of chemotherapy in the management of germ cell neoplasms.

Risk Assignment and Prognostic Factors

Critical to clinical investigations is the ability to define prognosis for complete remission and subsequent cure based on pretherapy characteristics. In the 1990s, the goal of therapy for patients with a "good" and "poor" prognosis differs. In the former, where CR proportions are high, therapy is directed towards minimizing the toxicity of treatment; while in the latter, for whom response is unlikely, an increase in cure fraction is the primary goal. The need to standardize the definitions can not be overemphasized.

Most reported criteria include an assessment of tumor bulk by incorporating some combination of the following: the size of the metastatic lesion(s), the degree of marker elevation, the number of metastatic sites and/or the presence or absence of visceral involvement. Criteria such as the site of primary tumor and histopathology have also been included [18]. Prognosis is then defined on the basis of a clinical grouping of defined disease characteristics or based on statistical modelling. The three classifications minimal,

Table 1. Indiana classification of germ cell tumors. (From [19])

Minimal
1. Elevated markers after retroperitoneal lymph node dissection as only evidence of disease
2. Cervical ±retroperitoneal nodes
3. Unresectable nonpalpable retroperitoneal disease
4. Minimal pulmonary metastases:
   A. <5 per lung field
   B. Largest <2 cm in diameter

Moderate
5. Palpable abdominal mass (or mass >10 cm on abdominal CT scan) with no supradiaphragmatic disease
6. Moderate pulmonary metastases:
   A. 5–10 per lung field with largest <3 cm
   B. Or solitary mass of any size
   C. Or mediastinal adenopathy <50% intrathoracic diameter

Advanced
7. Advanced pulmonary metastases
   A. Mediastinal mass >50% of intrathoracic diameter
   B. More than 10 pulmonary metastases/lung field
   C. Pulmonary metastases with largest >3 cm (±nonpalpable retroperitoneal disease, ±cervical nodes)
8. Palpable abdominal mass plus pulmonary metastases
9. Hepatic, osseous or CNS metastases

moderate, and advanced used by the group at Indiana University [19] are an example of the former. These criteria are listed in Table 1.

Alternatively, multivariable statistical techniques can be used. The specific methodology employed varies with the desired endpoint. For example, when considering the dichotomous endpoint of a CR vs. an incomplete response (IR) to therapy, logistic regression is utilized. The CR considered may be to chemotherapy alone (CT), to chemotherapy plus surgical resection of residual disease (CT + S) or the total number of complete responses (CT + CT + S). At MSKCC, patients with residual teratoma in post-chemotherapy surgical specimens are considered complete responders to CT. However, if the endpoint of interest is the time to an event that has not occurred for an individual patient, such as the time to relapse or survival, the Cox proportional hazards model ("Cox regression") is used. Endpoints considered can include time to relapse, disease-free survival, overall survival, or event-free survival. In some cases, patients who relapse can be rendered disease-free with salvage therapy. In these cases, the time to relapse may suggest an inferior outcome, while disease-free survival is not jeopardized. This is important when considering patients treated on observation protocols for stage I disease [7] or in the adjuvant setting following retroperitoneal lymph node dissection for stage II disease [11]. Event-free survival was defined as an endpoint to increase the statistical power of treatment comparisons. In this type of analysis, for example, the number of relapses and deaths are combined to give a higher number of total events.

At MSKCC, a multivariable analysis of factors predictive of complete remission was performed on 171 patients with germ cell tumors. Three pretherapy variables were shown to be significant: the initial lactate dehydrogenase level (LDH); the human chorionic gonadotropin level (HCG); and the total number of metastatic sites. A mathematical equation was derived to predict the probability of achieving a CR to CT + CT + S and was defined as:

$$\text{probability of CR} = \text{EXP H}/(1 + \text{EXP H})$$

where $H = 8.514 - 1.973 \log(LDH + 1) - 0.53 \log(HCG + 1) - 1.111$ TOTMET. TOTMET has a value of 0, 1, or 2 depending upon the number of sites of metastases. The value 0 is for patients who have marker elevations as the only manifestation of disease, while 2 refers to patients with two or more metastatic sites. Patients with a probability of CR greater than 0.50 are considered to have a good prognosis and are prospectively treated on trials designed to minimize toxicity. Patients with a probability of CR less than 0.50 are considered to have a poor prognosis and are allocated to trials designed to improve efficacy [20]. For these, the primary outcome is the total number of CR to C + CT + S. All other responses are classified as incomplete. This mathematical model pertains only to patients with nonseminomatous neoplasms of gonadal origin. Patients with extragonadal nonseminomatous germ cell tumors are considered to have a poor prognosis, while patients with pure seminoma are classified in a good risk category.

The unique features of the MSKCC model are (1) the use of serum markers as continuous as opposed to discrete variables, (2) the absence of a discrete size of a lesion such as ≥10 cm vs. <10 cm to discriminate a "poor" vs. "good" prognosis patient, and (3) that an individual site of disease alone such as liver or brain metastates, does not result in a poor risk classification. This model has been used to prospectively assign patients for chemotherapy protocols since 1982.

The criteria used to assign risk differ substantially among various institutions and the specific criteria utilized can influence reported results. In a study comparing the criteria used by MSKCC, the Indiana University, the National Cancer Institute and the EORTC, to classify 118 patients treated with germ cell tumors, concordance in the classification of poor risk was observed in only 56% of cases. Depending on the criteria used, 31%–72% of the population were classified as poor risk, while CR proportions (CT + CT + S) ranged from 38% to 62%, depending on the criteria utilized [18]. Similar differences were seen in the survival distributions. In addition, a wide range in CR proportions was noted for patients for whom risk classifications did not agree. Thus, patients considered poor risk by less stringent criteria but good risk by more rigorous criteria had CR rates as high as 87%. The study also demonstrated that the classification of patients with a good prognosis in a poor prognosis category could lead to the conclusion that a particular therapy was more beneficial with respect to a higher overall response proportion and improved survival when, in fact, no improvement existed.

Several groups have reported prognostic models and prior models are continually being refined. The MRC bases treatment allocation on the basis of discrete values of tumor markers and tumor bulk. Three prognostic groups were defined with 3-year survivals of 91%, 74% and 47%, respectively [21]. The EORTC developed a model using specific disease sites, tumor markers and pathologic features [22].

The Indiana group has updated their original classification to incorporate serum tumor markers [19]. Until these criteria have been standardized, randomized comparisons will be required to show the superiority of one treatment regimen over another.

More recently, the prognostic model in use for patients with testicular primaries at MSKCC was evaluated on poor risk patients and patients with extragonadal primaries. The results showed a difference in prognosis for patients with mediastinal and retroperitoneal primaries with CR proportions in 38% and 61% of cases, respectively [23]. The model [20] was poorly predictive of outcome for patients with extragonadal primaries. Using Cox regression, the presence of brain or live metastases were shown to be predictive of subsequent event-free survival [22].

Clinical Trials

Toxicity Reduction and Good Risk Disease

The adverse effects of chemotherapy programs employing combinations of cisplatin, bleomycin and vinblastine include myelosuppression, intestinal ileus, Raynaud's phenomenon, pulmonary fibrosis, nephrotoxicity and second primary tumors [1–3,24–27]. With the demonstration of activity for etoposide (VP-16-213) in relapsed patients [12,13,28], several groups undertook randomized comparisons to evaluate the role of this agent as primary therapy.

The Southeast Cancer Study Group (SECSG) compared four cycles of cisplatin, vinblastine and bleomycin (PVB) to four cycles of bleomycin, etoposide and cisplatin (BEP). Extent of disease was classified as previously described [19]. A total of 244 patients, 121 on PVB and 123 on BEP, were matched with respect to primary site and extent of disease. Considering tumors of testicular origin, 80% and 84% responded. Equivalent response proportions were observed for patients with minimal and moderate disease, while there was a trend toward a higher CR proportion for advanced disease patients, 30% vs. 63%, respectively, treated with the etoposide regimen

[29]. Overall, 95% with minimal, 81% with moderate and 50% with advanced disease achieved CR. The toxicity was significantly less in the etoposide arm with respect to paresthesias ($p = 0.02$), myalgias ($p = 0.0002$), and abdominal cramps ($p = 0.008$). The degree of myelosuppression was similar between the two groups. Based on this study, BEP became the new standard regimen. In an attempt to further reduce toxicity, the group subsequently compared four vs. three cycles of BEP in patients with a favorable (minimal or moderate disease) prognosis. At last follow-up, 92% of patients treated with 3% and 92% of patients treated with four cycles were continuously disease free [30]. As expected, toxicities were significantly reduced by deletion of the fourth cycle.

At MSKCC, to evaluate a less toxic regimen in patients with good risk disease [20], a randomized comparison of the five-drug regimen, VAB-6, which contains cyclophosphamide, dactinomycin, bleomycin, vinblastine and cisplatin and the two-drug regimen of etoposide and cisplatin (EP) was performed. The study was based on prior experience with VAB-6 [31], showing efficacy but with substantial nephrotoxicity, nausea, emesis, and myelosuppression, and the activity and toxicity profile of EP as front-line [32] and as salvage therapy in a relapsed population [13]. One-hundred and sixty-four patients randomized to receive either four cycles of EP or three cycles of VAB-6 were matched with respect to the probability of achieving a CR. Overall a CR to CT was observed in 68 and 72 cases, respectively, and to CT + S in 11 and four cases, for an overall CR proportion 96% (79/82) and 93% (76/82), respectively. IR were observed in three and six cases, while a relapse from CR was documented in nine (11%) and nine (12%) of patients treated [33].

While the proportion of patients who required surgery to resect residual disease did not differ, fewer patients in the EP arm had residual viable germ cell tumor (8%) compared to the VAB-6 regimen (20%). Toxicities were significantly lower for the etoposide and cisplatin arm with respect to the granulocyte nadir ($p = 0.006$), degree of thrombocytopenia ($p = 0.01$), nadir magnesium level ($p = 0.0001$) and quantitative emesis. No drug-induced deaths were observed and relapse-free and event-free survival distributions similar between the two arms similar [33]. In a subsequent report of the EP-treated patients after a median follow-up of 61 months, nine (11%) relapses were observed between 6 and 17.5 months, of whom 3 were salvaged. No late relapses were observed. Overall, 71 patients (87%) remained disease-free [34]. Thus, at MSKCC, four cycles of EP is considered standard therapy for good risk patients.

In an attempt to further reduce toxicity in patients with good risk disease, a randomized comparison of EP and E + the platinum analog carboplatin (EC) was recently reported. The rationale was based on the reported activity [35,36] and toxicity profile of carboplatin. Further, the ability to administer carboplatin on an outpatient basis offered the potential for a significant reduction in cost. Two hundred sixty three patients were randomized to receive four cycles of EP given every 3 weeks, or EC given every 4 weeks. Treatment groups were balanced with respect to the median elevations of LDH, HCG and alpha-fetoprotein (AFP), number of metastatic sites and probability of CR. With 192 evaluable patients, 95 or EP and 97 on EC, efficacy was equivalent although the rate of hematologic toxicity was significantly higher in the EC arm [37]. However, with longer follow-up, a higher proportion of relapses was observed in the EC arm [38]. It was therefore concluded that EC could not replace EP as standard therapy for good risk patients.

The role of bleomycin in patients with good risk disease, for whom long-term toxicities are of great importance, has been questioned [15]. In addition to acute toxicities such as a bleomycin-induced pneumonitis, a reduction in vital capacity or diffusion capacity has been observed in approximately 20% of cases treated with agent [39]. Cardiovascular complications such as Raynaud's phenomenon [24], myocardial infarction and hypertension [39] have been observed. Based on the reported comparison of VAB-6 and EP, bleomycin has not been utilized for induction therapy in good risk patients at MSKCC.

The first trial comparing BEP and EP was reported by the Royal Marsden in patients with "low volume" disease [32]. The CR proportion was lower using the two- vs. the three-drug regimen, 75% (12/16) vs. 89% (16/18) respectively. The study results were questioned because of the small number of patients treated, and because the total etoposide dose was only 360 mg/m^2 as opposed to the more conventional 500 mg/m^2 that is utilized more commonly.

More recently, the Eastern Cooperative Oncology Group (ECOG) reported a comparison of three cycles of BEP, the previously defined standard [30], to three cycles of EP in favorable prognosis (minimal or moderate disease) patients using the Indiana Classification. The dose of etoposide was 500 mg/m^2 per cycle. A total of 166 evaluable patients, 111 with minimal and 55 with moderate disease, were evaluated. The study was terminated prior to a scheduled interim look due to a higher proportion of adverse events defined as either relapse, persistent or progressive carcinoma, or death, in the EP arm. Overall 31.3% (26/83) of patients on EP and 15.6% (13/83) on BEP had adverse events ($p = 0.03$). Toxicities were similar. The authors concluded that bleomycin remains an essential component of therapy for favorable prognosis patients [40].

While the previous two studies would appear to reach opposite conclusions with regard to the role of bleomycin, a direct comparison is not possible for several reasons. First, the classification schemes are not directly comparable. Second, the number of chemotherapy cycles differs between the two programs. The studies, however, do suggest that there is a minimum amount of chemotherapy that must be administered to patients with a "good risk" or "favorable prognosis" without compromising cure: four cycles of EP or three cycles of BEP.

The Southwest Oncology Group (SWOG) recently reported a trial comparing two three-drug regimens – PVB and the three-drug combination VP-16, cisplatin and vinblastine (VPV). Of 160 patients treated, 46 had minimal and 116 had maximal disease using SWOG group criteria. Patient demographics were comparable between the two arms. No difference in the proportions of patients achieving disease-free status was noted, 77% for PVB and 73% for VPV. However, a significantly lower incidence of mucosal ($p < 0.001$), neurologic ($p = 0.006$), dermatologic ($p < 0.001$) and pulmonary toxicities ($p < 0.001$) was observed in favor of the VPV regimen. The authors concluded that bleomycin could be substituted with VP-16 in this three-drug regimen [41].

Poor Risk Disease

The controversies regarding categorization of patients as poor risk has been previously discussed. As a result, reported "improvements" in single arm studies must be considered in the context of the differences in risk definitions and interpreted cautiously [18,19,42,43]. At MSKCC, nonseminomatous testicular cancer patients with a probability of CR less than 0.5 and all those with extragonadal nonseminomas are treated on poor risk protocols. The results of three sequential studies are outlined in Table 2. The first evaluated patients treated on VAB-6 alone, the second six cycles of the potentially non-cross-resistant VAB-6 alternating with EP [44], and the third utilized an induction regimen with carboplatin/etoposide/bleomycin [45]. Overall, no difference in the overall CR proportion, durability of CR or overall survival was observed between the three studies.

The SWOG and SECSG group trial evaluated 157 patients in a randomized comparison of high ($40\,mg/m^2$ per day $\times$ 5) vs. conventional ($20\,mg/m^2$ per day $\times$ 5) dose cisplatin plus VP-16 and bleomycin. Of 141

Table 2. Poor risk trials at MSKCC

	VAB-6 [31]	VAB-6/EP [44]	EBC [45]
Patients (n)	29	39	32
Primary site (%)			
testis	62	67	72
extragonadal	28	33	28
Response (%)			
complete	43	51	57
durable	29	33	44

Regimens: VAB-6, vinblastine, cisplatin, bleomycin, cyclophosphamide plus dactinomycin; EP, etoposide plus cisplatin; EBC, etoposide plus bleomycin plus carboplatin.

evaluable patients, 68 on high and 73 on conventional, disease-free status was achieved in 44 (65%, 33 CT, 10 with resection of residual teratoma, and 1 to CT + S) and 51 (70%, 34 CT, 12 with resection of residual teratoma, and 5 to CT + S) cases, respectively. The number of relapses, two (5%) and four (8%), the proportion of patients continuously disease free, 57% and 62%, and overall disease-free survival, 72% and 75%, respectively, were comparable between the two groups. Toxicities were significantly higher in the high dose arm. This study shows that there is a plateau in the dose–response curve for cisplatin in this disease and that escalations of cisplatin are unlikely to improve outcome [46].

As risk classification becomes standardized, single arm phase II trials can be utilized to provide leads for investigation in the phase III setting. Only the latter methodology can definitively prove a survival benefit for one approach over another. However, as a finite proportion of patients with poor risk disease are cured using standard therapy, an alternative approach is to use the posttreatment decline in serum tumor markers to select those patients destined to fail primary therapy. This will be discussed below.

Salvage Therapy

Patients who are referred for salvage therapy also represent a heterogeneous group. They include patients who have relapsed from a previous complete remission and those who never achieved CR status. Within the former, the time to relapse has been shown to be a significant predictor for subsequent response. Patients who have never achieved a CR include those with an unresectable tumor in partial remission – residual abnormalities on CT scan that could not be resected – with or without elevated markers. The number of prior cisplatin containing regimens, 1 or 2 or more, can also influence results. Fortunately, the proportion of patients who are primarily refractory to all chemotherapy is small.

The initial trials of salvage chemotherapy demonstrated the role of EP in patients who had relapsed from VAB-6 or PVB like programs. At MSKCC, of 17 patients who relapsed from a prior CR, 8 (47%) achieved a second CR, compared to 0 of 28 patients in a refractory group who had not achieved a previous CR [13]. The results did not change in a updated report of 83 patients with relapsed vs. refractory disease, where CR was observed in 42% vs. 8%, while 3-year survivals of 36% vs. 8%, respectively [47]. Thus, cure can be achieved using second line therapy.

More recently, the cyclophosphamide analog ifosfamide, an oxazaphosphorine, has been evaluated. Early clinical use was limited by the development of hemorrhagic cystitis, due to the metabolism of the drug to acrolein and chloroacetic acid. Interest was renewed with the demonstration that thiol compounds such as *N*-acetylcysteine and mesna could ameliorate this toxicity [47,48]. Using *N*-acetylcysteine, the incidence of hematuria was

30% (24/80) vs. 2.3% (5/149) using mesna [49]. Thus mesna is now used as standard therapy.

In untreated patients, Schmoll et al. oberved 2 CR and 13 PR in 20 patients treated with 60 mg/kg for 3 consecutive days [50]. Subsequently, a phase II trial at Indiana University was performed in 30 patients with refractory disease. Using 2 g/day × 5 days, 23% (one CR and six PR) responded [51].

Ifosfamide has subsequently been evaluated in combination with cisplatin and VP-16. At Indiana University, three groups of patients were reported on separately: (A) patients with an unresectable tumor in partial remission as the best response to induction chemotherapy, (B) patients who achieved a previous CR but who relapsed within 2 months, and (C) patients who relapsed from primary therapy and who had subsequently progressed on second line cisplatin and etoposide [52]. Twenty-seven percent (6/22) of group A, 43% (3/7) of group B and 37% (7 of 19) of group C attained a CR, seven of whom remained disease free. Toxicities, primarily myelosuppression, were significant in this heavily pretreated population.

Additional reports have utilized ifosfamide and cisplatin in combination with either VP-16 or vinblastine, depending on the induction therapy used. Patients previously treated with VP-16 receive vinblastine and vice versa. The results from three centers evaluating only patients who did not achieve a complete remission, or IR patients showed the combination to be potentially curative in a small, but definite, proportion of patients. As shown in Table 3, CR was noted in 12 of 35 (34%) [53], 8 of 29 (28%) [54] and 10 of 40 (25%) of cases [55]. Of greater import was that the CR's were durable in 26%, 10% and 15% of cases, respectively. No responses were observed in six patients primarily refractory to PEB or PVB [56]. While toxicities were significant, the results show that durable CR could be achieved in 17% (18/110) of refractory patients. Currently, this combination is being evaluated in combination with a hematopoietic growth factor in an attempt to ameliorate the myelosuppression of this regimen in heavily pretreated patients.

Table 3. VIP/VelP as salvage therapy

	Indiana[a] University [53]	Institut Gustave-Roussy [54]	MSKCC [55]	Institut Tumori [56]
Patients (*n*)	35	33	42	6
Evaluable patients (*n*)	35	29	40	6
CR (*n*)	12	8	10	0
Durable (*n*)	9 (26%)	3 (10%)	6 (15%)	0

[a] Considers only patients with unresectable PR or incomplete responders to treatment (Groups A and C, see text).

A second approach involves an exploration of the dose–response curve of carboplatin using autologous bone marrow transplantation. In a dose-escalation study a fixed $400\,\text{mg/m}^2 \times 3$ dose of VP-16 was used with escalating doses of carboplatin. The dose of carboplatin ranged from 900 to $2000\,\text{mg/m}^2$ per cycle. In a heavily pretreated population with refractory disease, the dose-limiting toxicity was enterocolitis and hepatotoxicity. All patients developed granulocytopenic fever and seven (21%) died from treatment. The recommended phase II dose was $1500\,\text{mg/m}^2$. Of 32 evaluable patients, 8 achieved a CR which was maintained in 4 cases [57]. In an updated experience of 40 patients followed for a minimum of 2 years, 8 (20%) were alive free of disease [58].

An alternative approach currently under study at MSKCC utilizes the decline in serum tumor markers for early prediction of patients destined to fall short of a complete remission, a prerequisite for cure. These patients can then be changed to high-dose chemotherapy with autologous marrow support earlier in their treatment course. The serum half-life of HCG is approximately 30 h, while APH has a half-life of approximately 5 days. A posttreatment decline by half-life has been shown to correlate with both the attainment of a complete remission to CT or CT + S, and for subsequent relapse from CR. In a retrospective analysis of changes in marker levels, patients with an abnormal serologic response did poorly. For example, considering either AFP or HCG, only 1 of 9 (11%) patients with a prolonged decline in marker levels achieved a complete response to CT in contrast to 97 of 109 (89%) who showed satisfactory declines in half-life ($p < 0.001$). Similarly 8 of 12 (67%) CR with prolonged half-lives relapsed in contrast to 18 of 139 (13%) with satisfactory declines ($p < 0.0001$) [59].

Previously, experience has shown that approximately one third of patients with poor risk disease based on pretreatment characteristics can be cured of their disease with standard chemotherapy [30,44,45]. These patients show an appropriate decline in serum tumor markers post of chemotherapy. By determining marker levels at weekly intervals, nonresponders are identified early and referred directly for a treatment program designed to maximally exploit the dose–response curve for carboplatin and etoposide. In a preliminary report using this trial methodology, 23 patients were treated with VAB-6 for two cycles with weekly tumor markers. Those with prolonged half-lives, more than 3 days for HCG or 7 days for AFP, were then treated with high-dose carboplatin, etoposide and autologous bone marrow rescue. Of 16 evaluable cases, 3 (19%) achieved a CR to VAB-6 alone, while 6 (38%) became CR after transplant. All nine CR patients were free of disease for 6–24+ months. Three were no treatment-related deaths. While additional patients are being enrolled, the preliminary results suggest that a higher proportion of patients with poor risk disease achieve CR status using this approach [60]. The current treatment policy is to administer two cycles of high-dose therapy with bone marrow rescue.

Future Directions

The treatment results of patients with germ cell tumors remains one of the most gratifying in clinical oncology. With the increased understanding of the natural history of the disease in all stages, treatments are being tailored more and more to the individual patient and his prognosis. Those with good risk disease are receiving effective but less toxic chemotherapy, while those with poor risk disease are being offered salvage therapy earlier in the course of the illness, where both the chance of success is increased and patient tolerance improved.

Future developments must focus on the standardization of prognostic factors, to facilitate comparisons between treatment groups. In addition, the models in use need to be expanded to incorporate biologic parameters that have been or have yet to be identified. For example, the number of copies of the chromosomal abnormality i(12p) found in over 90% of patients with germ cell tumors was recently shown to have prognostic significance [61]. The treatment policies for postchemotherapy surgery need to be standardized [17], particularly for patients with poor prognosis disease to accurately assess the impact of a chemotherapy program on response. Further, as more is understood of the biology of these tumors, novel strategies such as growth factor inhibition [62], or differentiation through use of retinoic acid derivatives [63], can be explored. The clinical advances in the managment of germ cell tumors were realized through the design, execution, and completion of carefully designed clinical trials. The importance of continued participation in these protocols can not be overemphasized.

References

1. Einhorn LH, Donohue JP (1977) Combination chemotherapy with cisdiamminedichloroplatinum, vinblastine, and bleomycin in disseminated testicular cancer. Ann Intern Med 87:283–298
2. Stoter G, Vendrik CPJ, Struyvenberg A et al. (1979) Combination chemotherapy with cis-diamminedichloroplatinum, vinblastine, and bleomycin in advance testicular non-seminoma. Lancet 1:941–945
3. Vugrin D, Herr HW, Whitmore WF et al. (1981) VAB-6 combination chemotherapy in disseminated cancer of the testis. Ann Intern Med 95:59–61
4. Einhorn LH, Williams SD, Troner M et al. (1981) The role of maintenance therapy in disseminated testicular cancer. N Engl J Med 305:727–731
5. Einhorn LH (1981) Testicular cancer as a model for a curable neoplasm: the Richard and Hinda Rosenthal Foundation award lecture. Cancer Res 41: 3275–3280
6. Samson M, Rivkin S, Jones S et al. (1984) Dose-response and dose-survival advantage for high- versus low-dose cisplatin combined with vinblastine and bleomycin in disseminated testicular cancer. Cancer 53:1029–1035
7. Peckham MJ, Brada M (1987) Surveillance following orchidectomy for stage I testicular cancer. Int J Androl 10:247–254

8. Feinstein AR, Sosin DM, Wells CK (1985) The Will Rogers phenomenon. Stage migration and new diagnostic techniques as a source of misleading statistics for survival in cancer. N Engl J Med 312:1604–1608
9. Bosl GJ, Geller NL, Chan EYW (1988) Stage migration and the increasing proportion of complete responders in patients with advanced germ cell tumors. Cancer Res 48:3524–3527
10. Horwich A, Eastan M, Husband J et al. (1987) Prognosis following chemotherapy for metastatic malignant teratoma. Br J Urol 59:578–583
11. Williams SD, Stablein DM, Einhorn LH et al. (1987) Immediate adjuvant chemotherapy versus observation with treatment at relapse in pathological stage II testicular cancer. N Engl J Med 317:1433–1438
12. Fitzharris EM, Kaye SB, Saveryuttu S et al. (1980) VP-16-213 as a single agent in advanced testicular tumors. Eur J Cancer 16:1193–1197
13. Bosl GJ, Yagoda A, Golbey RB et al. (1985) Role of etoposide-based chemotherapy with refractory or relapsed germ cell tumors. Am J Med 78: 423–428
14. Einhorn LH, Williams S, Loehrer P et al. (1989) Evaluation of optimal duration of chemotherapy in favorable-prognosis disseminated germ cell tumors: a Southeastern Cancer Study Group protocol. J Clin Oncol 7:387–391
15. Bosl GJ (1991) Management of germ cell tumors: toxicity reduction and the use of bleomycin. In: Horwich A (ed) Testicular cancer. Chapman and Hall, London
16. Foster RS, Donohue JP, Bihrle R (1991) Stage A non-seminomatous testis carcinoma: rationale and results of nerve-sparing retroperitoneal lymphadenectomy. Urol Int 46:294–297
17. Toner GC, Panicek DM, Heelan RT et al. (1990) Adjunctive surgery after chemotherapy for nonseminomatous germ cell tumors: recommendations for patient selection. J Clin Oncol 8:1683–1694
18. Bajorin D, Katz A, Chan E et al. (1988) Comparison of criteria for assigning germ cell tumor patients to "good risk" and "poor risk" studies. J Clin Oncol 6:786–792
19. Birch R, Williams S, Cone A et al. (1986) Prognostic factors for favorable outcome in disseminated germ cell tumors. J Clin Oncol 4:400–407
20. Bosl GJ, Geller NL, Cirrindane C et al. (1983) Multivariate analysis of prognostic variables in patients with metastatic testicular cancer. Cancer Res 43:3404–3407
21. Horwich A, Brada M, Nicholls J et al. (1989) Intensive induction chemotherapy for poor risk non-seminomatous germ cell tumors. Eur J Cancer Clin Oncol 25:177–184
22. Stoter G, Sylvester R, Sleijfer DT et al. (1987) Multivariate analysis of prognostic factors in patients with disseminated nonseminomatous testicular cancer: results from a European Organization for Research and Treatment of Cancer multinstitutional phase III study. Cancer Res 47:2714–2718
23. Toner GC, Geller NL, Lin SY, Bosl GJ (1991) Extragonadal and poor risk nonseminomatous germ cell tumors: survival and prognostic features. Cancer 67:2049–2057
24. Vogelzang N, Bosl GJ, Johnson K et al. (1981) Raynaud's phenomenon: a common toxicity after combination chemotherapy for testicular cancer. Ann Inter Med 95:288–292
25. Bosl GJ, Leitner SP, Atlas SA et al. (1986) Increased plasma renin and aldosterone in patients treated with cisplatin-based chemotherapy for metastatic germ-cell tumors. J Clin Oncol 4:1684–1689
26. Redman JR, Vugrin D, Zrlin ZA et al. (1984) Leukemia following treatment of germ cell tumors in men. J Clin Oncol 2:1080–1087

27. Coleman MP, Bell CMJ, Fraser P (1987) Second primary malignancy after Hodgkin's disease, ovarian cancer and cancer of the testis: a population based cohort study. Br J Cancer 56:349–355
28. Williams SD, Einhorn LH, Greco FA et al. (1980) VP-16-213 salvage therapy for refractory germinal neoplasms. Cancer 46:2154–2158
29. Williams SD, Birch R, Einjorn LH et al. (1987) Treatment of disseminated germ-cell tumors with cisplatin, bleomycin and either vinblastine or etoposide. N Engl J Med 316:1435–1440
30. Einhorn LH, Williams SD, Loehrer PJ et al. (1989) Evaluation of optimal duration of chemotherapy in favorable-prognosis germ cell tumors: a Southeastern Cancer Study Group protocol. J Clin Oncol 7:387–391
31. Bosl GJ, Gluckman R, Geller NL et al. (1986) VAB-6: an effective chemotherapy regimen for patients with germ cell tumors. J Clin Oncol 4: 1493–1499
32. Peckham MK, Horwich A, Blackmore C, Hendry EF (1985) Etoposide and cisplatin with or without bleomycin as first-line chemothrapy for patients with small-volume metastases of testicular nonseminoma. Cancer Treat Rep 69: 483–488
33. Bosl GJ, Geller, NL, Bajorin D et al. (1988) A randomized trial of etoposide + cisplatin versus vinblastine + bleomycin + cisplatin + cyclophosphamide + dactinomycin in patients with good-prognosis germ cell tumors. J Clin Oncol 6:1231–1238
34. Bajorin DF, Geller NL, Bosl GH (1991) Two drug therapy in patients with metastatic germ cell tumors: long term follow-up. Cancer 67:28–32
35. Horwich A, Duchesne G, Dearnaley D et al. (1988) Single agent carboplatin (SAC) as initial therapy for advanced seminoma. Proc Am Soc Clin Oncol 7:117
36. Motzer RJ, Bosl GJ, Tauer K et al. (1987) Phase II trial of carboplatin in patients with germ cell tumors refractory to cisplatin. Cancer Treat Rep 71: 197–198
37. Bajorin DF, Sarosdy MF, Bosl GJ et al. (1991) A randomized trial of etoposide + carboplatin (EC) vs. etoposide + cisplatin (EP) in patients (PTS) with metastatic germ cell tumors (GCT). Proc Am Soc Clin Oncol 10(535):168
38. Bajorin DF (1991) A randomized trial of etoposide + carboplatin (EC) vs. etoposide + cisplatin (EP) in patients (PTS) with metastatic germ cell tumors (GCT). Presented at the American Society of Clinical Oncology, Houston, Texas, May
39. Stoter G, Koopman A, Vendrik CPJ et al. (1989) Ten-year survival and late sequelae in testicular cancer patients treated with cisplatin, vinblastine, and bleomycin. J Clin Oncol 7:1099–1104
40. Loehrer PJ, Elson P, Johnson DH et al. (1991) A randomized trial of cisplatin (P) plus etoposide (E) with or without bleomycin (B) in favorable prognosis disseminated germ cell tumors (GCT): an ECOG study. Proc Am Soc Clin Oncol 10(540):169
41. Wozniak AJ, Samson MK, Shah NT et al. (1991) A randomized trial of cisplatin, vinblastine, and bleomycin versus vinblastine, cisplatin, and etoposide in the treatment of advanced germ cell tumors of the testis: a Southwest Oncology Group Study. J Clin Oncol 9:70–76
42. Pizzocaro G, Zanoni F, Salvioni R et al. (1985) High complete remission (CR) rate with cisplatin, etoposide, bleomycin (PEB) and surgery in patients with poor prognosis germ cell tumors (GCT) of the testis. Proc Am Soc Clin Oncol 4:110
43. Ozols RF, Ihde DC, Lindchan WM et al. (1988) A randomized trial of standard chemotherapy vs. a high-dose chemotherapy regimen in the treatment of poor prognosis nonseminomatous germ-cell tumors. J Clin Oncol 6:1031–1040

44. Bosl GJ, Geller NL, Vogelzang NJ et al. (1983) Alternating cycles of etoposide plus cisplatin and VAB-6 in the treatment of poor-risk patients with germ cell tumors. J Clin Oncol 5:436–440
45. Motzer RJ, Cooper K, Geller NL et al. (1990) Carboplatin + etoposide + bleomycin for patients with poor risk germ cell tumors. Cancer 65:2465–2470
46. Nichols CR, Williams S, Loehrer P et al. (1991) Randomized study of cisplatin dose intensity in poor-risk germ cell tumors: a Southeastern Cancer Study Group and Southwest Oncology Group protocol. J Clin Oncol 9:1163–1172
47. Motzer R, Geller N, Bajorin D et al. (1990) Salvage chemotherapy (CT) for germ cell tumor (GCT) patients (PTS): prior complete response (CR) as an important prognostic variable. Proc Am Soc Clin Oncol 9(548):141
48. Brock N, Hilgard P, Pohl J et al. (1984) Pharmacokinetics and mechanisms of action of detoxifying low-molecular-weight thiols. J Cancer Res Clin Oncol 108:87–97
49. Loehrer PJ (1990) Ifosfamide in testicular cancer. Semin Oncol 17 [Suppl 4]:2–5
50. Schmoll HJ (1982) The role of ifosfamide in testicular cancer. Proceedings of the 13th international cancer congress 13:626
51. Wheeler BM, Loehrer PJ, Williams SD et al. (1984) Ifosfamide in refractory male germ cell tumors. J Clin Oncol 4:28–34
52. Loehrer PJ, Einhorn LH, Williams SD (1986) VP-16 plus ifosfamide plus cisplatin as salvage therapy in refractory germ cell cancer. J Clin Oncol 4: 528–536
53. Munshi NC, Loehrer PJ, Roth BJ et al. (1990) Vinblstine, ifosfamide and cisplatin (VeIP) as second line chemotherapy in metastatic germ cell tumors (GCT). Proc Am Soc Clin Oncol 9(520):134
54. Droz JP, Ribrag V, Ghosn M et al. (1990) Phase II trial with etoposide (VP16) plus ifosfamide plus high-dose cisplatin (VIhP regimen) in refractory germ cell tumors. In: Murphy G, Khoury S (eds) Therapeutic progress in urological cancers. Liss, New York, pp 739–747
55. Motzer RJ, Cooper K, Geller NL et al. (1990) The role of ifosfamide plus cisplatin-based chemotherapy as salvage therapy for patients with refractory germ cell tumors. Cancer 66:2476–2481
56. Pizzocaro G, Salvioni R, Piva L et al. (1990) High activity, moderate toxicity modified cisplatin, etoposide, ifosphamide (PEI) salvage therapy for germ-cell testicular tumors (GCTT). Proc Am Soc Clin Oncol 9(508):131
57. Nichols CR, Tricot G, Williams SD et al. (1989) Dose-intensive chemotherapy in refractory germ cell cancer – a phase I/II trial of high-dose carboplatin and etoposide with autologous bone marrow transplantation. J Clin Oncol 7:932–939
58. Broun ER, Tricot JK, Fox E et al. (1991) Long term follow-up of salvage chemotherapy in relapsed and refractory germ cell tumors using high dose carboplatin and etoposide with autologous bone marrow rescue (ABMR). Proc Am Soc Clin Oncol 10(533):167
59. Toner GC, Geller NL, Tan C et al. (1990) Serum tumor half-life during chemotherapy allow earlly prediction of complete response and survival in non-seminomatous germ cell tumors. Cancer Res 50:5904–5910
60. Motzer RJ, Gulati SC, Brown J et al. (1991) High-dose (HD) carboplatin (C) + etoposide (E) with autologous bone marrow rescue (AUBMR) in poor risk nonseminomatous germ cell tumor (NSGCT) patients (PTS) with slow serum tumor marker decline during induction with cyclophosphamide, bleomycin, actinomycin D, vinblastine, cisplatin (VAB-6). Proc Am Soc Clin Oncol 10(518):164
61. Bosl GJ, Dmitrovsky E, Reuter VE et al. (1989) Isochromosome of chromosome 12: clinically useful marker for male germ cell tumors. JNCI 81:1874–1878

62. Dmitrovsky E, Moy D, Miller Wh, Li A, Masui H (1990) Retinoic acid causes a decline in TGF-alpha expression, cloning efficiency, and tumorigenicity in a human embryonal cancer cell line. Oncogene Res 5:233–239
63. Strickland S, Smith KK, Marotti K (1980) Hormonal induction of differentiation in teratocarcinoma stem cells: generation of parietal endoderm by retinoic acid and dibutyryl cAMP. Cell 21:347–355

Subject Index

Printing: Mercedesdruck, Berlin
Binding: Buchbinderei Lüderitz & Bauer, Berlin